SOCIOLOGY OF MARRIAGE AND THE FAMILY

THE NELSON-HALL SERIES IN SOCIOLOGY

Consulting Editor: Jonathan H. Turner
University of California, Riverside

About the Authors

Randall Collins graduated from Harvard in 1963 and received an M.A. in psychology at Stanford University the next year. At the University of California, Berkeley, Collins received an M.A. and Ph.D. in sociology, and later taught at the University of California at San Diego and Berkeley and at the University of Wisconsin, Madison. He has contributed to sociological journals and is the author of several books, including *Conflict Sociology* (1975), *Sociology Since Mid-Century: Essays in Theory Cumulation* (1981), and *Weberian Sociological Theory* (1986). A member of the Crime Writer's Association of Great Britain, Collins also wrote *The Case of the Philosophers' Ring* (1979), a detective novel set in London and published both in this country and in Great Britain. Collins has been a Visiting Member at the Institute for Advanced Study in Princeton, and was awarded the Theory Prize by the Theory Section of the American Sociological Association in 1982. He was president of the Pacific Sociological Association in 1993. Now a professor of sociology at the University of California, Riverside, he lives in San Diego, California, with his wife, Judith McConnell, a superior court judge, and their three children.

Scott Coltrane completed his undergraduate studies at Yale University and the University of California, Santa Cruz, and received an M.A. and Ph.D. in sociology from the University of California at Santa Cruz. In 1988, he received the C. S. Ford Cross-Cultural Research Prize for a study of fathers' involvement in child care and the public status of women. Coltrane has published articles on family and gender in several scholarly journals, including *American Journal of Sociology, Journal of Marriage and the Family, Journal of Family Issues, Social Problems*, and *Gender & Society*. His latest book, *Family Man* (1995), summarizes the results of his research on fathers and gender equity. Now an associate professor of sociology at the University of California at Riverside, he lives in Riverside, California, with his wife, Wendy Wheeler-Coltrane, and their two children.

FOURTH EDITION

SOCIOLOGY OF MARRIAGE AND THE FAMILY

Gender, Love, and Property

Randall Collins
University of California, Riverside

Scott Coltrane
University of California, Riverside

NELSON-HALL PUBLISHERS

Chicago

Project Editor: Rachel Schick
Design/Production: Tamra Phelps
Photo Research/Illustrations: Nicholas Communications
Typesetter: E. T. Lowe
Printer: R.R. Donnelley & Sons

Library of Congress Cataloging-in-Publication Data

Collins, Randall, 1941–
    Sociology of marriage and the family : gender, love, and property
Randall Collins, Scott Coltrane. — 4th ed.
      p. cm.
    Includes bibliographical references (p.   ) and indexes.
    ISBN 0-8304-1392-8
    1. Family—United States.   2. Family—Cross-cultural studies.
   3. Family life education—United States.   I. Coltrane, Scott.
   II. Title.
   HQ536.C716   1995
   306.85'0973—dc20

94-15709
CIP

Manufactured in the United States of America

10 9 8 7 6 5 4 3 2

The paper used in this book meets the minimum requirements of American National Standard for Information Sciences—Permanence of Paper for Printed Library Materials, ANSI Z39.48-1984.

For Maren,
Anthony,
Lindsay,
Colin, and
Shannon

CONTENTS IN BRIEF

CHAPTER 2
THE SOCIAL INGREDIENTS OF THE FAMILY   29

CHAPTER 3
FROM KINSHIP POLITICS TO THE PATRIARCHAL HOUSEHOLD   61

CHAPTER 4
THE LOVE REVOLUTION AND THE RISE OF FEMINISM   101

PART TWO
DIVERSITY IN MODERN FAMILIES   133

CHAPTER 5
THE TWENTIETH CENTURY   135

CHAPTER 6
FAMILIES AND SOCIAL CLASS   169

CHAPTER 7
RACE AND ETHNICITY: VARIATIONS IN FAMILIES   207

CHAPTER 10
CONTRACEPTION, CHILDBEARING, AND THE
BIRTH EXPERIENCE   319

PART FOUR
FAMILY REALITIES   367

CHAPTER 11
HOUSEWORK, POWER, AND
MARITAL SATISFACTION   369

CHAPTER 12
PARENTS AND CHILDREN   409

CHAPTER 13
FAMILY VIOLENCE   455

PART FIVE
FAMILY CHANGES   499

CHAPTER 14
DIVORCE AND REMARRIAGE   501

CHAPTER 15
LIFE TRANSITIONS 539

CHAPTER 16
THE FUTURE OF THE FAMILY   563

PREFACE

This textbook is intended for use in several ways and in different types of courses. One important set of topics in the book covers the concerns of a course on contemporary marriage. It is designed to introduce students to what to expect in their own family life: what information is worth knowing, what problems may be expected, what practical techniques are available for dealing with issues that arise.

A second kind of course focuses more on the family as an institution in society. Here we are concerned with what a family is in general, why it takes particular forms in particular societies, and what forces are responsible for changing and shaping it as the world moves into a new phase. What makes this area of study especially timely is the fact that we are living through a period of major change in the history of the family. For a sociologist, it is an exciting time to be alive and observing. In particular, the issues raised regarding the positions of women and of men in society are deeply implicated in the changes that are remaking the modern family. However we may stand personally on the issues of modern feminism, they are a central reality that we must come to grips with if we are to understand what is happening to the family today.

Yet a third approach to the family is to see it as the locus of a series of practical issues that bear on the work of many different professions. Here we are dealing with a course that is taken by students preparing to become counselors, therapists, social workers, as well as educators, family law specialists, and law enforcement personnel. The family is deeply implicated in the problems of poverty, crime and delinquency, school achievement and career success, and, more specifically, in the controversies over child and spouse abuse, marital and extramartal sex, illegitimacy, abortion, divorce, child custody, remarriage, and aging. One might even argue that politicians, political activists, and citizens in general need to be concerned with the sociology of the family in order to act insightfully on these issues as they intersect with public policy.

It is possible to organize a course out of this book around each of these concerns. The course on contemporary marriage would emphasize the introduction (chapter 1), the material on the modern era (chapter 5), and the series of chapters on women and men forming marriages, running households, having children, and experiencing various problems and transitions (chapters 8 through 15). The course on family as a social institution, in addition to these chapters, would focus especially on chapter 2 (theory of the family) and chapters 6 and 7 (social class and racial differences), and on the historical and comparative materials in chapters 3, 4, and 5. A course especially organized for professionals who will deal with family-related problems would pay attention especially to chapters 6 and 7 (social class, race, and ethnicity), chapter 9 (premarital, marital, and extramarital sex), chapter 10 (contraception and childbearing), chapter 11 (family power and marital happiness), chapter 12 (children, mothering, and fathering), chapter 13 (family quarrels, conflict, violence, and abuse), chapter 14 (divorce and remarriage), chapter 15 (life transitions, adolescence, and aging), and chapters 5 and 16 (recent trends and the future).

These three different approaches to the family are not entirely separate. It is not possible to prepare oneself for one's own marriage by means of a sociology course without being exposed to some more general sociological principles explaining why things are as they are today; and social problems relating to the family are always manifestations of deeper social patterns. Conversely, a purely theoretical and intellectual interest in the family nevertheless should have as a by-product the greater ability to anticipate what will happen in one's own marriage, and to act intelligently in regard to the social problems connected with the family. Sociological theory is not merely an esoteric subject for intellectuals. Insofar as sociology is successful in explaining why certain social patterns exist under certain circumstances, it is presenting information that is of both theoretical and practical importance. This book attempts to show both the basic framework of our knowledge and what we can do with it.

Whatever approach you choose, the following supplements are available. The Instructor's Resource Manual includes outlines, summaries, key terms and concepts, class activities, discussion questions, suggested readings for instructors, and recommended films. The Test Bank contains multiple choice, true/false, and essay questions. It is available in printed format and on IBM-compatible disks.

Lest this sound too austere, we would like to add a word about enjoying the book. The family is not just a practical problem to be solved, a theory to be mastered, or a social issue to deal with. It is also a center of the emotional, the beautiful, and the dramatic in life. Love, sex, childhood, the phases of life, and the struggles of human history are all part of the life drama. We have tried to put into this book what is most interesting about these aspects of the family, and to illustrate, where possible, with some of the great literature and art of our culture. We hope the book will prove useful, illuminating, and socially progressive; but also we hope it will give the reader pleasure.

PART ONE

COMPARATIVE AND HISTORICAL PERSPECTIVES

1

EXPLAINING THE FAMILY

The family is a multisided reality. From one point of view, there is nothing more ordinary. Mom and Dad, brother and sister, sitting in front of the TV, doing the dishes, taking out the garbage. What could be more mundane, more routine? The family in that sense hardly seems worth paying attention to in a college course. It is merely what happens when you stay at home, away from where things are going on.

And yet the very same institution, the family, is incredibly dramatic. Most novels and an extraordinary number of dramas have been written about the family, certainly more than about any other subject. One of the oldest of all Greek plays, *Oedipus Rex*, is about a family. *War and Peace* is the story of a family; so are *The Brothers Karamazov*, *Hamlet*, and *Star Wars*. One can hardly go to the movies without encountering a story about a family, especially if the movie is meant to be "serious" (or comedy) rather than merely outdoor adventure. And even the latter usually has its family overtones, if we count romantic subjects. For in real life, love, dating, and sexual affairs all are part of the orbit of the family; they are either part of the courtship process through which families are formed or, more illicitly, part of the way they fall apart and become reshaped.

So, ironically, even if one leaves the humdrum of home for a night out "where the action is," one is more than likely to encounter some dramatics related to the family, either in the sphere of personal romance, or, even more likely, by sitting in a darkened theater watching the story of some other family.

What should we make of this? It is not simply a matter of literature versus life. Every family has these two characteristics. Great literature captures the conflict between them: it is what Flaubert's *Madame Bovary* is about—and also most of the modern feminist novels. And the contrast is there in everyone's own life. Falling in love, dating, courting, and getting married are among the more exciting things in life. Family life, too, provides a lot of negative excitement: family quarrels, jealousies, parents' and children's struggles with each other, extramarital affairs, the traumas of separation, divorce, and death. Power struggles in the home are a major theme, both in the psychiatric literature and in the feminist concern for women's liberation. Sigmund Freud was more than anything an analyst of the aftereffects of family life; most clinical psychology and counseling deals with repairing individuals and reorganizing the way they deal with their families. Even more extreme is the family as a favorite location for violence. Wives, children, and sometimes husbands have a better chance of being battered in their own house than out on the streets, and murders occur most commonly between spouses. You don't have to go out to see the fights; you can see them at home.

The family, then, is mundane and routine. And the family is ultradramatic, accounting for both our personal highs and our worst lows. Moreover, these two sides are connected. The dramatics of love and sex, the ideal image of the bride, the joyfulness surrounding a newborn baby, all of these lead to the routine, mundane side of family life. The transition from dramatic to mundane is often a shock; the letdown causes much of the disgruntlement and conflict that take place in the family. We could put it more strongly: the family is the subject of powerful ideals—or illusions, even ideologies—which constitute a major part of its attraction. But as family life gets under way, reality asserts itself: material work has to be done, the diapers changed, the bills paid. The mundane world has its structure too. It is not only alliance and support, but property and power, domination, and conflict.

We are not saying that the family simply consists of two phases, a naive, idealistic phase before marriage and nothing but mundane realities and conflicts thereafter. The ideals and the emotional high points keep reasserting themselves

FEATURE 1.1
Sex versus Gender: What's the Difference?

The words *sex* and *gender* are used in a special way in this book, and in the writings of many contemporary sociologists. Both are ordinary words, but their commonsense usage creates certain confusions that make it preferable to distinguish between them.

In everyday language, the word *sex* has two different meanings. It refers to (*a*) the characteristic of being male or female, but also (*b*) erotic behavior. It is in sense *a* that one might refer to "the sexual division of labor," while sense *b* comes out in such expressions as "How's your sex life?"

This double meaning of *sex* can be confusing. *Sexes* (sense *a*) are not necessarily *sexy* (sense *b*).

There is a further problem with an indiscriminate use of the word *sex*. The term has a strong biological connotation, in both sense *a* and sense *b*. Males and females are primarily distinguished by their genitals, and erotic behavior is an act of the physical body. But a good deal of the way in which the sexes (sense *a*) behave is not biologically determined at all, but is *cultural*. Biological males and females learn how to play the *social roles* of men or women. Just what these behaviors are depends on the society they happen to live in, and on each person's individual experience.

In order to distinguish between the *biological equipment* of males and females and the *social roles* built upon them, many sociologists now use two different terms. *Sex* refers to the biological person and his or her physical actions, while *gender* is used to refer to the social roles of being male or female. Thus, female and male styles of dress or talk would be seen as gender differences constructed in daily life; similarly, the fact that women earn less than men is a form of *gender* discrimination. (We should be aware, though, that many sociologists continue to say *sex* discrimination or *sex* differences; it is difficult, if not impossible, to completely standardize everyone's terminology.)

This distinction enables us to use the word *sex* in a clearer way. Its meaning now specifies the biological and erotic dimension. To talk about *sexual* property rights or *sexual* possessiveness, as we will do in this book, is to discuss the rights of erotic access that someone claims over another person's body. How people behave erotically, though, is not just "natural"; it has a biological component, but it, too, is strongly influenced by society.

In sum: *Gender* reminds us that the social roles of being male and female are largely produced by the culture. *Sex* refers us explicitly to biological characteristics and erotic behavior.

throughout the months and years: in some marriages, this happens quite a lot. But the mundane world of monthly bills, work that has to be done, power that some people wield and to which others must acquiesce—all this is always present, whether we are conscious of it or not. Families that live on a higher level of emotional success are those that have worked out their mundane family lives satisfactorily. Otherwise, when the mundane world intrudes, as inevitably it must, its hard realities always take priority and burst the bubbles of our ideals.

THE FAMILY IN CURRENT CONTROVERSY

The family has become a topic of public controversies, cutting in all directions. It is not just a source of drama in one's own life. One can hardly pick up a newspaper or turn on the TV without coming across an issue closely connected to the family.

Take crime, for instance. We hear a great deal about "crime in the streets"—conveying an image of youths away from homes, out of control of parents, scorning the activities that their parents would approve of. The rhetoric is no doubt exaggerated, but nevertheless there is a connection. Gangs are a kind of substitute for families and flourish where families are weakest. Another dominant issue today is drugs, and the gang violence that goes on to control the drug trade. Here again we find a public controversy tied to the family in a negative way. The issue, however, is not as simple as some politicians put it—that the family is breaking down, traditional values are disappearing, and therefore crime is rampant.

To understand the issue better, we could look at the intersection between the family, race, and crime. A major target of the antidrug crusade is the inner-city youth gangs, especially those composed of blacks. The inner city is precisely the place where conventional families are in crisis. The majority of black children now are reared in single-parent families, typically by unmarried or divorced women living at poverty level. Why are there so few marriages in this group? Part of the reason is that a considerable proportion of young black men are in jail or have criminal records, and hence do not make particularly desirable partners for young black women. In addition, murder is the largest cause of death among these young men, contributing to the shortage. The result is a vicious circle: poverty and crime greatly weaken the family and its ability to control and support children; and the inability of these families to do much for their children has the effect of perpetuating poverty and crime.

Here we have crime occurring outside the family, but related to what is happening inside. We are also greeted with a good deal of news about crime and controversy inside the family. Take a sample of headlines for just one week in one city. Two teenage boys are accused of murdering their wealthy parents in order to receive their inheritance. In defense, they claim to have been sexually abused in childhood. A divorced woman is found guilty of hiring two teenagers to murder her ex-husband. Another woman is on trial for murder; divorced by her husband who wanted to marry a younger woman, she broke into their apartment and shot them both in their bed. A demonstration against an abortion center turns violent, and protesters are arrested. On the positive side, a women's group demands that the city do more to provide child care for employed parents.

And then there are the politicians, campaigning for election. How do they treat family issues? Whatever their political stripe, they usually announce their commitment to family traditions in the most visual way. Right on the front page of their campaign literature, we see pictures of the candidates, posing with spouse and children, smiling at the camera. Apparently the way to get elected is to look like part of a conventional happy family.

Stratification: Inequality Inside and Outside the Family

Is there a more general pattern, a sociological insight that helps explain these different kinds of family issues? We propose that one thread that weaves them together is *stratification*—the pattern of social equality or inequality. Most sociologists, one way or another, deal with some aspect of stratification. It has several dimensions, which we may find both inside and outside the family;

Judge Nomoto's robe and chair are symbols that reinforce the status of her title and position.

families are part of the larger system of stratification in American society. In addition, we should pay attention to inequalities inside the family—in the relationships between husbands and wives, parents and children, and among sisters and brothers.

Stratification is usually analyzed in three dimensions: (1) *Economic class* is the dimension ranging from poverty to wealth, and includes the occupations and careers by which people get their income. (2) *Power* determines who controls whom. It ranges from the level of political elites at the most macro level of the state to the most intimate powers, such as who gets to decide when to have sex. (3) *Status* is the dimension concerning who is held in most respect. It is manifested in the way people talk about each other and in the way they display themselves, their homes, and their possessions. It is a realm of symbols and feelings, ranging all the way from admiring smiles to not even noticing someone is there.

We shall say a lot more about these aspects of stratification throughout this book. For now, notice what a stratification perspective tells us about the controversies we have already mentioned. We have seen that issues of crime, drugs, and gangs are woven together with family patterns, especially in the poverty level of economic *class*. In addition, the *status* dimension is involved when we examine the way in which racism is a part of the chain of causes that makes poverty, as well as gangs, so common in black communities. We can examine how youth gangs are a way in which youths can claim some *power* as well as status in a society in which they have very little advantage on any of the stratification dimensions.

FEATURE 1.2
How to Avoid Talking about Social Class

Most Americans do not like to talk openly about social class. In actuality, social class affects so many aspects of the world around us that we have to invent terms to refer to it. In the 1980s, a favorite newspaper cliché that referred to certain upper-middle class persons was the word *yuppies*. Other terms that describe the same group are *upscale* and *"life in the fast lane."* Back in the 1950s, the term *leadership* was a way of referring to wealthy and powerful people who controlled business organizations, government, and the professions. An earlier generation created the term *high society*, which is now abbreviated to just plain *society*, in the sense of the *"society pages"* of a newspaper—a peculiar terminology implying that only a few wealthy people count as members of society. For referring to the working class or to the lower class, we have such euphemisms as *low rollers*, or *the streets*—as in the phrases *"crime in the streets"* or somebody having *"street smarts."* Generations ago we spoke of *"the other side of the tracks."*

Create a game of "sociological trivia." See how many terms you can think of that people use when they talk about different social classes but wish to avoid the words *upper class*, *middle class*, *lower class*, and *working class*.

We should also take a close look at the position of black families in the entire class structure of American society. A sizable proportion of black families are not in the poverty sector, and their children are not involved in crime. The black community is split along class lines; we can avoid stereotyping blacks in general by taking class into account. By the same token, we should be aware that black poverty is only the most visible part of the poverty sector. The majority of poor families are white. Once again, we need to be aware of class if we are to understand the range of families in our society.

The instances of crime inside the family should make us aware of the ways in which stratification operates. Spouse abuse is a blatant example of the exercise of *power*, typically by men against women. In the same way, child abuse represents the power of parents over children taken to an extreme. Many parents see this use of power as normal, making no distinction between spanking and breaking a child's bones. Fortunately, most husbands and most parents do not push stratification to such an extreme inside the family. But abuse happens because there is a larger pattern of family stratification, usually taken for granted. Men who abuse their wives typically feel that men have the right to exercise power in the family, and they become angry when they feel their power is challenged. Violent abuse is only the tip of the power iceberg.

Looking at stratification gives us some sociological insight even into sensationalistic headlines. Take the story of the woman who murdered her ex-husband and his new wife. This is a woman in her mid-forties, once pretty and now overweight, who devoted her life to being a housewife and mother. Her husband, a wealthy lawyer, divorced her to marry a much younger woman, attractive and successful with her own career. The story could serve as a symbol for the conflicts in today's gender stratification. The murderess appears to be a rather traditional woman in a time when the traditional family is changing. She is doubly left behind: divorced by her husband when she is no longer attractive and upstaged by the nontraditional women of today who have economic independence and pursue their own careers. Although she went too far when she resorted to murder, we should notice that this traditional woman did not take her position passively. Behind the headlines, we see a woman fighting gender stratification in the family.

The Dual-Income Family Becomes the Center of Contemporary Stratification

Of course, you have every right to say, What does this have to do with me or with the families I know? Most families are not involved in murders, serious abuse, or gang crime. Nevertheless, a family's place in the pattern of stratification is the most important thing we can know about it. Externally, families are the main way in which each individual becomes placed in the class structure of society. Internally, how much stratification there is between husbands and wives and between parents and children tells us a great deal about what will happen in that family, ranging from housework to sex to family quarrels and personal happiness.

Since we will say a great deal more about these processes later, let us focus on just one key point here. The family has always been the building-block of the class structure. The upper class consists of wealthy families, who usually have managed to pyramid their wealth by inheritance across the generations. The lower class has had a weaker family structure, consisting of many isolated individuals and of families unable to rear children effectively for the competition for economic position. Between these extremes have traditionally been the blue-collar working class and the white-collar middle class. In both of these the traditional ideal was a stable family, with the husband employed as breadwinner and the wife taking care of home and children.

The main difference between traditional working-class and traditional middle-class families was that in the former, young women usually had to go to work to make ends meet. Traditional middle-class women, on the other hand, did not have to work for pay. The *high-status* thing for them to do was to stay home and take care of the house and children (if possible, with the help of a working-class woman as a servant). It is for this reason that working-class women have usually wanted to quit their jobs and become housewives, since this would have been a sign of success, of living like middle-class women.

Since about 1970, there has been a major shift in this pattern (see figure 1.1). The *status* system has drastically shifted for middle-class women. It has become high-status for women to pursue careers—not just jobs they hold before they are married, but positions in the professions and in business that are full-time and the equivalent of the best careers of men. A number of things have contributed to this shift—above all, the women's movement, but also the rapid growth in the proportion of women who are college-educated, which now surpasses that of men.

But this is not merely a *status* process—a shift in what is considered respectable and admirable for women to do. It has had a profound effect upon *economic class*. As we shall see, *the most important way in which a family gets itself into an upper-middle class level of income today is by having two middle-class incomes*. The combination of working husband and working wife has become the most important feature differentiating families by economic levels (see figure 1.2).

What we might call the "new upper-middle class" consists of two professionals or business executives married to each other: two incomes of $40,000 to $50,000 or more combining to permit an affluent lifestyle. Further down the class structure, two working-class incomes (say, a carpenter and a secretary) will combine to make a decent income at the middle level. In contrast, the traditional single-earner family usually pays an economic price. One middle-class income today can barely keep a family at an average level of consumption, let alone match the style of life of modern couples with two professional incomes. Further down the scale, one working-class income may be barely enough for survival. Many people in the poverty class are not unemployed or on welfare; they simply are trying to get along on one inadequate income. The upper class stands out as the only group in which the wealth of one individual (usually the husband) is so great that the traditional single-employed pattern

FIGURE 1.1: The Decline of Traditional Families: Labor Force Patterns from 1940–1990

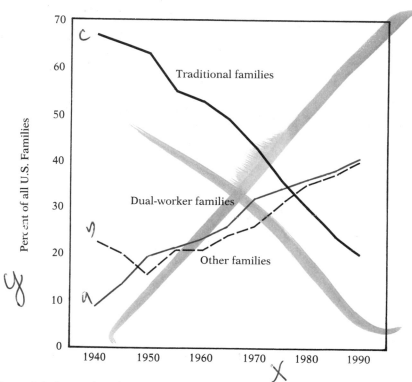

Figure 1.1 shows that the traditional family in which the husband is a bread-winner and the wife is a homemaker has declined significantly since the 1940s. Starting in the mid-1970s, families with two workers became the most typical family type. "Other families" include employed single-mother families, unemployed single-mother families, employed single-father families, unemployed single-father families, married couples with wife-only working, married couples with no workers, and other related people living together, with or without workers. These other family forms are increasingly common.

Source: Howard Hayghe, "Family Members in the Work Force," *Monthly Labor Review*, Vol. 113, No. 3, 1990, p. 16; and Dennis Ahlburg and Carol DeVita, "New Realities of the American Family," *Population Bulletin*, Vol. 47, No. 2, 1992, p. 25.

FIGURE 1.2: The Advantage of Two-Income Families over One-Income Families

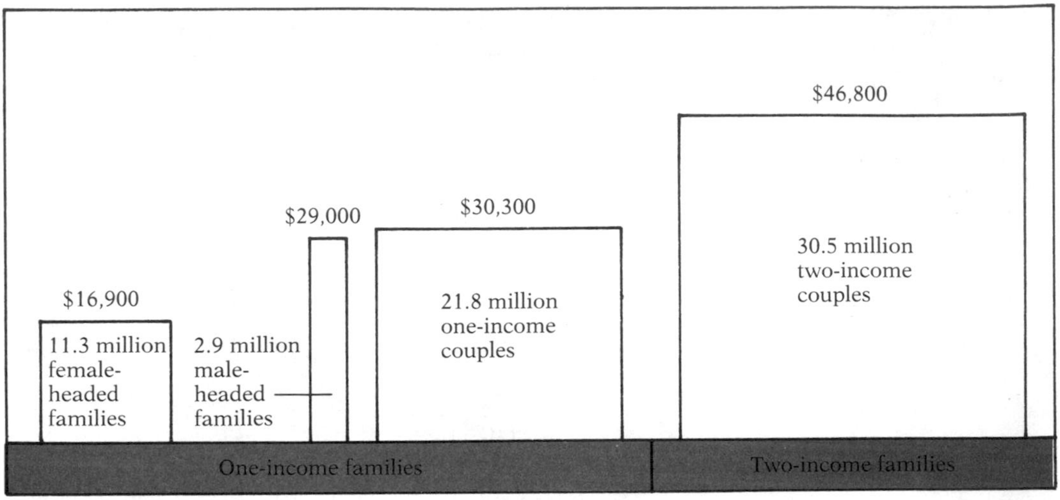

$46,800

$29,000 $30,300

30.5 million
two-income
couples

$16,900

11.3 million   2.9 million   21.8 million
female-   male-   one-income
headed   headed   couples
families   families

One-income families   Two-income families

Source: *Statistical Abstract of the United States,* 1992, No. 708.

still holds. But the upper class is only a small percentage of the population. Everywhere else, the difference between two-income and single-income families is crucial.

The economic class structure puts pressure on every family within it. Whether they want to show off their possessions or not, prices are driven up by the amount of money people have to buy things. This is most obvious in regard to homes. In most major urban areas of the United States, there has been a tremendous inflation in the price of housing. Homes have become so expensive that families without two incomes can scarcely afford to buy anything within many residential areas.

For this reason, American families have become locked into the dual-earner pattern. This will no doubt become increasingly true in the future. Older homeowners, who bought at a time when prices were much lower, could afford to live on one income. As these people die or move away, virtually all of the substantial homes, which used to be considered part of a middle-class life-style, will be taken over by dual-income couples. Although the pattern may have started with the women's movement, it has now become an economic reality. One can predict that, except for a small number in the wealthy upper class, families that want to maintain a middle-class lifestyle will have to be headed by dual-income couples.

The Intersection of Gender, Race, and Class

On the other hand, not everyone is going to be successful in this struggle for economic position. A large number of people, for instance, may never be able to afford to own a house. An alarming pattern that is emerging in the late twentieth century is the way that the class structure is becoming more polarized. Economic inequality, which declined between 1940 and 1970, has in recent years been growing again. Only dual-income families have kept up with inflation in the 1980s. The top 20 percent of the population, roughly speaking, the upper-middle and upper classes, has gotten an increasing proportion of the income, while everyone else's income has declined.

Several dimensions of inequality intersect. On the privileged side are women who have overcome the barriers of gender stratification and who are pursuing high-paying careers. When these women of the new upper-middle class are married to men of the same class, their families reap double advantages. On the less favored side is a large middle class struggling to hold on. Families in the poverty sector typically suffer from a combination of disadvantages. Race and class tend to reinforce each other at this level. In addition, gender stratification piles on another disadvantage for women who bring up their children as single parents on incomes that are already below the level of those of men.

The stock broker in the foreground has overcome some of the barriers of gender stratification inherent in the business world.

Equality and Inequality

What will happen to families in the 1990s and as we move into the twenty-first century? To get a good grip on this question, we will have to look at today's families at different levels of the class structure. It is apparent that upper-middle class families are moving one way, and lower class families are moving in the opposite direction. For the families between, there are conflicting tendencies to sort out. Only upper-class families, at the wealthy peak of the society, appear to be holding to the traditional patterns. To determine which way the overall pattern is shifting, we must look at the macrostructure of society and the sociological theories that explain its long-term changes.

In addition, we must look at the micro level of stratification inside the individual family. As sociologists, the most useful thing we can do is to take a good look at how the three dimensions of stratification apply to men and women who are married to each other—or who are thinking about getting married or divorced. (1) *Economic class.* What sources of income does each man and woman have? What effects do their occupations and careers, whether hoursework or paid jobs, have on how they relate to each other and to their children? (2) *Power.* What kinds of things give a man power over his wife or a wife power over her husband? And what kind of power do they have over their children? To what extent are the children able to evade their parents' power? (3)

The working-class position of these men may affect how they relate to their families.

Status. How do people regard each other in the family? For the family, status has a special set of meanings and emotions: love, affection, and admiration are part of it; so are jealousy, disrespect, and hatred.

Our sociological theories are designed to get at the causes of these patterns. We want to know what factors will make a husband or a wife more powerful on the micro level, what makes a marriage more happy or more disharmonious. On the macro level, we want to know what changes the pattern of dual-income families in the society and the answers to many other questions about the stratification of the overall structure.

In principle, stratification can range from extreme degrees of inequality on down to zero. Just because we pay stratification a good deal of attention does not mean that inequality is inevitable. Some societies, past and present, and some relationships among individuals are much more unequal than others. Among all these variations one possible pattern is equality. Men and women could be equal within the family; and conceivably all families could be equal in terms of economic wealth, power, and status. Not that equality is easy to achieve. Some forms of it, especially class stratification in the larger society, seem to be moving toward greater inequality rather than greater equality.

There is a particular kind of stratification that has traditionally been at the core of the family itself: gender stratification. On this level, there is reason to believe that equality is possible, and indeed it is sometimes achieved in families. We will be especially concerned with the causes of different degrees of gender stratification and will try to pinpoint the conditions under which there is gender equality in the family. As we shall see, the degree of gender equality or inequality has a major effect on the amount of conflict and of happiness within the family.

EXPLAINING GENDER STRATIFICATION IN THE FAMILY

A crucial question is how we can explain gender stratification in the family. This is a relatively new area in sociology. Before about 1970, few theories raised the question of whether there are differences between men and women in their power, economic position, and social status. A conventional, rather idealized family was taken for granted in the older theories. A husband who goes off to work each morning, a wife who stays home and takes care of the kids, a dog named Spot—this was referred to as the sexual division of labor (leaving aside the dog, of course).

In this model, men did the work and supported the family in the economic sphere. Their labor was regarded as *instrumental*—that is to say, practical, goal-oriented, unemotional. Women, on the other hand, were regarded as primarily *expressive*. Women took care of the emotions and the personal relationships. They nurtured the children and saw to their husband's emotional needs, providing relaxation and therapy when he came home from the male sphere of practical achievement (Parsons and Bales 1955).

What's wrong with this picture? Listing its defects has become quite an intellectual industry in the last twenty years. Here are a few of the more important points:

1. Is it really true that husbands work and housewives do not? This overlooks what actually has to get done in the home day after day. Cooking food, dusting furniture, cleaning floors, washing and ironing—all this is work too. Even taking care of children is far more than an "expressive" activity. There is a great deal of sheer physical, practical content to it, from changing diapers to getting toddlers dressed, undressed, washed, and fed. When children get older and are able to do some things for themselves, new forms of work appear, such as transporting them from place to place by what seems to many women to be "mom's taxi service." It is a gross misnomer to call men's employed activities "work" but not women's activities at home. The difference is not whether it is work or not but the way that it ties into the economic structure. The key point is that men's work is paid, whereas traditionally women's work has been unpaid.

2. Why have men usually worked in the better-paying sector of the economy and women in the unpaid sector? Is this simply a natural arrangement? As we look far enough across different societies and different periods of history, we see that the kinds of work women did varied a great deal. In some hunting-and-gathering or horticultural societies, for instance, women produced the bulk of the food that the whole society needed to survive. It would seem, then, that the "sexual division of labor" is not an inevitable arrangement.

In twentieth century societies, too, we can see a flaw in the theory that women and men are in different spheres because of purely functional reasons. In all modern industrial societies, including our own, a considerable number of women have worked for pay outside the home. Until the pressure exerted by the women's movement in the last few years, virtually all of these women worked at jobs that yielded low pay and little power and status. Women were almost totally excluded from positions in the higher levels. So-called "middle-class" or "white-collar" positions for women were largely restricted to secretarial and retail sales positions—jobs with virtually no power and a rate of pay lower than for most manual jobs for male workers. Women in the professions have been heavily concentrated in elementary school teaching or in nursing—both professions in which men tended to dominate the higher-level positions.

There have been plenty of blue-collar working-class women as well. This is especially true in the garment industries, where women operate sewing machines in the so-called "needles trades." We tend to think of a typical factory scene as consisting of armies of men pouring out molten metals in the grim ambiance of the steel mills. A better image today would be long rows of women processing plastic goods or computer chips.

Another sector in which women have always been in the majority is the service sector, for example, working as waitresses and beauticians. This area includes the humblest positions in our society: custodians, hotel maids, and cleaning women who empty the trash and tidy up after the rest of us. When this kind of work is done in a private home by a paid household servant or housekeeper, it is almost always done by women. Here we see a good example of how gender stratification crisscrosses the class structure. Female household servants holding one of the lowest-paying jobs in the working class are typically hired by women who are in the upper and upper-middle classes. These employers used to get their wealth from their husbands, but increasingly they themselves have high-paying professional careers. Class stratification and gender stratification are both hierarchies of privilege in their own right. Sometimes class and gender reinforce each other, as in the case of low-paid housekeepers who are at the bottom on both scales. But sometimes class operates independently of gender, as in the case of the upper-middle class professional women in a position to hire working-class women to work for them. Even though some women get to the top or near it, other women make up a large proportion of those at the bottom.

In short, women have always worked, either without pay in the home or with pay outside it. In both cases, women have typically received less material rewards than men and little power or status from their work. Perhaps this is why women's work has been "invisible" to most people, including the authors of the older sociological theories. The newer theories look for ways in which women are excluded from the higher paying jobs, rather than taking it for granted that these are automatically men's positions. Once we see that men and women have their own economic positions, however high or low, we recognize that the structure of the family is going to be shaped by these economic positions. For economic position outside the family brings resources that give power inside the family. If women have traditionally done household services for men at home, it is at least partly because of the power that has accrued to men who have more economic resources than their wives (see figure 1.3).

3. Is it really true that men are *instrumental*—hard, cold, practical—and women are *expressive*—emotional, personal, down-to-earth, uninterested in achievements or in abstract ideas? This is in some ways the most controversial question in the theory of gender. The traditional theories of the sexual division of labor assumed that this distinction was natural. Supposedly society needs both instrumental and expressive skills, therefore combining "male" and "female" roles in every family makes a perfect mix. This gave a functionalist justification for gender differences in society.

Newer critical theories are split on this point. On the one hand, theories that might be referred to as "egalitarian" or "socially constructivist" argue that males and females are essentially the same beneath the skin and that the way they behave is socially determined. If men have dominated the social positions in which practical activities were carried out—whether in economics, politics, science, or any other sphere—this has created the image of men as "instrumental," purposeful achievers. By the same token, if women have been channeled into subordinate positions and given no opportunity to control business, government, cultural institutions, and so forth, it is not surprising that women develop the image of being nonpractical, uninterested in achievement, and oriented away from the dominant spheres of society. What women were considered to be good at, then, were the virtues of the places where they happened to be. Women's status came from being lovers, mothers, friends, and confidantes; thus, women acquired an image of being subjective, oriented to the personal, the intimate and the immediate, and away from the harsh realities of the world of men. Egalitarian theories hold that these differences are socially constructed and that if men and women are put in the same positions, they behave the same way.

Another line is taken by various theories that have also emerged with

FIGURE 1.3: Women Still Earn Less than Men

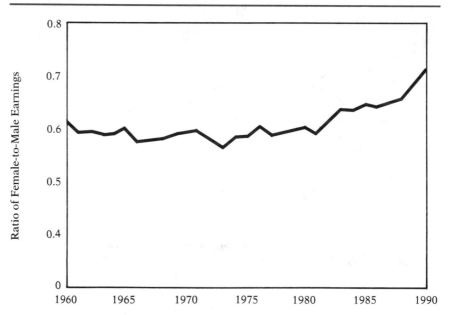

This figure shows the ratio of female-to-male earnings for year-round, full-time workers in the United States from 1960 to 1990. Women's earnings are on the rise, but they still make only about 71 cents for every dollar earned by men. Because women are much more likely than men to be unemployed, work part-time, or hold seasonal jobs, the gender gap in pay is even larger than this figure shows.

Source: U.S. Bureau of the Census, "Income, Poverty and Wealth in the United States," *Current Population Reports*, Series P-60, No. 179, 1992d.

the radical criticism of the last few decades. These might be referred to as "essentialist" theories of gender. In some respects, their position agrees with the traditional theories that assert that there are fundamental differences between males and females. However, the radical essentialist theories hold the "male" qualities responsible for most of the evils of society. They posit that the instrumental, impersonal, achievement-oriented approach has been manifested in the pursuit of economic profit, political power, and military power without regard for the values of human beings. In the realm of ideas, the instrumental attitude is linked to the "positivistic" attitude of science, which tries to reduce the world to purely objective elements, without room for the human side of things. This critique links feminism to Marxist and critical theories of society generally. The female qualities, emphasizing personal, intuitive, and humane attitudes, are equated with the ideals that are subordinated and even crushed by the dominance of male institutions: war, violence, the pursuit of profit, and the subjugation of nature.

In this text, we do not attempt to argue all of the wide-ranging issues that are raised by these critiques. In our view, the traditional theories that divided males and females into two different cultures are misleading, to say the least. What shall we put in the place of this reasoning? We can approach this from the point of view of *what kind of theory gives us the most ability to explain social differences*. The traditional, functionalist theory did not convincingly explain the pattern "male = instrumental" and "female = expressive." For one thing, there is a good deal of evidence that females also engage in many instrumental activities. Much housework is hard, practical labor as is much of the work that women have put into the paid labor force. Although women have been discriminated against and discouraged from entering positions of political power or economic control, or in the world of science, there is plenty of evidence that when women succeed in attaining these positions, they perform "instrumental" tasks in the same way as men. (Two important studies on this point include Rosabeth Kanter's *Men and Women of the Corporation* [1977], and Cynthia Epstein's *Deceptive Distinctions* [1988].)

Conversely, there is evidence that "expressive," personal, and intuitive activities can also be carried out by men. For example, the realm that has been traditionally considered the sphere of the personal, caring relationship *par excellence* is mothering. Yet there is evidence (see chapter 12) that when men take part in fathering, their personalities take on the qualities of being caring rather than aggressive and impersonal. Similarly, studies of artists and creative scientists—males as well as females—show that their personal style tends to emphasize intuitive judgment rather than the mechanical style of the "positivist" image (Bar-Haim 1988; for the opposite argument, see Keller 1985). We would conclude that what are traditionally considered to be "male" or "female" styles are really two parts of the range of possibilities within the abilities of all human beings.

In actuality, both men and women tend to have a mix of instrumental and expressive activities. Our task as sociologists is to show the conditions that push the mixture further in one direction or the other, in each particular case.

The question then is: How deep are the conditions that determine what is *gender*, that is, the culturally defined aspect of what is considered to be male or female in each society? The theories range all the way from conditions that are so primordial that one can doubt they would ever change, to conditions that are relatively superficial and can be pushed around with enough personal or political willpower. The continuum would look something like this:

- Male and female roles are determined biologically and cannot change, except as genetic endowments might change over the long course of evolution.
- Male and female are categories very deep in the culture. Everything else is seen through the "eyeglasses" of gender. This implies we can't get outside of our own gender's field of vision. The culture might mysteriously change at some time in history, but we can't say how or why. This position is more popular among anthropologists than among sociologists.
- Male and female are personality traits developed early in childhood. Relationships between small children and their mothers are crucial in setting boys and girls on the paths that will constitute their gender for the

By taking on nontraditional roles, fathers may change their children's perceptions of appropriate gender activity.

rest of their lives. This position allows a social determinism, but only rather deep in each person's life. Studies of how fathering affects personality, though, show us there is a possibility of variations even at this level.

■ Male and female are styles of behavior that are determined by the social structure of each society. As the social structure changes, so too does the content of gender.

Obviously, as sociologists we are likely to be partial to the last line of analysis. We put a lot of attention into showing the social conditions that make males and females act in particular ways, and that define what are the images of men and women. Since we think that child-rearing is also a social process, we focus on how much of gender can be explained by what happens in early childhood; and we put this in the larger social setting that affects how parents behave.

We want to be alert to the ways in which individuals do *not* live up to the prevailing ideals. Not every male is "machine man," not every woman is "earth mother." We want to be able to see more clearly not only how society puts people into slots; but also how people squirm in these slots, and how they struggle to change the shape of the society into other forms.

Explanation by Comparison

The method of comparison is central to our explanations. This applies to virtually every topic we will consider. If we want to understand how men and women act inside and outside the family, we need to make comparisons to see how much things can vary and under what circumstances. This is one reason why we need to take a long time-perspective, to look across the centuries and from one society to another. The view from someplace else lets us know when we are taking some feature of our own society as more permanent and fundamental than it really is. This is true all across the board. Every aspect of the family has at least some variations, and these will help reveal the social conditions that are causing them. Whatever we look at—how much divorce is taking place today, how much premarital sex, how much controversy over abortion, or any other topic—a look at the comparisons will help us understand it more deeply.

Comparison is such an important method for explanation that we often seek out unusual comparisons to make. In chapter 9, for example, we will spend some time looking at "open marriages," "swingers'" groups, and the types of communes that existed in the 1960s and 1970s. Should we pay attention to these just because we are curious about exotic forms of sexual behavior? That might be one reason, but there is a more important point here. The analysis in chapter 9 is concerned with marital sex, and the processes that maintain sexual fidelity, or put strains upon it, in the conventional family. But to understand conventional family sexual arrangements, we gain insight by comparing them with what happens when these arrangements are altered. Open marriages, swingers' groups, and communes provide us with kinds of real-life experiments that show what happens when sexual behavior is organized in different ways. As we will see, swingers' groups are more successful than the other forms of nonmarital sexuality, but they institute rather strict controls over couples and sharply separate sex from love. These features tell us something about the general processes that affect sexual ties within ordinary marriages as well.

The method of comparison is important for everything in sociology. At some points, we will compare heterosexual couples with homosexual couples.

FEATURE 1.3
Some Changing Family Principles

Tribal Marriage Exchange

On reaching puberty, the Konyak Naga boys begin to look for girls from the clan complementary to their own, and they exchange little gifts, the value and nature of which are strictly fixed by custom. These gifts are of such importance that a boy's first question to the young girl whose favours he seeks is as follows: "Will you take my gifts or not?" The answer being, perhaps, "I will take them," or "I have taken the gifts of another man. I don't want to exchange with you." Even the wording of these overtures is fixed by tradition. This exchange of gifts initiates a whole series of reciprocal protestations which lead to marriage, or, rather, constitute the initial transactions of marriage, *viz*, work in the fields, meals, cakes, and so on.

Claude Lévi-Strauss, *The Elementary Structure of Kinship*, 1949

The Patriarchal Family

By a girl, by a young woman, or even by an aged one, nothing must be done independently, even in her own house.

In childhood a female must be subject to her father, in youth to her husband, when her lord is dead to her sons; a woman must never be independent.

She must never separate herself from her father, husband, or sons; by leaving them she would make both families contemptible.

Him to whom her father may give her, or her brother with the father's permission, she shall obey as long as he lives, and when he is dead, she must not insult his memory.

Laws of Manu, India, ca. A.D. 100

Men have authority over women because Allah has made the one superior to the other, and because they spend their wealth to maintain them. Good women are obedient. They guard their unseen parts because Allah has guarded them. As for those from whom you fear disobedience, admonish them and send them to beds apart and beat them. Then if they obey you, take no further action against them.

Mohammed, *The Koran*, ca. A.D. 630

By marriage, the husband and wife are one person in law; that is, the very being or

Again, whether or not we are interested in the specific facts about gay or lesbian couples, the comparison will show us what processes are quite general and cut across all types of couples. The comparison also shows us what features seem to be specifically related to males and females and their interaction.

It is because of the importance of comparison that we will examine families at different historical times (chapters 3–5). Many of these family forms are irrelevant today, for they will never come back again. But comparing the conditions under which the different types of family systems exist gives us the key to the processes that produce variations and thus enables us to understand what is producing the family structures of today. Such comparisons also give us hints about the future.

legal existence of the woman is suspended during the marriage, or at least is incorporated and consolidated into that of the husband; under whose wing, protection, and cover, she performs every thing. . . .
Upon this principle, of a union of person in husband and wife, depend almost all the legal rights, duties, and disabilities that either of them acquire by the marriage. . . .
A man cannot grant any thing to his wife, or enter into covenant with her, for the grant would be to suppose her separate existence.

> Sir William Blackstone, *Commentaries on the Laws of England*, 1765

The Victorian Family

It is really a stillborn thought to send women into the struggle for existence exactly as men. If, for instance, I imagined my gentle sweet girl as a competitor it would only end in my telling her, as I did seventeen months ago, that I am fond of her and that I implore her to withdraw from the strife into the calm uncompetitive activity of my home. It is possible that changes in up-bringing may suppress all a woman's tender

attributes, needful of protection and yet so victorious, and that she can then earn a livelihood like men. It is also possible that in such an event one would not be justified in mourning the passing away of the most delightful thing the world can offer us—our ideal of womanhood. I believe that all reforming action in law and education would break down in front of the fact that, long before the age at which a man can earn a position in society, Nature has determined woman's destiny through beauty, charm, and sweetness. Law and custom have much to give women that has been withheld from them, but the position of women will surely be what it is: in youth an adored darling and in mature years a loved wife.

> Sigmund Freud, youthful letter to his fianceé, ca. 1885

The Egalitarian Family

Equality of rights under the law shall not be denied or abridged by the United States or by any State on account of sex.

> Equal Rights Amendment, proposed by Congress in 1972 and ratified by thirty-five states, 1972–1979

THE USES OF FAMILY SOCIOLOGY

There are a great many issues concerning the family that we will consider. Our aim is to get an overall picture of what causes what. Why has the family taken particular forms in the past? What is happening to it today? What can we expect for families in the future? These issues are important, for men and women as groups, for the larger society as a whole, and for each of us as individuals.

Social and Political Issues

The family has become an object of politics and public controversy. This has happened even though the family is private and most people want it to remain so. Issues of employment discrimination against women affect the family, whichever way they are decided. The effect is indirect but powerful nevertheless, because the outcome determines how much financial independence women have and thus their desire to get married and to stay married as well as their power within a marriage. Laws on illegitimacy and regulations on welfare support for dependent children affect the family too, especially at the poverty level, where many households are headed by single women. Laws regarding contraceptive information for teenagers or the even more vehement controversy over abortion can have a dramatic, and sometimes tragic, effect on the flow of many family lives. The politics of the future may deal with such issues as public child-care facilities; these affect the careers of married people and make a link between economics and the shape of the family. We live in a time when the family is not taken for granted any longer but is constantly being reshaped by public, political, and private decisions.

Many of the current trends in the family were being manifested before politicians and other commentators got around to advocating their proposals: the long-term increase in employment among married women, for instance, or the steady rise in the illegitimacy and divorce rates. The world has been changing, and politicians to a certain extent only tinker with it. Nevertheless, the politics of the family is going to be a live issue for the foreseeable future. One aim of a textbook on the sociology of the family ought to be to make us more aware of these realities: what effects different government policies are likely to have on the family, and also what aspects of the family have been moving along on their own momentum apart from political control.

Our Personal Life Experience

The theories and facts presented from this academic viewpoint have another relevance, too. We all experience families. Virtually everyone has been brought up in one, and most of us will have a family of our own. The family has, to a considerable degree, shaped what we are, and it will be a major part of

~~our life experience in the future.~~ The sociological theory of the family has a personal and practical relevance that virtually no other part of sociology shares. This is not to say that sociology is capable of drawing faultless guidelines on how to live a happy family life. Many of the issues are too powerfully embedded in the social world we live in for us to manipulate them easily. But the sociological perspective gives us some insight. If it cannot let us evade many of the conflicts, it can at least alert us to what they are and present some ways of dealing with them.

Careers in Family-Related Professions

The sociology of the family is also relevant from a third point of view. Increasingly, the family is a subject for professional practitioners. Most of the work of psychiatrists and clinical psychologists as well as counselors and social workers deals with the family or its effects on individuals. We would argue that, although some of the problems that one witnesses are *psychological*, in the sense that we can see them in individuals, they are not necessarily psychological in their origins. We need to understand the sociology of the family to see how many issues arise. For example, there was a time when women struggling for some autonomous roles in a male-dominated system were diagnosed as merely having psychiatric problems (see feature 11.6); today, a sociological perspective gives better insight into the causes of their situation and possible courses of action. The family also has a major effect on how children succeed in school, and indeed on important apsects of their subsequent careers. Professional educators as well as career counselors and personnel managers thus would also benefit from some lessons in the sociology of the family. The growing importance of political issues around the family makes it of practical relevance for all of us to know what aspects of the surrounding world affect the inside of the family and vice versa.

A NOTE ON SOCIOLOGICAL VERSUS PSYCHOLOGICAL APPROACHES TO THE FAMILY

The approach taken in this book is sociological. That is to say, it analyzes the family as a structure of relationships among persons, and as a part of the larger society. This differs from a psychological or individually based approach, which would deal with the family from the point of view of a particular person within it. This does not mean that we will pay no attention to individuals' experiences. The sum total of individual experiences and actions, after all, are what make up the family. But from a sociological viewpoint, we do not merely stop with the individual experiences or motivations, but try to show how they fit into a larger pattern.

The sociological approach to the family, then, is from the outside in,

FEATURE 1.4
Some Professional Careers Using Family Sociology

Professional Family Specialists
   Family planning counselors
   Family life educators
   Marriage and family counselors
   Psychiatrists and clinical psychologists
   Social workers
   Sex therapists
   Rape counselors

Professions Often Dealing with Family Issues	*Examples*
Lawyers and judges	divorces; child custody; juvenile crime
Teachers	learning problems; family motivation and cultural background
Medical doctors and nurses	child and spouse abuse; family stress
Law enforcement	spouse abuse; family background of crime; murder investigations
Politicians and social-change activists	gender equality; public education; child care; abortion; population policy

rather than the inside out. The goal is not merely to acquaint the reader of this book with what family life *feels* like, but to offer some insight into why people feel as they do, and to explain why certain kinds of situations keep turning up. Being in love, for example, is a feeling of warmth, tenderness, or passion. As sociologists, we respect that. But as sociologists, we go further, and attempt to get "behind the scenes" and explain why these feelings arise at certain times, with certain persons rather than others. Sociology looks for the general process that produces individual experiences.

Sociology does not deny that individuals are unique. Each of us, in certain respects at least, lives a life that is different from other people. But most of what happens to us has happened somewhere else, too—to at least some other persons. We, like other people, are embedded in the larger networks of social structure, and this has a powerful influence on our lives. What makes us unique as individuals is the fact that each of us is embedded in the larger society in different ways; we inhabit different social networks and are subject to a different combination of social influences than are other people. One might say that there is a crisscrossing field of social forces, and that each of us is moving around in that field in a different location than other persons.

Thus, each of us has somewhat unique experiences. Falling in love or bringing up a child is never exactly the same from one person to the next. Nevertheless, these experiences are made up of certain common social ingredients. What sociology attempts to do is to analyze how these social ingredients, these general social processes, operate. Though these general sociological principles never completely explain everything that happens in one's own individual experience, nevertheless they provide a crucial background for understanding what is going on. We stand out as individuals against the background of society; but without society, we would not be what we are.

SUMMARY

1. The family is routine and taken for granted, but it is also the source of major emotional experiences, both positive and negative. It has sexual, emotional, and economic sides, all entwined.

2. Many issues surrounding the family are connected to patterns of stratification, including the dimensions of economic class, power, and status. These affect the positions of men and women both inside and outside the family.

3. Deriving income from two wage earners is now the most important way families acquire a middle or upper-middle class standard of living. Only a small proportion of families reach high income levels with only the man working; if only the woman works, the income tends to be below average or even at poverty level.

4. There is no natural "sexual division of labor" between men and women, since all arrangements are socially determined. Women do a great deal of practical, "instrumental" work, and men do carry out "expressive" emotional activities. Gender stratification processes have made women more likely than men to be in the lower-paying and less powerful jobs throughtout the different sectors of the labor force, and women do the bulk of the work within the home.

5. Theories of the family focus on general mechanisms that explain why aspects of the family take various forms under different social conditions. Comparisons among different groups and different societies are an important method of establishing such theories.

6. The family has an important link to social issues because it provides a major basis for perpetuating economic inequalities and ethnic/racial distinctions, as well as the domination of male over female. The family is also central to individual life experience. It has a powerful effect in shaping children's personalities and beliefs; and it is a major determinant of individual happiness at different points during one's life. In addition, family sociology is relevant to careers in many professions.

2

THE SOCIAL INGREDIENTS OF THE FAMILY

What are the basic features of the human family? We all know a family when we see one. But what are its essential parts? What makes it tick?

The question of defining a family may seem like a simple one. What automatically comes to mind is a husband, wife, and children, living together in a house. But families often are both more and less than this. If we bring other relatives into the household, they too are part of the family. Moreover, there is a sense in which we say people comprise a family even if they don't live together. This is what we mean when we say, "The family all gathered at Grandma's for Thanksgiving."

A family, then, can be more than husband, wife, and children in a home. Can it be less than this? Apparently, it can. The common household does not seem to be necessary. A married sailor in the navy still has a family when he is away for a year on his ship. In some tribal societies of the matrilocal type, a wife lives with her mother, while her husband lives with his mother and only visits his wife in the evenings.

Are children necessary for a family? This is a matter of debate. Some sociologists confine the term *family* to a couple with offspring, using the word *marriage* to refer to the husband-wife relationship by itself. According to this definition, a woman living with her illegitimate children has a family but not a marriage. Notice, though, that the children do not have to be biological offspring in order to constitute a family. Adopted children are treated as full members of a family in our own society. In some traditional societies, such as medieval Japan, adoption was very common, especially in families that lacked sons of their own to inherit the family property and carry on the family name. The crucial matter that defines parents and their children, then, is not the sheer biological fact of procreation, but a social consensus of relationship between them. This is seen in the distinction that is usually made between legitimate and illegitimate children. Those born without the proper social ceremonies of marriage between their parents are often not considered members of the family (at least not in the full sense), even though the biological ties are there.

Is it then the male/female relationship of a husband and a wife that is the core of the family? There is good reason to argue that this is so. But still we need to be careful. There are a number of practices found here and there in the world that violate this definition (see Ortner and Whitehead 1981). Among some African tribes, for example, there are cases of marriages between two females. In our own society today, there are communities in which the prevailing form is polygynous marriage—one man with a number of wives (see feature 2.1).

What definition we choose, of course, is in a certain sense arbitrary. We can just decide on what word we are going to attach to what observable thing in the world. The word *automobile* was picked out around 1900 when motor cars were invented, but it would work just as well to call them *velocipedes* or whatever we wished—as long as everybody was clear on what the word referred to. So if we liked, we could arbitrarily define the "family" to mean husband, wife, and children living together. The price we would pay for this definition would be that all sorts of arrangements among men, women, children, and households would fall outside the definition. We might overlook these because we wouldn't know what to call them.

FEATURE 2.1
Two Forms of Nonstandard Marriage

An African Marriage Between Women

In the Nuer tribe in the Sudan, it was possible for two women to marry each other. A wealthy woman who believed she was sterile could marry a younger woman by going through the same ceremony with her that was ordinarily used to unite a woman with a man. The older woman would find a man to make her "wife" pregnant. The older woman was called the husband in this marriage, and their children referred to her as "father." In this society, women were important economic producers and were able to accumulate property of their own, thus allowing some of them to forego marriage to a man. (Evans-Pritchard 1951, 108–9).

Polygymous Marriage in an Arizona Town

In the town of Colorado City, Arizona, polygyny is the dominant form of marriage. In virtually all families, one man is married to three or more women. Some men have as many as sixty children, and households of thirty or more are not uncommon. Consequently, the houses in this town tend to be huge, rambling structures; they are continually growing as the occupants add on new rooms to accommodate their growing families.

The town is just across the border from Utah and is inhabited by members of a dissident Mormon sect. When the Mormon church (the Church of Jesus Christ of the Latter-Day Saints) was founded in the mid-1800s, some members practiced polygyny, although there was controversy over its religious basis. In 1890, the Mormon church officially rejected polygyny because the U. S. Supreme Court had declared it illegal; but dissident groups broke away to continue the practice. One of these groups survives in Colorado City, while there are rival groups in a few nearby towns. All the land in Colorado City is held in common trust. When a young man wishes to marry, he must receive permission from the trustees in order to build a house. Since most of the wives go to a few of the men, not all males can marry. Community members say social pressure exists toward boys who rebel against this system. They are urged or even forced to leave town. Men who conform to the system, though, are given extra wives as rewards. The wealthiest men possess the greatest number of wives and derive status from the size of the family they can exhibit.

Girls, on the other hand, are under pressure to marry before they are eighteen. Marriage is described as the only sure route to heaven. Women are subordinate to men, and their primary task is to bear as many children as possible. The males wear more modern clothing styles, while the females wear their hair in the styles of the 1930s and dress in prairie-style dresses that they sew themselves. Makeup, earrings, or other jewelry are not allowed.

Though there is some dissension within the system, it also has strong advocates. One man recalled that his father had three wives. "I didn't feel one bit backward about the way I lived," he said. "The kids from monogamous families were the ones that were funny to me." Children generally refer to all their father's wives as "mothers," although some of them refer to them as "aunts." Said the informant: "How can it do anything but strengthen a home if you have two mothers?" (*Los Angeles Times*, April 13, 1986)

We should try to define the family as broadly as possible, so that we can study as many phenomena of this general type as are of interest to us. There is even a philosophical position, taken by such thinkers as Ludwig Wittgenstein (1953), that definitions of words are always shadowy and elusive at their borders. Any word covers a variety of instances that are connected together like a net but do not actually have any single characteristic in common to them all. According to this approach, we all have a gut-level feeling for what we mean by "family." Instead of trying to arrive at a precise definition that covers all instances, what we should do is take our commonsense understanding as a basis for exploration.

It is still worthwhile, however, to push our investigation far enough to find out what makes a family system work. Exploring the basic concept of the family is a way of explaining why families take the forms that they do. It is the first step in the search for a theory of the family.

It is sometimes asserted that the family is universal, found in all societies. It has to be so, according to this line of argument. For without the family, societies would not be able to reproduce themselves, physically or culturally. On the other hand, utopian thinkers from time to time have tried to envision worlds without the family. There have been attempts to put some of these plans into practice, for instance in American communes or in the Israeli kibbutz.

An aerial view of a kibbutz in the Jordan River Valley. These Israeli farming cooperatives assume many traditional responsibilities of the family by providing members with food, housing, child-care services, and education.

FEATURE 2.2
What Counts as a Wedding?

Wedding ceremonies vary a great deal from society to society. Our own wedding customs include many ritual elements: the church ceremony itself, the bride's white dress, throwing rice on the couple as they emerge from the church, tying tin cans to their car, the bride's throwing her wedding bouquet to the maids of honor. In other societies, the rituals are quite different.

In medieval Europe a couple was not married within the church but outside the church doors. This was because the medieval Church put its highest value upon a life of celibacy, and marriage was regarded as a concession to the weakness of the flesh. The bride wore a red gown, the color of sensual love; the white wedding dress did not come in until the 1800s (Ariès 1962, 357–58).

In the modern Israeli kibbutz, meals are eaten in a dining hall, and children are brought up in a collective nursery. All that is involved in getting married is for a couple to put in an application to live in the same dormitory room (Spiro 1956).

In the matrilineal society of the Trobriand Islands, premarital sexual intercourse is perfectly normal, and nobody comments on it. What the Trobrianders find shocking, though, is the modern Western custom of dating, in which an unmarried man and woman casually go out to eat. For among the Trobrianders, it is sharing a meal that constitutes the wedding ceremony (Malinowski 1929).

In ancient Greek society, the wedding consisted of transferring the bride from her father's domestic religion to her husband's. First her father offered a sacrifice on his domestic altar and declared to his family gods that he was giving his daughter into another family. Then she was veiled, dressed in robes reserved for religious observances, and taken to her husband's house in a procession. Those around her carried a torch and sang a religious chant, with the refrain *O Hymen, O Hymenaie* (Hymen being the Greek god of marriage). When they arrived at their destination, the husband pretended to seize the bride by force and carried her over the threshold, taking care that her feet did not touch the sill. Finally, she touched the sacred fire in her husband's house and was initiated into her new family worship, under the protection of her husband's family gods (Fustel de Coulanges 1973, 44–46).

What all these ceremonies have in common is simply the fact that the couple publicly show the community that they are now going to live together. The wedding is basically a public announcement; all the special ritual is to let everyone know that an important change is taking place.

(We shall see later to what extent these forms really do eliminate the family.) Other thinkers have predicted, on the basis of current trends, that the family may disappear in the future. This doesn't seem to be happening yet, although it is true that the forms of the modern family are shifting. But the question remains open: Is it really *possible* that a society could exist without *some* kind of family form? This is one of the questions that a basic theory of the family should help us answer.

THE FAMILY AS A PROPERTY SYSTEM

There are three kinds of property involved in the family. *Rights of sexual possession* include the rights of sexual intercourse and prohibitions on intercourse with outsiders. Sometimes the rights also extend to claims over a person's emotions or affection, although this is mainly found in modern societies like our own. *Economic property rights* include the material household itself, the income that supports the family, and the labor that different family members put into making the household a living concern. *Intergenerational property rights* include the rights that children have to inherit the family's economic property, and also the rights that parents have over their own children, economically and otherwise.

We should bear in mind that not all societies have the same particular rights under each of these headings. In some societies, these rights are heavily concentrated in the power of men over women, parents over children, or a single head of the family over all the other members. In other societies, there are more rights for women and for children. In some, fathers are all-important, while in others a woman's brother has far more power. But all societies have some way in which these three kinds of property are arranged. If we look at every society, including our own, in terms of these three kinds of property rights, we will gain insight into the basic dynamics that determine what goes on in the family. Moreover, the comparisons help tell us under what conditions these property relations vary: when there is the greatest amount of sexist oppression, when we find equality among men and women, or even what conditions produce female domination of such property rights.

Rights of Sexual Possession

The first kind of possession in any marriage consists of erotic rights over human bodies. In all societies, a marriage establishes the right to sexual intercourse between a particular man and woman. Sometimes, although not always, this is an exclusive right. Often marriages have given the man exclusive sexual possession over his wife while leaving him relatively free to have intercourse with others. In our modern family system, the ideal is bilateral sexual possession, with both husband and wife confining their erotic activities to each other.

To call this arrangement "property" may sound crass. After all, we usually think of property as a physical thing that someone owns. To speak of someone's body as being the property of another makes us think of slavery, degrading the human person to the staus of an inanimate object. Nevertheless, there are good reasons in sociological theory for extending the concept of property to erotic rights in marriage. Kingsley Davis (1936, 1949), who originated the term *sexual property*, pointed out that property is not the thing itself that someone owns, but a *social relationship*. That is, the concept of property is a kind of agreement among people about how they will act toward particular things. If you own a car, for example, that does not mean there is a bond between you and your car. What it means is that (1) you have the right to use the car; (2) other people do not have the right to use your car; and (3) you can call on the rest of society to enforce your rights; e.g., you can call the police if someone takes your car without your permission.

Erotic rights over human bodies are property rights in almost exactly the same sense. A marriage may be socially defined in our own society as follows: (1) the husband and wife have the right to sexual intercourse with each other; (2) other people do not have the right to sexual intercourse with either of them while they are married; and (3) if violations occur, the aggrieved party can go to court and demand damages, as in the form of a divorce. These property rights are different in certain respects than the rights people have over their cars. Husbands and wives in our society cannot buy or sell their sexual possessions, for instance, and they do not have the right to destroy them. Exactly what the rules of sexual possession are varies from one society to another. Some societies (such as ancient Rome) have allowed the husband to sell or kill his wife and children. The fact that we do not allow this does not mean that there is no property relationship; rather, the state does not allow people to do certain things with their sexual possessions, more or less the same way that the state does not allow people to do just anything they wish with cars, such as drive them without a license or faster than the speed limit.

If it seems crass to define married rights to sexual intercourse as a form of property, the concept nevertheless helps us to see certain things that we otherwise might miss. In many traditional societies, marriages were arranged quite impersonally as a collection of property rights and duties and involved little or no affection between husband and wife. But even in today's society, when there have been major changes in the kinds of property rights that married people have over each other, we still have a hard, tough core of demands that married people make on each other. Sexual rights remain a basic aspect of family possessiveness, and even our ideals of love and affection are actually treated as exclusive rights, as we shall see in a minute.

How can we show that marriage always involves sexual possession? For one thing, in practice we take marriage as being established by sexual intercourse. Suppose a couple is married in a religious or legal ceremony. Are they really married? Our laws actually say no, in that the marriage license and the ceremony do not have full force until marriage is sexually "consummated," as the expression goes. If the couple never have intercourse, say because of sexual impotence, then this is grounds for annulment. An *annulment* is not a divorce; it simply declares that the marriage itself never came into existence because an essential ingredient was missing. Even in very traditional societies that do not allow divorce (except under extreme circumstances that we will examine), lack of sexual consummation is grounds for annulment.

An act of officially approved sexual intercourse, then, is what constitutes marriage in our society, and in all others. I say officially approved, because intercourse may occur among unmarried people, sometimes quite openly,

FEATURE 2.3
Brunhild's Wedding Night: A Medieval Battle over Sexual Possession

The Nibelungenlied was the most popular German poem of the Middle Ages. In this excerpt from the story, the German King Gunther has brought back his new bride, Brunhild, from Iceland, where he won her favors in a contest to the death. But Gunther's friend Siegfried had secretly helped him win the contest, and Brunhild is suspicious. After a great celebration in his castle with all the knights of the realm, Gunther goes to his room to consummate the marriage.

Listen to how gallant Gunther lay with lady Brunhild—he had lain more pleasantly with other women many a time.

His attendants, both man and woman, had left him. The chamber was quickly barred, and he imagined that he was soon to enjoy her lovely body; but the time when Brunhild would become his wife was certainly not at hand! She went to the bed in a shift of fine white linen, and the noble knight thought to himself: "Now I have everything here that I ever wished for." And indeed there was great cause why her beauty should gratify him deeply. He dimmed the lights one after another with his own royal hands, and then, dauntless warrior, he went to the lady. He laid himself close beside her, and with a great rush of joy took the adorable woman in his arms.

He would have lavished caresses and endearments, had the Queen suffered him to do so, but she flew into a rage that deeply shocked him—he had hoped to meet with "friend," but what he met was "foe"!

"Sir," she said, "you must give up the thing you have set your hopes on, for it

Reprinted with permission of Penguin Books Ltd. from *The Nibelungenlied*, trans. A. T. Hatto (Penguin Classics 1965, revised edition 1969) pp. 87–93. Copyright © A. T. Hatto, 1965, 1969.

will not come to pass. Take good note of this: I intend to stay a maiden till I have learned the truth about Siegfried."

Gunther grew very angry with her. He tried to win her by force, and tumbled her shift for her, at which the haughty girl reached for the girdle of stout silk cord that she wore about her waist, and subjected him to great suffering and shame: for in return for being baulked of her sleep, she bound him hand and foot, carried him to a nail, and hung him on the wall. She had put a stop to his lovemaking! As to him, he all but died, such strength had she exerted.

And now he who had thought to be master began to entreat her. "Loose my bonds, most noble Queen, I do not fancy I shall ever subdue you, lovely woman, and I shall never again lie so close to you."

She did not care at all how he fared, since she was lying very snug. He had to stay hanging there the whole night through till dawn, when the bright morning shone through the window. If Gunther had ever been possessed of any strength, it had dwindled to nothing now.

Gunther confides in his friend Siegfried, a knight who has magic powers. Siegfried agrees to make himself invisible and come to help Gunther that night.

Siegfried has gone to Gunther's chamber where many attendants were standing with lights. These he extinguished in their hands, and Gunther knew that Siegfried was there. He was aware what Siegfried wanted, and dismissed the ladies and maids. This done, the King quickly thrust two stout bolts across, barred the door himself, and hid the lights behind the bed curtains. And now mighty Siegfried and the fair maiden began a game there was no avoiding and one that gladdened yet saddened the king.

Siegfried laid himself close to the young lady's side. "Keep away, Gunther, unless you want a taste of the same medicine!" But Siegfried held his tongue and said not a word. And although Gunther could not see him, he could plainly hear that no intimacies passed between them, for to tell the truth they had very little ease in that bed. Siegfried comported himself as if he were the great King Gunther and clasped the illustrious maiden in his arms—but she flung him out of the bed against a stool nearby so that his head struck it with a mighty crack! Yet the brave man rebounded powerfully, determined to have another try, though when he set about subduing her it cost him very dearly—I am sure no woman will ever again so defend herself.

Seeing that he would not desist, the maiden leapt to her feet. "Stop rumpling my beautiful white shift!" said the handsome girl. "You are a very vulgar fellow and you shall pay for it dearly— I'll show you!" She locked the rare warrior in her arms and would had laid him in bonds, like the King, so that she might have the comfort of her bed. She took a tremendous revenge on him for having ruffled her clothes. What could his huge strength avail him? She showed him that her might was the greater, for she carried him with irresistible force and rammed him between the wall and a coffer.

"Alas," thought the hero, "if I now lose my life to a girl, the whole sex will grow uppish with their husbands for ever after, though they would otherwise never behave so."

The King heard it all and was afraid for the man; but Siegfried was deeply ashamed and began to lose his temper, so that he fought back with huge strength and closed with Brunhild desperately. To the King it seemed an age before Siegfried overcame her. She gripped his hands so powerfully that the blood spurted from his nails and he was in agony; but it was not long before he forced the arrogant girl to recant the monstrous resolve which she had voiced the night before. Meanwhile nothing was lost on the King, although Siegfried spoke no word. The latter now crushed her on to the bed so violently that she shrieked aloud, such pain did his might inflict on her. Then she groped for the girdle of silk round her waist with intent to bind him, but his hands fought off her attempt so fiercely that her joints cracked all over her body! This settled the issue, and she submitted to Gunther.

"Let me live, noble King!" said she. "I shall make ample amends for all that I have done to you and shall never again repel your noble advances, since I have found to my cost that you know well how to master a woman."

Siegfried left the maiden lying there and stepped aside as though to remove his clothes. And now Gunther and the lovely girl lay together, and he took his pleasure with her as was his due, so that she had to resign her maiden shame and anger. But from his intimacy she grew somewhat pale, for at love's coming her vast strength fled so that now she was no stronger than any other woman. Gunther had his delight of her lovely body, and had she renewed her resistance what good could it have done her? His loving had reduced her to this.

And now how very tenderly and amorously Brunhild lay beside him till the bright dawn!

(Continued next page)

FEATURE 2.3 (*Continued*)

A *sociological comment:* The story recalls the violent Viking age several centuries before *The Nibelungenlied* was written. Brunhild is something like the Valkyrie, the mythical armed women who accompanied the Norse god Odin. There must have been some reality behind these stories. Iceland, where Brunhild had reigned as a maiden queen, was the quasimythical Viking frontier where, according to the poet, there was a matrilineal system and female warriors of extraordinary strength. We notice that marriage is conceived of rather crudely, especially from the perspective of the poet's own age, when the Vikings had given way to a polite courtly society of knights and ladies.

There is no church ceremony, and the wedding consists only of a large feast and then the first sexual intercourse itself. Brunhild defends herself, in a kind of temporary triumph of the old matrilineal ideal. Then Siegfried shows that his masculine powers are even greater, and she is subdued. Notice that after she has finally submitted to intercourse, she suddenly loses all her strength; the Viking warrioress is turned into a soft and amorous lady of the new courtly style. One can interpret this as a fight between two different systems of sexual property, with a male-dominated system overcoming a female-dominated one.

without implying any consequences. The official sanction takes the form of some kind of public announcement, which brings the rest of society into the picture. Through its representatives, the courts and/or religious leaders, society thus takes a hand in recognizing that rights to sexual intercourse have been established, and thus is ready to back up these rights by appropriate punishments when they are violated. In all societies, a marriage takes the form of some kind of public ceremony that makes this sort of announcement to the community. In early medieval Europe, for instance, the marriage was not religious at all but took the form of a large feast at which relatives and friends were entertained on the night the couple had their first (at least "official") intercourse (see feature 2.3).

The fact that sexual possession is a key to marriage is brought home to us by the way in which people react to its violation. The violation of sexual possession is *adultery*. This is considered a crime in most societies, although some punish it much more severely than others. In conservative, Catholic-dominated countries such as Italy divorces were not allowed up until just a few years ago, although an exception was made in the case of adultery. This is because adultery was considered such a major violation of the central features of marriage that a marriage could not survive it. The importance placed on sexual rights in many societies is illustrated by the unwritten law that a man would not be convicted of murder for killing his wife's lover. (This unwritten law was usually rather sexist in operation, since it less frequently condoned a wife's killing her husband if she caught him in adultery.) The degree of such sentiments has varied at different times. In ancient Greece and Rome, a husband could kill his adulterous wife and her lover, and the law took no notice of it. Today, a court action would almost certainly be instituted, although juries are inclined to give relatively light sentences to murders committed out of this kind of jealousy.

The modern law also recognizes sexual rights in another sense. In most states, the husband has the legal right to sexual intercourse with his wife (and vice versa); hence, in these states, there is no crime of rape between married people, according to the law. This implies that the marriage contract itself gives away one's right to control one's own sexuality, once and for all; until a divorce takes place, the law has nothing to say over what happens sexually between the couple. As a result of the widely publicized Rideout case in 1978, in which a husband was charged with the rape of his wife while still residing with her, a number of states (including California, Massachusetts, and Minnesota) have passed laws prohibiting marital rape. These laws, and the cases that have arisen since their passage, are controversial, which indicates how strongly held are our beliefs about this kind of "property." Marriages are changing in the direction of sexual rights by mutual consent, but still recognizing married couples as sexual partners.

There are certain practices that might seem to violate this conception of marriage as sexual possession. Some societies allow institutionalized wife swapping or extramarital affairs, and others practice polygamy. But here is where our basic theory of property is very revealing. If you own a car, it does not violate your right of property if you let somebody else drive it. In fact, a loan or a gift actually emphasizes whose property it is. In order to borrow something, one person has explicitly to ask the owner and be granted permission; so the very act of borrowing involves a social acknowledgement of who actually owns the property. The same is true in societies that practice "wife swapping." This was once fairly common, for instance, among Eskimos (Hoebel 1954). Eskimo men took long hunting trips, and when one stopped in for hospitality at a distant house, the host would often lend the visitor his wife. In reciprocity, the host could expect the same courtesy when he traveled far from home. That does not mean that Eskimos had no concern for sexual possession. They certainly did, and Eskimos often quarreled when a man took someone's else's woman without asking. The Eskimo murder rate, in fact, has been one of the highest in the world. Similarly, when mate "swapping" is practiced in groups of "swingers" in our own society, there are always explicit agreements that the relationships are temporary and that each partner gets something in return (see chapter 9).

The Incest Taboo and Exogamy

Property rights, as we have seen, have both a positive side—who has the rights to something—and a negative side—who is excluded from it. One of the most important forms of sexual property is the negative side: who is *excluded* from sexual intercourse with whom.

The incest taboo is a restriction on sexual intercourse, and hence on marriage, among close relatives. But what has counted as "close" has varied historically. Medieval Christian societies prohibited marriage with cousins, as well as aunts and uncles, nephews and nieces, out to the seventh degree (Goody 1983, 44); whereas in many tribal societies first cousins are the preferred marriage partners. But mother-son and father-daughter intercourse has been condemned everywhere as incestuous; brother-sister marriage, although allowed in a few places and circumstances, has also generally been prohibited.

The incest taboo is usually thought of as a way to prevent inbreeding. It is not really useful to think of this as a biological phenomenon; clearly incest prohibitions are not instinctual, since they are violated often enough, whereas instincts are incapable of being violated. Nor should we expect that people throughout history and back into primitive times would have had a theory about the genetic effects of inbreeding, since genetics was discovered only within the last 100 years. The incest taboo has been so powerful because it is so central *socially*. As many theorists have pointed out, without the incest taboo human society would scarcely be possible. The incest taboo forces individuals to go outside of the families in which they were born in order to find sexual partners. Without this pressure, each family could stay isolated, continually inbreeding among its own children. The world would consist of little isolated families, with no contact or exchange among them. The incest taboo is so powerful socially because it lays down the rule that *no family can exist alone*.

How did the incest taboo come about, if it is not instinctual? One cannot simply invoke a psychological explanation. Some theories have suggested that there is a natural lack of sexual attraction among persons who are brought up together. But this certainly isn't true among many animal species, and Sigmund Freud claimed that the unconscious wishes of his patients showed they frequently had incestuous desires. Freud thought that some crucial event occurred early in human history that enforced the incest taboo. Freud's own speculation about what happened is not widely accepted, but he is probably right that the incest taboo did not emerge from inner psychological processes but had to be *enforced from the outside*.

The French anthropologist Claude Lévi-Strauss (1969) proposed a method of reconstructing what happened in unwritten family history. Lévi-Strauss argues that the incest taboo does not stand by itself but is part of a larger system of family structure. The incest taboo can be regarded as a kind of treaty between different families: each one promises that it will send its own children out to find marriage partners if other families will do the same. There are different ways in which this can be done. In some systems, the women leave home, and other women come in as wives for the men who remain there. (This is called a *patrilocal system*.) In other systems, the men live with their sisters, and husbands are temporary visitors in the homes of their wives. (This is called a matrilocal system.) In both cases, there is an exchange. Lévi-Strauss argues that families exchange women because the incest taboo requires that they cannot be sexual partners for those who stay at home.

Viewed in this way, the incest taboo is a political treaty among families. Every time a marriage is made, one family has given away a daughter or a niece. In most cases, there isn't a corresponding man or woman from the other family to be given back right away in a compensating marriage. If the Jones family (or the Black Eagle clan), marries a daughter to the Smith family (or the White Eagle clan), it doesn't usually happen that the Joneses (Black Eagle) have a son ready to marry a Smith (White Eagle) daughter at exactly the same time. But a bond is created between the two families because they have intermarried once and expect that when the opportunity arises there will be an intermarriage in the opposite direction. Lévi-Strauss points out that it is even useful that the exchanges do not happen in both directions at the same time. It is better that one family should be waiting in expectation, so that the alliance between the families always has something to look forward to. It is these *overlapping* exchanges that make for the strongest bonds.

Lévi-Strauss goes on to make a technical analysis of many of the different types of marriage exchange systems that exist in tribal societies. In some of them, a man is expected to marry his cousin on his mother's side and the woman marries her cousin on her father's side; in other systems, the preferred cousins are chosen in a different way. Each type of marriage system links together different sets of families. As a result, there are different types of social structures. For in tribal societies, there is no state, no police force, no stores or factories, no economic or political organizations other than what the family provides. The network of intermarrying families *is* the entire social structure. It is by working out these rules of intermarriage that larger units beyond each individual family are created. Moreover, these families do not merely exchange men and women as marriage partners; they also exchange food, help in building houses or clearing land for crops, and cooperate in hunting together. The group of families that has intermarried constitutes a political unit, to defend one another against attack and to mount expeditions attacking their enemies. Intermarriage is thus not just a matter of getting a sexual partner for each individual but creates the entire alliance structure of the society. It is not only a sexual activity but also an economic and political one.

We can see now why the incest taboo was so important in getting human society started. The taboo is part of a system of sexual exchange that makes possible all the other economic and political exchanges. The incest prohibition is the negative rule in this system; it says that families must not use their own members for sexual partners. Corresponding to it is a positive rule, which says that families must go outside to find new sexual partners. By going outside, alliances are made, and the larger society becomes possible.

The incest taboo, then, did not emerge from some kind of repulsion inside the family. It appears to have been negotiated between groups, as part of a package deal that they would begin to intermarry in a particular way. Lévi-Strauss implies that those families that made these treaties and began to prohibit incest became stronger because they gained allies. They were more assured of food and military power than families who did not prohibit incest. Hence the families with the alliance system would have survived better than the isolated, incestuous families that failed to make the transition. Eventually all families that remained had to abide by the incest taboo and make their children available only to outsiders. When they failed to do so, they were punished from outside. The surrounding society came to expect that children

FEATURE 2.4
Incest in Ancient Egypt

Marriage or sexual intercourse between brother and sister has almost everywhere been forbidden as incestuous. Nevertheless, there are a few instances historically where such marriages were allowed or even encouraged. In the royal family of native Hawaii, for example, brothers and sisters regularly married, and so did siblings in the succession of the Egyptian pharaohs. Cleopatra was married to her brother (in fact, to two of them, Ptolemy XII and Ptolemy XIII), and so were many other Egyptian queens. Both of these patterns can be explained by kinship politics. Since intermarriage is a way of making alliances between families, royalty attempts to marry only the children of royalty. Both Hawaii and the Kingdom of the ancient Nile Valley were quite isolated from other societies; hence there were no other kingdoms nearby with which their royalty might intermarry. Hence, brother-sister incest became a solution for keeping inheritance strictly within royal lines.

In ancient Egypt around the time of the Roman Empire, however, brother-sister incest was much more common even than this. Among the landowning classes in general, a very high proportion of marriages were between brothers and sisters. Jack Goody (1983, 43–44) points out that this was a strategy for keeping family inheritances intact. Under Egyptian law of this period, women had considerable rights of inheritance. At the same time, land holdings along the narrow irrigated strip of the Nile were extremely valuable, and the elite of wealthy families who owned them were very reluctant to break them up. Hence brother-sister incest, which had already been legitimized by the marriages within the family of the pharaohs, was adopted by the landowning classes in general as a way to keep the family property together.

After the Christian church began to become powerful, however, it strongly condemned these marriages as incestuous. Elsewhere in the Roman Empire and in northern Europe, too, the church fought against marriages among cousins, men marrying the wives of their deceased brothers, and other marriages among close relatives. Goody (1983) interprets this as an economic strategy on the part of the church. For the older marriage customs were all ways of keeping inheritance within the extended family, by ensuring that there would usually be an heir available. The church, though, benefited from bequests made to it by people who died without heirs, and this was one factor in its efforts to eliminate these "incestuous" practices. By the late Middle Ages, the definition of incest had been extended so that it included relatives no more closely related than seventh cousins.

would be reserved as sexual partners for someone outside the family and reacted angrily against violators who failed to keep their part in the social alliance system.

As we shall see in the next chapter, there are quite a few different ways that sexual exchange systems can be organized. In some societies the older generation does most of the exchanging, using their offspring as pawns in a game of marriage politics. In other societies, the younger generation is able to negotiate its own individual exchanges. The shifts among these types of exchange systems make up a major part of the history of the family.

There are also differences in how large a network is involved in exchanges. Some societies have enforced exchanges that go very far from home, whereas others exchange only among a sort of extended family. Continual interchanges among a group in which the cousins are constantly intermarrying would be regarded by us as rather incestuous, even if they did go outside the immediate nuclear family. Like everything else in human societies, the incest taboo is a variable, and its extent is a matter of degree. In one period in ancient Egypt, marriage between brothers and sisters was fairly common. It took special economic and political circumstances for this to happen, and the coming of Christianity soon took family policies to the opposite extreme (see feature 2.4).

Unilateral and Bilateral Sexual Possession

Sexual possession, like any other kind of property, can be distributed more equally or unequally. In many traditional societies, it constitutes what may be called one-sided or *unilateral sexual possession:* the woman is the sexual possession of the male but not vice versa. In medieval Arab societies, for example, women were strongly restricted as the sexual possessions of men. They were kept secluded in harems, and virginity at marriage was considered a prime mark of a woman's value. A man's honor depended upon being the only sexual possessor of a woman, and her marriage value was zero if she was not a virgin. But Arab men were not at all restricted. They could have intercourse with prostitutes and slaves and keep concubines without the wife's having any say in the matter. In such a society, we would say that there is one-sided sexual possession: of men over women but not vice versa.

Sexual possession is not necessarily organized in this extreme male-dominated form. In many societies women have sexual rights over their husbands. This, in fact, is often true in certain kinds of polygamous societies. In Madagascar (off the coast of Africa), wealthier men usually had several wives (Linton 1936). Each wife had her own hut as part of the family compound. The husband was supposed to rotate regularly among them, spending a night with each in her turn. If the husband violated the rotation by spending an extra night with a particular wife, the other wives considered that he had committed adultery. This was a type of horticultural society in which women had con-

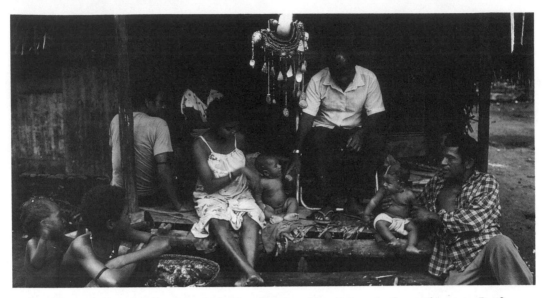

The paramount chief of the Trobriand Islands, off eastern New Guinea in the southwestern Pacific, visits with relatives, who sit respectfully at his feet, at the home of one of his wives. In the Trobriands, polygamy is practiced by a few men of high rank.

siderable economic importance as food producers, and hence women had a certain amount of power to assert some controls over men.

Our own society today is officially closer to a form that can be called *bilateral sexual possession*. This means that in principle, both husband and wife are sexual possessions of each other. Neither one has the right to extramarital affairs, and both have sexual rights over each other's body. This has been the official ideal in Western societies for at least the past few hundred years, although there has been some tendency for it to be violated in practice, especially by men. In recent years, though, there is some evidence that women are now committing adultery as much as men in some sectors of society (Blumstein and Schwartz 1983), although men still remain ahead on that score (Smith 1991).

The whole issue of sexual possession, and especially its unilateral or bilateral nature, remains a matter of controversy among sociologists and feminist theorists. Gayle Rubin (1975), using a combination of anthropological evidence and Freudian psychoanalytic theory, has argued that the basic pattern in all societies is for men to exchange women. That is, there is a basic taboo, the incest taboo, that requires that individuals must leave their own families to find sexual partners. Hence, one can say that families are actually exchanging sexual partners. In tribal societies, as many anthropologists have pointed out, these exchanges are deliberate political maneuvers by which families establish alliances with one another. In many such tribal societies (as well as medieval societies like the Arab example given above), the exchange is explicitly controlled by men: women are treated as sexual property whom men use to make exchanges for the purpose of political alliances. Rubin argues that although the family political alliances have disappeared, the basic psychological structure of *men possessing and exchanging women* continues to exist in our own society.

Thus, for Rubin one-sided sexual property still exists, and is the basis of gender inequalities throughout our society. In her analysis, the deep structure of the problem goes back to a second taboo that is implicit in the incest taboo: a rule of "compulsory heterosexuality" that requires male and female children to give up any erotic drives for members of their own sex. Rubin argues that small children, in fact, have no sexual preference until it is forced upon them by the conventions of culture. She calls for a revolution in the family, in order to overturn the foundations of the system of gender inequality. If males and females took equal part in caring for babies and small children, Rubin argues, the infant's original bisexual orientation would not be overturned by the imposition of heterosexuality. Gender distinctions would disappear, as it would make no difference whom one chose as a love object. Males would no longer be regarded as dominant and superior, and men would no longer control women and exchange them in a system of sexual property. The result, in Rubin's theory, would finally be to overturn the culturally based system of gender differences and gender inequality in all its aspects—economic, familial, and sexual.

Rubin's theory is one of the most challenging of radical feminist theories, and its implications continue to cause serious discussion. At the same time, from a sociological viewpoint, it seems to have certain flaws. The most basic one is the claim that the system of sexual property exchange in our modern society is basically the same as in tribal and traditional societies. Rubin is an anthropologist, and her empirical material comes entirely from studies of small tribal societies in which the family organizes the entire political and economic system. What she does not pay attention to are the large historical shifts that occurred in the family, first as the state emerged and broke down these family networks, and later, as families lost control over the sexual bargaining and marriages of their children. (These historical changes are described in chapters 3 and 4 of this book.) It no longer makes sense to claim that there is now a system in which men exchange women, although that was true in some societies at some times. There continues to be an exchange in the form of the modern marriage market, but this is a system in which *individuals*, both males and females, are doing their own exchanging. Gender inequalities may enter into this, insofar as males may often have more income or higher prestige occupations than females. But this does not seem to be inevitable. Nor is there any basic cultural rule in our society that says that *men* exchange *women*. It does not necessarily seem to be true that all persons would have to become bisexual in order for gender equality to come about. It is quite plausible that gender equality could be accomplished by greater equality in economic opportunities for women and men, and that this, in turn, would change the traditional gender roles in the family and the marriage market.

Love as Emotional Possession

The reason that this kind of analysis of sexual possession sounds somewhat improper to our modern ears (although it hardly would have in medieval society) is that we have added another element to the modern marriage. That is the ideal of love. In our society, unlike those of the past, marriages are supposed to be founded on love. Since we commonly distinguish between love and sex, defining marriage in terms of sexual possession sounds unduly cynical: a violation of the spirit of mutual affection and caring that makes up the bond of love.

Nevertheless, our ideal of love does not so much replace sexual possession as add another dimension to it. For one thing, love and sex are not really so separate in our own minds. Sex can take place without love, but our ideal says that they should take place together. Love without sex, between a husband and wife, would be considered rather strange by most of us, and a partner might very well doubt the other's love if she or he refused ever to "make love," in the telling phrase.

What helps clinch the argument is the fact that we tend to treat love as a form of property too. The very language of love reveals a distinct undercurrent of possessiveness. Probably the most common expressions of love are phrases like: "Will you be mine?" "I'm yours forever." "I've lost my heart." Popular love songs have more of this kind of terminology in them than the ritualistic "I love you." Words like *mine* and *yours* are probably more common in love talk than the word *love* itself.

In our modern system of marriage, we have simply created a new form of property, *emotional possession*. Love is a right that people acquire over each other's affections. According to the modern marriage ideal, this exchange of love vows should take place between every married couple, and traditionally this love bond was supposed to last for a lifetime and be inviolable to any outsider. This is even recognized in the law. The phrase "alienation of affection" describes the violation of emotional property rights that has been recognized as a basis for divorce and damage suits. Ideally, love-property is supposed to be bilateral between both partners; both love each other equally. In practice, of course, this is not always the case. In the love affairs of courtship, men and women often play a psychological game over who is more in love with the other. Like other forms of property, possession of the emotional rights of love can lead to a good deal of maneuvering for domination. In fact, this is one of the traditional points made by most writers on the subject of love for centuries. Love itself is not at all a cynical sentiment, but it is nevertheless implicated in the struggles of a property system. Loving someone and wanting their love is usually a form of possessiveness that works in much the same way as traditional sexual rights.

Sexual Possession or Legitimacy of Childbearing?

The anthropologist Bronislaw Malinowski (1929, 1964) pointed out that sexual intercourse is widely available outside of marriage in many societies, but that bearing illegitimate children is prohibited or disapproved of everywhere. Hence, the purpose of the family is to regulate not sex but childbearing. Malinowski's principle of legitimacy states that all societies demand that every child have a sociological father as an official male link between the child and the community. When Malinowski formulated his principle of legitimacy, one of his main arguments was that marriage is not a license for intercourse. This seems to flatly contradict the theory that marriage is sexual possession. Malinowski observed that the Trobriand Islanders, a tribal horticultural society, were shocked by illegitimacy but allowed wide-open premarital intercourse. He pointed out that many other tribal societies put no restrictions on premarital sex. Hence, sex is not what constitutes the family, since that can be procured outside of it. This left Malinowski's candidate theory, legitimacy, as the explanation of the core of the family.

Several points can be raised against Malinowski's argument. Malinowski happened to have studied the Trobriand Islanders, whose culture is similar to many other Melanesian, Polynesian, and Oceanian peoples. These cultures are among the most sexually permissive ever known. Most societies, on a world scale, have been a good deal more restrictive of individual sexual rights, at least as far as women have been concerned. The great agrarian empires and states that have existed in the last two or three thousand years in China, India, and Eurasia have comprised the great bulk of the world's population. There, women were usually very restricted, while men were allowed nonmarital carousing. The segment of tribal societies that practiced wide-open premarital intercourse by both sexes includes only a tiny proportion of all the people who have ever lived on the earth. Malinowski was basing his argument on a fairly narrow base. It is true that our own society also has recently moved into a rather widespread pattern of premarital intercourse by both sexes. Does this mean that sexual possessiveness is disappearing? The evidence that we will examine in later chapters indicates that it is not.

Malinowski is confusing sex per se with sexual possession. Just because sex occurs outside of marriage doesn't mean that marriage isn't a contract of sexual possession. Possession means that somebody has the perpetual right to intercourse until this right is terminated by death or dissolution of the marriage. It also means, almost always, the *exclusive* right to sexual intercourse. As we have seen, this right has usually been exercised by men over women more than by women over men. Men have almost always wanted their women to be "faithful" and "chaste" at least while they were married to them, without applying the same standard to themselves. This is what we have called one-sided or unilateral sexual possession. But some kind of sexual possession is always there in a marriage. What varies is how widely premarital intercourse is allowed. But this is just part of a larger family system in which sexual possession is a strong feature. Malinowski's own Trobriand Islanders believed a married woman should reserve herself for her husband and regarded her violation as adultery.

The comparative evidence is on the side of the property theory. Marriage is an exclusive right to sexual intercourse on the part of a man, a woman, or both. The universal existence of the incest taboo supports this; a mother cannot marry her son, for example, because of the taboo on intercourse between them. Where sexual possession is not allowed, marriage is not allowed. The mother is the exclusive sexual possession of the father; the incest taboo on the son is simply the flip side of the father's property right. Similarly, the incest taboo on father-daughter intercourse upholds the mother's sexual right over her husband, at least from rivals among her own dependents.

The property argument is wider and more inclusive than the legitimacy argument, for the legitimacy argument can be better handled within it. Sexual possession automatically carries with it legitimacy of any children that are born out of the relationship. In effect, the parents announce to the community that they are establishing a long-term sexual relationship. In return, the community regards their child as legitimate. Sexual property is more basic and comes first. Once there is sexual property, you automatically get legitimacy.

From the point of view of the property theory, Malinowski's legitimacy theory is too narrow. It only deals with the third kind of property, intergenerational property. A legitimate child is one who is entitled to inherit the family property and its social status in the outside world. Legitimacy, then, is not merely society's way of connecting children with fathers and hence providing them with a link to the social structure; it is also another facet of the family property system. It is not surprising that the highest illegitimacy rates in the world occur in the poorest societies and in the lowest economic classes. Ghetto blacks in urban America or slum dwellers in Haiti have little concern for legitimacy because they have no property, and—a related fact—women have more autonomy from men.

Sexual possession links together the other two aspects of the family, intergenerational property and economic property (see figure 2.1). Intergenerational property (children and their inheritance) is a result of sexual possession. And economic property is always implicated in any permanent form of sexual possession, that is to say, in any marriage. A marriage always has some

FIGURE 2.1: Sexual, Economic, and Intergenerational Property

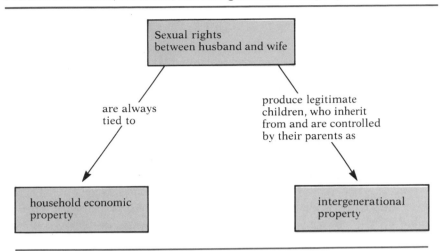

provision for family housekeeping, and it almost always transfers some economic rights among the man, the woman, and their families. It is for this reason that the sexual rights of women have followed the ups and downs of their economic position. Marriages are an arrangement in which sexual rights and economic rights are always involved. Other things may be involved too, but these form the core that makes the institution work. The sexual rights that each partner gets depends on how much power they have, and economic resources have been a key to the balance of family power. When women have had proportionately more control over their own livelihood, and even over that of the man, they have had correspondingly greater sexual rights (Blumberg 1984). When men have held all the economic resources, one-sided or unilateral sexual possession has prevailed.

Economic Property and Household Labor

The modern family is not just a sexual arrangement. In day-to-day life, the most obvious thing about it is its economic activities. In fact, these are so obvious that they are often taken completely for granted. The family is a kind of business, engaged in the enterprise of running a household. It is an enterprise for cooking food, cleaning clothes, providing lodging, and taking care of children. We might say that the modern family is a mixture of hotel, restaurant, laundry, and baby-sitting agency.

This has always been true. Just how the household economy was arranged has varied among different societies in history, but all of them have taken care of these basic economic needs.

In agrarian societies, like medieval Europe, China, or the Arab world, or in ancient Greece, Rome, and Egypt, households pretty much made up the entire economy. Food was produced by family farms, whether large manors or peasant holdings; cloth was woven by women in their quarters, and most of the necessities of life were produced in the household itself. Even the specialized trades were carried out in households. A medieval merchant did not have business hours, the way that stores today are open between 9:00 A.M. and 5:00 P.M. (or some other fixed time). The medieval merchant was always open, because he lived in his shop—or, to put it differently, he worked in his home. The family and the business had not yet been separated.

In the tribal societies studied by anthropologists, the situation is different once again. But the family is still very central as an economic institution. Tribal societies have much less stratification into wealthy and poor classes than agrarian societies do; what they have instead is a widespread horizontal linkage among families. The family network is the main basis of the tribal economy, its politics, war, and religion. In fact, virtually no other organizations besides the family exist in such societies. Thus, a family that has taken a man or woman in marriage from another family might be obligated to deliver yams or other kinds of food at certain times of the year. In return, the other family might have to help in joint fishing expeditions or other collective economic activities. The family and kinship network *is* the economic system, comprising much of what we would consider the market.

There are three things we should notice from these comparisons:

1. In all societies, the family is economic. In fact, our word *family* comes from the ancient Roman word for "household," including property as well as people (see feature 2.5). In our modern society, the family is much less important in the total economy than it once was, but it is still very important for the family members themselves. Children, of course, demand economic support from their family, but so do adults. Legally, the husband and wife hold their household property in common. If the husband alone works, the wife has the right to economic support. If she works, her husband shares her income. In both cases, there is also the nonmonetary side of the family economy: the labor put into running the household. Usually, most of this labor is put in by the woman, in the role of housewife.

2. The family economy is always connected to sexual possession. In tribal societies, this marriage market is recognized fairly explicitly as part of the system of economic exchange. Hence, many tribes have specific rules about whom children should marry. For example, some tribes want cousins on their father's side to intermarry (patrilateral cross-cousin marriage), while others prefer intermarriage of cousins on the mother's side (matrilateral cross-cousin marriage). These rules keep certain branches of the family tied together by repeated exchanges generation after generation.

FEATURE 2.5
Where Does the Word *Family* Come From?

Our modern term *family* comes from the Latin. But the ancient Romans did not use *familia* to mean blood filiation or kinship. It meant rather the household property—the fields, house, money, and slaves. The Latin word *famulus* means "servant." In Rome, the plebian form of marriage consisted of a man buying his wife, and she became recognized by the law as part of his property—his *familia*.

The ancient Greeks used the word *oikos*, which is translated "family." But again, it meant "property" or "domicile." *Oikos*, in fact, is the Greek root for the words *economy* or *economic*: the first economy that the ancient civilizations knew about was the family economy, the property that was produced and consumed in the household. Aristotle said that the family (*oikos*) was composed of three elements: the male, the female, and the servant.

The Latin word *pater*, which is related to our word *father*, also originally had an economic and political rather than a sexual connotation. The term *paterfamilias* could be applied to a man who had no children and was not even married—for instance if a youth inherited the family property because of his father's early death. The *pater* was essentially the domestic authority. Slaves and clients applied the term to their masters. A man might address a god as *pater*, and poets used the word for anyone they wished to honor. It was synonymous with *rex* or *basileus*, which we translated as "king" or "chief."

In Arabia at the time of Mohammed, the word for marriage was *Nikah*, which literally meant sexual intercourse. In the Koran it was also used to mean a contract. Marriage thus was conceived of as a contract for sexual intercourse. (Fustel de Coulanges 1973, 89–90, 107; Bullough 1974, 67, 141; Snodgrass 1980, 79)

Many agrarian societies had the institution of the dowry, wealth that a bride brought to her marriage. Without a good dowry, a woman (at least of the higher social classes) was not very marriageable, and families went shopping rather coldly for a woman with a proper dowry attached. Some tribal societies went to the opposite extreme with the institution of the bride price. A woman was so valuable economically that a man paid to marry her, trading such goods as cattle (Goody 1976).

In our own society, the economic trade is almost entirely between the individual husband and wife, not between their families. But the trade remains important. Some theorists have argued that the basis of gender inequality in our society is economic inequality: men have higher incomes, while women have to compensate by providing the menial household labor. In all societies, the marital deal seems to be some version of material property traded in return for an arrangement of sexual possession.

3. Sexual possession and economic property thus always go together to make up the family system, but in varying combinations. As we've seen, there are different kinds and degrees of sexual possession, and the same is true of family economic property relationships (see figure 2.2). There are many different combinations possible and hence we find a variety of family types around the world and throughout history. Our own typical family arrangement is only one among many. It is not always necessary to divide the bundle of

property rights among family members in the same way. In matrilineal/matrilocal societies, for example, the husband has sexual rights but little or no economic property in the marital arrangement. His wife keeps her property in her own lineage, often under the control of her brother. Economic property and sexual possession are thus split in this kind of system. Intergenerational property rights are also split off from the husband's control, as the matrilineal principle dictates that children inherit from the female line, not from their father. But even in

FIGURE 2.2: Some Patterns of Family Descent and Inheritance

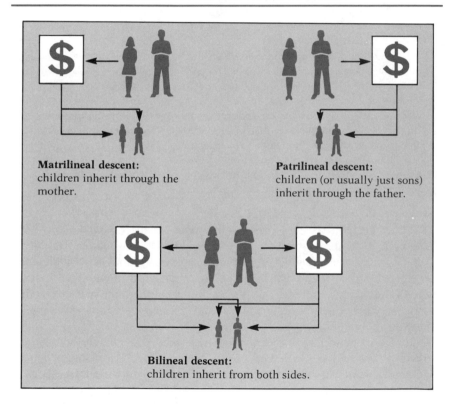

Matrilineal descent: children inherit through the mother.

Patrilineal descent: children (or usually just sons) inherit through the father.

Bilineal descent: children inherit from both sides.

Interestingly enough, there has been a tendency for beliefs about impregnation to reflect the lineage structure. Many matrilineal societies deny that sexual intercourse with the father has anything to do with pregnancy; the child is viewed as produced entirely by the mother (Malinowski 1964, 7). In patrilineal societies, there is a tendency for men to assert that the child is entirely formed by the father's sperm, while the mother's womb provides only the receptacle (Bullough 1974, 62). Sociologically, we would regard these beliefs as *ideologies*, attempts to justify the prevailing kinship system.

TABLE 2.1   Some Combinations of Family Ingredients

	Sexual possession	Economic property unit	Childbearing and rearing
Monogamous marriage	yes	yes	usually yes
Cohabiting couple	yes	yes	usually no
Single-parent family	no	yes	yes
Gay or lesbian couple	yes	partly	usually no

this case, in order to comprehend the family institution of the matrilineal society, we have to know how *both* sexual possession and economic property are organized.

Cohabiting, Single-Parent, Gay, and Lesbian Couples

In our own society of the late twentieth century, the various ingredients of the family seem to be coming apart again. It is as if we were experimenting with all sorts of ways of combining different pieces of family structures (see table 2.1). Many men and women live together as *cohabiting couples* without legally getting married. For all intents and purposes, this is like a marriage as far as the sexual aspect goes: the couple has sex regularly (see chapter 9) and excludes other persons sexually, just like a marital rule of fidelity. They are also sharing a joint household, and in that sense have set up something like the common economic property that constitutes a family. The legal standing of this joint property is still rather iffy. But some courts are treating cohabitation as if it implied the same kind of property contract as a regular marriage, so that there are property rights to be divided if the couple breaks up (see feature 14.1). What cohabiting couples usually lack that a marriage has, is not property relations but children.

On the other hand, there are many families now in which there is only one adult, but there are children. Here there is no sexual possession (sexual relations between parents and children are prohibited by the incest taboo, of course!); and there continues to be economic property. The children have the right to inherit from their mother or father. But suppose this is a mother living in poverty (rather typical of this type of family), with nothing to inherit. There is still an economic relationship implicit in the requirement that parents must support their children. This right of support is one of the most important "property rights" of children today.

For yet another arrangement, consider gay and lesbian couples. If they live together, there is usually some degree of sexual fidelity or possession. (For some data on this, see Blumstein and Schwartz 1983, and feature 9.2.) While they do not usually share custody of children in the same way that heterosexual

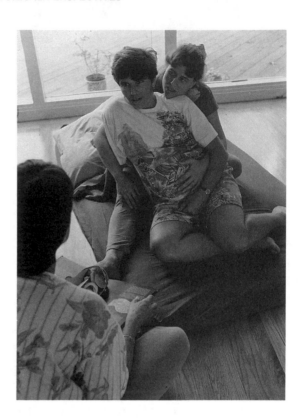

A lesbian couple at a birth preparation class.

couples do, gays and lesbians sometimes have children from previous marriages, adopt, or conceive through artificial insemination (see chapter 10). Approximately one in three lesbians are mothers and one in ten gays are fathers. Because our society has strong taboos against homosexuality, the presence of children in such households is extremely controversial and most adoption agencies are reluctant to place children with lesbian or gay couples. What the public seems to be beginning to accept is an ideal-type of gay or lesbian couple, which has some sexual possession, and no childbearing or childrearing. Would they have joint economic property? In practice, such couples share the economy of a household, as much as cohabiting heterosexual couples do. So far, economic rights of inheritance among homosexual partners have not been legally sanctioned by the courts. This is an area in which gay rights activists have struggled to change the law, so that when one partner dies, for instance, the survivor would have a right to inherit the deceased one's pension fund.

Looking at the various kinds of property, then, turns out to be a good way of understanding what sort of ingredients make up family relationships. If the borderlines of what constitutes a family are fuzzy, that is because there are several different kinds of property. Sexual, economic, and intergenerational property do not all neatly overlap. That is why it is necessary for us to look at all three of them.

HOW TO EXPLAIN FAMILY DIFFERENCES

A theory is useful to us when it explains why things sometimes take one form, sometimes another. In the following chapters, we are going to trace the implications of these theoretical principles. If the family involves three kinds of property relationships, we should ask: What determines the kind of sexual possession there is in a family? What determines how a family's economic property is structured? And what determines parents' relationships with their children? The rest of this book will address these questions in detail, but let us briefly anticipate what kinds of answers we will encounter.

One type of theory argues that economic property is the key. First, we would want to know what kind of economy characterizes the society in which we find the family. Is it a hunting-and-gathering band, a society of peasant agriculture, an industrial capitalist society, or any one of several other types? Then, within this society, we would look at its economic stratification. Are all families economically equal, as is the case in some primitive tribes? Or do some families own most of the property, and other families work for them, as in agrarian societies dominated by land-holding aristocracies? Is there a range of social classes through upper, middle, and lower levels of property?

We should expect that the shape of the family will vary with its economic position. Moreover, since a crucial feature of the family is the relationship between the men and women within it, we want to examine the ways in which the two genders are connected to property. A theory of this type attempts to show that the power of men and women within the family, and hence the kind of family roles they play, is determined by their economic positions.

A second type of theory argues that variations in sexual property are the key. In this line of theory, we would examine: How exclusively is sexual intercourse confined to marriage? Who is sexually eligible for a marriage, and who is tabooed? How permanent are the rights of sexual possession? Does one sex dominate in the degree of sexual possessiveness, or is the relationship egalitarian? Does sexual possession involve emotional possessiveness—feelings of love, including rights over possessing someone's love—or is the emotional side regarded as unimportant? We especially want to know what causes these different kinds of sexual property arrangements.

This kind of theory proposes that sexual possessiveness is really a question of power. Hence, the variations in sexual property that make the family so different at different periods of history are determined by the political structure of the society. As we will see, a key feature of politics is the extent to which the state is organized through the family or independently of it. Before the rise of

The Medicis, a powerful Italian family, dominated Florence during the Renaissance and, through kinship politics, extended their influence throughout Europe. Catherine de Medici, pictured here with her son Charles IX by her marriage to Henry II, became virtual ruler of France in the latter half of the sixteenth century.

modern bureaucracy, politics consisted of alliances among families. Sometimes particular families of the aristocracy dominated the rest of the society, and their households—their palaces, fortresses or country estates—were the centers of government. There were a variety of arrangements of this type. In all of them, politics was essentially *kinship politics*. Marriages were controlled by political motives, and sexual possession was a matter of social honor and represented the standing of the whole family in its political networks. Individuals did not control their own sexual lives as private possessions until the point in history when a political shift severed the connection between the family and the state.

Kinship systems are ways of organizing sexual possession so that their members are bound together into alliances. These variations mesh with the historical varieties of political organization. Even in our own society, in which the state is no longer directly connected to the family, there remain some important indirect connections. For the state always upholds the existing marital property system. It is the state that issues marriage licenses and also authorizes the breaking up of families through divorce. Although family politics no longer determine who is supposed to marry whom, there are still a great many legal rules controlling marriage. These include restrictions on who is too young to marry or to have intercourse and penalties for violations of sexual property such as adultery. There is also the willingness of the courts to allow spouses to enforce their own sexual property rights through personal violence. State controls also affect child-bearing, ranging from prohibition of contraceptives or abortion to their promotion. Under some governments there are subsidies for having children, such as welfare payments or tax deductions; elsewhere we find the opposite, such as tax penalties for children in countries trying to limit the birthrate. We need to be constantly aware of the power of the state, then, in analyzing why one or another of these policies is in effect, and how it affects the shape of the family. In our own times, political movements for and against the legal rights of women are among the most important influences on the way the family has been changing.

These two lines of theory are not mutually exclusive. Both of them focus on some aspect of the property system. Focusing on economic property leads us to consider the predominant technology, and its connections to such things as the kind of work that men and women do, their activities in the household and in the labor force. It also implies that the third kind of property—intergenerational property, involving relationships between parents and children—is mainly determined by economic conditions. How many children there are depends largely on economic conditions affecting the birth and death rates. Key factors are whether children can do economically useful work, and on the other hand, how expensive it is to raise children.

The theory focusing on sexual property is a little more exotic. It leads us into the area of highly charged emotions, including pride, jealousy, and violence. But these, too, are socially structured. These emotions are connected with the status dimension of stratification, as family members show off their position in the networks of society. And they are connected to the power struggles of politics. To modern ears, "sex and politics" sounds like a scandalous headline in a newspaper. But the connection is deeper than that, for both sex and politics are fundamentally concerned with making alliances; and both sex and politics are overlaid with attitudes about morality, claims about rights which, if necessary, people will act quite violently to defend. Relations of parents and children are political and emotional as well: children were part of the political maneuvers in societies based on family alliances. As the family/state connection has become more remote, children, like their parents' sex lives, have become a more private issue, yet they are still a force in political efforts at control.

In chapters 3, 4, and 5, we will trace these processes as they have shaped the family across history, from the most primitive societies through the family changes of the twentieth century.

SUMMARY

1. The family is a set of social rather than merely biological relationships.
2. The property theory sees the family depending on three kinds of property: rights of sexual possession, economic property, and intergenerational property. The amounts of each of these kinds of property held by men and women vary among societies; these different combinations make up the various types of the family.
3. Kingsley Davis's theory of sexual property points out that property is not a thing but a social relationship among an owner, a possession, nonowners, and the rest of society, which the owner can call upon to back up his or her rights over possession and to keep nonowners away. Marriage is sexual possession involving the exclusive right to have sexual intercourse with a person.
4. The incest taboo is found universally in some form in all societies. The incest taboo is regarded as cultural rather than biological, since it establishes a system of exchange between families. Each family must give up its isolation by sending its offspring to find sexual partners outside. In tribal societies, various forms of marriage exchanges made up economic and political networks and thus established the larger society.
5. In some societies sexual possession is unilateral: the husband has exclusive sexual rights over his wife but not vice versa. In our society, the ideal of sexual possession is bilateral: both husband and wife are supposed to be sexually faithful to each other. The degree to which sexual possession is unilateral or bilateral depends upon the relative power and economic resources of men and women.
6. The modern ideal is that marriages take place not merely for sex but because of a bond of mutual love. However, the prevalence of jealousy indicates that love is also a form of socially sanctioned possession. It can be analyzed as an emotional possession that goes along with sexual possession.
7. The family also consists of economic property. This includes both the family's income and the household labor involved in running an unpaid domestic business of cooking, cleaning, lodging, and child care. Our word *family* itself comes from the ancient Roman word denoting the physical household.
8. The family economy is always connected to sexual possession. Marriage systems in different societies have varied depending on how men and women have traded both economic and sexual rights. In our own society, critical theorists have argued that the weaker economic position of women

outside the family has brought about both sexual subordination and the custom that the wife provides most of the unpaid household labor.

9. The third aspect of the family, intergenerational property, is the result of the first two kinds of property. It includes both economic and sexual rights between parents and children. Economic rights include inheritance, and the parents' control over children's labor. Parents traditionally have had the rights to control their children sexually, especially in deciding who they marry. This has declined, but a negative sexual rule continues to apply: the incest taboo.

10. In the late twentieth century, the different forms of family property are being combined in various ways. Cohabiting couples are similar to marriage partners in sexual possession and in maintaining a joint household, but usually lack intergenerational property. Single-parent families lack sexual possession; gay and lesbian couples may have sexual possession and joint household property.

11. To explain variations in families across historical periods as well as differences within our own society, two main background factors are important: the economic structure of the society, especially the economic property of men and women; and the system of political alliances, which determines the patterns of sexual possession.

FROM KINSHIP POLITICS TO THE PATRIARCHAL HOUSEHOLD

In this chapter we look at the family through the long sweep of history. The family varies a great deal across these societies. There are tribal systems with patrilineal, matrilineal, and many other forms, and complicated rules for marriage exchanges. Some of these forms are among the most egalitarian between men and women that have ever existed, while others resemble a ferocious war between the sexes. In technologically more advanced societies, the so-called ancient and medieval civilizations, the family form changes again, typically into the patriarchal household. Here society is rigidly stratified, and male domination over women is pushed to an extreme. Yet even here women have some resources for gaining power and status, pointing toward changes to come later.

SOCIETIES: AN OVERVIEW

Let us look briefly at all the major types of societies existing in different parts of the world and historical periods. Figure 3.1 categorizes the main types of societies according to their principal economic base. Lenski (1966) showed that these economic structures, and the resulting system of power, affected the overall degree of stratification. Since economic and political structures vary among these societies, we should not be surprised to find that their family structures are also quite different.

Hunting and Gathering

Hunting-and-gathering societies are small wandering groups of a few hundred people or less. This is the type of society that prevailed during the hundreds of thousands of years of the Stone Age. It has survived almost until today in remote areas like the deserts of Australia, the Kalahari in Africa, in the Arctic among the Eskimos, and in the Philippine jungles.

The popular movie *The Gods Must Be Crazy* gives a good depiction of one of the few hunting-and-gathering societies that survive in the late twentieth century. The family tends to be a simple unit, in many cases consisting of no more than the monogamous nuclear family made up of a single couple and their ungrown children. These families are often grouped together into larger bands of people who intermarry. But connections are loose. People wander in and out of the various campsites, so there may be different people sleeping there at different nights. There is little property to tie anyone down. Divorce is common, even more so than in our own society.

Primitive Horticulture

Primitive horticultural societies rely on the earliest form of agriculture. Usually the women cultivate the plants by hand and make baskets and

FIGURE 3.1: Typical Family Structure Varies with the Economic and Political Organization of the Society

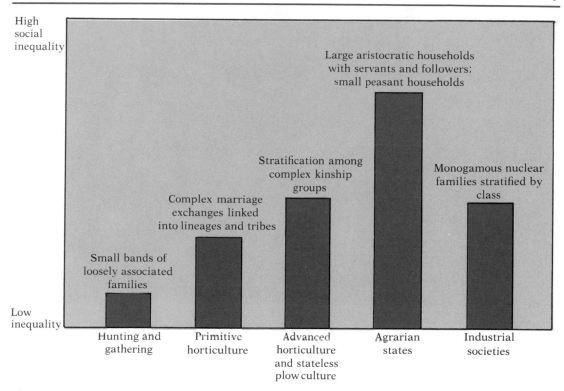

Inequalities of wealth and power are lowest in hunting and gathering societies, highest in state-organized agrarian societies, and intermediate in industrial societies.

pottery, while the men engage in fighting or clearing the land. The society may be as large as a few thousand people, linked together by exchanges, periodic councils, and religious ceremonials. It is here that we are especially likely to find complex marriage rules of the kind described in Lévi-Strauss's theory. The whole society is connected together by marriage exchanges. Smaller and larger kinship networks are nested inside each other like concentric rings. At the smallest level, homes may be matrilocal or patrilocal. Beyond these are lineages which may descend on matri or patri sides. Often there are larger clans or tribes, which in turn may be divided into various sections. Men and women may have their own ceremonial groups and centers, often taking the form of secret religious societies. With all these options to choose from, horticultural societies have tremendous variety in their forms of kinship.

These types of societies have been extensively studied by anthropologists

Dressed in the colors of the Bahian goddess of the sea, three women of a fishing community on the east coast of Brazil offer thanks for the morning's first catch.

in the Americas, Africa, Southeast Asia, and the islands of the Pacific. Somewhat similar forms of organization are found in *pastoral* societies, which live by herding cattle, sheep, goats, or other animals; and in *fishing* societies, which formed in abundant coastal areas like the Pacific Northwest.

Advanced Horticulture and Plow Culture

Through the use of metal tools, *advanced horticultural societies* were able to support larger and denser populations, ranging from 10,000 up to a few million. Here stratification began to become pronounced, with a hierarchy of chiefs, priests, and military aristocrats exacting tribute from the lower classes.

Examples of advanced horticultural societies include the Mayans and other societies of early Mexico, the Incas in Peru, and some of the societies of sub-Saharan Africa such as the Azande in east Africa or other large societies in Dahomey and Nigeria in west Africa.

Agrarian plow culture is a more advanced form of technology. These societies tend to emphasize property in terms of land, plowing the same ground over and over again for many years, whereas *horticultural tribes* usually shift to new land fairly often when the old soil is exhausted. In both these types of societies, the nuclear family form is rare, and a large proportion of them have complex family systems. Because there is more stratification than in *primitive horticulture*, social class is beginning to become a rival to family structure as the key basis of social position.

Agrarian States

Agrarian societies are those in which the large-scale state emerged. This is because plow culture used animal power rather than exclusively human labor, and was able to produce a considerable surplus that enabled other aspects of civilization to be built up. The great historic civilizations of Egypt and the Middle East, China, Japan, India, ancient Rome, and medieval Europe fall into this category. They had a variety of technological developments, including the use of irrigation for extensive agriculture, wheeled carts pulled by oxen or other animals, metal weapons, large-scale buildings, and the use of writing. These states might include millions of people over huge territories, organized into a feudal system or a centralized empire. Here stratification was extreme, with more wealth concentrated in the hands of a small aristocracy (about 2 percent of the population) than in any other type of society (Lenski 1966).

We should bear in mind that not all agrarian societies developed large-scale states. Plow-farmers lived in the so-called "barbarian" tribes in northern Europe at the time of the Roman Empire, for instance, and there have been plow-cultures and irrigation-cultures in Asia and in sub-Saharan Africa that did not have strong state structures. In these societies, complex kinship systems tended to remain in effect. Where the state structure did appear, however, it had an important effect on the family system. The households of the military aristocracy became the most important family units. In these societies, the complex marriage rules that tied together horticultural kinship systems tended to fade away. Instead, we find the aristocrats bringing large numbers of nonkin into their households. This was the era of palaces, castles, and fortified buildings, where powerful families employed large numbers of servants or sometimes slaves, and surrounded themselves with military followers and political dependents. In these societies, the most important influence on the family was its class position; large aristocratic families were at the center and dictated most of the conditions under which peasants and servants could have families of their own.

Industrial Societies

The use of machinery run on inanimate energy sources such as coal, electricity, and oil characterizes *industrial societies*. Although cities are found in agrarian societies, urban population outnumbers rural residents for the first time in industrial societies. The family system simplifies dramatically as we move toward the wealthier industrial economies. Today, less-developed societies that have taken steps toward industrialization show a drop in the prevalence of complex family systems, while among heavily industrialized societies monogamy and the nuclear family have become almost universal.

This comparison has its limitations. Not all societies fit clearly on the chart. Some have been mixtures of one type with another, such as pastoral societies that borrowed some of the techniques of agrarian civilizations (for example, Genghis Khan's Mongol empire). Nor do societies necessarily go through all these "stages"; they may skip one or more. We have concentrated only on a crude measure of family complexity that leaves out the variations in details of kinship systems found among societies of the same general type. But the chart does give us some idea of the typical forms of society that have existed, and we can now look more closely at some of their family structures, and at the conditions that caused them.

SOME THEORIES OF GENDER INEQUALITY

How can we explain why family systems have taken different forms? In recent times, a number of sociological theories have tried to deal with this question by analyzing the family as a system of stratification. The most striking differences among types of families lie in the degree of equality or inequality one finds among their members. Some of this inequality may be between members of the same gender: the rights of the oldest brother against the younger brother; of father versus son; of the mother over her son's new wife who moves into their household; or the first wife in a polygamous system versus lower-ranking wives or concubines. But it is most convenient to begin by analyzing the sources of stratification between men and women. Theorists look for the sources of power that make a society male dominated in various respects, female dominated in certain spheres, and (all too rarely) egalitarian. A successful theory of gender stratification should not only be able to explain the varying degrees of inequality between the genders but go on from there to show how these same resources lead to other features of the family.

Notice that theories must always be comparative. In order to highlight the conditions for inequality, we must compare them with the conditions in which relative gender equality occurs. More precisely, we look at the variations across a continuum, comparing situations in which gender inequality is high with those in which it is moderate and those in which it is low.

There are two main approaches to the theory of gender inequality. One version emphasizes *economic* causes, especially the forms of work that men and women do, and the environments in which their societies attempt to make a living. A second version emphasizes *political* causes, especially the ways the kinship system produces alliances by marriage exchanges, and the various kinds of military situations and state organization that modify the power of men and women over one another. The two theories are not mutually exclusive. In fact, each theory fills gaps left by the other in explaining the family's historical changes.

Economic Theories

Some version of *economic determinism* has been implied in many of the oldest theories of the family. It was once held (and still is by some sociologists as we have seen in the previous chapter) that there is a natural division of labor between the sexes: males evolved as the hunter and later the breadwinner, while females specialized in having babies and child rearing and caring for the home. We know now that this is not necessarily true. In hunting-and-gathering societies, women typically produce more of the food by gathering than men do by hunting (Blumberg 1978, 6). In fact, women may produce as much as 60 to 80 percent of the family's diet. Moreover, in primitive horticultural societies, women not only grow almost all the crops but also do most of the manufacturing of their ceramics and basketry. The idea that men alone provide the income and women merely care for the home is fallacious, based on the experience of the middle and upper classes in comparatively modern times. But even in Victorian England, the anthropologists who theorized about "man the hunter" (or breadwinner) ignored the millions of working-class women who had always worked outside their homes.

Economic theories have become more sophisticated. They recognize that the kind of work that women do has varied a great deal from one type of society to another and that the form of *ownership* in that society is crucial for how much benefit women reap from their work. For working does not automatically pay off in control, as we can easily see by comparing the incomes of factory workers today with those of persons who do nothing but invest in the stock market. The hypothesis is that the status of men and women depends upon how they fit into the larger economic system rather than simply upon how much work they do. This form of analysis goes back to Marx and Engels. Friedrich Engels deserves to be called the originator of the *theory of gender stratification*, although his precise hypothesis is not much followed today. Engel's book *The Origins of the Family, Private Property, and the State* (1884/1972) proposed that the family originally was a form of "primitive communism," in which equality prevailed and sexuality was promiscuous. After this came a stage of matriarchy and finally monogamy and patriarchy. The last was truly the fall out of the Garden of Eden, since it brought about the subjugation of women and was caused by an economic change, the institution of private property. Along with this, Engels argued, women were guarded as exclusive sexual property, while men reserved sexual freedom for themselves in the form of sexual access to slaves, concubines, or prostitutes.

Engel's theory is mistaken in various respects. There was no stage of "primitive communism" and sexual promiscuity, but he was right in the general sense that hunting-and-gathering societies are much more egalitarian than most subsequent forms of economy. There was also no stage of matriarchy; like other early anthropologists, Engels mistook evidence of *matrilineal descent* for a system of female property and power, which did not necessarily follow. And the origins of private property cannot be neatly correlated with the rise of patrilineal and male-dominated families. But Engels inspired later theories by his emphasis on how an economic system can determine the power of men and women.

There are a number of modern versions of economic theory of gender stratification. Some early models, such as Boserup (1970) and Friedl (1975), concentrated on the relationships between women's work, kinship structure, and gender stratification in various kinds of preindustrial societies. More comprehensive models covering the whole range of societies were formulated by Rae Lesser Blumberg (1978, 1984), Joan Huber and Glenna Spitze (1983), and Janet Chafetz (1984). Figure 3.2 depicts some of the main variables in these models.

The theory centers on the fact that a woman's power depends upon how much control she has economically. Working is not enough; it must be supplemented by control over the means of production (a point also made by Karen Sacks 1979). Blumberg elaborates this across the entire range of human societies. In many hunting-and-gathering and horticultural societies, women are not only important economic producers but often are able to translate production into economic control and hence into power and status. In agrarian societies with a powerful state, women tend to be excluded from the major areas of work and from ownership as well (although there are exceptions in the upper classes), and hence their status generally becomes very low. In industrial societies, a woman's relative status depends upon her income or other property. This is even true in the supposedly communistic and egalitarian Israeli kibbutz, where women's status fell when they were excluded from the productive heavy agriculture and manufacturing upon which the kibbutz economy rested (Blumberg 1976, Buber Agassi, 1991).

The question then arises: What determines women's economic control? Blumberg points to three sets of factors: the strategic indispensability of women's labor, the effects of the kinship system on female property, and the stratification of larger societies.

Strategic Indispensability of Women's Labor

Women get the most economic power when they represent a crucial labor force. This happens, for instance, where women produce most of the diet (as in hunting-and-gathering societies); where women control the technical expertise for their work (as in horticultural societies); where women do their work in

FIGURE 3.2: An Economic Theory of Gender Stratification

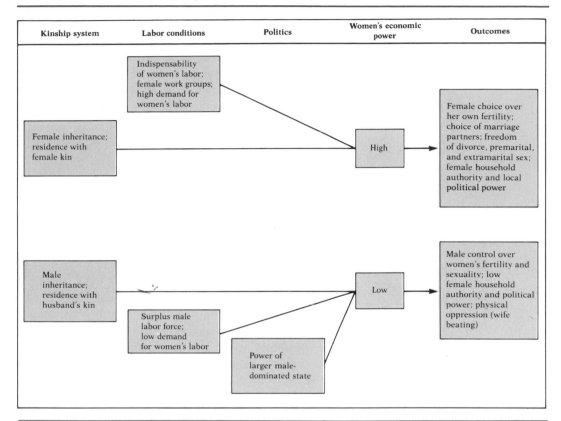

all-female groups independent of male supervision (as in many African societies). It also happens when there is a shortage of male labor, such as in wartime; hence there were women in previously all-male jobs during World War II. Conversely, a surplus of labor undermines women's economic position (Chafetz 1990). One reason that women's status tended to be low in agrarian civilizations was that there was always a surplus population of landless peasants, ready to fill any position, and hence forcing women out. Similar processes have happened during economic depressions in more recent times.

The fact that women's labor is *sometimes* indispensable shows, incidentally, that social rather than biological factors determine what work women do. It is not simply a matter of strength. Women in peasant or tribal societies were probably a good deal stronger than women—or many men—in our white-collar industrial society. In ancient China, during a wartime siege all the men were called to military duty, while the women not only did all the work but also built the fortifications (Griffith 1963, 38). Similarly, because World War II killed off a sizable proportion of the male labor force in Russia, the women did most of the heavy labor.

As we reach modern societies, the demand for women's labor takes on a new importance (Chafetz 1984). With the rise of the industrial factory system, large numbers of women were pulled into paid employment outside the household. At first, these were usually working class women, laboring under harsh conditions in the mills and receiving relatively low wages. This gave them little economic leverage to change the system of gender stratification, but it did establish a pattern of women in the labor force. Labor unions became organized, primarily among men, and generally attempted to restrict the kinds of jobs women could hold. The rationale was partly to protect women from hard and dangerous work, but also to save their jobs for men. Thus during the last century there has been a tendency for working-class women to go into the labor force only to be pushed out again by men.

An even more important change came about as women entered white-collar work in the late 1800s. Office work expanded with the rise of modern bureaucracies. The typewriter was invented, and the job of typist became a female occupation. Soon the position of office secretary became a female job as well. With the growth of commercial business, large numbers of women became clerks and cashiers. Again these tended to be relatively low-paying jobs, but the pattern was now established in which middle-class women often worked for part of their lives—usually when they were young and unmarried, but perhaps longer if their family needed to supplement its income. At first, the highest level job for women was school teacher, an occupation that expanded rapidly in the late 1800s and early 1900s with the spread of mass education. Thus the expanding demand for female labor gave women at least a fingerhold on some of the economic resources that give women more freedom and power in their lives.

Throughout the last century, the movement of women into the paid labor force has been upward. But women's jobs have usually been badly paid compared to men's jobs. Men control the economic balance of power, which has reinforced a male-oriented family structure. In modern societies, women's labor has never been the crucial, indispensable economic factor. Nevertheless, the gradual expansion of women's jobs is one of the conditions which has slowly exerted some pressure for family change.

Effects of the Kinship System on Female Property

When economic goods are passed along by *inheritance* from women and to women, female economic power is maximized. Also contributing is the *residence* pattern: where a woman resides with her female relatives (a form of *matrilocal residence*), the female group is most likely to control its own property. This power is lessened when she resides with her brothers and is lowest of all when the woman leaves home to live with her husband's relatives *(patrilocal residence)*. Compared to these, the official *descent rule (matrilineal, patrilineal, bilineal descent)* is relatively unimportant. In other words, whether children officially belong to a lineage traced through their mother or their father (or both) may not have very much effect on actual economic power. As Alice Schlegel (1972) has shown, in many matrilineal systems property is passed from mother's brother to sister's son, and husbands or brothers end up dominating their wives or sisters. In some patrilineal societies in West Africa, women still have de facto control of the land, which they work by horticulture. Descent rules thus appear to be a kind of ideology that can mask the actual economic situation between the genders.

The Stratification of Larger Societies

Women do best in their own local communities and in societies that do not extend far beyond the community. Larger societies tend to have male-dominated networks of warfare and politics, and the state tends to intervene to limit or eliminate local female power. In the great agrarian civilizations, women in the cities close to the upper classes were veiled and secluded, while peasant women in the countryside had much more freedom. In modern times, as Peggy Sanday (1981) has pointed out, the British colonial administration in Africa destroyed the local political institutions of native women and replaced them with an all-male native hierarchy. For this reason, women have had greater economic power in smaller and more isolated tribal societies.

The economic model can be visualized as a kind of chain. Labor indispensability comes first, modified by the kind of kinship system. These in turn can be overruled by an unfavorable *stratification system in the larger society*. All these affect economic power, which in turn determines women's status. In real-life terms, this includes the degree to which women make their own decisions about when to have children, contract their own marriages, divorce, have premarital or extramarital sex, have authority within the household and wield political power in the local community. When female economic power is low, women have few of these rights; men control their fertility and their sexual lives; household authority and community participation are low. Under these conditions, women are most likely to be subject to male physical abuse (wife beating, rape), to be second-class subjects politically, and even to be viewed as unholy according to the prevailing religious beliefs.

There is a self-reinforcing loop here. When women have low power, they can be forced to be baby-producing machines, which in turn makes them less able to participate in productive labor and to acquire economic power. This is what tends to happen in agrarian societies with a strong state, where women's power is lowest. The men who head these families usually want large numbers of children, since they are useful as farm workers or sometimes as fighters. In addition, a high birthrate tends to produce a surplus population, which further undermines a woman's possibility of wielding economic power. The situation thus becomes a vicious cycle.

Childbearing and Population Conditions

Huber and Spitze (1983; Huber 1991) point out that there is a loop connecting overall population conditions and the amount of their lives that women have to devote to childbearing; and this in turn affects how free women are to gain economic resources (see figure 3.3). In societies with primitive technologies, the amount of pressure depends upon the environment. When conditions are harsh and the death rate is high, women may be almost constantly pregnant or nursing small children. Their fertility is crucial to offset the high death rate. But this in turn restricts the kinds of work they can do. Pregnancy and nursing prevent them from hunting, herding animals, or

FIGURE 3.3: Demographic Conditions Affecting Women's Economic Position

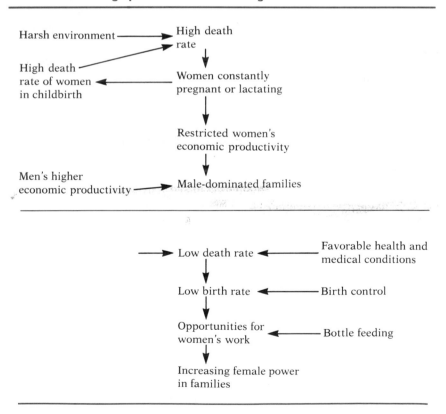

growing crops any distance from home. These economically key activities are dominated by men and gives them dominance in the family system.

On the other hand, in some of these societies providing food is relatively easy. In simple horticulture, where women grow most of the crops in small gardens, their work is easily compatible with breast-feeding children. In such societies, the position of women tends to approach that of men (who specialize in warfare farther away from home). Both sexes have means of supporting themselves, and as a result they are not very committed to marriages. Divorce rates are sometimes extremely high, higher even than in our own society (Friedl 1975). In addition, because women produce so much, men regard them as valuable, and where possible they marry as many wives as they can. This is the type of society in which *polygyny* is most common throughout the populace (see feature 3.1).

In this perspective, the most important effect of industrialization is the way it affected births and deaths. With industrialization, the death rate declined, especially the level of infant mortality, because of generally higher levels of health and nutrition. As a result, the pressure for large numbers of pregnancies also declined.

At the same time, children stopped being economically valuable. Fewer households ran family farms or shops where the children were useful hands. Instead, children became more of an economic burden, especially as it became the custom for them to be supported through a long period of schooling. This change in children's status was another factor contributing to a decline in the birthrate. With the development of contraceptives and the use of birth control, the number of children women were expected to bear dramatically decreased in the early twentieth century.

On top of this, around 1910 sanitary baby bottles came into use. This new convenience drastically reduced the demands on women to care for their small children, and opened the way for mothers to devote themselves full-time to the labor force. To be sure, even in earlier conditions, it was not always necessary for women to spend much time breast-feeding. As Blumberg (1984) indicates, *when there was a lot of opportunity and demand for women's labor,* women would cut down on the length of breast-feeding, space out their childbearing, and otherwise adjust so that child care did not interfere with their work. In agrarian and early industrial societies, some women specialized as wet nurses, breast-feeding other women's children. In Paris around 1800, for example, only one woman in twenty was breast-feeding her own child (Sussman 1982). Wet nurses were hired not only by women who were employed, but also by upper-class women to relieve themselves of the care. The down side of this practice was that care by wet nurses was not very good, and the death rate among children was high. At any rate, in the twentieth century the practice disappeared.

By the twentieth century, then, in industrial societies, it was possible for women to work permanently in the labor force, and pressures to bear and rear large numbers of children had dropped considerably. This potentially opened the way to increasing women's economic power and to other changes in the family. But as long as men controlled the good jobs, this change remained largely potential.

Criticisms

These economic theories capture an important part of the explanation. Some shortcomings, though, may be suggested. First, there is the unanswered question as to how women gain control of inheritance. Certain kinship systems seem to favor this, but how those kinship systems arise is not explained. Second, the point that a surplus of male population will push women out of the desirable jobs assumes that there is some reason, unexplained, why men

FEATURE 3.1
Polygyny in West Africa

Note: *polygyny* comes from *poly-gyny*, meaning "many women" (or wives). It should not be confused with *polygamy—poly-gamos*, meaning "many marriages." *Polygamy* is a broader term, including both **polygyny** *and* **polyandry** (*"many men"* or *"many husbands"*).

Many West African societies have a system of polygyny. Most men, if they have any wealth at all, will have several wives. Usually each wife lives in her own hut, inside a walled compound, taking turns feeding the husband and having sex with him. But this is not exactly a masculine paradise. The women have a good deal of economic power, since they produce and market the crops. Women are frequently richer than their husbands, and some women today own whole fleets of pickup trucks, known in West Africa as "mammy wagons." The women are in a sense better organized than the men, since the co-wives work and live together, while the men are split up. Even sexual rights are carefully regulated, so that the situation is not so much a despotic harem as a group of women sharing a man among themselves. And eventually, since the men tend to be much older than their wives, the women outlive their husbands and remarry; so there is a kind of polyandry that

occurs as well, but spread out in time. Divorce is easy, and there is no dual sexual standard, with women having a great deal of leeway in the area of premarital and extramarital sex.

The senior wife has a great deal of power in this situation. She dominates the junior wives, as well as carrying much weight in the community, and she can get considerable deference from the children of the polygynous compound, including grown sons. She takes an active part in the politics of marriage bargaining that goes on in acquiring junior wives. For one thing, any additional wives will have to live with her, in a sense even more closely than with her husband, and their compatibility or suitability as a group of women is to be considered. Also, contrary to our usual notions of jealousy, the senior wife may actually push quite strongly to acquire some junior wives. Her prestige, as well as her domestic authority, depends upon how many wives there are ranking below her. And the prestige of her husband in the community, which she shares, depends upon how many wives he has. To be married to a man with only one wife is somewhat demeaning, a sign of social failure. It is something that one woman might taunt another with. (Lloyd 1965; Clignet 1970) will be able to get those jobs. Finally, there is the observation that the larger political system can intervene to overturn women's local power. This is true, but why does it happen that men have controlled the networks of long-distance politics and stratification?

These problems are a useful introduction to the second type of theory of gender stratification, the political approach. This analysis looks at kinship precisely as a form of politics and sees the larger political system as something that emerges out of the shifting alliances of different family systems.

Political Theories of the Family

Political theories of the family go back to Max Weber (1922/1968, 356–84; 1923/1961, 38–53), Collins (1986a), Claude Lévi-Strauss (1949/1969), and subsequent *alliance theories* of kinship. Here is another version of a political conflict theory (Collins 1971; 1975) (see figure 3.4).

FIGURE 3.4: A Political Theory of Gender Stratification

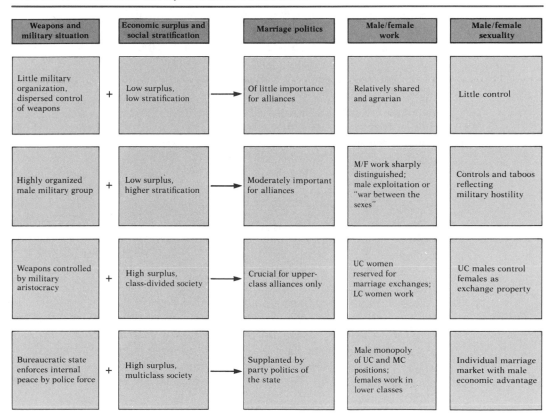

Weapons and military situation		Economic surplus and social stratification	Marriage politics	Male/female work	Male/female sexuality
Little military organization, dispersed control of weapons	+	Low surplus, low stratification	Of little importance for alliances	Relatively shared and agrarian	Little control
Highly organized male military group	+	Low surplus, higher stratification	Moderately important for alliances	M/F work sharply distinguished; male exploitation or "war between the sexes"	Controls and taboos reflecting military hostility
Weapons controlled by military aristocracy	+	High surplus, class-divided society	Crucial for upper-class alliances only	UC women reserved for marriage exchanges; LC women work	UC males control females as exchange property
Bureaucratic state enforces internal peace by police force	+	High surplus, multiclass society	Supplanted by party politics of the state	Male monopoly of UC and MC positions; females work in lower classes	Individual marriage market with male economic advantage

It is convenient to begin with the question, What are distinctive traits of males and females? What is there about men and women that could explain differential advantages they have in family systems but at the same time could account for *variations* in gender stratification in different systems? We are looking for some factor that is not exactly gender neutral—as that would make it impossible to explain why men and women occupy any different positions at all—but that translates into different outcomes in different social situations.

There are three possible candidates: differences in sexual characteristics, childbearing, and differences in size and strength. The first of these, differences in genitals and secondary sexual characteristics, is obvious enough, but it does not seem to make much of a social difference. Male and female genitals are complementary to each other, so to speak, which is one main reason why the family exists in the first place; but in themselves they explain nothing about patterns of inequality or domination, and why men should so often dominate the property or political system. Male genitals are not a qualification for work or political positions. As one successful modern woman remarked, the only job that requires a penis as a necessary qualification is that of male prostitute.

A more powerful candidate is the uniquely female capacity for childbearing and breast-feeding. It is better to phrase it that way than to speak of female child rearing, since it is yet to be proved that women alone are capable of rearing children once they are born. But childbearing is significant enough, and a number of feminist (as well as nonfeminist) theories have made this the crucial basis for gender stratification. Shulamith Firestone (1970), for example, argued that women have been subordinated since the beginning of the human race because pregnancy made them unable to fight and hunt with men, while childbirth and prolonged breast-feeding extended this vulnerability throughout most of their adult lives. For Firestone, the greatest shift in the status of women was the modern invention of birth-control devices, which finally freed women from the tyranny of birth and made it possible for them to seek careers outside the home—although in an uphill struggle because of the late start.

But we have already seen, in the discussion of the economic theories, some reasons why the importance of childbearing need not be so overwhelming. Women have actually done quite a lot of nonhousehold work in most societies throughout history. Moreover, when they have had enough social power they have been able to control their own fertility; the image of women historically as uncontrollable baby-producing machines is inaccurate. It is also an exaggeration to regard pregnancy as a continuous period of physical debility; we know that today many women work into their ninth month of pregnancy and return not long after childbirth. It is also more than likely that hard-pressed tribal or peasant women did not have the luxury of taking a great deal of time off. A Japanese film on a New Guinea tribe actually shows a woman about to go into labor building her own hut in the jungle, where she is required by taboos to be ritually secluded for the childbirth.

After babies are born, moreover, it does not necessarily follow that mothers become specialized primarily in child care. Long before modern experiments with taking babies to work, tribal mothers were carrying infants with them in slings and baskets while they produced what might be the bulk of their family's food supply (Blumberg 1984). In more stratified societies, children were often cared for collectively in some fashion, by nurses (including wet nurses for infants), relatives, or others; and this was done not only by the upper classes but often out of necessity by peasant women who could not interrupt their work. These are instances of *some* women specializing in child care to free other women from it, but that it be done by women does not seem mandatory. Conceivably it could be done by males as well, under a different social arrangement; and in fact the extent to which men have shared parenting tasks does vary from society to society. In some egalitarian tribal societies (Sanday 1981, 60–64; Coltrane 1988), male parenting is considerable. Child care is also sometimes done by older siblings, including brothers. The point is that the female's biological specialization in childbirth is not necessarily a social institution but is made so only under particular circumstances.

This brings us down to the third difference: the fact that on the average males are somewhat larger than females and have a heavier musculature. The difference is not great compared to some of the primates; it is only about 10 percent (Van den Berghe 1973, 15, 42), and since these are averages, some proportion of women are bigger and stronger than some proportion of men. But why should this make a difference socially? Despite some common assumptions, it does not turn out to make much difference in the kinds of work men and women can do. As shown in the discussion regarding Blumberg's theory of the strategic indispensability of women's labor, women have been able to do all kinds of heavy work when necessary. Perhaps stronger men might be able to do it more easily, but that should not necessarily make a difference unless there is some kind of competition going on. If it is just a question of what women can accomplish, there is no reason why women should drop out of heavy work (like plowing in agrarian societies) unless there is some other social pressure, as indeed there is in such societies. Moreover, we should not assume that male work automatically requires more strength. Metalworking and certain other high-prestige crafts have usually been monopolized by men, such as by the secret societies of iron forgers found in tribal societies; but this is a case of social control of a skill rather than of biological fitness. This is even more true of male success in monopolizing positions involving writing after it was invented in the ancient civilizations; this was certainly light labor in terms of muscular effort, but since it was the pathway to power and privilege, males defined it as their own province.

We seem to be running out of alternative explanations for male advantage, but there is one possibility left. Size and muscles are useful for fighting, and this is a sphere that is inherently competitive. Here it is not just a matter of being able to do the job adequately but of being able to beat someone else. Fighting is in this sense unlike any productive human activity, since there is no intrinsic standard about when the job is well done: it is always a matter of who is stronger relative to someone else. This may be the factor we are looking for; although differences in size between men and women are relatively constant, the importance of fighting is a *variable across societies*. This is what we need to focus on if we are to explain why male/female status varies from one social situation to another.

Fighting brings us into the realm of politics. Politics originally concerned little more than the organization of coalitions for warfare; later, after the rise of the bureaucratic state, violence was monopolized at the state level (this is in fact Weber's definition of the state), and hence peaceful politics emerged. Males gained power primarily by monopolizing the political realm. But again, this is a variable. In some societies (such as our own), women have some political power, and potentially they could have quite a lot. But this depends upon particular kinds of conditions. Partly this is a matter of how closely politics is connected with warfare; and secondarily it is a matter of how the military is actually organized. In general, the closer that fighting comes to the immediate community and to the family, the more male power is enhanced. Conversely, peaceful societies, and those in which fighting is done by specialized troops and in distant places, are usually favorable to female power.

There are a lot of variations here, and the different shape of the family across human history is to a considerable extent affected by the amount of military threat and by political organization. Let us look briefly at some of the ingredients of these situations.

Men and Weapons

Although men have tried to monopolize the weapons with which fighting is done, there are some exceptions. In the West African Kingdom of Dahomey, the king had a bodyguard of 5,000 armed women (Sanday 1981, 86–87). These female warriors had to be celibate during their period of service, which shows one way child rearing might be dealt with: it was made a specialized role for nonmilitary women. This part of Africa in general has particularly strong political and economic participation by women. There are also ancient stories (as in Homer) of Amazon societies consisting entirely of women warriors. Blumberg (1984) suggests that these may not be entirely mythical but reflect real situations in which the male population was decimated by war. It sometimes happened that all the able-bodied males went off on a foreign military expedition and were then wiped out; under these circumstances, women not only took over the economy but also may have armed themselves and taken control. There are similar stories among the Vikings, especially in such frontier areas as Iceland. There the god of war, Odin, is depicted as surrounded by female warriors, the Valkyrie, who hover over the battlefield;

The legends and myths of many cultures include references to female warriors. The Valkyrie, female warriors in Icelandic mythology, are depicted in a production of Richard Wagner's opera *Die Walküre*.

and real Icelandic history also contains accounts of ferocious queens who ruled by wielding their battle axe. (The story of Brunhild, recounted in chapter 2, depicts some carryovers from these events.)

In more recent times, Israeli women fought in the 1948 war, when the crucial question of establishing the state of Israel hung in the balance. With modern machine guns and other technological weapons, the factor of strength becomes less significant in warfare. This is one reason why women have recently been entering the armed forces, although their role in combat remains controversial. Both men and women have been on both sides of this recent issue; but whatever the merits on either side, it is clear that physical differences in strength are no longer the crux of the matter, but rather social beliefs.

The significant part is that men have usually gone to great lengths to monopolize weapons. Weapons offset advantages in sheer strength, especially as they become technologically more advanced. Whatever physical strength one starts out with, one's capability for threatening others goes up rapidly if one is better armed. Men have not only been somewhat bigger than women (on the average) but have made great efforts to accentuate that advantage by making sure that they have been armed and the women have not.

There is a very revealing myth among the Ona tribe, a group of Indians living on Tierra del Fuego at the tip of South America (Cooper 1946; Lothrop 1928). The Ona were hunters who warred among themselves, usually over the capture of women. Bows and arrows, the principal weapons, were forbidden to women, and it was taboo for women to enter the house of men's secret society where the weapons were kept. Periodically the men would carry out a ceremony in which they emerged from the men's houses armed and wearing terrifying masks, representing male spirit powers, to bully the women and children. Adolescent boys, when they were ready to be admitted to the secret society, were first whipped and then finally shown the chamber where the masks were put on. They learned the secret that the terrifying spirits were really the men, the group they were now part of themselves. And they were told this myth: women were witches, responsible for all misfortune and death. At one time in the past, the witches had ruled the Ona. But the men banded together and killed all the women, except for some small girls, who were brought up henceforth to respect men forever. That is why only the men could enter the men's house, don the masks, or handle the weapons. Perhaps the story includes some vague memory of real events in which the tribe changed from a form in which women had more power. Whatever the historical truth behind the myth, it shows that these men realized that their domination over women depended upon their organization and monopoly of the means of intimidation.

It is not only weapons that are power multipliers, but organization as well. Weapons used in a group are much more effective than those used alone. Hence, when males have been organized as a group, their domination over women has been most extreme. The same principle works the other way, of course. The general hypothesis, then, is that military factors (the kinds of threats from outside, the kinds of weapons) determine how males and females are organized and hence the degree of domination in the family system. There are many possible outcomes.

Where there are few weapons and not much military threat, there is little incentive for males to organize in military groups, and relationships between men and women are quite egalitarian. This occurs in some primitive gathering tribes in environments where there is an abundance of food. For example, the most recently discovered tribe in the world is the Tasaday, first noticed in the 1970s in a remote part of the Philippines (Nance 1975). The men did not hunt, and both sexes took part in gathering food. The group was unstratified, but the person with the greatest influence was an older woman. Other very gentle and egalitarian societies existed earlier in the tropical forests of the Malay peninsula, among the pygmies in the forests of central Africa, and in the tribes of the Kalahari desert (Sanday 1981, 19–24; Draper 1975). Similar gender equality is found at a more advanced level of technology, for example in some

TABLE 3.1   **Matrilineal and Matrilocal Societies at Four Levels of Economic Development**

Percent of societies with matrilineal kin groups	
Hunting-and-gathering societies	10%
Primitive horticultural societies	26
Advanced horticultural societies	27
Agrarian societies	4
Percent of societies with matrilocal residence	
Hunting-and-gathering societies	3
Primitive horticultural societies	15
Advanced horticultural societies	5
Agrarian societies	1

Source: Lenski and Lenski, 1974, 190–91.

peaceful rice-growing societies in Southeast Asia (Bacdayan 1977; Boserup 1970). Under these circumstances, male or female lineages tend to be relatively unimportant.

Where there is constant fighting and all the males are armed, men tend to be organized very strongly and females are subordinated. This particularly takes the form of *patrilineal* family (inheritance is in the male line) and the *patrilocal* household or community (the woman leaves home to live with her husband's kin). This family form is predominant in agrarian state societies, especially in the military classes, but it also occurs in many horticultural and pastoral societies.

On the other hand, even in relatively militarized societies the women may be in a stronger position. This occurs whenever the kinship system is *matrilineal* (inheritance passed through the mother's rather than the father's line) and especially *matrilocal* (wives stay at home with their kin and are joined by their husbands). Such matrilineal/matrilocal systems are not very common (see table 3.1), but where they exist they split up men's resources: their kin and property are in one place and their home is in another (Murphy 1957). (In a matrilineal/patrilocal system, women go to live with their husbands, but men inherit through their mother's house, which also splits up men's resources.) These forms seem to be good at knitting local groups together, so that not too much domestic fighting takes place. The result is that women often hold the balance of power. Matrilineal and matrilocal forms also seem to be correlated with prevalence of *foreign* wars, in which the men of fighting age are away from home a good deal (Ember and Ember 1971; Divale 1984).

Neither matrilineality nor matrilocality is very common, even at the primitive horticultural level (where we find 26 percent of societies to be matrilineal and 15 percent to be matrilocal; see table 3.1). But these dimensions still capture only part of the complexity of family organization. Tribal societies frequently have a larger kin group beyond the individual family; there can be secret societies, clans, totemic groups, and other forms of organization in which both men and women are variously enmeshed. Even patrilineal/patrilocal societies may be offset by other kin-group organizations outside the household, so that in some of these societies women may end up being fairly well organized compared to men. For instance, in the Igbo society of Nigeria, the women had their own political and religious hierarchy separate from the men, culminating in a female monarch who lived in her own palace and who adjudicated all affairs relating to women. In parts of the Igbo realm, each village had a woman's council attended by all married women. Persons, including men, who offended village rules were punished by a mob of women who "trashed" the offender's property and sometimes administered corporal punishment (Sanday 1981, 88–89, 136–40). In general, the better women are organized, and the more men are split up, the more favorable the position of women in that family system (see feature 3.1).

Finally, there are situations in which both males and females are relatively well organized in separate groups, but in which men monopolize the use of force while women control the economy. For instance, in some parts of New Guinea or the South American jungle, most of the food is produced by women's horticulture, while the men engage in head-hunting or cannibalistic expeditions against rival tribes. Both sexes are well organized: the men in secret societies or other groups, the women because they are left alone in their work by the men's incessant preoccupation with warfare. The result is that the men actually depend upon the women for livelihood but give nothing substantial (in the form of livelihood) in return. Hence the men attempt to terrorize the women, through violence as well as ceremony. At the same time, they are afraid of the women; these societies hold beliefs that women are polluting, place taboos on contact with them, and fear that they are dangerous witches. Divale and Harris (1976) and Murphy (1957, 1959) refer to such societies as exhibiting the "male supremacist complex," but a more accurate description might be a genuine "war between the sexes." In societies in which men have little trouble getting their way over women (such as in most agrarian state societies) there is relatively little use of violence; power and property are so overwhelmingly in male hands that there is little reason to resort to force. It is because women in these head-hunters' realms have relatively great leverage that the men are so afraid of them and resort so often to force and vituperation to gain control.

Work and Class Stratification

What kind of work men or women do tends to be an offshoot of the military situation. Where the men are primarily fighting, they will leave the work of producing food to women, as is the case in many horticultural societies. (In some horticultural societies, though, both men and women grow and harvest plants; just as in some hunting-and-gathering societies both men and women gather.) Men tend to monopolize work that is closely connected to their fighting groups: for example, hunting or herding horses or other large animals are activities that foster groups that can also conduct warfare.

When plow argiculture and irrigation were invented, the men usually ended up taking over farming. But this was not, as is often claimed, because male strength is needed for these tasks. Rather, it took place because such societies produced a considerable surplus and became very socially stratified. This did not happen overnight, and some agrarian societies—people living off plow culture—remained tribal and relatively unstratified. Even these societies usually developed military groups, out of which came a new form of stratitifcation. The stratification involved a very important shift in the military structure; the lower class of men was disarmed, and only a small aristocracy of warriors (knights, samurai, or men with some other honorific title) carried weapons. This aristocracy forced everyone else, female *and male,* to work for them. Nonaristocratic males were transformed into peasants and set to work in the fields, plowing or digging irrigation ditches. Women, too, continued to do a great deal of work, especially women in the peasant classes, where they often worked in the fields alongside the men. But women were no longer the primary work force, because another change set in as well: there was considerable social pressure to turn them into baby-producing machines, indoor servants, and, in the higher social classes, erotic objects and pawns in the game of marriage politics.

KINSHIP POLITICS

We have already been introduced to Lévi-Strauss's theory that marriage is a form of exchange among families that creates alliances. In chapter 2, we saw how the incest taboo established a link between each isolated family and the outside world and thus made human society possible. There are many different forms of exchange; some are local, while others are far-reaching. There may also be formal rules of exchange—such as the rule that a man should always marry his mother's brother's daughter if possible (what is called *matrilateral cross-cousin marriage)*, or exchanges may be negotiated informally. The formal rules tend to be most elaborate in stateless horticultural societies, while informal exchanges are more likely in societies based on both less advanced and more advanced technologies.

Lévi-Strauss's theory has been criticized for depicting the world of kinship in an overly male-oriented way. He describes how men exchange women (their daughters or sisters) in order to form advantageous political alliances. But it is also possible to turn the model around; instead of putting the male in the center, we could start from the woman and ask to whom she is connected. The rule quoted above then would state that a woman should always marry her father's sister's son (providing her father has a sister and she has a son) and we could call this *patrilateral cross-cousin marriage* (which it is from the female viewpoint). This is just a matter of anthropological terminology, but there is a real difference that is often overlooked. We should not assume that women have no initiative or powers in these marriage exchanges. Karen Sacks (1979) points out that women in matrilineal societies often have power as *sisters*, just as even in ostensibly male-oriented patrilineal/patrilocal societies women may exert considerable influence over marriage negotiations and other family matters. In the latter case, both men and women are moved around as pawns, but primarily because they are members of the younger generation, while older women (as well as men) may have considerable influence in this political game (see feature 3.1). In the Viking period, which included the invasion of Normandy in northern France around A.D. 850–1050, women were very active in pursuing marriage politics to advance their families politically. Eleanor Searle (1988) goes so far as to call this a system of "predatory kinship."

Just how much initiative and power men *or* women have in the way this game is played depends upon the power situation sketched out above. In general, we might say that the more that military factors favor male power, the more that men will control the marriage exchange system. Conversely, where those factors favor female power, women will have a corresponding influence in exchanges. Lévi-Strauss's male-centered view may be more appropriate for certain situations, while a female-centered view is more accurate for others. There are also mixed cases, in which both genders are taking part in the maneuvering, and yet another type, in which neither men nor women as a group are particularly influential, but individuals choose their own partners by personal preference.

In the more egalitarian and peaceful gathering and primitive horticultural societies, for example, there is not much need for political alliances, and individuals find their own mates. Sexual and even romantic love is not found only in technologically modern societies but is common, for example, in some unmilitarized societies in the Polynesian islands. Again, in our own society, families are no longer politically very important, so they have little choice in the matter of whom their members marry. It is still true that we fall in love within the constraints of a kind of invisible exchange system, but it is based more on social class than on explicit political negotiations.

The more advanced horticultural societies are where formal rules of the "marry-your-father's-brother's-son" type are most likely to be found. Rules like this, according to Lévi-Strauss, are strategies for linking different families together. Some strategies produce what Lévi-Strauss (1949/1969) calls a "short cycle," in which the same two families constantly intermarry, like a braid of hair across the generations. Other strategies produce what he calls a "long cycle," in which a long set of families is linked together, like a chain. Lévi-Strauss (1949/1969) theorized that it is riskier for families to follow the strategy of the "long cycle," because one family might always pull out and break the chain. But if the chain is successful, it is like a chain letter in which every person sends a dollar to the ten persons on a list of names, crossing out one name at the top and putting his or her own at the bottom; the end result is that whoever started the chain becomes rich. Some families thus end up with lots of alliances; they become "marriage rich," while other families who do not succeed at this strategy end up "marriage poor." Lévi-Strauss argues that at one time in history this could have produced a "family revolution," creating unequal societies with upper and lower classes, and thereby bringing to an end the form of marriage politics played by these formal rules.

Whether or not Lévi-Strauss's theory of the "family revolution" is correct, it is clear that more stratified societies emerged at some point. In these, the game of marriage politics was still played, but no longer in the same way. The rules no longer specified which cousins, or kinship groups were preferred as marriage partners and which were prohibited: such rules did not give individual families much freedom for action. In the more stratified agrarian states, aristocracies and wealthier merchants or landowners wanted to be able to marry off their sons and daughters to form alliances with whomever served their purposes. Permanent linkages between certain families, repeated over and over again across the generations as in tribal systems, gave way to a more flexible and in many ways more cynical use of marriages for immediate political advantage.

Comparing all these different kinds of exchange systems, from the most individualized to the most family dominated, the generalization seems to be: the more militarized the family, and the larger the economic surplus and resulting social stratification, the more emphasis there is on marriage politics. There is a tendency, then, for marriage politics to become more explicit as societies become more technologically advanced, from hunting-and-gathering societies up through the agrarian empires. Beyond that point, there is an important reversal. As we shall see in chapter 5, marriage politics suddenly begin to fall away as we get into industrial societies, because there *the family is no longer the center of politics.* The modern bureaucratic state takes power away from the maneuvering families of aristocrats and their retainers, and that suddenly puts the family in a different context: the private family, a newcomer in world history, is born. This shift away from marriage politics will be the subject of the next chapter.

The Politics of Virginity

One pattern that can be observed is that, the greater the emphasis upon marriage politics, the more sex becomes treated as property. In unstratified

FIGURE 3.5: The Varying Emphasis on Premarital Virginity

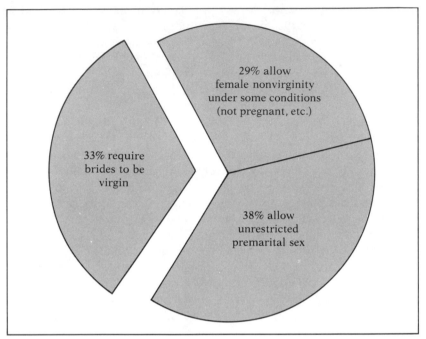

29% allow
female nonvirginity
under some conditions
(not pregnant, etc.)

33% require
brides to be
virgin

38% allow
unrestricted
premarital sex

Only a minority of societies require women to be virgins at marriage. The stratified agrarian states were the most restrictive, the horticultural tribes most permissive.

Source: Murdock, *World Ethnographic Atlas* (1967); data courtesy of Rae Lesser Blumberg.

gathering societies and simple horticultural societies, there is very little control over sexuality (see figure 3.5). Premarital intercourse is frequently allowed with any choice of partners, and sometimes marriage is a trial arrangement that can be easily broken up if unsatisfactory. Not all primitive societies are this sexually libertarian, but few of them require that a woman be virginal at marriage; at most it may be expected that she not be pregnant by someone other than her husband-to-be. As we shift toward societies with greater degrees of stratification, the proportion requiring brides to be virgins goes up.

When we reach the highly stratified agrarian societies, restrictions on sexuality become much tougher. Not only is virginity usually required, but there also tends to be a practice of early marriage, such as found in medieval India. A girl may be betrothed long before the age of puberty, so that she is the

FEATURE 3.2
The Politics of the Female Circumcision Ritual

Female circumcision, also known as clitoridectomy, infibulation, or female genital mutilation, is a traditional practice that is widespread in more than twenty African nations and is also practiced in the Middle East, Malaysia, and Indonesia. The practice also occurs in the United States and Europe, where immigrants from Africa and the Middle East continue to circumcise their daughters, often surreptitiously. It is difficult to estimate the number of girls and women worldwide who have undergone some form of female circumcision because of the secrecy that surrounds the practice. The United Nations reports that the practice affects up to 70 million women, while the Women's International Network estimates that 110 million girls and women in continental Africa alone have been circumcised. The practice of circumcising women dates back to the Phoenicians of the fifth century.

Female circumcision is usually performed on girls aged seven or eight, although some African tribes perform it on infants and other societies on young adult women. It is usually performed by midwives or elderly women of the community, who frequently do not have formal medical training. Circumcision is often performed under poor hygienic conditions, using an unsterile knife, razor, or sharpened stone, without anesthetic.

There are four basic types of female circumcision:

Type 1 is analogous to male circumcision and consists of cutting the clitoral prepuce circumferally to remove it. This is the least drastic type. Type 2 involves removing the glans clitoris or even the entire clitoris; part or even all the adjacent tissues (the labia minora) may be removed as well. Type 3, infibulation or "pharaonoic circumcision," involves removing not only the clitoris and adjacent tissues (labia minora), but the external labia as well; the raw edges of the

wounds are then sewn together leaving only a tiny opening for urination and menstruation. Type 4, which is rarely practiced, is referred to as introcision and involves enlarging the vaginal opening by cutting the perineum. (Rushwan 1990: 24)

Short-term medical consequences of female circumcision include bleeding, shock, infection, urine retention, and damage to surrounding tissues, all of which can result in death. Long-term complications include cysts, menstrual problems, recurring urinary tract infections, incontinence, infertility, complications during pregnancy and childbirth, sexual problems, and psychological problems such as anxiety, depression, neuroses, and psychoses. Many women undergo a painful series of de-circumcisions and re-circumcisions after each child is born. Scars from circumcisions are cut open before delivery and stitched together after. In addition, there is evidence that the unsanitary conditions that surround the practice of female circumcision have contributed to the spread of AIDS, which is rampant in all areas where female circumcision is practiced.

Considering these circumstances, why is female circumcision practiced? The crucial point to bear in mind is that it is a ritual. Rituals are tied in to the social structure of the group, and usually are connected to maintaining the pattern of stratification. Rituals reaffirm a traditional sense of identity; for that reason, the admonitions of outsiders—including the host society that surrounds migrants from an area with this tradition—often are regarded as attacks upon the integrity of the group. Rituals are founded upon the dynamics of group emotions, and are surrounded by ideologies. Thus, female circumcision is sometimes defended as a requirement of Islamic law. Strictly speaking this is

(Continued next page)

FEATURE 3.2 (Continued)

untrue; in this respect the ritual is similar to the custom of wearing veils, which is not enjoined in the Koran, but has become a tradition over the centuries as it has become a mark of high status in the group. Other ideologies supporting female circumcision are the beliefs that it promotes cleanliness, increases a girl's chance for marriage, improves fertility, protects virginity, and prevents immorality. Older women often feel pressured to have their daughters undergo the practice; they fear that their daughters will be ostracized or will not be able to marry if they are not circumcised. Many women accept the procedure because they are taught to believe that sexual pleasure is the exclusive right of men.

Efforts are being made to eradicate these long-held beliefs, including educating midwives, enlisting the support of religious leaders and women's groups, and enacting laws to punish those who practice female circumcision. Strategies to abolish the practice must take into account local customs and belief systems. Efforts to publicize the medical risks are generally more effective than those that emphasize sexual issues, which are highly controversial. The World Health Organization has implemented several programs at the grassroots level, including the Childbirth Picture Book Program. These picture books graphically depict how female circumcision is damaging and impedes normal childbirth. They are designed to be used by community health workers and are effective with an illiterate population. This program has been somewhat successful in decreasing the incidences of female circumcision.

In the United States, legislation has been introduced in Congress to make female circumcision illegal. In 1994, a Nigerian woman successfully appealed deportation from the United States on the grounds that her two daughters would be circumcised by her family upon return to Nigeria.

Sources: C.A. Baker, G.J. Gilson, M.D. Vill, and L.B. Cureto. 1993. "Female Circumcision: Obstetric Issues." *American Journal of Obstetrics and Gynecology* 3 (December): 1616-18; "Men Control Female Sexuality with Circumcision." 1993. *AIDS Weekly* (March 29): 10; "Puberty Rite for Girls Is Bitter Issue Across Africa." 1990. *New York Times International*, January 15; Hamid Rushwan. 1990. "Female Circumcision." *World Health* (April-May): 24-25; Women's International Network. 1993. "Progress Report: Campaign to Stop FGM." *WIN News* 19 (Autumn): 29.

exclusive sexual property of her husband-to-be. Any sexual experience she may have thus becomes a grave insult to him and a violation of the alliance between the families contracting the marriage. Under these circumstances, women are often secluded, veiled, or locked up in their homes. Any violation makes her completely dishonored and useless for exchange politics. Premarital sex, like adultery, is often punished by death of both man and woman or by the sale of the woman into slavery or prostitution (Vieille 1978). To prevent such immorality, many societies encourage the practice of female circumcision (see feature 3.2).

This extreme form of sexual property in stratified agrarian societies is very one-sided, however. Women are expected to be virgins at marriage, but men are not. Men may even gain glory by their sexual exploits, and prostitution in such societies is usually quite open and legitimate. There are certain sexual rituals that may be enforced upon men, however (see feature 3.3).

Another limitation is that sexual property rights are most strongly enforced in the upper classes and in respectable urban society. Peasant women cannot be locked up in women's quarters, because they are needed to do the work, outdoors as well as indoors, and they may not even wear veils, though upper-class women in the same society would be dishonored to have a man see their face. The wearing of veils grew up particularly in the Islamic societies of the Middle East. At first it was a mark of social distinction, since it meant that a woman could remain secluded from contact with men other than her own husband—which is to say, it implied that she did not have to work in the open like the common people (Hodgson 1974, 342).

In some societies, widows in the upper classes may be prohibited from remarrying (because they are not virgins); in a very status-conscious caste society like India, they were even encouraged (or forced) to commit suicide on their husband's funeral pyre. But a widow of lower rank, especially if she had some property or could make a living, would be more valuable to another husband. Thus in highly stratified societies lower-class women had more freedom than women of the higher classes. As peasant or village families gained

The wearing of the veil persists in Islamic societies. This custom represents the traditional way of life in which women are secluded from contact with men other than their own husbands. It is also a mark of high status in the society.

some wealth and respectability, though, they usually attempted to raise their status by imposing upper-class restrictions upon their women. In some agrarian societies modernizing in the twentieth century, change has occurred in opposite directions at once as the wealth of the society has grown. For instance, in contemporary Iran, traditional puritanical restrictions were spreading in the peasant villages, even at the same time that women in the cities were discarding the veil. After the revolution led by Muslim fundamentalists in 1979, the wearing of veils was imposed on all women. Strictly speaking, this custom is not a law of the Koran (Hodgson 1974, 343); but it had become part of the traditional way of life that the fundamentalists were attempting to restore.

FEATURE 3.3
The Politics of the Male Circumcision Ritual

Circumcision, the cutting back of the foreskin from the glans of the penis, is carried out unceremoniously in hospitals today as a hygienic measure. But historically, circumcision had religious significance; for example, it divided the believers from the nonbelievers in both the Jewish and the Moslem faiths. Karen Ericksen Paige and Jeffery M. Paige (1981, 122–66) carried out a cross-cultural analysis of circumcision and found it widespread in a certain type of tribal society.

In these societies, circumcision was a ritual, carried out in public and in the presence of a large crowd. Often the subject was not an infant but a larger boy or even a teenager, who had to undergo an ordeal something like a hazing. There were feasts and elaborate ceremonies, which might include camel races or dancers brandishing swords to drive away evil spirits. It was especially important for all the male relatives to attend, although in some cases females attended as well.

The types of societies in which circumcision rituals of this sort were carried out consisted especially of warlike tribes that ranged widely over grazing land with their camels, goats, sheep, or cattle. They were most likely to be found in the Middle East and North Africa, which are the classic homelands, of course, of ancient Judaism and of Islam. Karen and Jeffery Paige analyze the circumcision rituals as a device used in the political maneuvering of these tribes. They have no regular state but only coalitions of warriors from different families, who band together to defend their grazing territories and to conquer others. Feuds are common, and the whole family lineage is bound to avenge the death or injury of one of its members. Hence the extended family coalition is very important. These are extremely male-dominated societies, and the main way that a man can become powerful is to have many sons, who in turn might have many grandsons. Thus, by the time he reaches old age, a patriarch might turn out quite an extensive army of descendants and relatives. In order to do this, men who can afford it acquire as many wives as possible (polygyny), usually by paying a bride price in the form of so many camels, cattle, or other animals (which are the principal form of wealth).

How does the circumcision ritual fit into this situation? Paige and Paige argue that it is a way of monitoring the strength of a coalition. Not only

PATRIARCHAL DOMINATION IN THE AGRARIAN STATES

In the world's great ancient and medieval civilizations, family organization took a new form. No longer were complex kinship groupings as important as they were in tribal societies. Society was starting to emerge beyond the level of kinship.

But it only emerged halfway. For the agrarian state societies were the era of the *household*. Both economically and politically, almost everything was organized around households. The places where people lived were the same places as

are these societies violent, but they are also unstable. Family coalitions often fall apart, because it is always possible for a son to gather his own wives, descendants, and followers and move off with their herds of animals to graze somewhere else, or to raid someone's grazing lands. There are no fixed political boundaries to hold things together. Moreover, it is often brothers or cousins who get involved in these feuds and splits. The Old Testament gives many examples from early Hebrew history, when the Jewish tribes wandered with their cattle throughout the Middle East. Abraham's followers fought with those of his relative Lot (Genesis 13:6–7); Isaac and Ishmael, who were half-brothers (their father Abraham being polygamous), became rivals and Ishmael was driven away into the desert like the Arabian Bedouin (Genesis 21:9–21); Jacob and Esau fought over their inheritance and split the tribe once again (Genesis 28:1–5).

The purpose of the circumcision ritual is to bring together all the male kinsmen and to try to bind their loyalty to the military coalition. The ceremony is an occasion on which all who consider themselves a member of the extended family show up to be counted, and to stay away is a mark of disloyalty. Moreover, the ceremony centers on the penis of the son, which is the instrument through which a patriarch will acquire more grandchildren and hence a larger coalition. But the penis is, so to speak, a two-edged weapon: it can increase the coalition, but it also raises the danger of splits between the sons when they gather their own followers. Hence the undertone of hostility and the infliction of pain in the ceremony. Moreover, given the low level of sanitation and of medical practice in these societies, circumcision brings the danger of infection and even of castration. The ceremony shows the willingness of the father to symbolically sacrifice something—the foreskin—and even more to risk his son's penis in the presence of the entire military coalition. It is, in short, a kind of display of bravado, simultaneously boasting about the potential reproductive power of the patriarch's lineage and reminding the group that he is indebted to them and will make sacrifices to hold the coalition together.

they worked, and these were simultaneously the places where people had their political loyalty.

If one had visited the bazaar (market) in an Arab city not long ago (or still today in some places), one would have an idea of what the patrimonial household economy meant. There are the merchants, with their goods lined up for display on the street in front of their shops. Step inside a shop, and you find more goods. But the back of the shop is not like an ordinary store. There are children playing around the goods, or more likely helping with the work, coming forward to sell the visitor something or running to do the parent's bidding. If one gets the idea that the father's word is law, that is not far from right. Somewhere in the back, too, may be the merchant's wife, helping with the store, but also doing the usual domestic duties of cleaning, carrying water, cooking. For the store is a home, and the family lives there in the back, or perhaps upstairs. That is one reason why the bazaar keeps such long hours, as opposed to our conventional nine to five; the store is almost always open, because the proprietor never goes home: he is already home, available whenever anyone comes in.

Multiply this example by thousands or millions, and spread it backwards over the history of medieval Europe, the Middle East, India, China, Japan, and you will have a fair idea of what life was like in a society of *patriarchal households*. Most such households were agricultural and rural, rather than mercantile like the bazaar example given above; the key factor is that there is no difference between the place of work and the residence. What is different from a tribal form of kinship is that the households are much more stratified; some are big, as large as palaces, others very small, mere one-room hovels. They are also much less organized around kinship. There is almost always a family at the center of every household, but there are now other people as well. Servants are found in every household that can afford them; large households might have dozens or even hundreds, dressed in a livery (uniform) that shows off the status of the master, but even the modest middle-level households try to have at least an old woman or a young servant girl, quite possibly a slave.

Slaves, in fact, tend to be very much in evidence. These are rather callous societies, among the cruelest known. Ancient Rome lived off the work of slaves, and they existed in medieval, Christian Europe as well, and in the American colonies (Patterson 1982). Households of the aristocracy often preferred to own eunuchs, men who had been castrated. Eunuchs were even trusted with considerable authority. This was partly because they could have no families of their own, hence did not pose a threat to the family succession, and partly because they could be used to guard and attend the women without fear of adultery. Eunuchs were especially popular in the Byzantine and Ottoman (Turkish) empires but also elsewhere in the Middle East, the Roman empire, and in China, especially where there were harems.

Somewhat overlapping with the category of servants were apprentices, pages, and retainers. It was quite common, especially in England and northern

Europe, for children in their early teens to live in the household of someone of higher rank and wealth. A peasant girl might be the housemaid for a yeoman farmer; a boy might go off to the town to apprentice with a baker or some other craftsman. This was partly a way of making a living, partly a form of education, of learning the trade, with the hope of amassing enough money to set oneself up in business later on. We have a description of a typical bakery in London in 1619, just after Shakespeare's death: it shows that the household included the baker and his wife, four journeyman employees, two apprentices, two maidservants, and three or four children of the master himself. The total is thirteen or fourteen persons, the vagueness indicating how little attention was paid to small children (Laslett 1971, 1–3). It was not a sentimental society; if everyone was treated as part of some big family, it was an authoritarian and businesslike one, not to be confused with today's private families with their emphasis on warm domestic togetherness.

One more important characteristic of these households should be mentioned: they were essentially *fortified households*. Generally speaking, there was no police force; everyone kept arms who could, although this generally meant that the aristocrats had most of the weapons. Shakespeare's *Romeo and Juliet* gives a good idea of what the situation was like. Romeo and Juliet belong to two wealthy patrician families of the Italian city of Verona. Each house is a kind of fortress, and Romeo has to scale the walls to get to Juliet's balcony to see her. Each family is full of armed men—sons, cousins, retainers, servants—and the two families are feuding because the two groups of armed men have battled on the street and Juliet's cousin has been killed in the fighting. That is the reason why the love affair between Romeo and Juliet is so illicit: the two families not only are not allies but actually are something like opposing armies.

We have seen that the family, with its servants and apprentices, was the economic unit. At a higher level, the aristocratic family was a political unit. How much power it had depended, first of all, on how much armed force it had within its own walls, and second, on what kinds of alliances it had made with other households. The ruler, originally, was the person with the largest household. Louis XIV of France overawed Europe with his palaces, which might contain ten thousand people besides his own family: courtiers, ladies-in-waiting, guards, and servants. The royal household was the state; below it were lesser households of the nobility and officials, each of these in turn with its own retinue of servants and followers. Just as a peasant boy might be sent to an urban bakery to learn the trade and make his fortune, the son—and also the daughter—of a minor nobleman would usually be sent to be a page or maid-in-waiting at the house of some more powerful lord, with the hope of establishing a connection that would make *this* young person's fortune.

The society was thus held together by a kind of household-to-household link (Girouard 1980; Mertes 1988). We pointed out that this was halfway beyond the kinship system, because most of the people in any one household were probably not relatives. In a strange way, there was no clear dividing line

FEATURE 3.4
Medieval versus Modern Houses

This Seigneur's dwelling room of the fourteenth century combines a "living room," bedroom, hearth, and shrine. It exemplifies the medieval Great Chamber, with no connecting hallways between rooms and little privacy.

between servants, retainers, and guests. A powerful lord built up his reputation and his following by keeping what we might call "open house": people could come and visit him, eat at his table, and stay with him for weeks or even years. In return, the guests were expected to be loyal followers, perhaps carry out political duties, fight in his battles, or in the meantime just act as servants. The nearer one could be to the lord, the more favors one could expect from him; hence it was much better to sit at his own table, though popular lords might entertain so many guests that one was lucky to get a bite at the far end of the hall. In the court of Louis XIV, high lords vied for the honor of doing humble services that might bring them near His Majesty; one was honored with the title of Gentleman of the Bedchamber, which meant that he had the honor of holding the king's robe while he stepped out of bed; another gentleman held a shovel with coals of burning incense to sweeten the air while His Majesty sat on the royal chamberpot (Lewis 1957, 1–61). As one can see, these were societies without much privacy in the modern sense (see feature 3.4).

The patrimonial household of the medieval civilizations was a kind of fortress. Some houses of great lords were literally castles, though even urban dwellings tended to be built for protection. Thus an old European city street typically has stone outer walls, with only a few high windows and great barred gates. The house does not face outward, but inward toward the courtyard, where people can be outdoors but under their own protection. Similarly, if one walks down an old street in Japan, one sees nothing but blank wooden walls; only if you are admitted inside can you see the light and airy verandas opening onto an attractive garden. Unlike our modern houses, these medieval houses were not built for external display; the display was for insiders only.

Another difference one would notice is that medieval houses did not have hallways. We tend to take for granted that one of our homes will have a central hallway, a corridor with bedrooms, bathrooms, and other kinds of rooms leading off it. But a large medieval house did not have these corridors or private rooms; instead one large room simply led onto the next. There was no way to get

to one room except by passing through all the other rooms leading up to it. This was part of the lack of privacy in medieval society. There were virtually no special purpose rooms, such as bedrooms or bathrooms. The "Great Chamber" meant a very large room, where the master of the house could entertain large numbers of guests for dinner. The composition of the household shifted, with different guests and retainers there at various times. Beds were trundles that were put away in the daytime and were shared by several people, or sometimes merely by cushions or places to sleep on the floor. The high-ranking lords and ladies would have beds, but their servants probably slept in the same room with them. As for bathrooms, there were chamberpots tucked away here and there, which is one reason why people who had servants kept them nearby. Excretion or sexual matters could not be kept very private, and people rarely tried to make them so. In the humbler ranks of society, houses were just as crowded and smelly as among their superiors, but much smaller. A peasant family usually had a hut with a single room, in which all the business of the household was conducted.

But we have not yet arrived at the modern type of workplace, where people live in a private home in one place and go somewhere else to work with people who are not relatives at all. As we shall see in the next chapter, this patrimonial household structure eventually began to break down. Political and military circumstances made it impossible to keep an entire army and an entire government bureaucracy in the household and feed them at the lord's personal table. Organizations began to emerge outside the household, and these giant castles and palaces began to give way to a more private dwelling even for the upper classes. The king's servants gradually became bureaucrats. Sometimes they retained their old names, though their functions changed. *Chancellor,* once the title of the butler in charge of the wine cellar, became the name for a high government official. The secretary took care of the lord's private correspondence (and at this time was always a man); eventually the title was applied to such elevated personages as the secretary of state.

The lower classes, on the other hand, had always lived in much smaller groups, unless they happened to be servants in some great man's household. Since the poor by far outnumbered the rich, one could say that the fortified patrimonial household was for a small but dominant minority, while the majority of the populace lived in small nuclear households. In some senses the families of the peasant majority were not too different from families today. But in other ways their lives were quite different: like the upper classes, they had little privacy; they tended to treat their children unsentimentally, as economic tools; where possible, the men were patriarchal and authoritarian; and even poor families emulated the style of the families above them on the rungs of the social ladder, with whom they were often connected by ties of feudal duties. Any family that could afford it tried to acquire servants and to copy the manners of their superiors. They also tried to establish ties with the larger households by going to work in them or becoming their retainers and followers. In these ways, the patrimonial households of the aristocrats dominated the surrounding society, even though this society consisted of a much larger number of small households inhabited by something more like the modern nuclear family.

The Era of Maximal Male Domination

The agrarian state societies had a wider gap between top and bottom of the social structure than any other type of society that has ever existed. It has been estimated by Lenski (1966) that the top 1 or 2 percent of the population owned over half the wealth, and that the king or emperor alone usually owned one quarter of the wealth all by himself. There was little in the way of a middle class, and some 90 percent of the population might have been peasants in tiny huts, living barely on the edge of subsistence. Stratification between males and females also reached an extreme (Michaelson and Goldschmidt 1971).

In these societies, women were most callously treated as mere property. Wealthier men, at least in many parts of the world, had harems or concubines. And even where men were restricted to a single wife, they tended to keep their wives and daughters secluded under lock and key, or veiled so that strangers could not look at them. The lower classes could not afford to do this; a man could have only one wife, and she was freer because she had to work. But the ideal of the patriarch was widespread, and peasant men lorded it over their women to the extent that they could. Legally and politically, women tended to be treated as minors. Because of the practice of giving a dowry with a woman when she was married, girls were considered an economic imposition and female infanticide was often practiced. Even the religions prevalent in agrarian societies (Islam, Hinduism, Confucianism, Buddhism, Christianity) generally treated women as second-class citizens and as temptations in the way of men's religious duties.

Why did men in these societies want to dominate women? It cannot be explained by invoking universal psychological or biological drives, since we see that this form of extreme domination occurs only in particular historical circumstances. Moreover, even in these medieval agrarian societies, men would not typically have said that they were dominating women just because they wanted to. These were indeed rather callous societies; servants, slaves, and members of the lower classes were treated more or less as dirt beneath the feet of the aristocracy. Upper-class ranking was so ritualized in everyday life that people felt they were divided into two distinct species of humans. Lower-class women were treated badly because they were members of the lower class.

Within their own families, men no doubt believed that they were honoring and protecting their wives and daughters by the tight controls they placed upon them. As we have seen, wearing a veil was considered a mark of high status in Muslim societies. Placing strict controls on women's sexual freedom was considered in agrarian societies as a way of protecting the honor of the family and of the woman as part of it. Even the Chinese practice of binding girl's feet, which made them unable to walk freely, was justified as displaying the high status of the family that could afford to have its women in this condition. The ideology of these societies assumed that women were being controlled for their own good, for whatever raised the status of the family raised the status of its women members too. Like all ideologies, this covered up the real power resources that enabled persons in the most favored circumstances to control those below them. Political and economic conditions put upper-class men in positions of almost tyrannical control, and this influenced gender stratification throughout the society.

Countervailing Forces for Female Status

Nevertheless, there were some conditions that favored women in these societies. The class stratification system itself meant that some women were of very high rank, even though their rank depended upon a man. Wives and daughters of important aristocrats or wealthy patricians had an important status in the society. Men of lower rank might try to make their approach to a high-ranking lady, in hopes that she could wield some influence on their behalf. And the politics of family alliances produced a good deal of legal maneuvering about inheritance, which became more rigged to favor the female line (Collins 1986b). This was perhaps not done in order to give more rights to women but because families wanted to receive or give away inheritances through their women. Nevertheless, it had the consequence that upper-class women, at any rate, acquired some property rights. Sometimes, through the death or absence of the males in the family line, a woman might actually wield power as a queen or noblewoman. Interestingly enough, this did not usually change the status of women generally in the society; the reigns of some of the most famous queens, such as Queen Elizabeth I in England or Catherine the Great in Russia, were conservative times when the majority of women were held under tight restraints. But even in an extremely male-dominated world like Ottoman Turkey, women acquired some powers of their own (Dengler 1978). Turkish women were supposed to be secluded from the eyes of men; but one effect of this was that they could move about in their veils more freely and anonymously than anyone else in this rather authoritarian society. The harems of the wealthy had large staffs of female servants, since the effort to exclude men meant that women for the most part were left to themselves. Wives and daughters of wealthy men might supervise huge numbers of servants and dispose of considerable economic property, even acquiring property in their own names.

Especially in Europe, the aristocratic women were pioneers. They exerted some pressure, perhaps very gently, on the property system. More significantly, as we shall see in the next chapter, they began to change the culture of male-female relations. In medieval France and then elsewhere, aristocratic women began to create the ideal of courtly love. In the midst of a society in which women were largely treated as pawns in the game of marriage politics, or ruthlessly exploited if they were members of the lower classes, an ideal began to emerge that demanded deference toward women on the part of men. It was only one step, but it was to have important consequences.

SUMMARY

1. Family systems rise and fall in a bell-shaped curve as societies become technologically more complex. Relatively simple families are found in hunting and gathering societies, while complex kinship systems reach their peak in advanced horticultural and stateless agrarian societies. The kinship system begins to simplify again in the agrarian civilizations, then rapidly loses complexity to form the nuclear family unit in industrial societies.

2. Gender inequality is explained by both economic and political theories. Economic theories propose that women have the greatest power inside and outside the household, and the greatest freedom to arrange their own marriages, divorces, and sex lives, where women make the greatest economic contribution *and* are able to control economic property. This happens when women's labor is indispensable because of female expertise, work organization, and sex ratios in the population. Also contributing is a kinship system that allows women to control inheritance, and it is also important that stratification of the larger society does not interfere on behalf of men.

3. Male strength and female breast-feeding are not very important in determining

what work women can do. Where women have the social power and their labor is desirable, they have been able to control their own fertility and work out satisfactory child-care arrangements.

4. Political theories of gender inequality argue that the male size advantage has been used mainly to monopolize fighting and hence politics, and that men have accentuated this advantage by monopolizing weapons. Male dominance thus depends upon how much fighting there is and how the male military group is organized.

5. Where societies are relatively peaceful and men are not organized in all-male groups, the family system tends to be egalitarian. Societies with patrilineal and patrilocal families isolate women amid strongly organized male groups and are very male dominated. In tribal societies where fighting takes place far from home, however, there is a tendency for organization to be matrilineal or matrilocal and to give women a somewhat better position. In some societies in which both genders are well organized, there is a great deal of antagonism between the sexes and frequent use of male violence.

6. Men tend to monopolize work (such as hunting and herding) that can most easily be transformed into military activity, leaving gathering and horticulture to women. In class-stratified societies, aristocratic males shift the burden of agriculture to male peasants, and women are forced into a more specialized role as baby producers.

7. Tribal societies often practice complex forms of kinship politics, in which families exchange marriages according to certain rules (especially rules emphasizing cousin marriages). In stratified agrarian societies, these formal rules tend to disappear, so that families can maneuver more flexibly to make marriage alliances for their political convenience.

8. In societies that emphasize marriage politics, women are treated as sexual property and their virginity is closely guarded. Premarital and extramarital sex, however, is allowed in many societies with less political pressure.

9. The ancient and medieval agrarian civilizations were highly stratified between an aristocracy and the peasants, servants, and slaves that they dominated. The upper classes lived in large patriarchal households, in which the family was surrounded by servants and armed followers. The lower classes lived in much smaller units, but they tried to emulate the large upper-class household whenever their income rose.

10. These agrarian civilizations represent the peak of male domination in world history. Especially in the upper classes and in the cities, women were kept secluded or veiled and were legally treated as minors. Female infanticide was often practiced, and women had low religious status. But countervailing forces existed in some aspects of property systems and marriage politics, so that some women acquired economic rights and, occasionally, political power.

THE LOVE REVOLUTION AND THE RISE OF FEMINISM

It used to be believed that the modern family structure emerged quite recently. A shift was supposed to have occurred from the *extended family* to the *nuclear family* within the last 100 years or less. Traditional families were supposed to have been large, with several generations living together in the same household. Along with this went a lack of individualism. All the members subordinated themselves to the good of the family as a whole, contributed their earnings when they could work, and were cared for and respected when they were too old. The counterpart of this lack of individualism was that men and women did not get to pick their own spouses; instead, marriages were arranged by the family, often at a very early age. The family encompassed the individual from the cradle to the grave.

Then, according to this analysis, along came industrialization, urbanization, and (in some versions) capitalism. The self-sufficient rural household gave way to the small urban home. Each man had to go out and seek his own work in factories or offices. The craft worker was alienated from the means of production and subjected to the impersonal labor market of the city. No longer did each contribute to the family kitty; no longer were the old people sheltered by their loved ones. With this, family controls over marriage broke down, and each person chose his or her own marriage partner. For better or worse, the modern family was born.

This picture of the modern family transition hangs together nicely, but it has one main flaw: it simply isn't true. Until recently, sociologists, anthropologists, and historians had not looked closely at the families of traditional agrarian Europe. Instead they had relied upon a comparison between the nuclear family system that exists now and "snapshots" of the kinds of tribal societies studied by anthropologists. Somehow these tribal families were supposed to be an approximation of the families of our own ancestors. Since industrialization happened quite recently, it was assumed that the modern nuclear family emerged recently, too.

WHEN DID THE MODERN FAMILY BEGIN?

In the 1960s, a number of British and French historians began to search through local village records, church documents, and other surviving information about medieval and early modern communities. Their aim was to reconstruct what family life was actually like—how many people were living in houses at what time, how often children were born, and so on. The results were quite a surprise, for these western European families did not resemble the "traditional" picture at all. The method has been extended to the early American colonies and to a few non-Western societies such as Turkey and Japan. The picture that emerges shows our former beliefs about the rise of the modern family to be a myth.

Myth 1: The Extended Family

When we think of the traditional family, we imagine several generations gathered around the same hearth: father and mother and their children, some of them perhaps grown-up sons who have brought their wives to live in the family home, and grandfather dozing contentedly by the fireplace. But in fact this type of family was nowhere to be found, at least not in sixteenth and seventeenth century England, northern France, or America (Laslett 1971, 1977). Instead, what we find when we look back is something that looks very much like the modern nuclear family: father and mother and their young children. When the children were big enough to work, they left home; when they were married, they acquired their own house. Nor were old people taken care of by their children; one rarely finds evidence of a grandmother or grandfather living in these homes, but one does find evidence of old people living by themselves in their own cottages.

How could we have been so mistaken about the family? One fact that was overlooked was sheer demography. Most people in these traditional societies did not live very long; life expectancy was about 35 to 45 years. (In Plymouth Colony, for instance, it was 45.5 years for men, who lived longer than women: Laslett 1971, 98.) These figures, however, are for life expectancy at birth, which was reduced by the fact that many infants and children died. Once individuals reached adulthood, their chances of reaching old age were improved. But it is clear that the proportion of old people was much lower than in our times. Since people did not marry early, it was not so common either for men or women to live to see their grandchildren, or even to live much beyond the age at which their own children would marry.

Myth 2: Early Marriage

We also have the image of a society in which people married very young. We can think of Romeo and Juliet in Shakespeare's play, Juliet was thirteen, and her mother says to her:

> Well, think of marriage now; younger than you,
> Here in Verona, ladies of esteem,
> Are made already mothers. By my count
> I was your mother much upon these years
> That you are now a maid.

But this, too, is a myth. Laslett (1971, 86; 1977, 40) found that in Shakespeare's own day (the early 1600s in England) the typical bride was about twenty-four to twenty-seven years old, and the typical groom was about twenty-seven to thirty. In our own century, during the 1950s, the median age of marriage was much lower (twenty for women, twenty-two for men), and teenage brides were much more common than in any other period for which records were kept. (Since the median is the point at which half the marriages are above that age and half below, in the 1950s half of American brides were teenagers.)

Again, what is going on here? For one thing, our image of medieval child marriages is based on a skewed sample: until historians started looking at records of the common people out in the villages, all we had to go on were literary records (like Shakespeare), which are concerned only with the aristocracy. It is true that at the very highest level of society, children might have been betrothed at an extremely early age, with negotiations beginning as soon as a child was born. A study of all the English princesses born between 1035 and 1482 (Boulding 1976, 429–31) found that half of them were married by age fifteen, with some married off by the time they were nine. But these were dynastic marriages, carried on for the sake of political alliances with the royalty of other states. Lower down in the aristocracy, nobles' daughters married a little earlier than peasants' daughters, but not a lot earlier: Laslett (1971, 86) found a mean age of nineteen to twenty-one for noblewomen. Noblemen, on the other hand, tended to marry about as late as other men, around age twenty-six.

Why did most people wait so long to marry, especially considering that their life span did not leave them much time afterwards? Primarily, the reasons look economic. The household was an economic enterprise: a farm, a bakery, a mill, or whatever it might be. The work was done right in the home, and the entire family (together with its servants) constituted the work force. To become the legitimate head of his own business, a man had to be married (and, conversely, a woman could become the head of a business, with very few exceptions, only by taking it over after the death of her husband). So the decision to marry was an economic one. A man could not run such a business without being married, and in fact was not even legally allowed to do so in England (Laslett 1971, 12).

On the other hand, a couple could not marry unless there was a house or a cottage they could acquire, a business opening that could be taken over. This operated as an automatic safety valve against overpopulation, since the age of marriage adjusted to economic conditions. In good times, people could marry earlier; in bad times, they had to wait, and some of them might never marry. Thus, the number of children born would go down during bad times, decreasing the pressure to find farms and household establishments for them when they became of marrying age. Under these conditions, children tended to be born much later than they are now: in an English town in 1700, for instance, the average age of mothers at the birth of their children was thirty, and the age of fathers was thirty-five (Laslett 1977, 41). Not until an industrial labor force began to emerge after 1750 were many people able to marry and at a relatively younger age (Goldstone, 1986).

What typically happened was that children left home when they were teenagers, to become servants or apprentices in someone else's house. This provided labor power to run these household economies: workers had to be gotten from somewhere, since it was likely that children in most families were too young to help out seriously. Servants were kept under family discipline, and they could save up their wages until an opportunity came, in their twenties, to leave and acquire a household of their own. There was a typical life phase of working as a servant, at least in England and northwest Europe, experienced by about a third to a half of the population (Laslett 1977, 43). It was not a lifelong position but rather something that people commonly went through in their teens and early twenties.

The Western European Family Pattern Compared with the Rest of the World

It looks as if there was (and is today) a distinctive western European family pattern, characteristic of England, the Netherlands, and northern France, which was found already in the late Middle Ages. It was this family structure that was transplanted to the North American colonies that became the United States. Just exactly when it began no one yet knows. Researchers have found it as far back as the 1200s in England (Macfarlane 1979); possibly, it goes back even farther. This "modern" family pattern did not have to wait for industrialization and a modern capitalism to develop; it was there in these rural farming societies. (Conversely, there is evidence now that industrialism did not cause the breakdown of the family when it became ascendent in the late nineteenth and early twentieth centuries: Tamara Hareven [1981] shows that the same New England families had, for several generations, sent their members into the factories at a certain age, a practice not unlike the older system of apprenticeship. This was especially true of single women.)

The western European-style family did not exist everywhere, however. It had some distinctive characteristics not found elsewhere in the world (Hajnal 1965; Laslett 1977, 12–49). The household was nuclear, consisting only of parents with their young children. In eastern Europe, southern France, and many other places, by contrast, there was much more likely to be a complex household, with several generations living together. There, married brothers might share the family property, which could not be divided or sold off in shares. Or one brother (usually the oldest) might take over as head of the so-called stem family, living with his parents while they were alive (Berkner 1972). This is very much the same today.

In the western European family, there was a relatively small age gap between husbands and wives. Typically the husband was about two or three years older, although in as many as one-fifth of all marriages the wife was older. (This was probably due to the strong tendency for a widow who inherited her husband's business to be a prime target for economically motivated marriage.)

In eastern Europe, or Asian societies like China and Japan, husbands were likely to be much older than their wives. This made the relations of spouses especially remote and put women at a special disadvantage, since they were young and had acquired few economic and cultural resources compared to their husbands. But in western European families, spouses were more likely to have a closer relationship.

In the western European family, women in particular put off marrying and childbearing until they were older. This is one reason that the complex, extended family is more often found in non-Western societies. In Russia, for instance, three-generational depth was found in most families. Women married young and bore children early, thus making it possible for old people to see their own grandchildren grow up. This practice also provided a greater source of labor right within the family, eliminating the need to go outside the family for servants. (The sons' wives who came into the family, though, were the equivalent of such servants: especially before she had borne a child—preferably a son—the bride ranked the lowest in the household and was given all the menial tasks.)

Finally, the western European family was unique in having a high proportion of servants and in typically sending its members out to work as servants during their early adult years. In non-Western European societies, servants were much less common. Typically only upper-class households would have servants, whereas in England and western Europe, even small peasant proprietors were likely to have them. Where servants did exist in eastern Europe or Asia, they were more likely to be lifetime servants who were devoted (or confined) to someone else's family until they died. In western Europe, being a servant was not a permanent condition, but a life stage confined largely to the young and unmarried.

If by "the modern family," then, we mean the nuclear household with relatively close and companionate ties between husband and wife, it is certainly not a recent development. Why it should have appeared in western Europe already by the Middle Ages but not elsewhere in the world is a puzzle that has not yet been solved. But this is not to say that *no* important changes have occurred in the Western family in the last few centuries. The most notable changes are those that have affected sexual behavior, family sentiments, and the position of women. The ideals surrounding marriage shifted drastically; marriage became viewed much less as an economic arrangement (though it still was) and more as the expression of mutual love. Sexual puritanism was enunciated very strongly, especially in the 1800s during what might be called the "Victorian revolution." And connected with both of these, the position of women was changed, although in somewhat contradictory directions. On the one hand, women gained a much more exalted position, at least in the popular ideology and in polite society. On the other hand, a doctrine of "separate spheres" grew up that confined women (at least above the working classes) to a role of housewife and mother much more exclusively than before.

This combination of changes is best seen as the result of a conflict. Women were struggling to get out from under the legal and other disadvantages that oppressed them, and the new form of marriage market that was emerging gave them some new opportunities. The sentimental revolution of love and the Victorian revolution of prudery are part of this, but they are also part of a counterattack that gave women a very confined role along with their new gains.

THE LOVE REVOLUTION

The basic principle of the modern love revolution was to connect love with marriage. Love existed in previous societies, to be sure. But it simply was not expected to be the reason for marrying someone. In tribal societies, kinship rules that specified, for instance, that one should marry one's cousin on the father's side but not on the mother's side obviously excluded love as a motive for marriage. Similarly, where marriage was a matter of family politics or economics, the sentiments of the individual counted for very little. By the mid-twentieth century, though, marrying for love became *the* dominant ideal, so much so that one is embarrassed to admit to marrying for any other reason. This change, which came about gradually in the 1700s and 1800s, may be referred to as the *love revolution*.

The Ancient Ideal of Love

It is not that love did not exist in other societies. It did, and ancient societies treated it clearly in their mythologies. In ancient Rome, Venus was depicted as the goddess of love, and her son Cupid was depicted as a cherubic figure whose arrows made men and women into helpless lovers. In Greece, love's name was Aphrodite (from which comes the term *aphrodisiac*), and her son's, Eros (the origin of our word *erotic*). Especially in the later phase of Greek civilization called the Hellenistic period (ca. 300 B.C.–A.D. 300), popular literature was full of romantic plots about orphan boys who ran away with girls whom their fathers had betrothed to someone else (Hadas 1950). (In the end it usually turned out that the boy had been exchanged for another when he was a baby, and hence was really the son of a good family, so he could marry the girl.) In Asia, too, love poems from that period speak of yearning and passion.

But love was never depicted as the normal path to marriage. On the contrary, marriage was something quite separate. The Greek and Roman goddess of marriage was not Venus but Juno (Hera), the wife of Jupiter (Zeus), and she was always depicted as a rival and enemy of Venus. Zeus had many legendary love affairs, but his adulteries were an embarrassment to Hera and the object of her revenge. In real life, too, this seems to have been the case. Greek men kept their wives locked up inside their houses while entertaining themselves with cultivated prostitutes called *hetairai* (Pomeroy 1975). The

Greek philosophers, including Plato, wrote a good deal about love, but for them it was an almost exclusively masculine passion. The prevailing form of love-infatuation in the classical period (600–400 B.C.) was homosexual. Typically this involved an older man falling in love with a beautiful adolescent boy (Dover 1978). Obviously it could not be the basis for a family, nor did it even involve the two-sided, mutual love that is the modern ideal of heterosexual love (and of modern *homosexual* love as well). Homosexual love affairs apparently also occurred among women; our *lesbians* are named for the island of Lesbos, where the poetess Sappho lived and wrote of her affection for younger women. But women (at least in the respectable class) were usually severely restricted in this society, and it appears that they had little chance for love affairs, heterosexual or homosexual (Dover 1978, 171–84; Pomeroy 1975).

Such Greek philosophers as Plato spoke a good deal about love. But this kind of love was a combination of the purely physical and the extremely spiritual; what it lacked were exactly the modern elements of mutual closeness and sympathy between lovers. Plato's image of love was passionate yearning and admiration for someone (usually a young boy, although the object could also be a woman, as in the myths of Venus, Zeus, and the other gods and heroes). Plato spiritualized this emotion by declaring that erotic love was the worship of beauty, which was a reflection of the ideal Forms upon which Plato believed the earthly world was patterned. Thus, one loves somebody for his or her physical beauty, which in turn represents the unearthly Beauty of a higher, philosophical realm. Nowhere is the love reciprocal and emotionally intimate. Later Roman writers such as Ovid (43 B.C.–A.D. 18) wrote on *The Art of Love* but confined it to an amusing game of chases and adventures. The goals were sex and beauty, never intimacy—and certainly never *marriage*.

With the arrival of Christianity, love was placed even more in the background. This seems a bit incongruous, in that Christianity began as a religion of love. But the love it exalted was completely spiritual: not the love of ideal Forms as in Plato and later Neoplatonic mystery religions, but the love of Jesus Christ for the world he was to save, and reciprocally the love that was expected of Christians for God and their Savior. Christianity was hostile to the Greek and Roman myths of Zeus and his affairs, which they considered immoral. So, too, were regarded love stories like those of Ovid or the risque Hellenistic novelists. The eroticism of ancient pagan culture was one of its features that Christianity regarded as most abominable and that it tried hard to suppress.

Early Christianity placed a very strong emphasis upon asceticism (Queen and Habenstein 1967, 181–201). Its ideal person was an anchorite or monk, who denied all the desires of the flesh in order to attain holiness. Accordingly, marriage was not looked upon as a desirable state at all, although the church leaders made concessions in the form of accepting marriage as a second best, lest one who could not contain the sexual appetites commit the far worse sin of premarital fornication. Saint Paul wrote to the Corinthians: "For I would that all men were even as myself [i.e., a bachelor]. . . . But if they cannot contain, let them marry; for it is better to marry than to burn" (1 Corinthians 7:7–9). This doctrine was to last far into the Middle Ages. Priests and monks were celibate (or at least ought to have been) and regarded anything less than complete celibacy as a condition of semimitigated sin. Marriages had to take place outside the door of the church, since they were not holy enough to occur inside. The only kind of love that the Church advocated was purely spiritual and religious.

Economic and Political Restrictions on Love

In all social classes in medieval society, marriages were contracted for reasons other than love. Among the kings and the higher nobility, marriages were a form of diplomacy. Sons and daughters were married off, sight unseen, to the heirs of other states with whom one wished to form an alliance. Diplomats might be sent to negotiate as soon as a prince or princess was born; the same child might be "marketed around" to various kingdoms, to find the best deal that could be arranged. Given the fact that kingdoms and other realms like duchies, baronies, and the like were hereditary, it was extremely important what sorts of arrangements were made. A queen might bring as her dowry the right to inherit an entire kingdom. And since not every marriage produced a son, it happened quite often that a throne could pass from one royal family to another via the female line. As late as 1714, for instance, the throne of England was inherited by the House of Hanover, in Germany (which brought to England the first series of kings named George, which lasted through the time of the American Revolution). This was due to a complicated political agreement decades earlier that declared that, in the absence of male offspring to the current king, the throne would pass to the descendants of Princess Elizabeth of the Palatine. The agreement was due to the religious politics of the period, since the Protestant faction in power in England wanted to make sure that the throne descended to a Protestant family and not back into the hands of the Catholic kings whom they had just overthrown.

In the lower ranks of the aristocracy, similarly, families arranged marriages to ensure the prestige of their line and the growth of their property holdings. But in the middle class, too, marriages did not happen for love. In England (although not necessarily elsewhere in Europe) individuals were generally able to choose their own marriage partners rather than have them chosen by their parents. Such families were too unimportant to make alliances, and the nuclear household structure, as we have seen, tended to put children out on their own by their late teen years. But even so, this did not lead to love marriages. Marriages were just too important economically for other considerations to enter in very much. The man who wanted to be a property owner running his own farm or small business had to be married, and he chose a wife who could help him with the work, oversee the servants and apprentices, and, if possible, bring in some money or property to get started. The woman had even less choice in the matter. She needed a husband in order to have an economic establishment in which she could share; if she wanted any kind of career or social status, she would have to do it by helping in her husband's trade. After his death, she might even be admitted to his guild (as a weaver, butcher, chandler, smith, or in many other trades: Power 1975, 53–75) and run his business.

The fact that people did not live long contributed to a somewhat mercenary attitude toward marriage. Since men were usually older than their wives (though not necessarily by much), many women were left widowed. If a widow inherited a substantial business or farm and her children were still too young to run it, she would very likely remarry in order to have a man to help work the enterprise. On the other hand, the life expectancy of women was shorter than that of men, mainly because many women died in childbirth. When this happened, a widower married again as fast as he could, since a wife was needed, not for sentimental reasons, but in order to have someone help run the household business.

The same kinds of economic pressures affected marriage far down into the peasant and working classes. Peasant children were especially likely to hire out as servants during their young adulthood and only married when they could afford to acquire a cottage and a little plot of land of their own. Getting married depended on finding a woman and a man who could coordinate this economic arrangement as the opportunities turned up. This pattern was especially found in England. In countries where the extended family dominated, family pressures regarding the marriages of children were so strong as to exclude even the English pattern of individual choice. In England and northwest Europe and in the American colonies, there was individual freedom to choose, but one had to choose primarily on an economic basis. But in eastern and southern Europe and Asia, the choice was in the hands of the family rather than the individual. In both places, when economic times were bad, many people at the bottom of the class structure were condemned to live their lives single.

The Cult of Courtly Love

The so-called High Middle Ages (1100s) saw the rise of a new ideal of male-female relations. This was the cult of *courtly love*. It did not exist across all ranks of medieval society but was confined at first to a small section of the aristocracy, especially in France.

The main carriers of this new ideal of love were the troubadours (Hauser 1951, 202–31; de Rougemont 1956). These were minstrels who composed poems and accompanied their own singing; but unlike other entertainers at the

FEATURE 4.1
A Courtly Love Poem

To Anthea

Ah, my *Anthea!* Must my heart still break?
(Love makes me write, what shame forbids to speak.)
Give me a kiss, and to that kiss a score;
Then to that twenty add a hundred more;
A thousand to that hundred: so kiss on,
To make the thousand up a million.
Treble that million, and when that is done,
Let's kiss afresh, as when we first begun.
But yet, though love likes well such scenes as these,
There is an act that will more fully please:
Kissing and glancing, soothing, all make way
But to the acting of this private play:
Name it I would; but, being blushing red,
The rest I'll speak when we meet both in bed.
  Robert Herrick, ca. 1620

courts of the nobles, the troubadours were not lower class but were themselves knights. The first famous troubadour was William IX, Count of Poitiers in western France. Their stylized songs were always about the same subject: the poet's love for an aristocratic lady. The lady is highly idealized, superior, and inaccessible: the lover continually complains and suffers, while the lady always says no.

This manner of love is very different from that depicted in ancient Mediterranean and Asian literature, where the lover's demands are pure and simple passion for physical possession, which is soon enough accomplished. Here, for the first time, we find love that exalts the woman in a spiritual sense, and a main part of her charm is the fact that she is unattainable.

The troubadour's love ideal was very much a part of the system of feudalism. The lover swears fealty to his lady, in exactly the same way in which a knight becomes the bound vassal of his military lord. In real life, too, knights at this time began the custom of wearing a lady's token (a scarf or other garment) when they fought in a tournament or battle. Love was made a part of the political hierarchy of the day. But in fact this jump was not so very large. "Love" in medieval society originally meant, not a male-female relationship, but the fealty between lord and follower. Kissing was originally an emblem of loyalty: a new knight knelt to kiss his master's ring, and we are a little surprised to read in medieval chronicles that a king would kiss his knights on the cheeks when they met after a harrowing escape or a victorious episode in battle. Some of the flavor of that custom still exists in European countries, where the official who bestows a medal will give the recipient, male or not, the ritual kisses on both cheeks. Similarly, Shakespeare's sonnets sometimes seem to avow his love for a man, although this did not indicate a homosexual relationship but simply the polite way of addressing one's patron. We ourselves continue to speak of love in this nonsexual sense in referring to the love that we feel (or are required by custom to feel) toward our parents and children.

The medieval knight was thus doing something unprecedented when he vowed subordination and fealty to a woman. Such love was vowed only to aristocratic women, preferably of the highest rank. The same authors who speak of the rules of courtly love make it clear that low-ranking people cannot feel or be the object of love; if a knight feels a physical passion for a peasant woman, he is advised simply to rape her, since she cannot possibly appreciate his poetic sentiments. The cult of courtly love was part of the status of the aristocratic knights. It was more than a literary convention, for it was just at this time that the crude medieval warriors were beginning to live at a higher standard of comfort, and codes for polite conduct were being established. Knights had originally been recruited from the peasant class based on their qualities as brutal fighters; but during the 1100s, the knighthood began to establish itself as a hereditary aristocracy. The requirements of courtly love and the courtesies of chivalry that went along with it were ways that these newly established aristocrats tried to distance themselves culturally from the crude soldiers they had so recently been. Bowing and deferring politely to ladies was one of the marks of their new refinement. Whatever its motive, courtly love began a new era in Western society and marked the first time that women had deference paid to them precisely because they were women.

Courtly love was not entirely platonic, however. In the romances that became popular at this time, the predominant theme is adultery. The most popular romance, which set the pattern for the others, is the story of Tristan and Iseult. Sir Tristan is a knight who is sent by his king to negotiate for the hand of a neighboring princess and to bring her home to be his queen. On the way they fall in love and consummate their passion. The rest of the story concerns their further adventures as the king finds out about their affair, while Tristan and Iseult try simultaneously to keep their loyal vows both to their lord and to each other. (A similar theme, you may remember, can be found in the German tale of Brunhild and Siegfried, quoted in chapter 2.)

The adultery could be real enough. The troubadours and other knights were often attending ladies in their castles while their husbands were away fighting. The noble lady might be left in charge at home, which meant in military charge as well, since the home was fortified and garrisoned by knights. So the oaths of fealty in the courtly poetry might well be serious. Moreover, the high-ranking lords and their ladies had not married for love and had not developed a personal attachment to each other; often enough, they had been

The cult of courtly love, which appeared in Europe in the twelfth century, exalted women in the spiritual sense, but love still had little to do with marriage. For aristocrats, marriages were arranged to cement alliances and to insure prestige and wealth. For peasants, marriage was a matter of economic necessity. (*Scene of Courtly Life, anonymous*. France, ca. 1490. Philadelphia Museum of Art. Purchased: Subscription and Museum Funds.)

negotiated for by diplomatic missions exactly like that which begins the story of Sir Tristan and Queen Iseult. Poetic love affairs were a game to pass the time at these courts, nicely fitted inside the status hierarchy of the current military and domestic situation (see feature 4.2).

The more open avowals of the cult of love were rather daring and only

FEATURE 4.2
The Medieval Courts of Love

In the courts of medieval France, England, and Spain, love was a game with elaborate rules. Courts were convened, modeled on law courts, to decide such questions as: Can true love exist between married persons? The answer, handed down by the comtesse of Champagne (near Paris) at a Court of Love held in 1174, was no; true love can only exist when the lovers are under no constraint, whereas married people are bound by duty. Another favorite question was: Is marriage a sufficient excuse to refuse a lover? The answer again was no; the rule was set out with concurrence of the assembled ladies of the court, with the admonition that to disobey the law of love was to incur disgrace before every woman of gentle birth.

The laws of love, published by André, the chaplain to the King of France, went on to detail such judgments and to lay out a series of laws. They included the following points:

1. that love could only take place in the aristocratic class;
2. that multiple affairs could be carried on at once, and that jealousy was part of the game;
3. that affairs should be kept secret;
4. that lovers were to be emotionally obsessed with seeing each other, adoring tokens of their clothing, and so forth;
5. no pleasures were to be taken by force;
6. a woman should refuse nothing to her lover, including the "most intimate embraces";
7. and most emphatically, that love did not lead to marriage and was not to be constrained by any economic motives. There was, however, a kind of pseudomarriage, in that if a lover died the remaining lover must remain faithful to his or her memory for two years.

Courtly love, although openly spoken of in many places, was nevertheless risqué. It was scandalous in the eyes of the serious church (although it should be kept in mind that religion for many people at this time, including many of the clergy themselves, like André, the king's chaplain, was more of a formality than a personal moral code). And of course the real or even the literary adulteries undermined the sexual property rights of powerful lords. This is what makes the medieval courts of love unique in world history, for in other agrarian states upper-class men enforced a very rigid dual sexual standard and kept their women locked up in harems and women's quarters.

What made this cult possible in western Europe at this time was the political situation. Kings were not very powerful, as they depended upon the military forces raised by their noblemen. France itself was split into a network of little feudal states, which shifted alliances constantly. The king of England at various times in the 1100s held more than half of France, although never very firmly. This situation gave the local nobles room to maneuver, especially through dynastic marriages. And this in turn gave noblewomen a great deal of importance. Eleanor of Aquitaine (1122–1204) was the heiress to the territory of southwest France; she was married first (at age fifteen) to the king of France but later divorced him and took Aquitaine into alliance with England by marrying King Henry II. It was her own daughter, Mary Comtesse of Champagne, who held the famous Court of Love in 1174. Another daughter became queen of Spain, while a granddaughter became queen of France. This reliance on intense marriage politics gave aristocratic women comparatively great freedom and set the stage for the cult of courtly love.

existed where the political situation permitted. They depended upon a decentralized feudal system in which military power rested upon the armed knights. Wherever possible, the kings attempted to gain personal control by replacing the knights with paid armies of non-noble soldiers. This became increasingly possible as guns were introduced into warfare in the 1400s and the following centuries. Military chivalry gradually disappeared, but the courtly practices of politeness and idealization of upper-class women remained. In fact, the rise of professional armies meant that the nobility often had less to do and therefore spent their time in idleness at the courts (Dickens 1977). Hundreds or even thousands of people might live in a king's palace, like the one Louis XIV built at Versailles. To pass the time, card games had been invented in the 1400s, along with gambling, hunting for the men, lawn games like croquet and tennis for the ladies—and love affairs, as often as not adulterous. Courts acquired a reputation for dissoluteness, which sometimes caused their political downfall. The complaints of the English Puritans against the sexual practices of the courts of James I and Charles I helped fuel the revolution that cut off Charles' head in 1649; and the court of the French kings, with their official mistresses and hedonistic atmosphere, was one of the factors leading up to the execution of Louis XVI on the guillotine in 1793 during the French Revolution.

This type of courtly love was very much confined to the upper classes. It was neither an ideal nor a reality for the bulk of the population. The peasants and the small craftsmen not only continued to base their marriages on practical economic necessities but were also often forced to maintain sexual continence. A woman simply could not afford to become pregnant until she was married (or just before), and illegitimacy in the bulk of the population was kept under strong controls until the rural family structure began to break down in the 1700s and 1800s (Laslett 1971; 1977). Morals were fairly rigidly enforced in small rural communities, where church attendance was not only compulsory but easy to check up on, unlike in the cities and the courts. Throughout the medieval period, then, there were two sharply distinguished sexual cultures: one for the aristocracy, and another for the rest of the population. Only later would the two cultures come together in the modern ideal of love.

The End of the Patrimonial Household and the Rise of the Private Middle-Class Marriage Market

In the 1600s and 1700s, the structure of medieval society began to change. The patrimonial household began to give way to the modern home (Girouard 1980). The major difference between the two was the emergence of privacy. The medieval house, we may recall, was a public place as well as a dwelling. The house of a king or high lord was also the seat of government. The hundreds of knights and courtiers congregated there because it was their place of employment: the place where law courts were held, orders given, taxes collected, and all the other business of government carried out. It was also a fortified castle, serving among other things as a military barracks for its defenders.

Two of the main reasons for the decline of the patrimonial household were the rise of the state bureaucracy and the advent of the paid professional army (Stone 1977). Whereas medieval lords had a few hundred or at most thousands of troops, the kings of the 1600s had armies of foot soldiers numbering in the tens or hundreds of thousands and large numbers of government officials spread out over their domains. Government was separated from the aristocrat's home. Similarly, at the lower level of small businesses (although happening a hundred or so years later than the rise of the bureaucratic state), enterprises began to move into separate factories or office buildings. For the first time, the private home, reserved for the family alone, began to appear.

Along with this development came a similar change in the nature of marriage. Marriages became more of a private affair between the couple themselves, less of a political or business arrangement (Macfarlane 1986). At the aristocratic top of society, dynastic marriages became less important with the rise of bureaucratic government and the gradual loss of royal power. The great national revolutions in particular, such as the English revolutions of 1642 and 1688 and the French Revolution of 1789, transferred power to parliaments or other strictly national bodies of politicians and officials. It was no longer possible that a whole state might pass into the possession of a foreign power merely by marriage. (This is so even though royalty today still marries, for the most part, foreign royalty. Thus the present royal family in England is descended from the royal houses of Greece and Germany; but since the English monarchs are figureheads, this has no effect on international relations.)

At the lower level, too, marriages became less a matter of practical economic necessity. As business moved out of the home, the work of wives (at least in the middle class) became less crucial economically. As the private domestic sphere emerged, women found themselves confined within it. At least on men's part, the incentive for getting married became less an economic one and more a matter of acquiring a sexual and domestic partner. The modern marriage market began to emerge, and with it came the ideal of love. But this was a new and different ideal of love than that found in previous societies. It was not simply the erotic passion of ancient times, which went along with the dual standard of male-dominated patrimonial households. Nor was it the courtly love of the aristocracy, which was a pastime with adulterous overtones. For the first time, the ideal emerged that love was a mutual sentimental bond between a man and a woman, a relationship of caring that was supposed to last a lifetime and was connected to marriage. Love was no longer extramarital but at the very core of the marital relationship (Flandrin 1979; Shorter 1975).

This sentimental ideal of love as a marital bond arose as part of the new marriage market. Since marriages were no longer held together by the larger political and economic structure, whatever bond there might be had to come

FEATURE 4.3
A Victorian Love Poem

How do I love thee? Let me count the ways.

I love thee to the depth and breadth and height
My soul can reach, when feeling out of sight
For the ends of Being and ideal Grace.
I love thee to the level of everyday's
Most quiet need, by sun and candle-light.
I love thee freely, as men strive for Right;
I love thee purely, as they turn from Praise.
I love thee with the passion put to use
In my old griefs, and with my childhood's faith.
I love thee with a love I seemed to lose
With my lost saints—I love thee with the breath,
Smiles, tears, of all my life!—and, if God choose,
I shall but love thee better after death.
    Elizabeth Barrett Browning, *Sonnets from
    the Portuguese,* ca. 1830

from within. The basis for this bond, as suggested elsewhere (Collins 1971), comes above all from the motives of women in their marriage-market situation. For although men now needed little from marriage besides sex and domestic help, women still found themselves economically dependent upon a husband. To ensure their own economic well-being it was important to attach men to themselves personally, and with a strong and lifetime tie if possible. This is what the new love ideal did. It lifted love from the status of a game, a sideline amusement that the wealthy aristocracy could afford to play, and made it the emotional insurance that kept a woman and her loyal breadwinner tied together "until death do us part." In order to do so, it was necessary to evolve a new attitude toward sex, to keep it highly connected with sentimental love and confined to marriage.

THE VICTORIAN REVOLUTION AND THE SEXUAL DOUBLE STANDARD

Besides the new emphasis upon marriages for love, the transition to the modern family involved a sharp conflict over sexual practices. The older double standard prevalent in patrimonial societies was challenged by a new, more equal but puritanical sexual standard, which strongly confined sex to the married couple. The earlier practice enforced ideals of premarital virginity and chastity upon women but left men free to have concubines, slave women, and courtesans (if they could afford them) and even to practice rape against lower-class women. After considerable struggle, the ideal was established that both men and women ought to adhere to the same sexual standard—and that the female ideal of marital fidelity should be the standard. Needless to say, the official ideal was often violated. The Victorian period, during which this ideal was most strongly upheld in public, was also a time when men frequently consorted with prostitutes and mistresses. Some Victorian women pursued sexual adventures as well. It would be more realistic to say that the official sexual standard did not come in without a struggle. Various forces opposed it or favored it at various times. And no sooner did official puritanism reach its height than it began to give way to a sharp reaction, culminating in the sexual permissiveness of the twentieth century.

This whole process may be referred to as the *Victorian revolution*, because it was during the reign of Queen Victoria in England (1837–1901) that sexual prudishness reached its height. Absolute propriety was the imperative for all respectable people. The sight of a woman's ankle was considered extremely risqué; it was even considered improper to use the word *leg* in reference to a woman. Nevertheless, this "Victorianism" goes back much further into European history. The rules of sexual behavior that Victorianism strove to enforce were reflected in ancient Christianity and were especially heavily touted during the High Middle Ages (A.D. 1000–1400). Even during Victoria's own lifetime, the high point of prudery was the earlier part of the century (around 1800–1870), while in her old age even London high society was rent by numerous scandals of adultery and homosexuality, and her own son the Crown Prince Edward was the center of a notorious "fast crowd."

The "Victorian revolution," then, is only a metaphor for something larger and more complex. Just as love has a history going back before the "love revolution" of early modern times, the battle over sexual standards seen in "Victorianism" goes back to the 1700s, with roots even earlier. It was with the rise of the private household and the romanticized marriage market that sexual prudery became prominent. Already in the 1700s, Englishmen of the middle and upper classes were referring to women as "the delicate sex" and attempting to keep "improper" topics from their ears (Watt 1957). In 1818, the English physician Thomas Bowdler brought out a censored edition of Shakespeare, since the bawdy talk of the Elizabethan era was now considered too obscene for women and children to read. By the mid-1800s, as the new privatized family and the sentimentalized conception of women had spread widely throughout society, sexual prudery had reached an extreme. Even references to pregnancy and childbirth were considered obscene, to be avoided especially by women themselves.

A few days before a baby calf had been born and I had seen it . . . but then my father and mother forced me to keep out of sight of the field where the mother and calf were, and where I had been a few moments before. The thing I had seen I dared not talk about or ask about without "deservin' to have my ears boxed."

Even when my little brother was about to be born, we children were hurried off to another farmhouse, and secrecy and shame settled like a clammy rag over everything. At sunset, a woman, speaking with much forced joy and a tone of mystery, asked us if we wanted a little brother. It seems a stork had brought him. (Wertz and Wertz 1977, 79)

The Battle Against the Double Standard

The extreme prudery of this period capped off the separation that had been emerging between male and female spheres. Men and women were to make no suggestive allusions in each other's company, nor were women, as the "purer sex," to discuss such subjects among themselves. Sex was left in a male backstage of private clubs (for the upper class), saloons (for the middle and lower), smoking parties, hunting trips, and the "sporting world" of the theater and prostitutes (Chesney 1970).

Prostitution had existed, of course, since ancient times. Medieval European lords did not have official concubines or practice polygamy like their counterparts in Moslem, Hindu, and Chinese societies (there are exceptions: Charlemagne, who reigned early in the Middle Ages, ca. A.D. 800, had four wives). But they did allow themselves considerable license to have mistresses. Some of these, such as Louis XIV's favorite at Versailles, Madame de Maintenon, or Louis XV's Madame de Pompadour, had official titles and residences and were persons of greatest importance in court politics. Prostitution existed quite openly in ports and military towns for the sailors and soldiers. The acting profession was considered very close to professional prostitution. In India and China, female entertainers were identical with prostitutes and often were admired for their cultivation and admitted into high society—among men only, of course. All this continued in the Victorian period, only now it went underground. Cities like London and Paris had huge districts where prostitutes walked the streets, as well as hotels and rooming houses ranging from squalid dives to luxurious houses of prostitution for the upper classes (Chesney 1970; Marcus 1964).

A battle went on between the male and the female spheres, or rather between the cult of domestic purity on the one hand, and the male backstage with its "bad girls" and "fallen women" on the other. Some of the moves in this battle were legal (see the discussion below on the feminist movement), but much of it was waged in the moral tones of everyday life. Women, at least in the middle class, worked hard at making men behave "decently," and a central

FEATURE 4.4
Two Famous Fictional Affairs

Samuel Richardson's *Clarissa* (1749) was the most popular novel of the eighteenth century. It tells the story of Clarissa Harlowe, the daughter of a family attempting to climb into the ranks of the nobility. Her father wants her to marry a wealthy but unattractive boor named Soames, while she prefers a rakish young nobleman, Lovelace. In defiance of her family's wishes, she runs off with Lovelace, hoping to reform him by her own example of sexual purity. Lovelace tries all his wiles to make her yield to him and finally has to resort to raping her while she is drugged with opiates. Having won his conquest, he offers to marry her, but Clarissa declares she has been eternally dishonored, falls sick, and dies. Lovelace thereafter repents and seeks his own death in a duel.

Richardson expresses an early version of the idealization of women and the fight against the double standard, themes that were to become dominant in the Victorian era. The novel is not yet full-fledged Victorian; although Richardson is very moralistic, he nevertheless is willing to talk openly about an explicit sexual theme. In the next century, the Victorians would present the sentiments but censor any reference to sexual issues.

In contrast to the middle-class center of moral gravity in the English novel, Pierre-Ambroise-François Choderlos de Laclos's *Les Liaisons Dangereuses* (1782) reflects the courtly love games of the French aristocracy at its most decadent. It tells of a plot by two sophisticated and amoral aristocrats, the Vicomte de Valmont and his female friend the Marquise de Merteuil, to arrange the seduction of a naive young girl just out of convent school, Cecile Volanges. Their motives are partly revenge against the girl's mother for a previous slight, and partly sheer diabolical interest in entertaining themselves through their leisure hours. The plot goes off badly, and the characters all come to a bad end; but unlike Richardson, Choderlos de Laclos presents a cynical and

De Laclos's novel, *Les Liaisons Dangereuses*, reflects the courtly love games of the French aristocracy at its most decadent. In this still from the 1988 film version, *Dangerous Liaisons*, the scheming Marquise de Merteuil (Glenn Close) and her co-conspirator, the Vicomte de Valmont (John Malkovich), plot a virtuous woman's seduction.

completely unidealized portrait of both villains and their victims. The novel caused a sensation in France. It was condemned as immoral by official opinion, but the first edition was sold out within days, and subsequent editions were eagerly snapped up. Parisian ladies retired behind locked doors to read it. After the French Revolution a few years later, a copy was found in the library of the executed Marie Antoinette herself. Choderlos de Laclos, an army officer with time on his hands, wrote no more novels. He ended up as a general in Napoleon's army. In the nineteenth century, when Victorianism hit France, his book was banned and condemned to be destroyed as "dangerous." In 1988, it became a popular film, *Dangerous Liaisons*.

tactic was to desexualize proper conversation entirely. Nancy Cott (1978) calls this the tactic of "passionlessness," which women used to remove themselves as much as possible from being subject to men's sexual desires. "The belief that women lacked carnal motivation was the cornerstone of the argument for women's moral superiority, used to enhance women's status and widen their opportunities in the nineteenth century" (Cott 1978). Barbara Welter (1966) refers to this as "the cult of True Womanhood," characterized by "piety, purity, submissiveness, and domesticity. Put them all together and they spelled mother, daughter, sister, wife—woman." The one thing that they did not spell was sex object.

One can also say that women now had to be especially careful to control sex in the new individual marriage market. They could no longer count on being married off by their parents or making an easy match based on their economic potential as workers. Instead, they expected to have someone fall in love with them, and that because of purely personal attractiveness. Sexual prudishness had the effect of confining sex within proper marriage bonds. At the same time, there was the widespread sentiment that love led to marriage. One can easily interpret the feelings of love that large numbers of people were beginning to experience for the first time in history as the form that sexual passion took when it had to pass through a filter of refinement and idealization. And since true love is forever (another point in contradiction to courtly love), it neatly adds up to a lifetime marriage vow.

By the early 1800s, this doctrine was strongly in place. It had become the formal moral code of society, even though the old courtly love game might still be played in the salons of the wealthy (Stendhal 1967), and the workers and peasants still lived for the most part in the traditional hard-fisted business-enterprise family. The history of the next hundred years was to include the gradual spreading out of the middle-class ideal until it encompassed virtually all of society. The "Victorian revolution" that it represented was the result of a conflict over male and female status. In one respect it was an important historical victory for women, since it raised their status (at least officially) to a very idealized level. At the same time, it guaranteed most middle-class women the economic support of a marriage at a time when they were excluded from independent careers of their own. The price that was paid was to confine women very strictly to the domestic realm and to make this realm almost the polar opposite of the male world outside.

THE FIRST WAVE OF FEMINISM

Until the end of the nineteenth century, women were distinctly second-class citizens. In England, the United States, and elsewhere, women had virtually no legal rights. The husband was entitled to collect his wife's wages, if she worked, and to dispose of her property, if she had any when they were married. He was the head of the household and represented his wife in all public and legal matters. He could decide how their children were to be educated and in what religion they were to be brought up. A dying husband could will his children, even unborn, to other guardians; in the case of divorce (not easy to obtain in those days) he had control of the children. A wife was the humble subordinate of her husband; if she disobeyed him, he had the legal right to chastise her physically, and could even hand her over to the law for punishment.

Similar restrictions held in public life. Women were not allowed to preach in church (except in radical sects like the Quakers). They could not vote (although in England it is true that men of the lower social classes could not vote either, until 1884, and in the early days of the United States various states restricted the vote to property-owning males). Even speaking in public was a scandal for a woman.

It was this segregation that first gave rise to the organized women's movement (Chafetz and Dworkin 1986; Flexner 1959; Sinclair 1965; O'Neill 1970). The mid-nineteenth century was a period of liberal social reform, and the most important of all the moral crusades was the campaign to abolish slavery. In the 1830s there were hundreds of antislavery societies in the United States, a large number of them separate organizations of women. Women made up more than half the signatories of the huge petitions periodically sent to Congress on the slavery question. In the 1830s, Angelina and Sarah Grimke, two wealthy South Carolina sisters, were among the first women to speak in public. Their theme was the abolitionist issue, but the furor roused against their public participation (especially among the conservative New England preachers who supported abolition) led them to defend the rights of women.

The Grimke sisters were eventually persuaded to drop the feminist issue lest it compromise antislavery. It was a type of political bind that was to plague the feminist movement continuously. But not all women reformers were willing to give in. In 1840, when the World Anti-Slavery Convention was held in London, the two leading American women abolitionists, Lucretia Mott and Elizabeth Cady Stanton, were excluded from the meeting and had to sit in the spectators' gallery behind a screen. To his credit, the great abolitionist orator William Lloyd Garrison furiously withdrew and sat with the women. Out of this incident was organized the first women's rights organization, which held its opening convention in Seneca Falls, New York (the home of Elizabeth Cady Stanton), in 1848. Its members passed a resolution calling for women's suffrage, though only narrowly: votes for women was regarded as so extreme a demand that it might compromise everything else.

Nevertheless, the movement began to spread. At first the women's movement largely relied on the fervor of larger reform campaigns. In America, most of the early feminists were abolitionists. Others, like Susan B. Anthony, came to feminism from the temperance movement, where women at first were arrogantly subordinated to male crusaders against alcohol (mostly ministers). Even more women were involved in missionary associations, drives to organize philanthropy for "the deserving poor," and the very popular crusade to suppress vice and reform prostitutes. These were all middle-class movements; a smaller number of women were involved in efforts to improve the conditions of the working class, by outlawing child labor and establishing protective legislation for women to ensure minimum wages and maximum working hours.

These different reform crusades had varying fates. Slavery was abolished in the United States in 1863 (having been stopped in England in 1807). The temperance movement, fueled by women's votes in some states where local suffrage existed, succeeded in having a constitutional amendment passed in 1919, although Prohibition was repealed again in 1933. Protective labor legislation for women and children was finally achieved early in the twentieth century, although surrounded by controversy (some of the more radical feminists declared that it restricted the chances of women to work). Charity organizations, which were relatively uncontroversial, became firmly established.

The politically most popular crusades were those for the suppression of vice. In the 1870s and 1880s in particular, widespread campaigns cracked down on prostitution, which had been officially condoned especially in large cities. In the United States, these movements were connected with what later became called Progressivism, which was especially concerned with overturning "boss rule" in the urban immigrant communities. A disproportionate number of the immigrants were men, and they tended to come from countries with patriarchal traditions and a definite double sexual standard. The political and social centers of the immigrant communities were usually the saloons, which was one reason why these were special targets for the Anglo-American women of the temperance crusade. The various movements tended to overlap; the WCTU (Women's Christian Temperance Union) had its own section for "Social Purity" and another devoted to eradicating obscene literature, while all of these movements generally supported votes for women as a means of implementing their political programs.

In England, similar movements developed, although with some differences: temperance was never a very strong movement in England, while working-class reforms were much more strongly backed. Slowly some of the women's legal disabilities were removed. In 1857, a bill gave women some rights of divorce; in 1884 Parliament gave them the right to their own earnings and abolished the penalty of imprisonment for women who denied their husband his conjugal sexual rights. In 1869–70, women property owners were allowed to vote in municipal elections and to serve on school boards.

In the United States, the territory of Wyoming in 1869 gave the vote to women, partly as an effort to attract them to this sparsely settled territory; when it became a state in 1890, it was the first state whose women could vote in national elections. A few years later, Utah, Colorado, and Idaho followed suit.

Because the women's suffrage and anti-immigration movements became aligned, eastern states with large immigrant populations tended to be anti-suffrage. Their organized resistance to women receiving the vote helped defeat many state suffrage bills. Western states, however, were pro-immigration and also hoped to attract women settlers.

But progress was hard and slow. More than 480 campaigns were waged before 1910 in different American states to have the issue of the female franchise brought up, but only a tiny proportion were successful. Just after the Civil War, when the vote was guaranteed for Negro men by a constitutional amendment, the demand of women to be included was turned down by the abolitionist leaders, who argued it would jeopardize the passage of the amendment.

In England, too, there was very slow progress. In the 1860s, liberals like John Stuart Mill unsuccessfully introduced a women's suffrage bill into Parliament. In 1869, Lady Amberly (who was to be the mother of the philosopher and peace-movement leader Bertrand Russell) caused a furor by breaking the taboo on women speaking in public when she addressed a suffrage meeting. Queen Victoria was so outraged by this breach of decorum that she declared that Lady Amberly ought to be horsewhipped (O'Neill 1970, 31). But although some prominent political leaders and intellectuals supported women's suffrage, the political parties were either opposed to it or regarded it as an issue that could be sacrificed to more important things. Benjamin Disraeli, the Conservative prime minister, favored suffrage, perhaps out of his perception that women would likely support the Conservative party. The Liberal party, on the other hand, although it tended philosophically to believe in equal rights, dragged its feet because it felt women's votes would go to the Conservatives.

At the turn of the twentieth century, the pressure finally began to build. New Zealand in 1893 became the first country to give women the vote, and Australia, Sweden, Norway, and Finland soon followed suit. In the United States, the growing wave of belief that non-Anglo immigrants were swamping the country gave support to the idea that native women's votes could help turn back the tide. World War I brought the goal of women's suffrage in sight at last. The process here was much more peaceful than in England (see feature 4.5), although in 1913 a group of women protesting at the inauguration of Woodrow Wilson was attacked by a mob, and in 1917 women pickets at the White House were jailed and abused. But in fact the movement had been building momentum just before the war. In 1910 the state of Washington passed a women's suffrage bill, followed by narrow victories in referendums in California, Arizona, Kansas, and Oregon. Suffrage was winning the West, although this was followed by a reaction in the East as Ohio, Michigan, and Wisconsin defeated suffrage bills; further defeats followed in 1915 in New York, Pennsylvania, Massachusetts, and New Jersey. These were the states with large immigrant populations, which tended to be antisuffrage. But the suffrage movement was large and well financed, and it kept up its pressure. The confrontations at the White House in 1917 brought the matter to an emotional crisis point. In 1918 President Wilson called for women's suffrage as a war measure, and by 1920 the thirty-sixth state legislature had approved the constitutional amendment.

Elsewhere around the world, progress was much slower. Although German women received the vote in 1919 with the new Weimar constitution that followed the overthrow of the Kaiser, the civil code continued to give husbands control over family place of residence, behavior of children, and most economic issues in the home. In France, women did not receive the vote until the late 1940s (Tilly 1981). And this was despite the very notable participation of French women in the war effort in World War I; it is apparent that the war provided an emotional opportunity to get the franchise through in England and the United States, but that it could not have been done without the strong pressure of an organized movement. In Switzerland, women could not vote until the 1960s.

After the winning of the franchise in England and the United States in

1918–20, the feminist movement in both countries collapsed. Despite great expectations, few significant changes followed. As it turned out, women voted more or less the same way as their husbands, and neither the predicted era of peace nor the hoped-for nonpartisan reform happened. Nor did women's legal conditions improve greatly. Some modest victories had been achieved decades earlier, when women won rights to divorce and to keep their own earnings. But as late as the 1960s, women in some states were still required by law to take their husband's name and to live in the residence that he chose (which implied

FEATURE 4.5
The Radical Suffragettes in England

In England, the movement for women's suffrage seemed to have come almost to a standstill in the early twentieth century. After decades of legislative rebuffs at the hands of both the Conservative and Liberal parties, a radical group of suffragettes was formed in 1903. This was the Women's Social and Political Union (WSPU), led by Mrs. Emmeline Pankhurst. She was described as a strikingly beautiful woman; perhaps this asset, together with the fact that she was the widow of a former member of Parliament, gave her the courage to challenge the male establishment as it had never been challenged before.

Together with her daughters Christabel, Sylvia, and Adela, Emmeline Pankhurst advocated a militant approach. In 1905 Christabel and other women were arrested for interrupting a political meeting; by 1907, the suffragettes' marches were being broken up by the police and by violent attacks by male opponents. The confrontations attracted widespread support and sympathy for the WSPU, including huge financial contributions from wealthy supporters and thousands of new demonstrators. As the persecution became more severe, the women escalated their tactics. They were being arrested and sentenced to months in jail for speaking at illegal rallies; in response they picketed and chained themselves to public buildings to keep from being hauled away. The Liberal government, which had taken over from the Conservatives in 1906, was embarrassed, since its own principles called for women's suffrage, but its

reaction was primarily to impose yet longer jail sentences. The confrontation reached its climax in 1912–14, when hundreds of militant suffragettes were arrested for smashing windows in a demonstration in downtown London, during which the house of the Liberal cabinet minister Lloyd George was burned down. Christabel Pankhurst escaped to Paris, where she established an underground command post. Emmeline Pankhurst, in and out of jail, continued to generate support for her endurance of brutality.

The outbreak of war in 1914 brought the drama to its conclusion. The Liberal government had already been looking for an excuse to grant women's suffrage without appearing to back down. The widespread participation of women in the British war effort would challenge the idea that women's place was only in the home, and the suffragettes, under the leadership of the Pankhursts, called a truce and threw themselves into the war effort. In 1918, British women were finally given the franchise, although in a graduated package that at first gave votes only to women over age thirty, with full equality deferred to 1928. The military tactics of the British feminists were widely criticized, but they set an example that was later copied by the nonviolent movement for Indian independence led by Gandhi in the 1930s and '40s and later by the black civil rights movement in the United States and by peace demonstrators, antinuclear activists, and others.

that his job legally took precedence over hers); she also had no legal rights over his income nor a legal voice in spending it nor a right to independent financial credit (Kanowitz 1969).

Instead, the family as it emerged after the suffrage movement looked very much like a renewed version of the traditional model with its sharp separation of male/female spheres. After some initial inroads into the professions and the opening up of new clerical positions, women made no further occupational progress; in the early 1960s the degree of occupational segregation was about the same as it had been in 1900 (O'Neill 1970, 93). In education, women actually lost ground: in 1920 they accounted for 47 percent of all college students, but by 1950 their proportion had fallen to 30 percent. In advanced degrees, women fell from one out of six doctorates to one out of every ten. As educational degrees became more important for jobs, men rushed in to snap them up, especially with the aid of the G. I. Bill after World War II. Not until 1970 did women get back to where they had been before the war (*Historical Statistics of the United States* 1965, 384–85).

In the larger view, there were two main reasons for the failure of the first wave of feminism to achieve its goals.

The Victorian Strategy

One weakness was the extent to which the women's movement relied upon moralistic arguments to generate support. The movement began as an offshoot of various reform groups, and it gained a good deal of strength from the shame it could impose upon Victorian men for the vices they condoned. The campaigns against prostitution were among the most successful, and they held up the image of women as morally superior—"purer" than men. This attitude carried over into sexual matters generally. Thus, although feminists often recognized that unrestricted childbearing was one of the things that kept them tied to the home, nevertheless they generally opposed birth control and joined in the campaign against abortion that swept the United States in the late nineteenth century (see chapter 10; also Gordon 1982). The obscenity laws passed in the 1870s cut off information on birth control, besides silencing the few militant feminists who spoke out against the restrictions inherent in puritanism and who advocated free love or other utopian experiments (see feature 4.6).

The result was to draw the boundaries even more sharply between men's and women's spheres. The women's campaigns against alcohol, prostitution, and obscene language did not make any of these things go away but rather removed them into an even more sharply delimited "man's world" that was not even supposed to be heard of by their wives at home. Some feminists wanted to limit sex as much as possible not only outside of marriage but in it; in the 1890s they caused a sensation by advocating separate bedrooms for husbands

FEATURE 4.6
The Free Love Scandal

The most flamboyant of the early American feminists was Victoria Woodhull. A stylish and intellectual woman, she made friends with the robber-baron millionaire Cornelius Vanderbilt and broke into the male business world as a stockbroker. Together with her sister, Tennessee Claflin, she established a magazine in New York City in the late 1860s; it advocated radical causes, including not only women's suffrage but *free love*. Her argument was that marriage was the basis of women's second-class citizenship and that exclusive sexual possession was its main prop. This prop she proposed to overthrow. Her argument should be seen against the background of the various utopian communities that had sprung up in the 1840s and 1850s, including the Oneida Community, which actually practiced group marriage.

For a time, Victoria Woodhull acquired a leading position among New York suffragists, and was the first woman to run for president of the United States. In 1871, she was allowed to address the House Judiciary Committee, the first time that Congress actually recognized female suffrage as a legislative issue. Stirred by her success, in November of the same year she announced from the stage of a public hall in New York City that she believed in free love. She was immediately denounced from all sides, not only by clergymen and the press but by other feminists. Landlords turned her out of lodgings.

Not to be easily disposed of, Woodhull struck back by making public an instance of free love among her respectable compatriots, an affair between Henry Ward Beecher, a famous abolitionist preacher, and Mrs. Elizabeth Tilton, the wife of a liberal editor. Woodhull declared that they, too, were believers in free love but were hypocritical about coming out and supporting her on the subject. The result was a lawsuit by Theodore Tilton, Elizabeth's husband, against Beecher, who was acquitted by a sympathetic jury, although probably guilty of the charge. In the ensuing scandal, Theodore Tilton, Victoria Woodhull, and her sister all had to flee the country. The American women's suffrage movement was smeared by its opponents and for decades thereafter had to live down the taint of free love. The movement's reaction was to emphasize sexual puritanism even more strongly.

and wives; one even called for separate residences (O'Neill 1970, 41). On the other hand, it should be borne in mind that these women were fighting against a situation of rather extreme sexual exploitation; in the absence of other weapons, one of their stronger tactics was to place restrictions upon sex, in conjunction with the love revolution that led to the first great rise in women's status. If it also resulted in an idealization of women in a separate sphere, confined to a decorous courtship and then a restricted home life, that was part of the price in this battle of unequal forces.

The Neglect of Working-Class Women

The other weakness was that the feminist movement was essentially a movement of the middle class (Chafetz and Dworkin 1986). Working-class women were not much involved. Such women had no need to fight for rights to seek work outside of the home, since most of them were forced to work anyway (Roberts 1986). In the countryside, peasant and farm women were used to doing heavy work in the fields. Throughout the early industrial period, they helped their husbands at their trades. Coal miners' wives would drag carts of ore through the mine shafts, while craft workers' wives picked up and delivered heavy loads as intermediaries between their husbands and their masters (Scott and Tilly 1975). With the progress of industrialism, women workers were concentrated in textile factories. Many women worked for other women in private sweatshops as piece workers in the needle trades, in return for miserable wages and a confined place to stay. A large proportion of women workers were servants, living under conditions of patriarchal discipline. Virtually all these women either worked directly under the control of their parents or husbands or else had to give up their wages to their family. The fate of women workers was to be exploited, whether by men or other women.

Where working-class women had any political consciousness, it was generally in connection with a male-dominated labor union or (in Europe) a radical socialist movement. The unions were generally unfavorable to organizing in the women's occupations themselves; instead they joined in the push for protective legislation that would keep women from competing with men. "Some socialist newspapers described the ideal society as one in which 'good socialist wives' would stay at home and care for the health and education of 'good socialist children'" (Scott and Tilly 1975, 63). Middle-class feminists generally did not appreciate the position of working-class women, although one

Women clerical workers in a typical pre-World War I office: a large, American mail-order house.

wing of the feminist movement campaigned for social legislation to eliminate child labor and restrict women's hours and regulate their working conditions. The main effect of such laws, where they were actually applied, was to separate women even more into segregated occupational spheres or to take them out of the labor force entirely.

While all this was happening, though, an occupational revolution was coming about that had been anticipated by no one. After 1880, women began to appear in the labor force in large numbers as clerical workers: secretaries, typists, file clerks, bookkeepers, retail clerks, as well as schoolteachers and nurses. These were the first important inroads women had ever made into working in nonmanual labor. Previously, the working woman was always a member of the lower classes, working as a domestic, farm laborer, or industrial employee. Now, at last, there were career opportunities open for middle-class women, and for working-class women attempting to break into the middle class on their own. There was a price to be paid: occupations that had formerly been held by men, and that led upward into higher careers, now became sex-typed and segregated. Secretaries and clerks had formerly been men, and their jobs could lead on up into the higher management hierarchy. The new women secretaries had no place to go up; they became an occupational caste restricted by gender and separated from male white-collar workers whose careers could still take them higher. Moreover, those jobs were confined almost exclusively to single women; married women were expected to quit their jobs.

The growth of the female clerical "ghettos" was an ambiguous legacy of the nineteenth century. On the one hand, it institutionalized the Victorian doctrine of separate spheres and cut off any further progress that women might make in gaining alternative sources of support for themselves so as to become less reliant upon the patriarchal home. On the other hand, it did get middle-class women out of the house, at least for part of their lives, and opened the way to changes in cultural and sexual matters that gradually transformed the ethos of male/female relations. Out of this would eventually come the family changes of the twentieth century.

SUMMARY

1. The traditional family in western Europe and North America never was an extended, multigenerational household, nor did it practice early marriage (except in the highest aristocracy). Men and women tended to leave home in their teens to become apprentices or servants in other households, and to marry only in their late twenties when they could afford to start their own homes. By the time their children were grown, the parents were likely to be dead, since people on the average did not live much past their forties.

2. In eastern and southern Europe, and in Asia, however, traditional families tended not to have servants, and had a pattern of older men marrying younger women. These families were more likely to have an extended, multigenerational structure, whereas families in England, northern Europe, and the American colonies tended to be more individualistic.

3. In ancient societies, love was generally regarded as purely spiritual or physical and not at all associated with marriage. Early and medieval Christianity was not favorable even to marriage, since its ideal was the celibate priest or monk. In traditional Europe, marriages were arranged either as diplomatic alliances (among the aristocracy), or to procure work partners to run the household business or farm (among the middle class and peasants). In neither case was love a consideration.

4. The first period of idealization of women came in the 1100s in France, when the cult of courtly love developed among the knights and ladies of the feudal castles. Only women of high rank were subjects of this love game, which was always adulterous. The chivalric code had the long-term effect of introducing politeness toward women as part of upper-class culture.

5. In the 1600s and 1700s, the professional army and the bureaucratic state began to displace the armed patrimonial household and to separate work from the private home. Along with this came a new, individually based marriage market, based on personal attractions alone. The new ideal, for the first time in Western history at least, was that marriages should be based on a lifetime bond of personal, mutually sympathetic love. The new love ideal reflected the individual freedom of the marriage market but also the fact that the middle-class woman was now cut off from the economic world that had moved outside the home, and had to attach herself permanently to a husband who could support her.

6. During the Victorian era of the nineteenth century (actually it began a little earlier in England and America) society became extremely puritanical about sex, at least in its official standards. This may be interpreted as a struggle by women against the double sexual standard, which had given comparative sexual freedom to men, both in the old patrimonial household and continuing in a Victorian "underground" of widespread prostitution. It also reinforced the new private family by attempting to confine sex strictly to marriage, for both males and females.

7. The first organized feminist movement began as an offshoot of the movement to abolish slavery in the 1840s. It made slow progress, but did eventually succeed in giving women certain legal rights, such as the ability to sue for divorce and to keep their own property after marriage, and gained some protections for women factory workers. The right to vote was not fully won in the United States and England until the political

upheavals of World War I, eighty years later. Nevertheless, the winning of suffrage had little effect on the family, which continued to feature a very sharp separation between male and female spheres. The most important source of change at the end of this period was the opening up of white-collar employment to middle-class women for the first time, although in newly sex-segregated positions.

PART TWO

DIVERSITY IN
MODERN FAMILIES

5

THE TWENTIETH CENTURY

The American family has recently been going through a period of major change. It has been suggested in some quarters that even more significant changes will occur in the future. Some have gone so far as to predict that the family as we now know it is in the process of disappearing entirely.

WHAT IS HAPPENING TO THE MODERN FAMILY?

Sex, Marriage, and Divorce

There is no doubt that some important shifts have been taking place. For instance, there has been a big change in sexual behavior. Since the early 1960s, premarital sex has greatly increased (see figure 5.1). Once considered largely taboo, especially for women, sex before marriage is now experienced by a rather considerable majority. Living together before marriage, which not so long ago would have been considered scandalous, at least in polite middle-class society, is now widely taken for granted. Movies and magazines are much more explicit about sex than ever before. Even the rate of *marital* intercourse has gone up. Some observers have described this as an erotic revolution. The

FIGURE 5.1: The Rise in Premarital Sexual Experience, 1920s–1980s

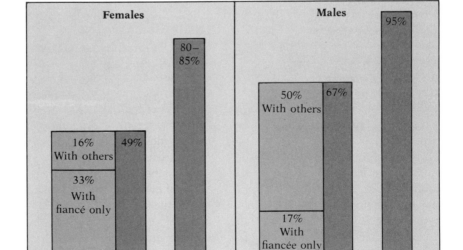

Source: Terman, 1938, 321; Hunt, 1974, 15; Tavris and Sadd, 1977, 34; *Family Planning Perspectives*, 1985; Mosher, 1990.

growth of sexual activity appeared to have leveled off by the late 1970s, however. By the late 1980s, the revolution was finished.

At the same time, marriage rates have fallen, especially for young people. Once it was common for working-class women to marry in their teens and for middle-class women to marry about the time they graduated from college. Now both groups are waiting longer before marrying. By the 1990s, most people were delaying marriage until their mid-twenties, and one out of four adults was still unmarried past age 30. As the marriage rate has gone down, the divorce rate has gone up, though there has been little change in the last decade (see figure 5.2).

A generation ago divorces were considered scandalous by many people. There was a time when a politician who had been divorced would have given up hope of public office. Legal restrictions on divorce required people to go through painful public trials to prove extreme cruelty, neglect, or adultery. In the 1960s the laws changed, and in many states divorces became much easier to get on a "no-fault" basis. But it would not be accurate to say that easier divorce laws are what has brought about the big increase in divorces. The rising trend in divorces has been going on for a long time, and what we have now is only the culmination of a trend at a very high level. During the 1980s, five out of ten marriages ended in divorce. In the 1990s, we can expect continued high levels of divorce, perhaps dipping down a bit so that four out of ten marriages will dissolve (U.S. Bureau of the Census 1992e).

Nonmarital Childbirth and Zero Population Growth

The combination of a falling marriage rate and a rising divorce rate is one reason why some observers have predicted the end of the family. Another is the sexual explosion of recent years. Such observers can also point to the massive increase in unmarried childbirth rates in the last decades (see figure 5.3). Where once illegitimacy was a relatively minor phenomenon, it has now reached considerable proportions. In certain groups of the population, such as poverty-level African-American women, it accounts for more than half of all births. Bearing an illegitimate child was once the most scandalous behavior of all, much more so than divorce or premarital intercourse. Now, some say, it is on its way to being accepted as normal, and this is happening not only in the poverty-stricken lower class. Among whites, one of every five births is now out of wedlock. Some educated middle-class women and their intellectual spokespersons have taken the stance that it is a woman's right to bear her own child without having to undergo a conventional marriage with a man. This is even reflected in television situation comedies; Murphy Brown decided to give birth without having a husband and without planning to marry.

Another important trend upsetting the traditional family has been the shift in the birthrate. A few years ago there was a lot of concern with the problem of overpopulation. At the rate people were reproducing themselves, it looked as if we were heading for a twenty-first century in which the United States would have to support a huge population. This image of wall-to-wall crowds of people has faded, at least for the advanced industrial countries. Birthrates began falling in the late 1950s and by the mid-1970s reached a level at which Americans

FIGURE 5.2: Long-Term Trends in Marriage and Divorce Rates

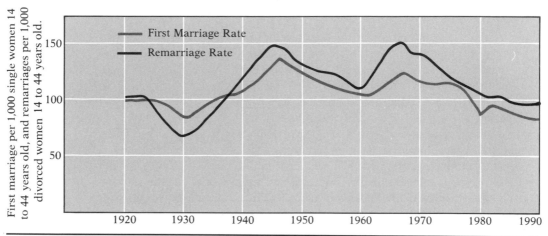

Sources: U.S. Bureau of the Census, 1992e, *Current Population Reports*, P-23, No. 180.

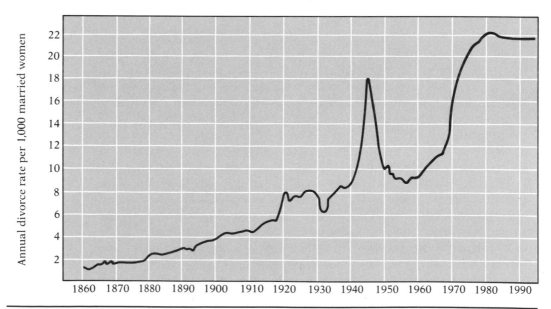

Source: Cherlin, 1992, 22; *Statistical Abstract of the United States*, 1992, No. 127.

FIGURE 5.3: The Rising Proportion of Births to Unmarried Women, 1940–1990

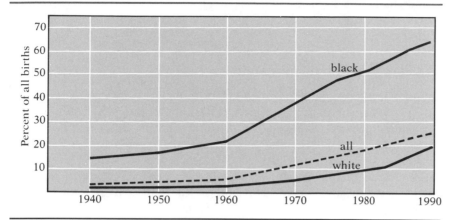

Source: *Statistical Abstract of the United States*, 1992, No. 89.

FIGURE 5.4: The Falling Birthrate, 1860–1990

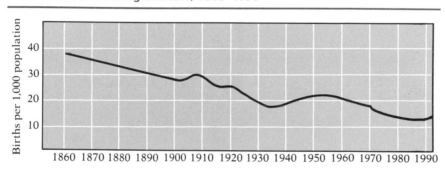

Source: U.S. Bureau of the Census, 1978a, 3; U.S. Bureau of the Census, 1992a.

were not quite reproducing themselves. By the early 1990s, the birthrate had crept back up to near replacement level (see figure 5.4).

This trend, in a sense, is the opposite of the trend in nonmarital childbirth. That is, on one side we see more sexual activity spilling over outside of marriage, driving up the nonmarital birthrate. On the other hand, we have fewer children being born in general and a declining birthrate. Actually the two trends are not incompatible. One reason we have such a high nonmarital birth*ratio* (i.e., the ratio between nonmarital births and marital births) is that the marital birthrate has been going down, so that there are fewer marital births to compare with the nonmarital ones. In a larger sense, though, the two trends add up to another challenge to the traditional family. The old-fashioned ideal of the large family, with Mom and Dad surrounded by a brood of four or five children, has clearly become a subject of nostalgia. New families are mostly small, and many of the children are born into families that do not include a father at all.

Change, Not Disappearance

The prognosis that the family is going to disappear, though, is overdrawn. In our opinion, and that of many sociological experts on the family, what is happening now is a change in typical family patterns. But a change does not mean the whole family institution is disappearing. There have been important changes in the past (see table 5.1). At each period of change, conservatives have claimed that the family itself was disappearing and that society was being undermined. Instead, what had disappeared was simply a particular form of the family, one that people had been used to for perhaps hundreds of years. But the

TABLE 5.1   Some Major Periods of Change in the Family

Human nuclear family emerges	1–5 million B.C.?	Strong sexual bonding ties together males and females to cooperate in raising children. Necessary because of long period of immaturity among human offspring as compared to other species.
Complex kinship exchange systems in tribal societies	8000 B.C.?	Extended family networks with complex marriage and economic exchange rules. Some family systems are matrilineal and/or matrilocal, although most are patrilineal and patrilocal or take still other forms. Family networks make up all of social organization.
Emergence of nonkinship social organization. Decline of complex exchange networks. Rise of the patrimonial household.	*Ca.* 3000 B.C. in Mesopotamia and Egypt; *ca.* 600 B.C. in Greece and China	Rise of the state as an organization outside family networks. Tribal family systems decline and are replaced by the patrimonial household, dominated by property-owning males and containing many servants and armed warriors.
Rise of the bureaucratic state and capitalist economy. Emergence of the private household.	A.D. 1700s/1800s in Western Europe and North America	Work and defense shift out of the household. Dominant patrimonial households disappear. Rise of the nuclear family as ideal. An individual marriage market replaces politically arranged marriages. Ideal of male breadwinner and female housewife.
Diversified marriage market. Tendency toward egalitarian male/ female marriage bargaining.	Ca. A.D. 1950–2000 in wealthy industrial societies	Variety of family structures. Widespread divorce and remarriage, along with both male and female careers, produces a series of relatively short-term marriages and informal cohabitations.

newer forms that developed were also versions of the family. And hundreds of years later, when these "new" forms themselves were giving way to something still newer, the conservatives again complained that the family was disappearing and that society was in jeopardy. (For debates on this issue, see Cherlin 1992; Popenoe 1993; Stacy 1993; Cowan 1993.)

New Variants of Family Structure

The new family system we are now seeing differs from the older one in several ways. With the big increases in premarital sex, illegitimacy, and cohabitation, a good deal of what used to be reserved for legal marriage is now taking place outside it. But these nonlegalized sexual and parental activities are not chaotic. They exhibit a pattern that has social significance. They may even be called new versions of family structure itself. Persons who have premarital intercourse, for example, do not simply sleep with everyone who comes along. There is a fairly limited number of sexual partners, and premarital pairings are put together and come apart in definite ways. Living together, once one gets over seeing it through the old scandal-tinted eyeglasses left over from former days, actually looks a good deal like a conventional marriage. It lacks the formal ceremony, but most people in our large urban society do not personally know about other people's marriage ceremonies anyway. Without being told, it is often hard to tell the difference between an unmarried couple living together and one that is married.

Even in the more extreme case of women with illegitimate children, there is a certain family pattern involved. These relationships are at least one part of the traditional family structure: parent and child. And there is often a network among women kinfolk with illegitimate children that is very much a family situation, if perhaps one that makes us think of the tribal kinship networks studied by anthropologists rather than the conventional American nuclear family.

The newer forms of sexual and parental arrangements, then, should be seen as new variants of family structure rather than phenomena that have nothing to do with a family system. At the same time, we should bear in mind that the conventional nuclear family is still here. A large number of households still have a father, mother, and minor children living together, although this makes up only 27 percent of all households, a drop from 39 percent in 1970 (see figure 5.5). The most common type of household now consists of a married couple with no children. What is happening is that the family structure in the United States is becoming more complicated because there are more different kinds of families and households existing at the same time. The family is becoming more diverse, and it is harder for advocates of just one form to say it is the ideal or norm.

For instance, a steadily increasing percentage of Americans are living past

FIGURE 5.5: Changing Shares of Household Types

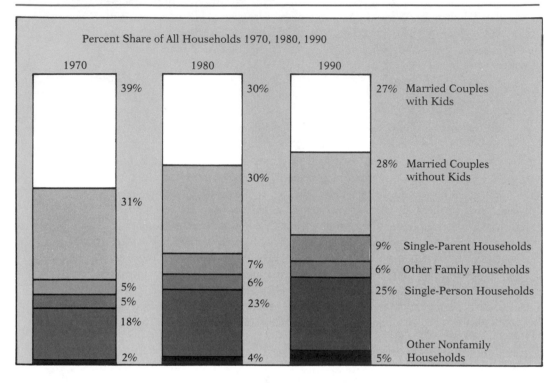

Since 1970 the share of households composed of married couples with children has decreased, while the share of households composed of single persons and single parents has increased.

Source: *Statistical Abstract of the United States*, 1971, 1981, 1992.

the time when their children have grown up and left home. So there are many households consisting only of husband and wife without children. As these people grow still older, one spouse or the other (typically the male) dies first, leaving a household of a single person. When we add these households together with those of young people who have not yet married and are living alone or with roommates, or who are cohabiting, then we find quite a substantial proportion of households that are not the conventional father-mother-children nuclear family. When we add in the people who have gotten divorced but have not yet remarried and women with illegitimate children, the total of nonstandard households goes up still further. And the number of smaller households is increasing, while large households are becoming rather scarce (see figure 5.5).

The Diversification of the Family

The stress we ought to place here is on the permanent *diversity* of types of families rather than on one form taking over entirely from the others (see figure 5.6). The nuclear family is not disappearing either, because it keeps on being

FIGURE 5.6: Types of Households Found Today

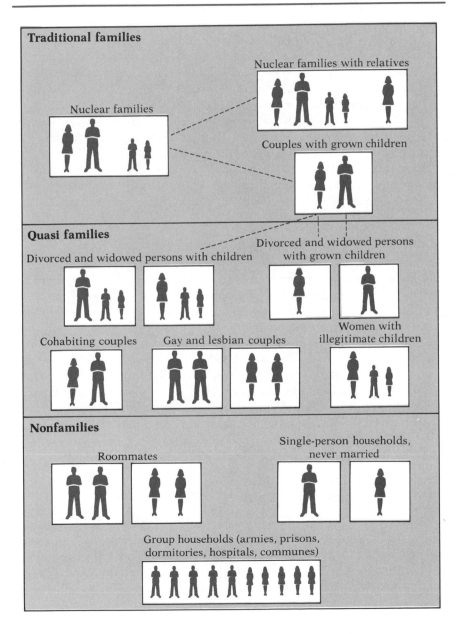

Traditional families

Nuclear families with relatives

Nuclear families

Couples with grown children

Quasi families

Divorced and widowed persons with children

Divorced and widowed persons with grown children

Cohabiting couples   Gay and lesbian couples

Women with illegitimate children

Nonfamilies

Roommates

Single-person households, never married

Group households (armies, prisons, dormitories, hospitals, communes)

reformed. People put off getting married longer these days, and hence there are more of them to count as single-person households, or shared-with-roommate households or cohabiting arrangements. But the rate at which people eventually marry is still high. Similarly, although there is a high divorce rate, there is also a high rate of remarriage. During the time when people are divorced, they add to the alternative-family total. Since there is a constant flow in and out between divorce and marriage, at any given time there is going to be a fair number of people in both kinds of living arrangements.

One interesting result of all this divorce and remarriage is that the kinship structure is getting more complicated. In the tradition of the nuclear family, each child has one father and one mother and lives in a single household until he or she is old enough to leave home and establish another household. With a high rate of divorce and remarriage, though, many children have two sets of parents and two households (or sometimes even more). Instead of merely having one set of brothers and sisters, they have two or more sets, living in different places. It is said that there is a whole new problem of etiquette at marriage receptions today; for when children of divorced people get married, a quite complicated set of families may have to be invited to the reception. In this sense we may actually have *more* family than before.

THE CAUSES OF LONG-RUN CHANGES

If we look at the way the family has been changing throughout the twentieth century, one pattern stands out. The period around 1950 is different from every other time. This is the time of the so-called *baby boom*, when suddenly the long-standing decline in the birthrate reversed itself, and the population shot up. Along with the baby boom went several other reversals in trends that had stood for decades: the marriage and divorce rates. After the 1950s, the baby boom disappeared, and the other trends in the family also reversed themselves once again. In retrospect the 1950s look like a strange aberration, a bump in the long-term curve. Andrew Cherlin (1992) has claimed that the real problem is not to explain why there was a liberalization of the family in the 1960s and 1970s (that was merely the continuation of a long-term trend), but why was there this anomalous countertrend, this revival of tradition, in the 1950s?

Why the Anomalous 1950s?

Let us examine the charts again. Figure 5.4 shows us that the birthrate had slowly been declining ever since the Civil War, a period of well over one hundred years (possibly longer, if it was declining in the early 1800s, too, which seems likely). The only exception is in the years from about 1940 to 1955, when the rate briefly rose again (the baby boom). In the 1920s and

The baby boom was definitely underway outside this Boston clinic in 1948.

1930s, the decline had been so striking that statisticians at the U.S. Bureau of the Census predicted that U.S. population would level off in the 1960s at about 140 million people. The baby boom caught them by surprise, and their prediction turned out to be about 60 million people short.

Along with the rising birthrate there was a rise in the marriage rate (figure 5.2). Actually this began a little earlier than the baby boom. The rate of marriage was lowest in the early 1930s, began to climb slowly throughout the war years, and then rapidly shot up in the few years just after the war, 1945–48. Thereafter it stayed at a steady high level throughout the 1950s and 1960s and began to drop in the 1970s. The divorce rate did just the opposite, mirroring the baby boom in a negative way. In the ten years before the baby boom really took off in the late 1940s, the divorce rate had been going up. There was a peak immediately after the end of the war in 1945, which is usually interpreted to mean that the disruption of relationships in the war years was being made official. Then the divorce rate dropped sharply, and it went down during the period when the birthrate was most sharply increasing. At the end of the 1950s, both rates reversed themselves: divorces began to go up again, and births to go down.

Finally, figure 5.7 shows us one more piece of the picture. American women were marrying earlier during the baby boom years. Early in the century, women typically married in their mid-twenties, and the level held

FIGURE 5.7: Women in Their Twenties Who Have Never Married

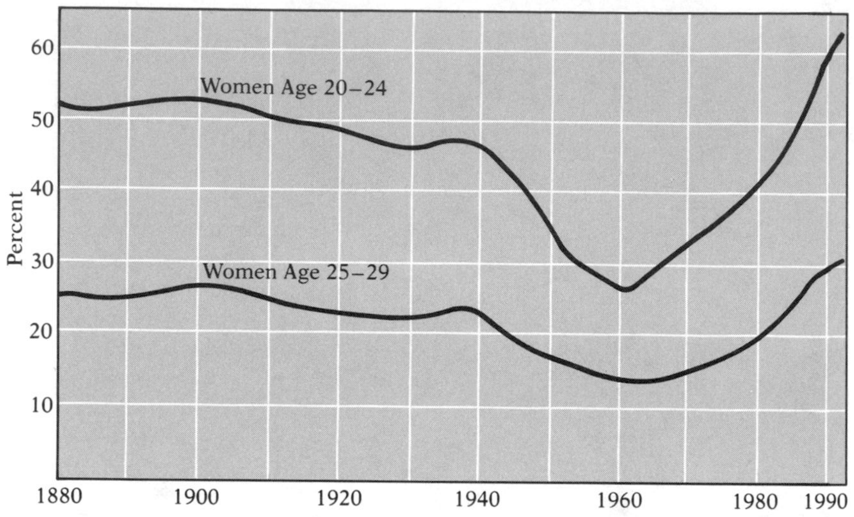

Source: U.S. Bureau of the Census, 1992h, P-23, No. 181.

steady for three decades. But in the 1940s, the age of marriages took a rapid drop, and half of all women were married by about age twenty-one. In the 1950s, it dropped even lower, and almost half were married before they turned twenty. Only in the 1970s did this pattern reverse itself, and the age of marriage began to rise again toward the pre-World War II pattern.

The different figures hold together. In the 1950s, marriage was especially popular. That generation had the highest percentage of married individuals on record: 96 percent of females, 94 percent of males (Cherlin 1992). And they married younger, divorced less, and had more children. No wonder the 1950s acquired a special reputation as the era of the "familistic generation," an era of tradition and conformity. Actually "the fifties" is a little inaccurate; the phenomenon began in the 1940s, with some trends beginning even before World War II. The rates of marriage and birth, for instance, were already inching up during that time, though they started from the low point of the previous downtrend, so that the full "baby boom" and "marriage boom" were not apparent until the late 1940s. Then in the late 1950s, most of the trends reversed themselves (although again some of them hung on longer, into the 1960s, and did not really shift until the 1970s).

How do we explain this fifties (really 1940s–1950s) bump in the trends? One argument often advanced is that the return to the traditional family was a

FEATURE 5.1
Nazi Germany and Stalin's Russia: Twentieth-Century Totalitarianism and the Family

The years between 1920 and 1950 were overshadowed by two great totalitarian systems. The Communist party came to power in Russia after the Revolution in 1917. Despite its idealistic beginnings, by the mid-1920s, following Lenin's death, the Soviet Socialist Republic degenerated into a dictatorship under Joseph Stalin. At about the same time in Germany, Adolf Hitler organized his Nazi movement around a militant core of disgruntled World War I veterans. After a long campaign of street demonstrations and battles, mostly aimed against the socialists but also against the Jews, Hitler's party was elected in 1933. Soon after, Hitler abolished the German democracy and set up his fascist dictatorship.

The Nazi program was committed to, among other programs, overthrowing the gains recently made by the German feminist movement. Women had won the right to vote in 1918; the Nazis officially excluded women from public life. The purpose of the German woman was "to minister in the home," devoting herself to "the care of man, soul, body and mind," from "the first to the last moment of man's existence." In *Mein Kampf*, Hitler declared that "the aim of feminine education is invariably to be the future mother." Contraception and abortion were made illegal, and the sole purpose of sex was to procreate new members for the German nation. Marriage was encouraged by taxing bachelors and spinsters, while taxes were rebated and interest-free loans given for each child born under a state-supported contract. (A woman took out the loan, but it was paid to her husband.) The measures were successful, at least in producing babies: the birthrate was raised 30 percent in the first two years after the Nazis took power.

But although the Nazi ideal was to keep the woman in the home, this policy failed. For one thing, the huge military losses of World War I had left Germany with two million more women than men, and these women had to support themselves at work. When military preparations began for World War II, the Nazis made both men and women liable for state labor and called out women in increasing numbers. At the height of the war effort in the 1940s, when the German army had conscripted virtually all the able-bodied men, most of the civilian work was done by women. As many as 80 percent of all adult women were working for the Third Reich. The state took a contradictory stance toward females: the police attempted to prevent women from smoking or wearing cosmetics, while at the same time houses of prostitution were maintained for the military and the Nazi leadership.

In the Soviet Union, the Revolution had begun with a principled effort to liberate women from the patriarchal Russian family. Women were given the right to choose their own domicile, name, and citizenship; to marry and divorce independently of the wishes of their parents or spouses; to pursue economic independence; and to control their own sexuality, with contraception and abortion on demand. Illegitimacy, adultery, and homosexuality were eliminated from the code of criminal offenses. In order that women should be able to participate economically and politically on a par with men, the plan was to establish public nurseries and a collective housekeeping service, along with maternity leaves at work.

But economic and then political forces intervened to prevent the carrying out of this liberal plan. During its early years the revolutionary government witnessed civil war and economic depression. In the early 1920s, there was widespread unemployment, which hit women especially hard in their campaign for economic equality. The

(Continued next page)

FEATURE 5.1 (*Continued*)

public child-care services could not be built, and mothers had to continue as before, often with job responsibilities on top of their other tasks. When the Soviet economy recovered, all resources were channeled into building heavy industrial equipment and armaments. By the 1930s, Soviet policy had shifted to support for the traditional family. The early revolutionary feminists like Alexandra Killontai were publicly censured. The purpose of sex, it was now declared, was solely to produce children. As in Germany, there was a great deal of rhetoric about "preserving the race," and an effort was made to build up the population. Women who had six or more children were given bonuses; mothers of seven were given medals. Abortion was again made a criminal offense. The old czarist legislation against homosexuality was reintroduced, providing for sentences of three to eight years, enforced by mass arrests. Illegitimacy again became a legal category, and both mother and child (but not father) were stigmatized. In 1936, couples who divorced were fined 30 to 50 rubles for "mistaking infatuation with love," and in 1944 the fine was raised to 500 to 2,000 rubles (which made divorce virtually impossible).

The rearing of children, which originally was planned on a libertarian basis, also took an authoritarian turn. An authoritarian youth organization was formed, with compulsory participation, under the leadership of Makarenko, a Secret Political Police official who had been in charge of delinquent boys. Progressive schools founded in the 1920s were abolished and replaced by traditional authoritarian ones. By 1943, co-education was eliminated, and the schools preached a puritanical doctrine of sexual abstinence in the name of commitment to the state. At home, parents were held responsible for teaching their children the correct ideological line.

Do the Soviet and Nazi cases show the impossibility of changing the traditional family? Although this conclusion has often been drawn from a superficial acquaintance with the facts, the lesson actually is somewhat different. The Nazis, on political grounds, attempted to reestablish the most sexist form of the traditional family. But their effort was undermined by economic realities: they needed women in the labor force and ended up undermining the traditional family more than supporting it. The Soviets started out with an extremely liberal plan but were unable to put it into effect for lack of economic resources. The key to equal rights for women was in the provision of state-supported child care and housekeeping, but these quickly went by the board due to the early economic crisis and then stayed low priority because of the subsequent military buildup. By the 1930s and 1940s, at the height of the Stalin dictatorship, the Soviet policy on the family and sexual behavior had become as authoritarian as the Nazis'. This provides a second lesson: there is a tendency for politically authoritarian regimes to invoke sexual puritanism and the authoritarian family as part of their general ideology of control over the individual.

These extreme instances of the authoritarian family are now in the past. The Nazi regime fell in 1945. After Stalin's death in 1953, Soviet policy gradually eased. The right to abortion was reestablished in 1955, and illegitimacy ceased to be registered in 1965. Women have moved closer to equality in educational access, although they are still concentrated in the lower-paid occupations. Women do make up a majority of Russian doctors, but even these are ranked lower than men in the bureaucratic medical system. Child-care facilities have expanded, but women continue to be responsible for most of the traditional housework and domestic child care. After the downfall of the Soviet Union in 1991, the situation of social services for women has gone into chaos. Feminist interests have been pushed into the background. Women there have a long way to go to catch up with their sisters in many parts of the West. (Millett 1970, 157–76; Neumann 1944; Geiger 1968; Fisher 1980) reaction to the postwar situation. According to this theory, after the end of wartime disruption with its temporary surge of divorces, people naturally turned toward reestablishing normalcy and the family. There are several problems with this explanation. One is that some of the trends (such as the rising rates of birth and marriage and the falling age of marriage) actually began before the war. Moreover, postwar situations do not invariably have this result. We can see this from figures 5.2 and 5.4, which show us that the rates of birth and marriage did not go up in the 1920s, after World War I. Nor was there a rise in the birthrate after the Civil War (which ended in 1865). World War II is thus more of a coincidence on the chart than an explanation of what actually happened in the 1940s and 1950s.

Part of the answer is that the war came on the heels of an economic depression. The war itself created considerable employment (the military boom), and after the war came a huge expansion of the American economy, which lasted into the 1950s. In general, people tend to marry when they feel they can afford it (Easterlin 1980). The better their economic prospects are, the younger they are when they decide they can get married. The converse of this is that during the Great Depression of the 1930s, there were strong reasons why many people did *not* get married, and why those who married felt they could not afford to have many children.

When the economic situation improved—culminating in the postwar boom, although there were smaller improvements earlier—it was as if a dam had burst. Suddenly there were two groups of people who had a chance to get married or to go ahead and have children if they were already married. There were those who put off getting married or put off having children in the 1930s, and soldiers returning from the war. Along with this, the generation just coming of age began to marry extremely early and to have children quickly. Familism had become something of a new cultural movement, and everybody was jumping on the bandwagon.

Working-Class Affluence and the Move to the Suburbs

We can understand this atmosphere better if we look at the way the 1950s were perceived at the time. It was the so-called age of suburbia, when mass housing developments sprang up for the first time outside the major cities. For a hundred years previously, the tendency had been for the population of the countryside to decline, while large city populations grew continuously larger. In the 1950s a countermovement developed: people were leaving the apartment houses of the city for single-family dwellings in the suburbs.

Culture critics of the time generally regarded the suburbs with horror. The houses stood in mass-produced developments like the famous Levittown, New Jersey; this was taken as a sign of uniformity invading our culture. The plethora

Mass-produced developments like Whitestone, in the New York City borough of Queens, aroused fears that the uniformity of suburbia was invading our culture.

of clubs, school teams, teenage gangs, and other organizations that went along with suburban life was also seen as evidence of growing conformity. David Riesman wrote in his famous sociological best-seller, *The Lonely Crowd*, that the traditional American inner-directed personality was on the decline and the other-directed conformist was proliferating. William H. Whyte described the impersonality of the suburbs as the habitat of *The Organization Man*, who was moved from place to place by gigantic bureaucracies. One folk song satirized the suburbs as "little boxes, on the hillside, little boxes made of ticky-tacky" that "all look just the same."

But the critics missed one crucial fact: the newly developed suburbs were not, in general, full of "organization men" working their way up the management ladder, nor did many doctors and lawyers live in the folksinger's "little boxes." The new suburbs were to a considerable extent working class (along with young professionals bound for better things) (Berger 1960; 1971). Houses were cheap, standardized, and prefabricated because that was the only kind of house the working class could afford. But from their point of view, at least they had a house rather than an apartment, and it was in the countryside rather than in the tenement district of a city. Americans in general have trouble seeing social class, and the upper-middle-class critics fell into the trap of assuming that the suburbs were built for people like themselves. Used to the luxury of old and grander houses, with more space between them and more investment in individualistic touches, they regarded the new suburbs with horror, as a decline in American taste. But for the working-class people, moving out to own their own houses for the first time was a tremendous step up.

The familism and "conformity" that were such popular topics of discussion in the 1950s, then, were apparently really to a large extent a movement of the working class (Berger 1960; 1971; Gans 1967). The immediate postwar decades have been described by a recent economist as an "economic miracle": for the first time in history, a middle-class standard of living seemed to be in sight for most or many of the working class. To own their own home, fill it with labor-saving appliances, own a car or even two: all these had been reserved for the rich, until this boom time in America. To keep one's wife home, like a respectable middle-class housewife, instead of sending her off to work with the masses—this was a new frontier of social status for the working class. The 1950s was one of the wealthiest periods in the history of the country. The United States, with its industrial plant running at full blast on the rich natural resources of North America, was producing the highest standard of living in the history of the world. The middle and upper classes got richer, but they already had their single-family dwellings. For the working class, this was something new, and it threw itself into the suburban lifestyle with enthusiasm. The "do-it-yourself" craze of putting in one's own room additions, plumbing, and wiring and doing all sorts of other jobs around the house made a splash in the 1950s. The traditional homeowner had been a white-collar worker who hired a handyman for these kinds of repairs; the new working-class suburbanites, though, could only afford to make these kinds of improvements by putting in their own manual labor.

This is one reason for the "conformity" charges made by the critics. For as we will see (in chapter 6), working-class culture is more group oriented and less individualistic and achievement oriented than upper-middle-class culture. The new suburbs looked roughly like middle-class communities on the outside, but culturally they continued to be working class and hence emphasized conformity (as Herbert Gans showed in his book *The Levittowners*, 1967; see also Berger 1960; 1971). Moreover, working-class culture is more familistic than middle-class culture. Working-class people marry younger, have children earlier, and confine more of their socializing to their immediate relatives and neighbors.

The rising rates of marriage and birth and the declining age of marriage, then, were to a considerable extent economic in origin. In that era of affluence, working-class people and younger, less affluent members of the middle class were more able to do what they preferred to do: to marry as soon as possible, and to raise a large family. Early economic constraints had kept them from doing these things; early in the century, the age of marriage was much higher, simply because most people couldn't afford to set up housekeeping on their own until they had worked for a number of years. In the economic boom that followed World War II, though, workers were finally able to indulge their familistic concerns more than ever before. It is probably this, more than anything else, that contributed to the unusual family pattern of the 1950s.

Why did this pattern come to an end, beginning in the late 1950s and the 1960s? In Cherlin's view (1992) the long-run trend just reasserted itself. The new suburban working class had a chance to emulate what it thought was a middle-class lifestyle, but actually it only moved the working-class style out to the better material conditions of the suburbs. The wave of familistic sentiment that went along with this transition began to be undercut by forces that had been operating in the middle classes for some time.

A Comparison: The Roaring Twenties and the Rebellious Sixties

If our analysis is right, the 1950s was a period when the working class captured the center of cultural attention in America. And since working-class culture is typically familistic and traditional, the family life of the 1950s looked like a trend back toward tradition. During most of this century, though, the cultural initiative has come largely from the middle and upper classes. Its thrust has been in a very different direction, especially regarding sexual liberalism. This is apparent in two of the more spectacular eras of change, the so-called "roaring twenties" and the rebellious sixties.

The 1920s

The 1920s was the era of our first great popular sexual revolution (see feature 5.2). Prior to this, the respectable middle and upper classes had attempted to live with Victorian propriety. Gradually, among the upper-class youths prior to World War I, a hedonistic culture was appearing. Drinking and sexual flirtation became the vogue. But this was confined to a rather small group and received little publicity until after the war. Then middle-class youth culture as a whole caught on to the new style, and it invaded the mass media.

Some of the changes were highly visible. One was the shift in clothing styles. Women's skirts suddenly went up, to the knee and even above. This does not seem like much to us now, but at the time it was a shock and a revolution. For the Victorian style had been long dresses, touching the ground and not revealing so much as an ankle. Legs had not even been a fit subject for proper talk, and now they were in full view. This was probably the single largest change in women's clothing styles in several thousand years of Western history. Women had worn a long robe or dress ever since Greek and Roman times, and the custom of covering the female body had scarcely varied throughout the centuries: style changes had been confined to the cut of the garment and to hair

In the 1920s, the youth culture began to experiment with new forms of music and dancing, more revealing clothing, and the use of alcohol, which was at that time illegal. Here, young New York City women begin a Charleston endurance contest at the Parody Club in 1926.

and head covering. Suddenly we were in a new era, and the style definitely announced a shift toward a sexier appearance.

In the 1920s, an independent youth culture sprang into being. Previously, middle-class parents had controlled their offsprings' socializing, which featured carefully chaperoned formal dances and gatherings of whole families with their guests at home. Now the youths went off and had their own parties, driving their own cars and frequenting roadhouses and dance halls. Prohibition of alcohol had just been enacted by constitutional amendment, but this seemed only to make the use of illicit alcohol more exciting. Dancing changed from the graceful ballroom traditions of the waltz to the new "hot" styles: the fox-trot, the Charleston, and many more. Jazz bands and their syncopated rhythms became the vogue; modern popular music was born. Jazz was considered somewhat immoral by traditional elders; as F. Scott Fitzgerald pointed out, the terms *jazz* or *jazzy* themselves first had the connotation of sex, and then wild dancing, before they finally came to mean a kind of music.

The upper-class youth started this new sexually oriented popular culture, but it soon spread. By the early twenties, according to Fitzgerald, it had infected the youths of the middle class in the smaller cities across the country. Within a few years, their elders had begun to catch on, and drinking parties, jazz, and the new styles had spread far and wide. The new dating style had set

FEATURE 5.2
F. Scott Fitzgerald on the Jazz Age

The first social revelation created a sensation out of all proportion to its novelty. As far back as 1915 the unchaperoned young people of the smaller cities had discovered the mobile privacy of that automobile given to young Bill at sixteen to make him "self-reliant." At first petting was a desperate adventure even under such favorable conditions, but presently confidences were exchanged and the old commandment broke down. As early as 1917 there were references to such sweet and casual dalliance in any number of the *Yale Record* or the *Princeton Tiger.*

But petting in its more audacious manifestations was confined to the wealthier classes—among other young people the old standard prevailed until after the War, and a kiss meant that a proposal was expected, as young officers in strange cities sometimes discovered to their dismay. Only in 1920 did the veil finally fall—the Jazz Age was in flower.

Scarcely had the staider citizens of the republic caught their breaths when the wildest of all generations, the generation which had been adolescent during the confusion of the War, brusquely shouldered my contemporaries out of the way and danced into the limelight. This was the generation whose girls dramatized themselves as flappers, the generation that corrupted its elders and eventually overreached itself less through lack of morals than through lack of taste. May one offer in exhibit the year 1922! That was the peak of the younger generation, for though the Jazz Age continued, it became less and less an affair of youth.

The sequel was like a children's party taken over by the elders, leaving the children puzzled and rather neglected and rather taken aback. By 1923 their elders, tired of watching the carnival with ill-concealed envy, had discovered that young liquor will take the place of young blood, and with a whoop the orgy began. The younger generation was starred no longer.

A whole race going hedonistic, deciding on pleasure. The precocious intimacies of the younger generation would have come about with or without prohibition—they were implicit in the attempt to adapt English customs to American conditions. (Our South, for example, is tropical and early maturing—it has never been part of the wisdom of France and Spain to let young girls go unchaperoned at sixteen and seventeen.) But the general decision to be amused that began with the cocktail parties of 1921 had more complicated origins.

The word jazz in its progress toward respectability has meant first sex, then dancing, then music. It is associated with a state of nervous stimulation, not unlike that of big cities behind the lines of a war. To many English the War still goes on because all the forces that menace them are still active—Wherefore eat, drink and be merry, for tomorrow we die. But different causes had now brought about a corresponding state in America— though there were entire classes (people over fifty, for example) who spent a whole decade denying its existence even when its puckish face peered into the family circle. Never did they dream that they had contributed to it. The honest citizens of every class, who believed in a strict public morality and were powerful enough to enforce the necessary legislation, did not know that they would necessarily be served by criminals and quacks, and do not really believe it to-day. Rich righteousness had always been able to buy honest and intelligent servants to free the slaves or the Cubans, so when this attempt collapsed our elders stood firm with all the stubbornness of people involved in a weak case, preserving their righteousness and losing their children. Silver-haired women and men with fine old faces, people who never did a consciously dishonest thing in their lives, still assure each other in the apartment hotels of New York and Boston and Washington that "there's a whole generation growing up that will never know the taste of liquor." Meanwhile their granddaughters pass the well-thumbed copy of *Lady Chatterly's Lover* around the boarding-school and, if they get about at all, know the taste of gin or corn at sixteen. But the generation who reached maturity between 1875 and 1895 continue to believe what they want to believe.

Even the intervening generations were incredulous. In 1920 Heywood Broun announced that all this hubbub was nonsense, that young men didn't kiss but told anyhow. But very shortly people over twenty-five came in for an intensive education. Let me trace some of the revelations vouchsafed them by reference to a dozen works written for various types of mentality during the decade. We begin with the suggestion that Don Juan leads an interesting life (*Furgen*, 1919); then we learn that there's a lot of sex around if we only knew it (*Winesburg, Ohio*, 1920), that adolescents lead very amorous lives (*This Side of Paradise*, 1920), that there are a lot of neglected Anglo-Saxon words (*Ulysses*, 1921), that older people don't always resist sudden temptations (*Cytherea*, 1922), that girls are sometimes seduced without being ruined (*Flaming Youth*, 1922), that even rape often turns out well (*The Sheik*, 1922), that glamorous English ladies are often promiscuous (*The Green Hat*, 1924), that in fact they devote most of their time to it (*The Vortex*, 1926), that it's a damn good thing too (*Lady Chatterley's Lover*, 1928), and finally that there are abnormal variations (*The Well of Loneliness*, 1928, and *Sodom and Gomorrah*, 1929).

The Jazz Age had had a wild youth and a heady middle age. There was a phase of the necking parties, the Leopold-Loeb murder (I remember the time my wife was arrested on Queensborough Bridge on the suspicion of being the "Bob-haired Bandit") and the John Held Clothes. In the second phase such phenomena as sex and murder became more mature, if much more conventional. Middle age must be served

and pajamas came to the beach to save fat thighs and flabby calves from competition with the one-piece bathing-suit. Finally skirts came down and everything was concealed. Everybody was at scratch now. Let's go—

But it was not to be. Somebody had blundered and the most expensive orgy in history was over.

It ended two years ago [1929], because the utter confidence which was its essential prop received an enormous jolt, and it didn't take long for the flimsy structure to settle earthward. And after two years the Jazz Age seems as far away as the days before the War. It was borrowed time anyhow—the whole upper tenth of a nation living with the insouciance of grand ducs and the casualness of chorus girls. But moralizing is easy now and it was pleasant to be in one's twenties in such a certain and unworried time. Even when you were broke you didn't worry about money, because it was in such profusion around you. Toward the end one had a struggle to pay one's share; it was almost a favor to accept hospitality that required any travelling. Charm, notoriety, mere good manners, weighed more than money as a social asset. This was rather splendid, but things were getting thinner and thinner as the eternal necessary human values tried to spread over all that expansion. Writers were geniuses on the strength of one respectable book or play; just as during the War officers of four months' experience commanded hundreds of men, so there were now many little fish lording it over great big bowls. In the theatrical world extravagant productions were carried by a few second-rate stars, and so on up the scale into politics, where it was difficult to interest good men in positions of the highest importance and responsibility, importance and responsibility far exceeding that of business executives but which paid only five or six thousand a year.

(Continued next page)

FEATURE 5.2 (*Continued*)

Now once more the belt is tight and we summon the proper expression of horror as we look back at our wasted youth. Sometimes, though, there is a ghostly rumble among the drums, an asthmatic whisper in the trombones that swings me back into the early twenties when we drank wood alcohol and every day in every way grew better and better, and there was a first abortive shortening of the skirts, and girls all looked alike in sweater dresses, and people you didn't want to know said "Yes, we have no bananas," and it

seemed only a question of a few years before the older people would step aside and let the world be run by those who saw things as they were—and it all seems rosy and romantic to us who were young then, because we will never feel quite so intensely about our surroundings any more.

F. Scott Fitzgerald, "Echoes of the Jazz Age," in *The Crack-Up.* Copyrighted 1945 by New Directions Publishing Corp. Reprinted by permission of New Directions.

---

in, while the traditional family as the center for social life was now becoming a thing of the past, found only among the elderly and in rural sectors of America.

With the fall of the stock market in 1929 and the coming of the Depression in the 1930s, the wild and hedonistic Jazz Age was over. But the changes that it made in the family, and in sexual styles and mass culture, were not to be undone. The modern age was firmly launched.

The 1960s

"The sixties" did not quite fit that decade chronologically. The era of political and cultural ferment that is usually called "the sixties" started just after the assassination of President John F. Kennedy in 1963 and went on until about the resignation of President Richard M. Nixon in 1974. It was a time of the civil rights movement, with its demonstrations, sit-ins, mass arrests, and even murders of civil rights workers in the Deep South, which finally culminated in the overturning of legal segregation. (A vivid account is *The Sixties* by Todd Gitlin, an activist who became a sociologist.) It was the time of the antiwar movement, centered on United States involvement of half a million troops in support of a military dictatorship in Vietnam, which eventually resulted in U.S. withdrawal in 1973. Both the civil rights and the antiwar movements were extremely popular on college campuses, where they led to numerous demonstrations, sit-ins, and strikes, some of which resulted in massive confrontations with police. Though the student demonstrations generally adhered to the nonviolent-protest philosophy espoused by the civil rights movement, the authorities sometimes put them down with force (such as at Kent State University in Ohio in 1970, when the National Guard opened fire on a demonstration and killed four students).

Not surprisingly, the prevailing mood of American youth during this period was one of intense alienation from traditional "straight" society. Conventional society was regarded not only as racist and militarist but also as sexually and culturally restrictive. A "counterculture" emerged. For a while, extreme clothing styles abounded. For fifty years, the American style of grooming had been for men to be close shaven; now beards and mustaches became the vogue. The conservatism of the 1950s had been manifested in conservative, military-style crew cuts, while women also had short, conservative coiffures. Now naturalism was expressed in long hair for both sexes; the eye was jolted when men's hair began to be as long as women's. Part of the youth culture developed a "hippie" style that included beads, Oriental robes, and other bright and exotic clothing. Some went to live in tepees in the woods; others established communes. LSD and other "psychedelic" drugs were popular for inducing visions and altered states of consciousness, while many people became devotees of Eastern mystical religions.

The counterculture and the political rebellion of the time were the most apparent public manifestations of the changes that were going on. But they were only part of a deeper change. As we have already seen, the 1960s and early 1970s formed a period during which the conservative family trends of the 1950s reversed themselves. The baby boom was over, marriage was being put off to a later age, and the divorce rate was rocketing up (figure 5.2). The illegitimacy rate was taking off (figure 5.3), and spokespersons of the counterculture

On May 4, 1970, the National Guard, called in to put down anti-war protests at Kent State University, fired tear gas and bullets into a crowd of about 500 students, killing four and wounding several others.

argued that it was discriminatory to punish either mother or child for an illegitimate birth. Sexual relations had traditionally been closely confined to marriage, or else to the courtship period that quickly led to marriage. But now sex was being explicitly severed from family legalities. The most striking form of this was the upsurge in the practice of unmarried couples living together or "cohabiting." In every generation up through the 1950s, this would have been considered scandalous. Now it was widely practiced in the middle-class youth culture, openly and shamelessly avowed, and the traditional public was forced to back down in its judgment.

What was responsible for these changes of the 1960s? The "revolution" was partly a revolution in the family, comparable to that of the 1920s. In both cases, there was a popular movement, with the most attention going to changes in clothing style, dancing, music, and illicit "trips" (Prohibition alcohol versus psychedelic drugs). The 1960s did have an idealistic political aspect, the movements for civil rights and against the Vietnam War, which the 1920s did not share. But in both cases there was a spectacular public movement that lasted for about ten years and then died away as economic conditions changed. The Roaring Twenties ended with the stock market crash in 1929, which ushered in the Great Depression. The sixties movement died partly because it won some important victories: the Vietnam War did end, and laws were passed overturning racial segregation. In both the 1920s and the 1960s, the period of upheaval left a more permanent monument in the shape of a changed family system and a changed public culture.

The Second Wave of the Women's Movement

A militant, politically vocal women's movement appeared on the scene late in the sixties period—that is, especially in the years just after 1971. The earlier political movements of the sixties had been male-dominated movements; and the hippies, too—though they broke out of the conventional family structure and espoused sexual liberation—tended to have a traditional setup of male leadership and female subordination. The women's movement challenged all that. More significantly, it challenged male supremacist practices throughout ordinary society: the exclusion of women from important managerial and professional jobs, their segregation into female "work ghettos" of secretarial and service work, and their domestic segregation into the traditional careers of housework and child care.

The women's movement acted politically to overcome legal discrimination against women through such organizations as the National Organization for Women (NOW) and the National Women's Political Caucus (NWPC). This public part of the movement constitutes a second wave of feminism, following the long first wave that spent the years from the 1840s to 1920 getting basic legal rights and political suffrage for women. But if we look at the trends in the

FIGURE 5.8: Married Women with Children Have Become Working Women

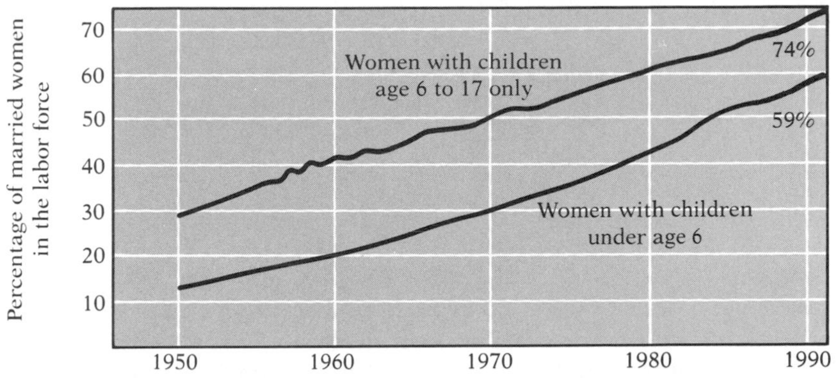

Source: U.S. Bureau of the Census, 1992h, *Current Population Reports*, P-23, No. 181.

family structure, we can see that the political aspect of the women's movement in the 1970s was just a manifestation of forces that had already been in motion, some for as long as twenty years (Chafetz and Dworkin 1986).

One crucial shift of this sort has been the trend for women to work outside the home. In the mid-1960s, only about 38 percent of women worked at paid jobs, and most of these were unmarried women, who often quit their jobs after they were married and almost certainly after they had children. Now 57 percent of all women aged sixteen or older work. And this trend is strong even among married women with children; three-fourths of women with school-age children now work, and even among women with infants less than one year old, over half are employed (see figure 5.8).

Moreover, the reasons why women work have changed. Formerly the bulk of working women were from working-class families. They worked at relatively low-paying and prestigeless jobs as waitresses, store clerks, domestic servants, and factory operatives. These jobs did not attract them; they worked out of economic necessity. But now we see large numbers of middle-class wives working in order to have independent careers of their own. This is happening despite the fact that women still have gotten into only a small proportion of the desirable managerial, professional, and technical positions, although there has been some improvement in this direction.

One of the main manifestations of the feminist movement at home appears to be an increase in the amount of dissatisfaction that women are showing with traditional domestic roles. This is one of the reasons the divorce rate has gone up, especially at the height of the recent wave of feminist consciousness in the 1970s. Women are putting off marriage longer, partly in order to pursue their

FEATURE 5.3
Breaking Down the Barriers of Gender Segregation

A successful woman lawyer describes conditions when she broke into the legal profession in 1970:

When I decided to go to work in [this city], I sent letters to all of the law firms in town that had more than six lawyers. Most of them responded that they had finished doing their hiring for the year, even though it was early. One of them actually answered, came out and said they didn't hire women, that they wanted to stick "with the boys." When I came out and interviewed, one firm said they might be interested in hiring me if I would do probate practice, by which they meant nothing but back-office research work. I also interviewed with a large company for a job as part of their legal staff, and the interviewer asked me about my birth-control practices. By the time that interview was over I didn't want the job.

The worst interview was one in which a government personnel board member said to me, "How are you going to choose between being a wife and being an attorney?" I said I wasn't married. He said, "Well surely a girl as attractive as you is going to be married sometime, and how are you going to handle it then?" I said, "When you were interviewed for your job, did they ask you how you were going to choose between being a husband and being an attorney?" He said, "Of course not." So I I declined to answer the question. I ended up getting the highest rating from the four-man panel. I finally took the job with a state agency, since none of the private firms or businesses would hire women.

own careers and partly because married life of the traditional sort is less appealing to them.

The women's movement has also taken a stand against many other areas of gender discrimination. Some of these have changed relatively rapidly once the pressure was turned on. In 1965 it was difficult or impossible for women to get credit cards in their own names or to get a loan from a bank. Now women have access to consumer credit. It used to be common for restaurants, clubs, and bars to publicly announce a "men only" policy. This kind of segregation, too, has changed, although it hangs on in some quarters. For example, the U.S. Junior Chamber of Commerce, which started admitting women in the late 1970s, reversed itself and attempted to become sex-segregated again in 1982. Cases of this sort remain in litigation during the 1990s.

Other areas of gender discrimination remain. Women still make up less than 5 percent of high-level government and business officials. At somewhat lower levels, the number of women holding elected office has gone up, from only one in twenty state and local officials in 1975 to about one in five today. As a result of elections held in 1992, dubbed "the year of the woman," the number of women who were U.S. Senators or Congressional Representatives nearly doubled from 30 to 53 (*Statistical Abstract* 1993). Women get paid a

There were a lot of restaurants where attorneys and business people gathered at lunchtime that didn't allow women to eat there. One of them was a restaurant that was located on land leased from the city, publicly owned land. The lawyers from my office and I went over there for lunch one day. There were five men and me. When we got there they wouldn't seat us in the businessmen's area, because I was along. They told us, though, it was because we didn't have reservations. So one of the men in our group phoned from the restroom and made a reservation for five minutes from now. Five minutes later we showed up again, and they couldn't avoid seating us. Afterwards I wrote a letter to the president of the corporation that owned the restaurant, pointing out they were on public property.

After some go-round on the legal matters, they changed their men-only policy.

Another place was a hotel that had a grill that was popular for lunch among business people and politicians. So I and two other female attorneys made a reservation for lunch. We walked in and the maitre d' told us we couldn't be seated because it was for men only. We told him if he didn't seat us he was violating the Civil Rights Act. He told us we wouldn't like eating there because of all the bad language men used. He actually put his hands on me and tried to push us out. We held onto the doorjamb. Then one of the other women said, "I'm with the attorney general's office and I want to eat now, because I have to be back in court." Finally he gave in. After that we didn't have any problems there. [From the author's files]

good deal less than men for similar kinds of work; across the board, working women make about seventy cents for every dollar a man makes (see figure 1.3). One might predict that if the feminist movement is successful in raising female wages up to the male level, women will have even less incentive for marrying and current trends in the family toward fewer and later marriages will become even stronger.

The women's movement obviously has a long way to go to reach its goals. But one reason to expect that it is going to continue having an effect over the long run is that many men support it. In fact, on some issues men are even stronger supporters of feminist goals than women are. Not all women favor the feminist version of equality. As we can see from figure 5.9, about 14 percent of women even recently would not have voted for a woman for president. But the opposition has been going down since the 1930s, when 60 percent of women and 75 percent of men were opposed to having a woman president.

Conflict over Family Change

Some of today's population not only does not support feminist goals but is actively opposed to them. This group is most heavily represented among the

FIGURE 5.9: Rising Acceptance for a Woman President: Percentage Who Would Vote for a Woman for President, by Sex, 1937–1991

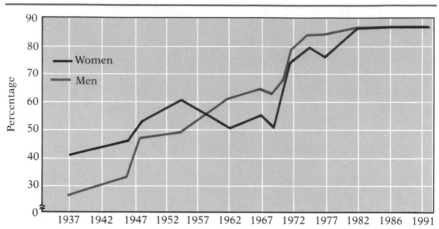

Source: 1937–1969, George H. Gallup, *The Gallup Poll: Public Opinion 1935–1971*, vols. 1 to 3 (New York: Random House, 1972); for 1972–1991, National Opinion Research Center, General Social Surveys.

older, less educated, and poorer people and in the rural areas of the country. Current estimates place it at about 25 percent of the population (Grigsby 1992; Liebman and Wuthnow 1983). It, too, has become politically organized in recent years, sometimes in religious form, such as the Reverend Jerry Falwell's Moral Majority. These groups were formed largely in reaction to the political and legal victories of feminists in the 1970s and to protest against changing trends in family structure. Just as the feminists have used politics to press for antidiscrimination laws, the traditional "profamily" movement has fought for legislation that would get women back in the home, reestablish the old ideal family form, and reinstitute some of the traditional taboos.

Social movements often operate in opposing pairs of this sort. The women's movement of the 1970s gave rise to a countermovement that favors an ultra-traditional family. In the 1920s, there was the Prohibition movement, which had its strength in rural and small-town America (Gusfield 1963), against which the Jazz Age culture rose in rebellion. Society is never uniform and homogeneous, and different classes and social sectors often go in different directions or openly conflict with one another. It is not surprising that today there are both feminist and antifeminist movements.

Chafetz and Dworkin (1987, 1989), comparing data on where feminist and antifeminist movements have emerged around the world, propose a theory that accounts for these movements. Feminist movements emerge when there are favorable economic opportunities for women. (This is in keeping with the economic theory of gender stratification presented in chapter 3.) Women's educational level has been rising, and so has the demand for white-collar labor. As women move increasingly into the labor force, there are more middle-class women with the resources to organize a social movement attempting to break down gender disadvantages and discrimination.

Status Conflict among Women

But not all women are equally well situated to take advantage of these new opportunities. Some women lack the education or the job chances to move into nontraditional occupations and to earn higher incomes. Other women have already organized their lives around the traditional pattern of wife, homemaker, and mother. To the extent that the feminist movement succeeds in redefining women's roles, this traditional group comes under pressure. Their roles are devalued by the new standard that defines a modern, successful woman as someone who has a professional or management-level job that was formerly reserved for upper-middle-class men. The entire status structure of stratification has changed, and traditional women are losing the idealized status that used to surround them as homemakers for their husbands and mothers for their children. This group is the source of an antifeminist backlash.

Antifeminist feelings thus are really a form of status conflict, arising from the changes in the structure of gender stratification. In the traditional structure, women's class position was seen by almost everyone as attached to their husband's position (or their father's, when they were younger). Now independent career women are pushing to define their own class positions in addition to those of men they might be related to. For some women, especially the group from whom the feminist activists are most likely to come, this means a push upward, into the upper-middle class. For other women who are unable to keep up, their class positions are reduced in the harsh light of the new standard. Hence conflict breaks out over the standards by which women are to be judged: on their own, according to their personal occupational success, or by the older tradition of making a home for a man. The antifeminist movement that recruits from the traditional group has become active on various political and legal issues. But its real concern is to defend the traditional status ideal; it fits the model of what Gusfield (1963) calls a "symbolic crusade."

Universalistic versus Traditionalistic Ideals

There is another way of looking at the kinds of ideals that are defended on both sides of this conflict. In each case, we should see the social movement as like a ship, floating on a much larger tide. The movement is the organized, self-conscious part of the female population; underneath are much larger numbers of women who are simply responding to the economic and social realities around them. On one hand, there are the women responding to the new economic opportunities, especially in the elite professions and in higher education. These women tend to be situated in social networks that are *cosmopolitan*, open to diverse experiences. As Chafetz and Dworkin put it (1987), they are in closer contact with men's occupations and take these as their reference group. Using the theory of interaction rituals (described in more detail in chapter 8), we would expect these women to have abstract and universalistic ideals. This is reflected in their belief in universal rights and abstract philosophies to justify the improvements they are seeking.

Antifeminists, on the other hand, are on board a ship floating on a very different tide. Their movement consists of women who are largely *encapsulated* in traditional female roles and spend most of their time in situations where they have little contact with more cosmopolitan occupations. Here we can say that the social structure is one of *high ritual density*; instead of being in cosmopolitan networks, these women are confined largely to the local routine of family, neighborhood, or gender-segregated job. The theory of interaction rituals (Collins 1988, 187–226) predicts that there is a high emphasis on conformity to traditional, particularistic symbols. Instead of dreaming of ideals of universal equality or the ascendance of feminist values, these women give their loyalty to their family or their community tradition. Since religion is one of the strongest examples of a set of traditional symbols and rituals, it is not surprising that antifeminist movements are frequently attached to the most traditional churches. In fact, such churches draw on the same social base as do antifeminist movements.

Who's Winning?

The 1980s were a period in which the women's movement, after its spectacular growth in the previous decade, appeared to slow down. Antifeminist movements mobilized around particular issues, such as opposing the Equal Rights Amendment (successfully) and becoming involved in the crusade against abortion. On these political and legal issues, particular events shifted from year to year, and no doubt will go on shifting somewhat unpredictably through the 1990s. It is important to bear in mind the image of two ships floating on two different tides. The female population, like modern society in general, is stratified in complex ways. Neither of the two big "waves," or segments, is going to disappear. But the long-term flow of the tide has been in the direction of expanding the well-educated, middle-class population of women and decreasing the segment that consists of full-time housewives, and perhaps even the portion in traditional gender-segregated jobs.

Between the feminist and antifeminist movements, we would have to say that the initiative has always come from the feminist side, while the opponents tend to react against them. Chafetz (1990) points out that the strongest determinant of an antifeminist movement anywhere in the world is the fact that a feminist movement already exists. And since the feminist movement itself emerges because the underlying economic tide has already started to flow, the existence of these movements means that changes are already under way. As Chafetz argues, women's movements act primarily to expedite a process already in motion.

What about the role of men in all this? On the overt level, men react in various ways. Some attempt to protect their existing advantages, holding on to male-dominated jobs. As we will see in subsequent chapters, men continue to use their balance of resources when they can to keep power in the home (chapter 11); at times, when violence is their only resource, they may use that in ugly outbursts of spouse abuse (chapter 13). Since men are the dominant members of traditional organizations, including religious and political ones, it is not surprising that men should be prominent in the leadership of organized antifeminist movements and other conservative groups (Chafetz and Dworkin 1987).

But even this is subject to social forces for change. We can cite several ways in which this happens. Men are also very prominent in the kinds of cosmopolitan networks, in the professions and educational institutions, that promote universalistic values. So there are many men who believe in the

One issue that has polarized women is abortion rights. Abortion is a particularly explosive issue because childbearing, an activity traditionally identified as a woman's primary role, can interfere with her ability to take advantage of economic opportunities. Pro-choice rallies such as this one held on Boston Common support the belief that women should be able to control their reproduction.

values of gender equality (to what extent they actually put them into effect in their own families, however, we shall soon see). Perhaps even more importantly, these egalitarian ideals are also reinforced by the way the class structure has been changing. The most important determinant of stratification now has become whether a family has one or two income-earners. Large numbers of middle-class men are strong supporters of women's rights, because these rights make a huge difference to their own life-style. The most important way that a family gets into the upper-middle class is to have two middle-class careers in it (as we saw in chapter 1). Hence, many men now have a stake in gender equality too. This is one more reason why we may expect that the move toward the modern family structure will continue.

SUMMARY

1. The present is a time of considerable changes in the American family. Rates of premarital intercourse, nonmarital childbirth, and divorce have gone up. People are putting off marriage to later ages, and the birthrate has declined. Some observers have suggested that the family is in the process of disappearing.

2. However, periods of major change in the form of the family have occurred before in history. We seem to be going through another important period of family change rather than seeing its disappearance.

3. Some new variants on family structure have become important, including the temporarily cohabiting couple, and the string of marriages and divorces sometimes referred to as "serial monogamy."

4. The main trend is toward diversification of family forms rather than a single standard form. The nuclear family of father, mother, and minor children now makes up only about a quarter of all households. The most common form consists of a married couple living with no children.

5. The 1950s, characterized by early marriages, the "baby boom" in birthrates, and a declining divorce rate, were an anomaly in the long-term trends of the twentieth century. One possible cause may have been the movement of working-class families, for the first time, into the *physically middle-class* single-family dwellings of suburbia, while importing working-class family styles.

6. The 1920s and the 1960s were comparable periods of liberalization in sexual manners and family formation. In both cases, the new style began in the youth culture and spread rapidly to older age groups.

7. Family changes of the last two decades are also connected with the "second wave" of the social movement for women's economic and occupational rights. The increasing mobilization of women through higher education, and their shift into full-time careers, are the major factors producing changes in the structure of the modern family.

8. Women in middle- and upper-middle-class careers have redefined women's status as depending upon occupational success. Traditionalist and antifeminist movements have emerged in reaction, to defend the status of traditional female roles. Women in cosmopolitan networks favor universalistic ideals of equality, while women encapsulated in gender-segregated roles conform to symbols of loyalty to family and community.

9. Feminist movements appear when economic conditions have already begun to change, and attempt to accelerate the change. The weakness of antifeminist movements is that they are reacting against changes already occurring. Men take part on both sides of these conflicts. Because contemporary stratification now depends on the economic advantages of two-income families, many men now have a stake in the gender-egalitarian family as well.

6

FAMILIES AND SOCIAL CLASS

One of the most important facts about families is that they are not all alike. There are considerable differences in family lifestyles, and among the most important of these are differences due to social class.

What do we mean by *class*? Classes are different categories of persons, usually thought to be socially unequal. It should be admitted right away that sociologists are not in agreement about what measures should be used as the bases for class ranking. Should they include income, education, social prestige or status, occupation, ownership or nonownership of property? Our belief is that all of these should be taken into account; in other words, we should deal with a multidimensional theory of social class. But for purposes of introduction, it is easy enough to reduce most of these to a single dimension, since persons possessed of one indicator of a certain status tend to have the other characteristics as well.

The basic idea of this chapter is that family life and family structure are profoundly affected by the economy. This is most evident when we look at economic disruptions such as the industrial plant closings in the American "steel belt" during the 1970s, the farm foreclosures in the American midwest during the 1980s, and the layoffs in the defense and aerospace industries in the 1990s. Extreme changes in the national or local economy place enormous burdens on families to survive on limited resources. But the economy also influences family structure and family interaction when it is running smoothly. How families react to localized economic conditions and how they are stratified by their positions in the larger economy are the subjects of this chapter. We will first take a look at the simplest means of ranking families—in terms of income. Then we will examine the lifestyles of two very different kinds of families: working class and upper middle class.

STRATIFICATION AND WEALTH

The most obvious basis of social stratification is how much money people have. This is surprisingly little analyzed by sociologists, who tend to regard money as too vulgar a measure. It is often pointed out, for instance, that money is not equivalent to status. The truck driver who earns $40,000 per year will still have lower status than the schoolteacher who earns $20,000, and will have a different lifestyle as well. Similarly, radical Marxian theories emphasize that the quantity of money is less important than how one earns it: wage earners, however much they make, have a different class interest and different class conflicts than business owners. These points are valid enough, but they are often overstated. The plain fact is that one of the most obvious differences among people is how much money they have, and money itself has major effects upon lifestyle, family crises, politics, and almost everything else.

Even though psychologists and other scientists try to identify generic patterns of family life and consistent stages in human development, all these processes are influenced by the availability of money. For example, how does a child's capacity for thinking develop? The amount of money a family has affects things like the quality of housing, the types of toys in the home, and the number of books available to a child. Because of money—or the lack of it—a child's perceptual, cognitive, and motivational development will be different, even when two children start out the same or when their parents use the same childrearing techniques (Elder and Caspi 1988). As we will see in this and later chapters, however, parents from different social classes also tend to act differently, primarily as a result of environmental factors linked to the availability of money.

Most of what we know about influence of money on families comes from studying the poor. Researchers have consistently found that the life chances of the poor are jeopardized by their lack of financial resources. Eitzen (1983, 65) reports that those with fewer economic resources are more likely to have premature or mentally retarded babies from prenatal malnutrition; they have a shorter life expectancy; and they are more likely to die from accidents, tuberculosis, influenza, pneumonia, and cancer of the stomach, lung, bronchi, and trachea. They are also more likely to get sick and stay sick longer because of deficiencies in diet, sanitary facilities, shelter, and/or medical care. In addition, people with limited incomes are less likely to go to college and are more likely to be arrested, found guilty, and given longer sentences for a particular violation. Finally, families with limited incomes are more likely to experience child abuse, spouse abuse, divorce, and desertion (Zinn and Eitzen 1987, 100).

Lack of resources in poor families clearly limits one's life chances, but an abundance of resources is equally consequential. It is an oversimplification to claim that the upper and middle classes derive their status more from culture than from wealth, for the marks of an upper- or middle-class lifestyle—taste in clothes, housing, and entertainment; educational attainment; familiarity with the arts; and savoir faire—can largely be acquired with money, especially over generations. The upper classes usually have a high level of culture because their families have had money long enough to spend it on education, art, the opera, and so forth (see feature 6.1). The truck driver who earns $40,000 a year has lower prestige partly because his (or her—but almost certainly his) income has been earned only recently, and he has not had time to translate it into a genteel lifestyle. Also, the income earned by the working class is generally much less secure; the truck driver may have a few good years, but he often has a short earning span that is vulnerable to sudden disruptions. Middle- and upper-class occupations have more financial security built into them. One of the main differences among social classes, then, is money, except that we should bear in mind that *long-term wealth has a much stronger social effect than short-term wealth.*

FEATURE 6.1
What Is an "Old Family"?

Members of the upper class like to pride themselves on belonging to *old families*. But stop and think about it: all families are equally old, since everyone has grandparents, great-grandparents, and so forth as far back as you wish to go. What they are really saying is that their families have had a lot of money for several generations. This is never stated openly, because claiming status on the basis of their money would be considered the sign of an upstart, a *nouveau riche*. Part of upper-class culture is the belief that talking about one's money is vulgar. Instead, they like to claim that their superiority is based upon their better taste, their contributions to charity, their "public service," and above all on their breeding and their families —as if they were the only ones who had families more than a generation back. Translated, this is a covert form of boasting, since families who have a great deal of money can invest it in education, travel, acquiring a taste for the arts, and other learning experiences that set them apart from people who have merely money and not "background." Of course, such an investment sets them apart even more from people who have neither money nor the kind of culture that one can buy with it.

Has Inequality Been Increasing?

For the most part, the world's wealth is concentrated in the hands of very few families. Unequal distributions of wealth are characteristic of democracies like the United States as well as monarchies like Saudi Arabia. In the United States, about one-half of one percent of the population owns 20 percent of the combined market worth of everything owned by every American, and the top one percent owns about a quarter of all wealth in America. Contrary to popular belief, the amount the wealthy own has remained about the same since the 1940s (Zeitlin 1978). For some forms of wealth, the concentration is even more marked. For instance, 2 percent of all families who own financial assets (such as savings and government bonds) hold over half of such assets, and the top 10 percent hold almost 90 percent of the total (Zarsky and Bowles 1986). We know something about the most visible of these families—the Rockefellers, DuPonts, and Mellons—but a look at the Social Register of every major American city would yield additional less affluent, but fantastically wealthy families. By controlling corporations and influencing the policy process, these families are able to maintain their economic and social position (Domhoff 1990; Parenti 1983). (Feature 6.3 presents a brief look at some aspects of elite upper-class families.)

In spite of the fact that so much wealth is concentrated in the hands of a few, we tend to assume that most Americans are sharing a bigger portion of the economic "pie." That seemed especially true when the size of the total pie was increasing. From 1940 to 1973, real earnings of Americans improved steadily, but since 1973, real earnings have stagnated and declined. As a consequence, many Americans are less well off than their counterparts born a decade earlier. When incomes are adjusted for inflation, the average weekly earnings of all American men have dropped significantly since the mid-seventies. Real weekly earnings for women also dropped during the 1980s, but in recent years they have increased (England and Browne 1992). During the last three decades we have seen marked increases in numbers of women employed, but also marked increases in unemployment rates for both men and women. For the first time since the Great Depression of the 1930s, American men face the prospect of lower lifetime real earnings than men born ten years before them.

The economic situation for working-class families seems to be diverging more and more from that of middle-class families as we move into the 1990s. Most observers agree that the life chances of working-class men have deteriorated over the past twenty years, even though working class women have relatively better job opportunities. As noted below, the new class divisions increasingly revolve around the number of workers in each family.

How Families Make Up the Income Hierarchy

In spite of popular rhetoric to the contrary, it seems that the rich keep getting richer and the poor keep getting poorer. In the 1980s and early 1990s, income inequality actually increased. And the changes were not even primarily the result of increasing unemployment, but stemmed from stagnant or declining incomes among employed people. In the 1980s, the number of full-time, year-round workers with low earnings rose significantly, and the real mean income of the majority of households either went down or stayed the same (U.S. Bureau of the Census 1992b; 1992c). The share of aggregate income earned by the top two-fifths of all American households rose between 1975 and 1990, whereas the share earned by the lower three-fifths stayed about the same or dropped (see figure 6.1). Look at figure 6.2. It shows the way income is distributed among families in the United States, and gives the occupations of both husband and wife. It also shows people who are not in the labor force (housewives and househusbands, the unemployed, and retired persons), and it indicates whether families consist of married couples or single family heads (widows and widowers, divorced persons, unwed mothers, etc.).

Several features should strike your eye. First, you might notice that occupations are distributed across the various income categories. Managers and professionals dominate the higher income brackets (and go way up to the top: this part of the income distribution goes up so high, relative to where most people are, that it would go far off the top of the page); but some managers and professionals are much further down, some even below the poverty line. This is partly due to the fact people at the beginning of their professional careers may barely eke out a living; it also reflects the fact that some business managers (like the owner of a small cigar stand) may not be employed in a very lucrative

FIGURE 6.1: The Rich Get Richer: Real Mean Income of Each Fifth of U.S. Households (adjusted for inflation)

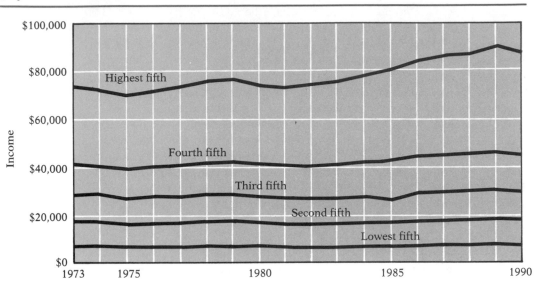

Source: U.S. Bureau of the Census, 1992b, *Current Population Reports*, Series P60, No. 183. Studies in the Distribution of Income.

establishment. Other occupations follow similar patterns. Each occupation clusters around some income level, with white-collar jobs higher on the average than blue-collar jobs. Some blue-collar families, however, especially those of skilled workers (plumbers, electricians, heavy-equipment operators), are in the higher income brackets.

A second feature to notice is that the income structure is not a triangle, widest at the bottom and narrowing toward the top. Instead, it is a little narrower at the bottom, bulges out toward the lower middle, then shrinks inward until it reaches the higher income brackets (upper-middle class and above: in the early 1990s, about $100,000 and up), when it suddenly becomes a tiny spike. One fact that this demonstrates is that the poorest people in the United States are not the majority. That is one reason—though there are others—why they have not been a revolutionary force, or even able to claim very much support in the form of welfare and other social reforms.

Third, notice how important it is for people to be in the labor force. The two poorest segments of the population are heavily made up of people who are retired, unemployed, or on welfare. Social security and welfare payments do not amount to much, despite some well-publicized cases that make it seem as though welfare recipients are driving Cadillacs and living in luxury. Whatever welfare fraud there may be, most people are better off working.

A fourth point worth noticing is that marriage can make a tremendous difference in economic standing, especially when both the husband and wife are employed. The poorest group on the chart consists almost entirely of families headed by a single adult (mostly women). Many of these people are also outside the labor force, and that makes them doubly poor. The next category up also contains many households with only one adult. This person is working, but there is only one income.

When we look above the medium budget line on the chart, a striking fact

FIGURE 6.2: The Stratification of Families by Income in the United States

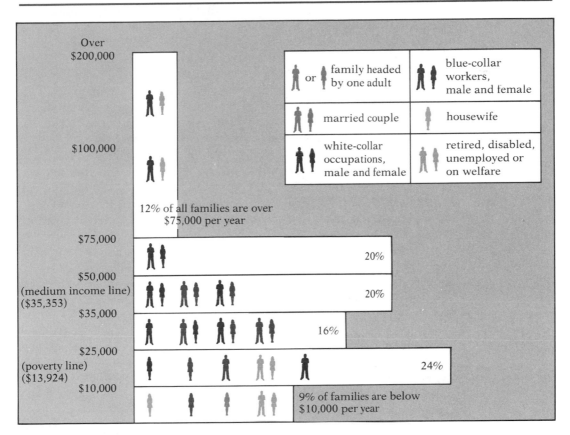

Figures in the chart show the most typical family pattern at each income level.

Source: *Statistical Abstract of the United States*, 1992, Nos. 724 and 697; U.S. Bureau of the Census, 1992h, *Current Population Reports*, Series P60, No. 181.

emerges: most middle- to upper-middle-income families in the United States have two incomes. Notice, for instance, that quite a few blue-collar workers (especially skilled workers) fall above the medium budget line. Virtually all of them are there because both husband and wife are employed, usually with the woman in a clerical job. But the same is true for white-collar families. Male professionals, managers, and clerical/sales workers are most likely to be in the middle to upper-middle range of income if their wives are employed. If both husband and wife are working in high-income white-collar jobs—managers and professionals—their combined incomes put them not merely into the middle class but into the affluent reaches of the upper-middle class.

Some families maintain the traditional pattern, with the man as sole breadwinner and the woman as housewife. Such arrangements are found in both white-collar and blue-collar families. But both of them pay a price. The blue-collar family in which only the husband has an income is almost always doomed to stay in the bottom half of the income distribution; and the white-collar family has a good chance of not making it out of the lower-middle-class income bracket. Of course, unemployed wives do make economic contributions that do not show up on income statements, and some husbands and wives place great value on at-home wives' emotional contributions to a family, especially during children's early years. One might say that traditional men who do not want their wives to work, or wives who prefer the traditional role, are paying a serious economic price for their tradition.

There is only one section of the labor force where this is not true. That is at the top. Up on the needle of income distribution, where it rises into the salary range from an affluent $75,000 to $100,000 per year on into the stratosphere of salaries in the range of $200,000 and higher, families revert to the traditional gender-role pattern. Most of the earners in this income range are men in the managerial and professional categories, but they are not in ordinary managerial and professional positions. These managers are the corporate executives, the managers of the few hundred giant corporations that dominate the American economy. The professionals are the most successful practitioners in a few of the most prestigious specialties: medical doctors, lawyers in the elite law firms (which work especially for corporate clients), and a few others.

In this group, the men earn so much by themselves that it makes little difference whether their wives add to the family income. They can afford a traditional family lifestyle; as we shall see, quite often they demand it. The difference between the top of the income distribution and the great bulk of the families that make it into the affluent upper-middle levels by having two family incomes is one of the most important splits within the structure of stratification in America.

Fifth, the stratification system, to a considerable degree, is based upon gender. This is true in the direct sense that women are much more apt than men to occupy the low-ranking and poorly paid jobs. It is also true that if women who head households are not employed—because they are on welfare, retired, or unwed mothers—they almost always end up in the poverty class. Men are less apt to head households if they cannot provide—one might say the women are left "holding the bag." And in retirement, men are much more likely to have reasonably adequate pension benefits.

Families in Poverty

The number of Americans living in poverty has grown substantially since the early 1970s. The Federal government calculates the minimum level of income necessary to meet basic subsistence needs for different family sizes and types, and adjusts this limit every year to reflect changes in the Consumer Price Index. In the early 1990s, the average poverty threshold for a family of four was about $13,000. Using the official guidelines for each year, about 23 million Americans, or 11 percent of the population, lived below the poverty level in 1974. By 1990 there were about 34 million people below the poverty line, representing almost 14 percent of the population. When welfare and other government money transfers are excluded from the calculations, over 50 million Americans, or over one-fifth of the population, are below the official poverty level (U.S. Bureau of the Census 1992d). What's more, the poor may be getting poorer: the average poor family has persistently had a yearly income further below the poverty line each year (Jaynes and Williams 1989).

Families headed by women represent the fastest growing poverty category.

Families headed by women represent the fastest growing poverty category. By 1990, more than half of the families below the official poverty level in the United States were female-headed households with no husbands present.

By 1990, more than half of the families below the official poverty level in the United States were female-headed households with no husband present. The growth of female-headed households is the primary reason why greater numbers of people are susceptible to poverty than in the recent past. What's more, because female-headed households are likely to include children, four out of every ten poor people in the United States are now under the age of eighteen (U.S. Bureau of the Census, 1992d). In fact, the poverty rate for most age groups has declined or leveled off since the early 1970s, but it has increased for children. In 1990, one of five American children under eighteen was being raised in a family below the poverty level (U.S. Bureau of the Census 1992d). Since about 60 percent of the children born since 1980 will spend some time living in a single parent household, the proportion of the nation's poor who are children will continue to increase (Kahn and Kamerman 1988). The increasing likelihood that women and children will live below the poverty level has been labeled "the feminization of poverty."

As we will see in the next chapter, a disproportionate share of families living in poverty are racial minorities. In 1990, one in three blacks and more than one in four Hispanics were below the official poverty level. The situation was even worse for children, with almost half of black children and over a third of Hispanic children living in poverty, as compared to fewer than one in five white children (Bianchi 1990). Even though blacks, Hispanics, and people of other races are more likely to be poor than whites, about two-thirds of all people below the poverty level in 1990 were whites (U.S. Bureau of the Census 1992d).

Family Events Cause Income Fluctuation

Researchers have found that when social class is defined in terms of income, movement between social classes is fairly common. For instance, only about half of families in the bottom fifth of the income pyramid remain there a decade later (Duncan et al. 1984). Differences in skill level or personality traits of family members accounted for little of the mobility, and events such as unemployment or disability accounted for only a portion of the change in individuals' incomes. The single most important factor accounting for changes in family well-being was a fundamental change in family structure: divorce, death, marriage, birth, or a child leaving home. In other words, changes in economic position of families are highly dependent on changes in the composition of families. When someone gets married, he or she often changes from living in a one-earner household to living in a two-earner household. When couples have children, parents often don't have time to work as many hours at their jobs. When people divorce, or become widowed, they typically lose the benefit of having the other person's income.

While family composition changes provide the best explanation of changing economic position, other events are also important, and reflect a certain fluidity in income levels. For instance, movement into or out of the labor force, such as being laid off from a job or taking a new one, has substantial effects on family income. Because women's hourly pay rates and employment hours are so much lower than those of men, women's movement into and out of the labor force produces an average change in annual income that is only about a quarter of that produced by men's movement into or out of the labor force (Duncan et al. 1984). For those who remain in the labor force continuously, large variations in economic situation reflect changes in work hours brought about by second jobs, overtime, job changes, and brief spells of unemployment.

Even though income changes occur frequently, most people maintain the same overall class position. That is, those who hold working-class jobs may go through bouts of unemployment or overtime work, but they are unlikely to move to upper-middle-class jobs. Similarly, entry into the elite upper class is severely restricted and the wealthiest families control about as much of the country's wealth as they did a half-century ago. For that reason, sociologists talk about stratification by social class as a relatively stable phenomenon. Economic changes for most families are the result of having one earner or two earners in the family, with the majority of families now composed of two earners. In addition, changes in family composition, particularly through marriage or divorce, are the most important precursors to changes in the economic position of families. As we shall see in later chapters, the downward mobility experienced by women after divorce, particularly if they have not been previously employed, reflects a profound change in class position.

THE MIXTURE OF SOCIAL CLASS STYLES IN AMERICA

We must face one important difficulty in dealing with social class in American society. Except for the very rich and the very poor, social class differences are not highly visible now, in the latter part of the twentieth century. Between the extremes, there are no sharp barriers between different occupations or income levels. The neighborhoods or suburbs in which people live are not usually divided into "working-class areas" and "middle-class areas," as was true earlier in our history and as is still likely to be the case in Europe. A majority of American manual workers own their own homes, and the areas in which they live include a proportion of white-collar workers as well, and very likely some business owners (perhaps of small businesses) and professionals and technical people of various levels (Halle 1984; Gans 1967). Only the upper-middle-class and upper-class neighborhoods are likely to be class segregated, consisting of expensive homes of the most successful lawyers, doctors, administrators, business managers, and others. Since neighborhoods are likely to be divided by race, Americans tend to be much more race conscious than class conscious.

Similarly, today's Americans tend to dress in much the same styles of clothes, no matter what their social class. At least this is so when they are not working. Factory workers, delivery truck drivers, or repair persons may wear special work clothes, not at all like the suits worn by business executives. But their leisure outfits are much the same. Moreover, some high-level personnel (scientists, professors, engineers, mass media personnel) tend to wear casual clothes at work, as do most college students. Though there are subtle differences in styles, it is no longer possible to tell people's social class at a glance, except when they are actually in their work situation. Furthermore, our leisure-centered society tends to have much the same culture shared by most people. The same professional spectator sports are followed by most fans; the same popular music, television shows, and movie stars are widely admired across social class lines (Halle 1984).

It is true that there is also a "high culture" style of reading "serious" books and going to the opera, the theater, art museums, and "art" movies. But even in the highly educated section of the upper and middle classes, only a *minority* of persons participate much in this "high culture"; the popular cultural style of TV, professional sports, and pop music is most widespread even in these social classes (Robinson 1977; Halle 1984; Gans 1974; Wilensky 1964; Csikszentmihaly and Rochberg-Halton 1981). At most, we can say that there is a distinctive "elite" culture only within a highly educated fraction of the upper-middle class, consisting especially of people who are themselves cultural specialists (teachers, musicians, aspiring artists, and the like) (DiMaggio and Useem 1978; 1982). For the rest of the population, blue collar, white collar, and managerial alike, there seems to be little in the way of class difference in cultural lifestyle. This is one reason why Americans are not very conscious of social class, except at the extremes.

Does this mean that social class is no longer of any importance in explaining how people will behave and think, and what kinds of family lives they have? That conclusion would be too hasty. Social class factors continue to operate, even if people are not aware of them as such. The economic pressures of different kinds of career trajectories have a strong effect on such things as the timing of one's marriage and the kinds of family crises that occur. And there are subtle cultural patterns that affect child-rearing styles, sociability, and the status concerns of the family members. We shall see, for example, that the higher social classes tend to be career-oriented and to subordinate everything else to their careers, while the working classes are more detached or alienated from their jobs, and stress the leisure and personal sides of their lives. The higher social classes tend to be involved in cosmopolitan social networks, while the working classes are more localistic and kin-centered; different styles of thinking tend to follow from those social patterns. We shall examine these differences in more detail in this chapter.

Are White-Collar Jobs Middle Class?

There has been a debate among sociologists about how "working class" and "middle class" ought to be defined. Our ordinary terminology distinguishes "blue-collar" and "white-collar" jobs: on one side are people who do manual work, usually wearing special work clothes because they get dirty on the job; on the other side are people who work in offices at clean jobs in their ordinary (if somewhat dressed-up) clothes. Nevertheless, the distinction may no longer be important. Skilled manual workers (electricians, mechanics, etc.) or those with strong labor unions often earn more than many white-collar workers. This is one reason why these manual workers can live in the same neighborhoods, wear the same leisure clothes, drive the same cars, and afford the same entertainment as white-collar workers. Moreover, the distinction between manual and nonmanual work may be more a matter of image than reality; many "white-collar" workers spend their time operating copying machines, typewriters, and computer keyboards and cash registers, while "manual" workers may be checking dials and switches and writing tallies. In short, the jobs may be different in their prestige and public image, but the actual work of both may be much the same: repetitive, boring, and subject to the authority of supervisors. Hence sociologists such as Erik Olin Wright (1979; Wright and Singleman 1982) have argued that the majority of lower white-collar jobs are really working class, while other jobs entail a mixture of class conditions.

If this is so, we should not be surprised that neighborhoods tend to mix white-collar and blue-collar workers, or that both of these occupations tend to wear the same styles of clothes and have the same culture. In other words, the "working class" in America is bigger than we would believe if we used the traditional criterion of manual occupations. Perhaps we need some new terms such as "blue-collar working class" and "white-collar working class." In fact, public consciousness has already arrived at this conclusion, except that it refers to all these people as "middle class" rather than "working class." This is an example of how status consciousness in the United States is skewed in an upward direction, allowing most people to inflate their relative position in their own eyes and in the public impression they make on others.

Husbands' and Wives' Occupations and Mixed-Class Marriages

Finally, we have the question of how we are to define the social class level of a marriage in which both husband and wife have jobs. The traditional method was to classify by the husband's occupation, as if the wife's position did not count in society's class rankings. This introduces a considerable bias: for example, if a male factory worker is married to a schoolteacher, it would be a mistake to consider the family as purely working class. And in fact this sort of combination is quite common. In the working class, wives typically have more education than their husbands, and many wives hold jobs in the lower professions such as teaching, social work, or nursing. Even more frequently, wives of blue-collar men have clerical jobs as secretaries, typists, or sales clerks. Hence the blue-collar/white-collar distinction is crossed in a great many families; a majority of what used to be considered "working class" families on the basis of the husband may really be only partly "working class."

Nevertheless, we have to consider this discussion in conjunction with the points raised in the previous paragraphs; that is, we must determine whether "white collar" occupations really ought to be analyzed as "middle class." We would suggest the following: that secretaries, typists, and most clerks are actually performing working-class types of jobs in office settings, and that we should consider them "white-collar working class" rather than "middle class." Wright (1979, 219) concluded that two-thirds of white women and almost three-fourths of black women actually hold working-class jobs. More recent census data on the occupations of employed women back this up (see table 6.1).

If we count this way, it turns out that most working-class husbands do have working-class wives, and there are not so many mixed-class marriages after all.

TABLE 6.1 Most Women Hold Working-Class Jobs

Occupational Classification in 1992

White-Collar Working-Class Women	
Clerical employees	27%
Retail sales	13%
Technical Support	4%
	44%
Blue-Collar Working-Class Women	
Blue-collar workers	10%
Service workers	18%
Farming, Forestry, Fishing	1%
Total Working-Class Women	29%
Middle- and Upper-Middle-Class Women	
Professional and technical	16%
Managers and administrators	11%
Total Middle and Upper-Middle Class	27%
Total employed women	100%

Source: U.S. Bureau of Labor Statistics, 1992. Employment and Earnings, vol. 39, no. 7, Table A-23.

Where mixed-class marriages do exist, though, we should recognize them as a separate category. For instance, there are still a substantial number of working-class men married to women who hold genuinely middle-class jobs, such as teaching or social work. To be honest, it should be admitted that sociologists do not have much precise information about these kinds of marriages, since most of the research (e.g., Rubin 1976; Komaravky 1962; Gans 1962; 1967; Halle 1984) has classified the family by the social class of the husband and has not systematically separated out wives who are in genuinely higher occupations. (As we have said, we do not think a bias was actually introduced by subsuming wives who were secretaries and clerks as working class.) A related problem, with which research has not specifically dealt, is *husbands* who are in what we have called the "white-collar working class"; they are more typically referred to as "lower middle class" and have usually been ignored in sociological studies of family styles. We suspect, though, that their lifestyle is similar to that of the manual working class.

Since clerical jobs are the single most common form of female employment, we should be aware of a rather different form of mixed-class marriage: middle-class husbands who are married to "white-collar working-class" wives. Although recently there has been a push for women to get into the higher-paid professions and into administrative and management positions, the occupational structure has been slow in opening up to women. Among employed women who are married to middle- and upper-middle-class husbands, clerical positions are still the most common work. Thus, there is a sense in which the class dividing line is most often crossed *downward* from husbands to wives, rather than vice versa. As we shall see, this is one reason why there tend to be strong power differences at home between upper-middle-class husbands and their wives and an underlying source of conflict within many families. And even when both husband and wife hold genuinely middle-class jobs, there is a strong tendency for the husband to outrank his wife, due to the general pattern of employment discrimination against women.

The structure of social class in the United States, then, is probably rather different than it appears on the surface. Several class levels and many possible occupational combinations exist for husbands and wives. There are poverty-level persons, those who are either out of the labor force or only sporadically employed. Then there are both blue-collar working-class and white-collar working-class, as well as middle-class (mixed authority positions), upper-middle-class, and upper-class occupations. Husbands and wives could be consistent—in the same class level—or inconsistent. Moreover, we should add cases where women are housewives—a position that sociologists have just begun to try to fit into the class structure. And finally, we should include single-adult families, where the household has only one source of class position.

Instead of diagramming all these possibilities, we shall use table 6.2, which shows only some of the most common class-types of marriages.

TABLE 6.2   Common Class Types of Marriages

Husband	Wife	Class Consistency
Poverty class	Poverty class	Consistent
Absent	Poverty class	Consistent
Working class (blue collar)	Working class (blue or white collar)	Consistent
Working class (blue or white collar)	Middle class	Inconsistent
Working class (blue or white collar)	Housewife	Consistent
Middle class	Working class (white collar)	Inconsistent
Upper-middle class	Middle class	Semiconsistent
Upper-middle class	Housewife	Consistent
Upper-middle class	Upper-middle class	Consistent
Upper class	Housewife	Consistent

WORKING-CLASS FAMILIES

It is not possible to describe all the different family types in this chapter. Instead, we have picked out two types from different parts of the spectrum: the working-class family and the upper-middle-class family. Moreover, we concentrate on the class-consistent types. These are working-class men (mostly blue collar, in the studies we have) married to women who hold blue-collar or white-collar working-class jobs or who are housewives. Furthermore, we should emphasize that the most vivid studies of family life (especially Lillian Rubin's *Worlds of Pain*) focus on the lower part of the working class, where there is more economic stress. For the other type, we deal with upper-middle-class families of the most traditional sort, where the man alone is employed. Since this group is similar to the upper class, we will add some materials about upper-class wives, who are one of the most culturally distinctive groups in the stratification of American families. We will also briefly discuss a relatively new, but increasingly popular upper-middle-class family type—the dual-career professional couple. The family types we have chosen are not at the extremes of the class system; they are all substantially employed, but they show the contrasting tendencies that exist beneath the superficial uniformity of cultural styles in late twentieth century America.

Early Marriages

What is family life like in this group? One notable fact about it is that working-class people tend to marry very young. Women marry around age seventeen to nineteen. Some marry even younger; most of the "child bride" stories that make for an occasional scandal in the newspapers are about working-class teenagers. Working-class males marry slightly older, but still a number of years before males of the middle and upper classes (Rubin 1976).

Why do they marry so young? For one thing, working-class youths do not usually go on to college; hence they have no special motive for putting off marriage. Moreover, they are usually not looking forward to an all-encompassing career. When they finish or drop out of high school, long-term career goals are often secondary, and going to work is not something they eagerly plan for (Rubin 1976, 155–84; Sennett and Cobb 1973). As compared to middle-class youths, working-class teenagers tend to name fantasy careers rather than realistic ones: boys want to be professional athletic stars, girls want to be singers or movie stars (Gottfredson 1981). When those dreams fade, and the excitement of the teen years is over, they are already full-fledged adults, and it is time to settle down and get married.

There is another reason for early marriage, an economic one. Working-class youths are eager to get out of their parents' homes (Rubin 1976, 49–52). Since their parents don't have much money, their homes are usually small and cramped; there is little privacy, and working-class parents often try to maintain strict control over their children. If working-class teenagers work, as many of them do, they are often asked to contribute their earnings to the household. (Working-class culture generally assumes youths will live at home until they

A young groom pours the first glass of champagne for his bride. For most working-class women, the wedding is the culmination of her dreams.

FEATURE 6.2
A Working-Class Woman's Image of Marriage

When I got married, I suppose I must have loved him, but at the time, I was busy planning the wedding and I wasn't thinking about anything else. I was just thinking about this big white wedding and all the trimmings, and how I was going to be a beautiful bride, and how I would finally have my own house. I never thought about problems we might have or anything like that. I don't know even if I ever thought much about him. Oh, I wanted to make a nice home for Glen, but I wasn't thinking about how anybody did that or whether I loved him enough to live with him the rest of my life. I was too busy with my dreams and thinking about how they were finally coming true. (Rubin 1976, 69)

marry, whereas middle-class culture allows and even expects more independence.) The solution is to get married. This is a legitimate reason for leaving home and spending one's income on oneself. It is also a reason for leaving school and any other adult controls over oneself. In short, it marks the beginning of independence. Among working-class women, in particular, marriage is a tremendous ideal; it is hardly an exaggeration to say that the single most important thing in a working-class woman's life is her image of herself as a bride. It fills her teenage imagination. After marriage, it is what she looks back on as the apex of her life (see feature 6.2).

The Early Marriage Crisis

After the wedding, there is usually a rude awakening, as Lillian Rubin (1976; see also Rainwater, Coleman, and Handel 1962) describes in detail. The couple may get their own home, but the early married years are almost always extremely hard. The couple experience a great deal of conflict, problems they never thought they would have to deal with. Middle-class marriage counselors have usually explained this as the result of marrying too young: the couple did not know each other well enough, and were not realistic or mature enough to cope with marriage. But such an analysis ignores the economic facts, for these are troubles especially found in working-class marriages. (It is fairly typical of middle-class psychologists to think that all problems can be resolved by adopting a middle-class lifestyle, without considering whether this is economically feasible.) The most serious source of the fights and unhappiness that are common at the beginning of working-class marriages is simply economic pressure.

This pressure comes mainly from two sources. One is that the husbands usually have low-paying jobs. The young man living with his parents who earns money at the gas station or factory feels that he has more than enough pocket money, and would be able to support himself if he didn't have to give it to his parents. But he finds that when he has to pay all the household bills (many of which are invisible to teenagers—insurance, loan payments, unexpected car and home repairs, and so forth), there is much less money than he thought. Added to this is the fact that working-class jobs are notoriously vulnerable to disruption by ups and downs in the economy; the young married man may find that his source of income is abruptly cut off when he loses his job.

A second source of economic pressure is that the wife's income is often rather quickly taken away from the couple. Most working-class young women have jobs before they marry, and the couple may expect that their two incomes together will make it possible for them to maintain their own home. The problem is that young working-class wives tend to have children very early in their marriages. In Rubin's sample of working-class California women, most were pregnant at the time they were married, and half of these had given birth by the time they had been married seven months. (In contrast, the group of middle-class women she studied had their first child three years after marriage.) Many of these young working-class mothers then left their jobs, since they hardly earned enough to cover the costs of child care and other work-related expenses.

This means that economic issues come to a head very soon. There is a tremendous feeling of letdown. Instead of living the happy, independent life they had imagined, the young couple feel trapped and may even fear that they are going under. It is this economic situation, not psychological problems, that sets off most of the fights and personality conflicts. Both husbands and wives experience a great deal of anger. Women attack their husbands for being poor breadwinners. The men are stung by this attack, since they accept the traditional role and feel demeaned that they cannot support their wives properly. Sometimes they respond by pulling "male prerogatives"—telling their wives to shut up, beating them, or going off to drink with their old friends of unmarried days. Such emotional escapes only reinforce the wife's viewpoint that her husband is irresponsible; and since he doesn't have the economic resources to back up his claim to male privilege, she continues to criticize him. Both sides feel done in, their dreams crumbled.

A baby's demands usually make matters worse. There are unexpected chores and pressures: getting up in the middle of the night to feed or quiet the baby, changing diapers, arranging the daily schedule around the new arrival. The parents' old independence is eroded even more. The baby also means added expenses, just at the time when the working-class couple is least able to afford them. To top it off, the wife spends much of her time and attention on the baby and tends to ignore her husband. In particular, their sex life tends to suffer, and the resulting sexual frustration creates more anger on the husband's part. Often, he is jealous of the baby. Numerous cases of child abuse arise out of such a scenario.

Fortunately, the hard times at the beginning of the marriage do not last forever. One solution is that the marriage simply comes apart. As we shall see in chapter 14, most divorces and separations occur soon after the marriage; the curve peaks right away and then gradually declines. Many working-class couples simply do not make a go of it.

The other common solution is for the couple to call on their families to bail them out. One of the most consistent findings about working-class families is that they rely on kin for economic and social support, a practice often labeled "pooling resources" (Bott 1957; Lein et al. 1974; Rubin 1976; Gans 1962; Rosen 1987; Stack 1974). Young working-class couples often live near parents and other kin and spend considerable time with them. This means that mothers, mothers-in-law, aunts, sisters, and other family members typically help out with tasks such as child care or shopping. In addition, they frequently will offer advice on the "best" methods of doing housework, disciplining children, or solving marital problems.

When a young working-class couple is forced to move back in with their

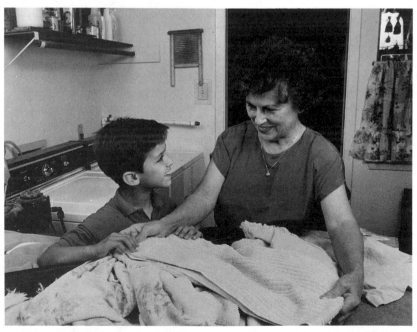

Working-class families tend to rely on kin for economic and social support. A grandmother, for example, may help out with tasks at home while the mother works at her daytime job.

parents, it is a bitter step to take, since they married in the first place in order to gain independence. But with a husband out of work, or a pregnant wife, or a young mother caring for a small baby, often the couple find they cannot afford to live in their own home or apartment. The young couple are a strain on their parents' household too, and carping over their money problems leads to many quarrels. Even without economic problems, the strain of depending upon in-laws can lead to a great deal of bitterness. The American folklore about the bad feelings between a man and his mother-in-law probably has a lot to do with this kind of situation, typical in many working-class marriages.

Eventually, the economic crisis passes. The husband finds work again, or moves to a better paying or more secure job. When the baby is older, the wife too returns to work and once more adds her income to the family's support. At some point, they are able to maintain their own home, and their sojourn with relatives is over. Nevertheless, the couple is still likely to rely on kin for ongoing help.

What we have just described, of course, is only a tendency. Not all working-class families go through this kind of economic crisis, though it is especially common in the lower-working-class families that Rubin studied. In the more highly skilled part of the working class (Mackenzie 1973; Halle 1984), there are fewer of these economically induced strains. As time goes on, many such families achieve relative affluence.

The Routine of Later Marriage

By their mid-twenties, a working-class couple has usually settled down. The pattern that emerges is a very localistic one. Working-class couples tend not to go out very much. Their social life together consists largely of visiting relatives. This is one very noticeable difference between the working class and the middle and upper classes (Fischer 1982; Bott 1957). One reason is, of course, economic: working-class people simply have less money to spend on entertainment. Young married couples often look back nostalgically to their teenage days, when they went to drive-in movies or to fast-food stands or listened to music more often. Once married, they spend most of their time at home, watching television.

Nor do they regularly invite other couples over to have dinner with them; the dinner party is a middle-class, not a working-class, custom. There is little male/female mixing. Both sexes tend to socialize predominantly with their own friends, and in different places. The house is usually the woman's sphere. She may drink coffee there during the daytime with neighbors or, more likely, female relatives. The men, on the other hand, tend to congregate outside the home. Sometimes they go to bars with their friends, or, even more commonly, spend their leisure time outside the house with other men, often fixing their cars. Sports consume much of the leisure time of working-class men. The men may be involved as participants in baseball, bowling, hunting, and fishing, or as spectators of professional teams. Often they are involved in gambling (Halle 1984).

Many working-class families, at least on the surface, are rather strongly dominated by men (Komarovsky 1962, 220–35; Rubin 1976, 93–113). In the families studied by Rubin, the men insisted on making the major family decisions themselves: what car to buy, when and where to go on vacation, whether to buy a house, and how to spend money. They often turned over the family checkbook to the women, but only because paying bills was a chore and there usually wasn't any discretionary money to be spent. Women were expected to care for the children and do the housework, without much help from their husbands, even when they were working at paying jobs. And, in fact, most such working-class wives did have jobs, usually part time. Some men, however, would not allow their wives to go to work. They felt that as long as a man could support his wife, it would reflect on him if he let her work. Some of these men would not even let their wives go back to school.

These working-class couples felt a great deal of dissatisfaction, especially the women. There was often a feeling of being cramped. Women said they were trapped in the house with their continuous housework. They complained that their husbands did not want to go out, even now that they were more secure and could afford it. They only wanted to get home from work, putter with their cars before dinner, and then watch TV before going to sleep.

In other cases the men did not come home after work but went out to drink with their friends. For the women in Rubin's study, this was worse yet than feeling trapped at home. Many of these women were fearful of alcoholism. They remembered their own fathers coming home drunk and abusing their mothers. They constantly repeated the gossip and folklore about men who took to drinking and "running around," who stopped supporting their families and ran off with another woman. And much of this may well have been true, although such behavior likely could have been a reaction to the economic pressures of working-class life, such as the shame of being unable to support one's family properly.

Most of the women in Rubin's sample were economically secure, although not prosperous. There was little reality to fears of their husbands' being unable to support them, at least at that time. But the "action crowd" at the bars was a constant image before their eyes. The women wanted to get out of the home, but they feared the milieu that their husbands went into when they got out. For them, the family circle was the only secure place. We find the same thing in other studies. Herbert Gans (1962) described working-class life as sharply divided between the routine of the self-enclosed family setting and an "action crowd," mostly male, that frequented bars and sporting events and went in for a culture of gambling, fights, and sexual prowess. Similarly, Halle (1984) found that about half of the oil refinery workers he studied (a rather well-paid

Working-class men often congregate at bars for "happy hour" after work.

working-class group) spent time at their favorite bars as part of their daily routine, instead of going directly home from work.

To an extent, working-class people pass through both milieus: the typical teenage experience is the "action scene," which is abruptly broken up by marriage. The women, in particular, get out of that environment and try to keep their husbands from ever returning to it, even though they themselves often look back nostalgically on a time when their lives were more interesting. For the men, the "action scene" is an escape from their wives and home life and the routine of their jobs. It is a place where they gather with their old friends and remember what it was like when they were free (Le Masters 1975).

Conflicts Between Male and Female Cultures

The settled family life of the working class, then, involves a rather severe split between men's and women's cultures. Rubin found there was little communication between husbands and wives, especially on personal and emotional matters. The children dominated the talk at the dinner table, with parents asking them about their school day or reprimanding them for their table manners. Then the family settled down to watch TV until bedtime.

Halle (1984), too, found this split between working-class men and their wives. (See also Locksley 1982.) He found that 40 percent of the men were unhappy with their marriages and another 25 percent said their marriages were somewhat happy; only a third said they had happy marriages. These figures should make us aware that the often-cited public opinion surveys tend to be superficial. At the time when Halle was making his study, about 70 percent of Americans were reported as saying their marriages were "very happy," and only 3 percent would admit they were "not too happy" (*Public Opinion*, 8 [1985] Aug./Sept.: 24, and Oct./Nov. 23). It appears that when asked a brief question by a stranger, most people tend to give the conventional answer, whereas Halle, who spent seven years getting to know his male subjects well, found much higher levels of unhappiness. (And, apparently, so did Rubin with her working-class wives, although she did not report statistics.) Moreover, the accounts of marriages by field researchers (like Rubin, Komarovsky, Rainwater, and others) frequently have not picked up the depths of the split between husbands and wives, because they have usually interviewed the husbands at home, on the women's "turf" (a criticism raised by Safilios-Rothschild 1969; and Halle 1984, 316). Hence, these husbands did not reveal much about their own lives in the masculine sphere, which often had an element of disreputable or even illicit activities such as gambling or sexual entertainment, activities they did not want their wives to know about.

Several typical sources of strain exist between working-class husbands and wives. One is the sharp split between the male and female social spheres. Working-class men spend most of their leisure time with other men participating in sports, hunting and fishing, or watching spectator sports; when they are home, they are likely to be absorbed in sports on TV. Husbands and wives share few activities except eating. To some extent, of course, this is also true of middle- and upper-class families, but it is especially severe in the working class. Moreover, what is distinctive about working-class marriages is that wives tend to be suspicious or even hostile to the male companionship groups. They view the male bar scene, male drinking and carousing, as a potential threat to themselves and their marriages. There is, no doubt, some reality to this, but we should understand that it has a symbolic dimension, for studies show that consumption of alcohol and the rate of alcoholism or "problem drinking" is approximately equal in all social classes (Calahan and Room 1974). Yet, it is working-class wives who feel particularly threatened by it; drinking is an emblem of the split between women and men. Thus, Halle (1984) found that the closer and happier marriages were those in which the men had rejected participation in the male bar culture, the marriages in which the husbands came home directly from work and spent more time with their wives.

Halle (1984) also reports that working-class marriages change over time. As blue-collar men grow older, they enjoy more economic security, their children grow up and move out, their wives often take jobs, and their home mortgages are paid off (or mostly paid off). Seniority on the job protects them from cutbacks, gives them choice vacation times, and ensures them a pension upon retirement. At the same time, working-class men often begin to develop physical ailments that slow their mobility, and most of them know someone who has died suddenly. A shift in leisure time activities often occurs at this point, with men who once spent most of their free time with men friends now spending most of their time with their wives. The wife becomes more of a companion, and couples travel together, particularly after retirement. As we shall see in later chapters, aging and maturity can also place special strain on families.

Earlier in the marriage, however, Halle (1984) reports that wives tend to be more socially ambitious than their husbands. Wives put on pressure to rise in the social class system, and they were sometimes critical of their husbands for the low status of their jobs. The husbands, on the other hand, were more likely to withdraw into their all-male sphere of sports, drinking, and the like, and to ignore the differences between themselves and the higher classes. Perhaps these women were status conscious because so many of them worked in white-collar positions, as secretaries and clerks. Even though their jobs tended to be relatively low-paying order-taking positions (and hence could be classified as "white-collar working class"), they had one distinctive feature that set them off from the men's blue-collar jobs: the women were in offices with a middle-class atmosphere and were usually in close contact with middle- or upper-middle class managers or professionals. Thus, they picked up a standard of status respectability from their bosses (who were usually men) by which they compared their own working-class husbands.

Halle found that many family disputes broke out because wives criticized their husbands for their crude manners or for embarrassing them in public. He quotes a working man:

> [My wife and daughter] wanted to get something from the store. So I gave them a lift, and Deirdre met a friend of hers and started chattering. You know, I was tired—I'd been working since 8:00 the night before. So I said, "Deirdre, get the *fucking hell* in here!"
>
> . . . [My wife] kept yammering on. I called her just now [on the telephone], and she still says I shouldn't have used that language in front of Deirdre's friend.
>
> So I'm in a bind. I know what's waiting for me back home. She'll start to yell at me and turn the children against me. (Halle 1984:60)

This is an example of the split between the standards of the male culture of the working-class group and the middle-class standards of public respectability upheld by the wife. Actually, we find when we turn to upper-middle and upper-class marriages (below) that wives often seem to specialize in keeping up the cultural image of the family. The main class difference is that in the higher social classes, the husbands go along with this status seeking, whereas in the working class, it tends to be a point of contention between the marital partners.

Finally, we should notice that working-class families experience a distinctive strain that comes from conditions that make them age faster psychologically.

This results, in part, from marrying earlier and reaching their peak earning capacity and on-the-job responsibility much sooner than professionals of the middle class (Aldous 1969). For working-class people, there is often a feeling that life is already over at the age of twenty-five—a time when most upper-middle-class people are feeling that their lives have scarcely begun. The working class idealize their teen years, and probably with good reason. It was then that they had the most sense of excitement and expectation, even if they were pursuing independence and expectations that turned out to be illusory. Among the women in Rubin's study, there was almost unanimous agreement that the fault lay in early marriage. They wanted to keep their daughters from making the same mistake, little mindful of the fact that they had disregarded their own mother's advice. One woman said: "I sure hope she doesn't get married until she's at least twenty-two or three" (Rubin 1976, 113). Her belief that this would be long enough for a woman to "see some more of the world" is an indication of the narrow horizons of the working class.

Working-Class Culture and the Family

Why does working-class family life have this structure? Part of the reason, as we have seen, is economic but not all of it is. For instance, although working-class wives are generally fearful and worried about their husbands' potential drinking, violence, and sexual infidelity, these problems are not unique to the working class. Studies of marital violence find that it happens surprisingly often in the middle class as well, as does marital infidelity (see chapters 9 and 13). Why then do working-class women react more strongly than middle- and upper-class women?

Part of the reason is that the working class tends to have a particular culture that might be called "localistic" and "group conformist." The British sociologist Basil Bernstein calls it the "restricted" code, as compared to the "elaborated" code of the middle class. The working-class culture, or code, sets a very strong line between insiders, whom one can trust, and outsiders, who are to be feared. Further, it has a strict sense of the moral rules incumbent upon insiders and makes heavy use of authority to enforce those rules.

The fear of outsiders manifests itself in numerous ways. Rubin (1976) found, for example, that when working-class mothers had to go back to work they left their children with relatives instead of taking them to a day-care center. When asked the reason, they stated very strongly that they did not want to leave their children with strangers. The same fear is shown when wives complain about their husbands going out to bars and getting in with "the wrong crowd." They are more comfortable when their husbands stay at home, even if they feel bored and restricted there.

It is the same with working-class sociability—there is a sharp barrier between those whom one will invite into one's home and the rest of the world.

The former consists almost entirely of relatives and a few old friends whom one has known since childhood. The contrast to middle-class sociability, which consists so centrally of dinner parties, is striking. But then, middle-class culture is different precisely because it does not have this same insider/outsider barrier.

The other feature typical of the culture is its moral code. The working class sees a clear, black-and-white dividing line between right and wrong, and it strictly punishes any deviations. The working-class parents in Rubin's study tended to be critical of the public schools and did not like having their children there because they felt that the schools did not enforce strict enough discipline. The psychological techniques used by middle-class parents in controlling their kids are not as evident in the working-class home. Working-class parents were much more likely to demand strict obedience from their children and to use physical punishment to enforce it (see also Bronfenbrenner 1958).

While other studies question whether working-class parents are any more likely to use physical punishment than middle-class parents (Erlanger 1974), most studies confirm that parental values and discipline practices differ according to social class. Research in industrialized countries from the 1950s to the 1970s shows that parents from all social classes discipline children for misbehavior, but working-class parents are more likely to value conformity to external authority, whereas middle-class parents are more likely to value self-direction. According to Melvin Kohn (1979), working-class parents are more likely than middle-class parents to value characteristics in their children such as manners, neatness, cleanliness, being a good student, honesty, and obeying orders. These traits are similar to those needed for success in working-class jobs. In contrast, Kohn finds that middle-class parents tend to emphasize consideration, curiosity, responsibility, and self-control; traits that are required for many managerial and white-collar jobs.

The same kind of difference between middle-class and working-class cultures has been found in many surveys of attitudes (summarized in Collins 1975, 61–79). The working class is much more likely to be moralistic and to adhere to traditional standards. This is found especially in attitudes about sexual behavior (there is much more intolerance of homosexuality, for example, or of premarital sex on the part of women), political dissent, and religious conformity. For violations, the working class is more likely to favor strong and violent punishments.

UPPER-MIDDLE-CLASS FAMILIES

The most important basis of social class differences is occupation. We have seen some of the reasons why it makes a difference in families' cultures: it affects the sheer amount of money people have, as well as their financial security or insecurity, and hence has an important effect on their lifestyles. It also helps produce the cultural codes that derive from a high-social-density/localistic or a low-social-density/cosmopolitan life experience. A further dimension of occupations is the way in which people experience power (see Collins 1975, 61–79).

Working-class people are generally *order takers*: they have no power on their jobs but take orders from others. Since taking orders is a psychologically demeaning experience, working-class people try to distance themselves from it by not identifying very strongly with their jobs. They tend to be cynical about the official ideals that they hear coming out of the mouths of their bosses, since these are used mainly to justify ordering them around. Working-class persons emphasize the leisure side of their lives and support the "cool," rather alienated themes of popular culture. Child-rearing practices tend to reinforce these values and prepare working-class children to assume jobs in which they must take orders from others.

People who are *order givers*, on the other hand, have a very different psychological perspective on their work. High-ranking managers and professionals tend to be proud of themselves and to identify strongly with their work roles. The experience of telling other people what to do gives them these qualities, both because it is more psychologically gratifying to give orders and because they have to take their work roles seriously if they are going to make other people respect their authority. Upper-middle-class childrearing values tend to emphasize taking responsibility for others' actions and feelings, as well as one's own, thereby preparing children to occupy positions of authority.

Members of the higher social classes have more incentive to identify with the ideals of their organizations and of the official, "front-stage" part of society in general. While the working class tends to be alienated from the larger society and to find refuge in their private lives, the upper classes live more in the larger world and are the main carriers of its various ideologies.

Order givers and order takers are not two mutually exclusive classes, and they do not exhaust all the levels of occupational classes. In between the higher social classes and the working classes are various intermediate groups (e.g., middle-level bureaucrats and officials, supervisors, and foremen) that take orders from people above them and pass them along to others below. These kinds of workers usually share parts of both the upper- and the lower-class cultures: they identify with rules and regulations in some ways but are not as committed to the larger ideals as high-ranking persons. There are also many white-collar workers who tend not to experience much order-giving *or* order-taking but who work in an office or professional setting surrounded by people more or less of the same rank. This is quite common among engineers, many scientific researchers, and other technical workers. In these white-collar groups, there is a typically "middle-class" style of being casual and informal on the job. This style is characteristic of the United States rather than of other countries, where the middle class is much more likely to be rather formal and concerned about the rituals of deference and power.

All these aspects of class cultures influence family life. Here, we will pick out just one type of class culture, that of the upper middle class, which contrasts most sharply with the family life of the working class. (See also feature 6.3, on the very highest social class.)

Upper-Middle-Class Cultural Traits

We already know some of the cultural traits found in the upper middle class. It is the class of high-ranking order givers: high-level managers in large corporations, the most important professionals (successful lawyers, professors in major universities, doctors, notable scientists, renowned architects, and so on), and higher government officials. Their culture has a number of things in common. As we might expect, these are the individuals who identify most strongly with the official ideals of their jobs. It is the successful scientist who cares most about the intellectual accomplishments of science; the doctor who goes on at greatest length about the need for proper medical care; the business executive who most extols the values of free enterprise in general and the merits of his (or her, though this is less common) corporation in particular. These people get the most out of their jobs in terms of prestige, power, and income, and hence they have a very favorable attitude toward their work. Many of them put in quite long hours, far more than the workers below them (Wilensky 1961). This is partly because they enjoy their jobs and tend to be highly satisfied with them, and partly because the higher levels of big organizations and the professions are very competitive places, and people need to put in long hours to stay ahead of their peers.

Members of the upper middle class also are more concerned about general and abstract ideas than other social classes. They are more likely to have taken a position on most issues, less likely to answer "don't know" to a public opinion poll (Glenn and Alston 1968; Mann 1970; Collins 1975, 67–77). Of all social classes, they are most likely to be active in politics—not only to vote, but to run for office, campaign for other candidates, or contribute money to their campaign funds. It is not much of an exaggeration to say that the upper middle class (together with the upper class, which has been much more widely studied) runs politics in America. That is to say, the upper middle class runs both the liberal and the conservative political parties. It is true, of course, that the business upper class, and its higher-ranking managers, tend to be the bulwark of the Republican party and to support conservative policies. But the Democratic party as well, including its liberal wing, is largely run by members of the upper middle class: not so much the business upper class but professionals such as lawyers, academics, and media professionals. Political campaigns in America are almost always between rival factions of the upper middle class. The reason for this is partly that this class has many more resources (money, contacts) to put into politics. But it is also that upper-middle-class culture is more concerned than working-class culture with general, abstract ideas (e.g.,

FEATURE 6.3
Women of the Upper Class

The family lifestyle of the wealthiest class can best be seen by looking at the wives of those men who control the major businesses. This is what Susan Ostrander (1984) did in her book *Women of the Upper Class* and Arlene Daniels did in her book *Invisible Careers* (1988). The husbands of these women are the presidents or chairmen of the board of major corporations, or they own old family firms, or they are partners in the law firms that handle corporate business, or they have major roles in stocks and finance. The upper class is not difficult to pick out; they announce their own membership in the *Social Register*, a volume published for most big eastern cities (and a few western ones). They belong to exclusive clubs. They usually attend exclusive private secondary schools, and their colleges tend to be the traditional Ivy League ones.

The daily routine of these women is not quite what the popular image of the rich would have us believe. "The life of the upper-class woman is not all champagne and roses, trips to Paris and Palm Beach. The upper-class woman is also not very interested in high fashion, nor is she a jet-set party goer. Her days are not spent lying around the country club pool or attending elegant ladies' luncheons" (Ostrander 1984, 3). They are much more conservative. They believe that they are morally superior to other people, that they are more responsible parents, that they contribute in essential ways to the community, and that they uphold the standards for the rest of society. They also have a distinct sense of being different from other people, and have a preference for associating with people like themselves. In the generally un-class-conscious U.S. society, upper-class women are among the most class conscious, and they have the sharpest sense of class boundaries.

"The club is a place to go where you can have lunch and discuss business without having to wade through the mass of people. You want to have people who are congenial, all of the same

social group," said one woman (Ostrander 1984, 97). There is an emphasis on belonging to clubs in which membership is by invitation only; one must be known and acceptable to the existing members, and new members must be sponsored by one of them. Many upper-class women are members because their families have been members. Women whose husbands have made it to the top of the corporate hierarchies can be invited to join. One of the main activities of established upper-class women is socializing these wives into the standards of the upper class and of picking out those who are to be sponsored into club membership. "Whenever a new executive wife comes along, I give her quite a lecture on ceasing to be an emotional person. She should never be nervous or cross. These men can't take that. She should always be prompt and ready to go" (Ostrander 1984, 47).

The same applies to the women's activities as mothers. They take their family roles seriously and believe they should protect their children from exposure to gangs, drugs, blatant sex, and the rest of what they perceive as the popular culture of the day. This is one reason why they want their children to attend private schools. But they are also concerned with maintaining upper-class homogeneity. Upper-class mothers are expected to prepare a new generation of the elite for its future responsibilities. These women generally consider it important for their children to marry someone of the same social class. For this reason, they want them to participate in dancing schools, debutante balls, and other sociable activities that are carefully arranged and controlled. These activities are supposed to be pleasant, but they are also viewed as a duty. One woman from a long-standing upper-class family, for example, commented: "My mother was one of the women who started [the debutante ball] so I felt the children should do it whether they wanted to or not" (Ostrander 1984, 89). These social events serve as

A "Baby Boutique" in Costa Mesa, California, caters to upper-class mothers. These women take their family roles seriously and believe they should provide their children with a well-protected environment that serves all of their needs.

rites of passage into membership within upper-class circles, and hence they are a barrier that status-seeking individuals would like to overcome from below. Upper-class women are aware of this and regard it as their duty to carry out the "dirty work" of separating the "acceptables" whom they will sponsor from the "unacceptables" whom they regard as mere "social climbers" (Ostrander 1984, 93–94; Daniels 1987, 224).

Another aspect of upper-class women's lives is participating as a community volunteer in charitable organizations. This does not so much mean the front-line activities of actually sitting at tables or ministering directly to the needy, but of serving on boards that plan functions and raise money.

They themselves are expected to be big contributors. This can be time consuming, and these women sometimes will describe it as a burden. But their charity work is part of their self-image as responsible persons and community leaders, and they regard volunteer work as superior to working for pay. Since virtually none of these upper-class women has a career of her own, charity work is a kind of substitute activity. For women whose husbands are in the upper middle class, charity work is a way of becoming known and associating with the upper-class network. On the other side, upper-class women see these relationships as yet another aspect of their "sponsorship" tasks and are suspicious of "newcomers" who they feel are pushing too hard. A moderate proportion of upper-class women regard their charity work as an obligation that they took on reluctantly. Some complained that the real power to make decisions was still held by the men of these charity boards; others suggested that participation was only symbolic, and that it accomplished little to alleviate real problems. One woman said she had "quit the Junior League because it seemed to her that 'the most important thing was how neat your hat was'" (Ostrander 1984, 125). In short, charity participation is essentially a ritualistic activity, felt to be an obligation ("noblesse oblige") but nevertheless an important way of displaying one's class position. While upper-class women feel obligated to participate in such activities, they generally do not stress the importance of maintaining their class position. In general, they find "class" an "ungenteel matter for discussion" (Daniels 1987, 227).

The lives of upper-class women are both formalized and constrained. Despite their wealth, they have little freedom. At home, they have servants to take care of the housework. But their husbands are the dominant figures in their lives; these are the men who control the major organiza-

(Continued next page)

FEATURE 6.3 *(Continued)*

tions of society, and their business activities take priority over everything else. "My husband never asks me what I think. He just tells me how it's going to be," said one woman (Ostrander 1984, 37). Wives make social arrangements for their husbands; they travel with them, when requested, even at short notice; they put up with their husbands' frequent trips and meetings away from home, making sure the home the men come back to is well ordered. Their task is to shield their husbands from having to be distracted by any domestic problems or issues.

Although upper-class women's volunteer activities help to support their husbands' careers, they typically minimize the importance of their work for this purpose. Because the women hold positions on powerful and prestigious governing boards, their husbands have the opportunity to make professional contacts and acquire "inside" information about future changes in the community that can affect their business or professional practice (Daniels 1987, 228).

Many of these women have inherited considerable money of their own, but this typically gives them little power. They usually put the money into their husband's care. These men, after all, are in the financial power centers of the world. It is their social networks at work that are the core of all this privilege. Living in intimate relationships with the sources of upper-class power makes these women aware of how dependent they are upon the power structure.

Does the environment need to be saved from pollution? Is racial inequality an important problem? Does free enterprise need to be defended?). They admire more abstract forms of art, where the working class prefers more sentimental and concrete forms of artistic "prettiness" (Bourdieu 1984).

The upper middle class is the cosmopolitan class par excellence. It is a class of social joiners and group members. The upper middle class also participates more in churches, professional associations, and clubs of various sorts. It is the wives of upper-middle-class men who support the various women's clubs and organize charity exhibitions, luncheons, and balls; it is the upper middle class, not the working class, that makes a point of belonging to the country club and the tennis club and takes part in the golf tournaments. (The same is true of more recent athletic fads such as long-distance running; the people who turn out for marathons and ten-kilometer races are largely upper middle class, almost never working class.)

One big contrast between the upper middle class and the working class, then, is the way their social lives are organized. The working class is organized very locally, around the home, relatives (for women), and nearby groups of male friends (for men). The upper middle class belongs to special-purpose organizations, with members who come together only occasionally and for one special activity only. Upper-middle-class people thus encounter many others in the course of their activities, although they interact with them much less closely. This is what is meant by the terms *high social density* (in the working class) and *low social density* (in the upper middle class). The same is true on the level of purely sociable relationships. The higher social classes tend to have many more friends than the lower classes, but they see them less often and are less close to them. Business executives, for instance, often send out Christmas cards to hundreds of their "friends"—persons they have socialized with at clubs, over business dinners, on trips, and so forth. Some higher executives even keep a secretary busy full time just taking care of files, so that the executives can be sure of who their "friends" actually are.

All this, of course, has an important effect upon upper-middle-class marriages. Upper-middle-class people tend to get married a good deal later than working-class people. Their lives tend to be oriented toward the future, not the present, and they are trained to wait for things. Not only do they marry later, but they put off having children longer. When children do arrive, they are more likely to have been planned through the careful use of birth control, so that they do not interfere with the husband's (and sometimes the wife's) career.

Wider Social Networks

The upper middle class is less family oriented than the working class. Whereas working-class people don't see many other people besides their own family members and relatives, the upper middle class see families, relatives, *plus* lots of other people. Their lives are less exclusively oriented toward the family network (Collins 1975, 79–80; Gans 1962; Fischer 1982).

Upper-middle-class careers, for reasons given earlier, are both more attractive and more demanding than working-class jobs. The husband in particular is likely to work long hours or be away for business trips or professional meetings rather frequently. And if he is not gone physically, he is often absent psychologically: his body may be present at home, but his mind is elsewhere, preoccupied with his work.

At the same time, since members of the upper middle class are strongly oriented toward the more intellectual media and keep up with ideas coming out of the technical professions (science, medicine, law, psychology), they tend to have well-defined opinions about family life. They may not devote so much time to their family, but they have well-articulated beliefs about it, just as they do on everything else. In the conservative part of the upper middle class, there is likely to be strong agreement with religious and political leaders who assert that the traditional family is the backbone of America; such persons will have strong opinions about the necessity for women to be full-time mothers, or the need to protect the sanctity of family life, and so forth. On the liberal side of the upper middle class, there is a strong concern with the psychological dimension of relationships between husband and wife, parents and children. Books are read, specialists sometimes consulted, family discussions are deliberately held, and avant-garde styles of relating are explicitly practiced. Yet all of this tends to be crammed into lives that are also busy with many things outside the family.

Corporate Career Families

Let us look a little more closely at a particular type of upper-middle-class family—the families of executives who work in the large corporations, the massive businesses that control 65 to 75 percent of the industrial economy. Because of its importance in our society, this group has been fairly heavily studied (Kanter 1977; Cuber and Harroff 1968). It is one sector of American society where wives generally are not employed. In a sense, both husband and wife are in the same job: both of them are "married" to the corporation.

Big corporations, in fact, have usually been explicitly concerned about their managers' marriages. They have tended to assume a rather strait-laced, even puritanical attitude. There is usually a belief that a good manager *should* be married and settled down, so that he can dedicate himself to business (Kanter 1977). Papanek (1973, 93) identified this "two-person career" as a distinctively middle-class career pattern involving "the induction of the wife into the husband's work orbit." (This attitude refers only to men; married women managers, on the contrary, have usually been regarded as not dedicated to the company.)

Many corporations tend to be all-inclusive, regulating every aspect of their employees' lives. Managers who are moved around the country from one corporate branch to another are expected to uproot their families, separate their children from their friends, and establish them in new schools as often as every few years. In this respect the corporation is rather like a career in the military. And like the military, corporations often compensate by arranging for housing, with some even providing company housing. Corporations also typically pay for their managers' membership in country clubs as well as city business clubs. Partly this is a matter of picking up business expenses, since executives often transact discussions and negotiations in these informal settings. It also is part of the corporation's interest in being a "total institution" surrounding its employees' lives and ensuring their loyalty. Corporations tend to assign their executives to take part in community charities, such as the United Way. Although not strictly a business contact, the participation of the company in such community affairs is regarded as an important part of its public relations (Greiff and Munter 1980; Fowlkes 1987).

The husband's life, then, tends to be totally absorbed in his career. This even extends to his leisure time, which may involve vacations paid for by the company in connection with company meetings. The wife is expected to be part of the "team." Some companies continue to look over the wife of a new manager to make sure she is the kind of person they want, and one who will support her husband's dedication to the company (Kanter 1977). Marrying the proper woman can thus be an important step in a male executive's career, and marrying the wrong one (or failing to marry) can be a fatal obstacle to his advancement.

Even today high-ranking careers such as corporate executive, medical doctor, lawyer, and scientist still tend to be male-intensive and male-dominated. The standards for training, mobility, and success in such careers contain a host of assumptions about what is primary and what is secondary. Most of these assumptions, including the expectation that one's career takes precedence over all other commitments, reflects a male-centered view (Epstein 1971). As more and more women attempt to travel the road to professional success, they are often heard to say, "What I really need is a good wife" (Fowlkes 1987, 358). With proportions of women and men in medical and law schools now near equality, there will surely be adjustments and conflicts lying ahead.

DUAL-CAREER FAMILIES

As more women seek education and training to enter professional careers as corporate managers, medical doctors, lawyers, and scientists, researchers have turned their attention to "dual-career couples" (e.g., Rapoport and Rapoport 1971; Rapoport and Rapoport 1976; Epstein 1971; Pepitone-Rockwell 1980; Poloma 1972; Holmstrom 1973; Garland 1972). Most studies have described the unique stresses and coping strategies common to families in which wives as well as husbands have professional careers. While dual-career couples represent only a fraction of families in which both wives and husbands are employed, the tendency has been to assume that what goes on in these "upscale" families is equally applicable to other middle- and working-class families (Ferree 1987). Using dual-career couples as models suggests that paid employment is the major source of personal reward for most people and that family work, like cooking, cleaning, and child care, is burdensome.

While dual-career marriages may be "more equal than others" (Hertz 1986), most researchers find that there is significantly more sharing of egalitarian *ideals* than sharing of domestic *tasks* such as child care and housework (see chapter 11). Most studies also show that having two careers provides professional couples with the resources to hire someone else to perform housework and child care, and that two relatively high incomes allow the couple to enjoy a high standard of living that includes frequent entertainment and dining out as well as expensive homes, cars, and clothes. While many researchers focus on the egalitarian attitudes of such couples, Hertz (1986, 415) suggests that "it is not an ideology of marital equality that determines careers; instead it is because of the two careers that equality becomes an issue." What appears to happen is that as the woman's career becomes as demanding as her husband's, there is no time for her to be a typical wife. Consequently the "invisible" work of the home comes into plain view for some dual-career husbands.

Hertz (1986) claims that the dual-career couple should be characterized as a situation with two husbands and no wife. She suggests that in this situation, both people feel pressure to limit their career aspirations and goals (see also Hunt and Hunt 1982). If men and women in demanding professional careers cut back on their employment hours, however, they are often perceived as unserious about their careers (Bourne and Wikler 1978). This tendency is reflected in an informal two-tiered system of job tracking in the corporate world: those who limit their time commitments to career because of family obligations are placed in a so-called "mommy track" or more recently "daddy track." Such unofficial categorization often leads to slower and fewer promotions and puts dual-career professionals who spend time with their families at a disadvantage in relation to their "fast-track colleagues (Schacter 1989). It appears that major adjustments in the requirements of professional careers are yet to come.

SUMMARY

1. The class structure in the United States is relatively stable, with most of the nation's wealth controlled by very few families. Individuals tend to move up or down the income pyramid because of changes in family composition—things like getting married, having a baby, or getting a divorce. Since 1973, real earnings for the average American have declined, and most families now have two wage-earners.

2. The poorest group in the United States consists of households of people who are out of the labor force because they are retired, unemployed, or on welfare. Households headed by women are especially likely to be at the bottom.

3. Families are most likely to be in the prosperous upper parts of the income distribution if both husband and wife are working full time, especially at white-collar jobs. Only in the very highest income groups is there a preponderance of families with the traditional structure of an employed husband and a nonemployed housewife.

4. Working-class people are especially likely to marry young, although this frequently results in severe economic strains early in the marriage, which are intensified by the early arrival of children. If the marriage survives this crisis, it tends to settle down into a routine in which the family sticks close to home for most of its nonworking hours. The woman tends to socialize with her relatives, the man with a local group of friends. There is usually a sharp split between male and female cultures.

5. Working-class culture tends to be highly localistic, with strong pressures for conformity within the group and distrust of outsiders. These conditions make working-class people typically moralistic and traditional in their beliefs, favoring strong discipline to control children and punish deviancy.

6. Upper-middle-class people emphasize delayed gratification and future

planning and tend to both marry and have children later than lower-class persons, so as to pursue their careers. They are involved in numerous networks of professional and business contacts and do much more nonfamily socializing than working-class people.

7. Upper-class families generally separate themselves from other classes by attending private schools, belonging to exclusive clubs, and acting as patrons of the arts. These families socialize their children to accept the responsibilities and obligations of the elite and ensure their class position by passing on their enormous wealth.

8. Corporate executives have typically been men whose wives contributed to their careers by remaining unemployed and running the home. Corporations regulated most aspects of their managerial employees' lives and corporate wives were expected to entertain business clients and perform other services for their husbands and for the company. High-ranking careers continue to be male-intensive and male-dominated.

9. Dual-career couples are a new family form that has received much attention from family researchers, even though they are less common than dual-earner couples who have working-class jobs. Dual-career couples can usually hire other people to perform child care and cleaning, and men and women professionals who limit their work hours to take care of children are still considered less serious about their careers than others.

7

RACE AND ETHNICITY: VARIATIONS IN FAMILIES

I n this chapter we will look at some of the ways in which family life is affected by membership in different racial and ethnic groups. Drawing on material from the last chapter, we will investigate how stratification results from economic factors and analyze the ways in which life experiences and family composition are similar and different for majority and minority families. In one sense, all Americans are members of an ethnic minority because no single ancestral group constitutes over half of our population. Sometimes age, sex, and religion are also used to designate minorities, but race and ethnicity are the most common ways of defining minority status.

Over the years, our government has defined different groups as minorities and used different labels to designate them. All this switching around has made it difficult to describe precisely what is happening to American minorities. The government currently collects statistics for four major minority groups: (1) African-Americans or "blacks"; (2) Asian-Americans or "Asians and Pacific Islanders"; (3) Native Americans or "American Indians, Eskimos, and Aleuts"; and (4) Latinos or "Hispanics." Because of higher-than-average birth and immigration rates among minorities, their populations are growing at a rapid rate. By the middle of the twenty-first century, today's minorities will comprise nearly half of all Americans (O'Hare 1992).

AFRICAN-AMERICAN FAMILIES

Although in many respects black families in America reflect the general culture, certain unique pressures upon them, coupled with their own individual cultural heritage, have caused these families to devise some distinctive structures for surviving and getting ahead. Yet despite such commonalities, there are also important economic and social differences among black families. For example, blacks whose families came from different regions of Africa or the Caribbean have very different linguistic and cultural backgrounds. Blacks raised in the American South have different experiences and traditions from those raised in the North, and those living in urban areas differ from those living in rural areas or those residing in the suburbs. One thing African Americans typically are not, however, is a reflection of the stereotype commonly applied to them.

The Myth of the Black Family

It is widely believed that the black family deviates greatly from white family structure in the United States. The general image is of marital violence, broken homes, large numbers of children, and a resulting cycle of poverty, illegitimacy, delinquency, welfare, and unemployment. During the racial upheavals of the 1960s, Daniel Patrick Moynihan wrote a report for the U.S. Department of Labor, the so-called Moynihan Report, which argued that the economic problems of blacks were basically due to the weakness of their family structure. He described it as a "tangled pathology. . . the principal source of most of the aberrant inadequate, or antisocial behavior that . . . perpetuates the cycle of poverty and deprivation" (Moynihan, Barton, and Broderick 1965, 47). Moynihan, now a U.S. senator from New York, argued that the source of this family structure was the historical experience of slavery, which separated fathers from their children and left the black family with a heritage of dependence upon mothers. "The Negro community has been forced into a matriarchal structure which, because it is so out of line with the rest of American society, seriously retards the progress of the group as a whole" (Moynihan, Barton, and Broderick 1965, 29).

The key attribute of the so-called broken family structure seen by Moynihan was the absence of the father and the unusual amount of power wielded by women. Allegedly, the psychological results of this include a low self-image on the part of male children; lower IQ scores with resulting low school achievement; high dropout rates; delinquency; and unemployment (Moynihan, Barton and Broderick 1965).

This image of the black family, however, has a number of faults. It ignores social class and especially overlooks the large number of black families that do not fit the stereotype. Less commonly noticed but quite apparent once one thinks of it, the indictment assumes a rather sexist and privatized model of the "normal" family.

Social Class Differences

One important fact missed by the broken/matriarchal image of the black family is that a large number of African-American families do not have this structure. About half of all African-American family households were married couple households in 1990 (O'Hare 1992) and most adult black men were working and supporting their households. More than a fourth of all black families have incomes above the median for the United States, and in this group 90 percent have two parents present (Willie 1981). Still, African-Americans are much more likely than members of other racial/ethnic groups to be unmarried, to live in single-parent families, and to live below the poverty line. In 1992 fewer than half of all black adults were married and fewer than half of all black children were living with two parents (U.S. Bureau of the Census 1992f).

Racial Discrimination and Its Paradoxical Consequences

Until the 1950s, the United States was a thoroughly and even openly racist society. Segregation laws were enforced in southern states, while de facto discrimination prevailed in the North, and racial stereotypes were an accepted

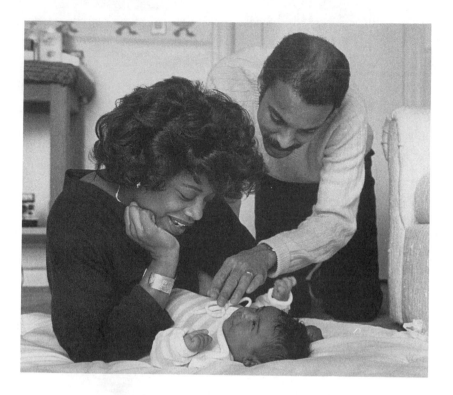

Two parents are present in about half of all black families.

part of popular culture. The civil rights movement of the 1950s and '60s, after both nonviolent demonstrations and violent uprisings, brought a rapid change in the legal structure, although informal racial discrimination has proved more resistant to eradication. One effect of this discrimination has been a serious financial penalty paid for being black.

Because of racial discrimination, the employment pattern of blacks has taken a peculiar form. They have been able to acquire decent white-collar jobs, or even stable blue-collar positions, mainly by working in government rather than private business. Many black women have been able to pull their families up to the middle-class level by working as schoolteachers. This has given a special importance to education in the black community, where women have frequently had more education than their husbands. It has been more difficult for black males to find comparable middle-class jobs. Where they have found them, it has usually been through government employment, such as the post office. Cities with large-scale government employment have thus tended to attract a large black population; one of the largest percentages of blacks in any city (65 percent) is in Washington, D.C.

Such economic pressures have also affected forms of family organization typically adopted by both the black middle class and the black working class. Neither fully resembles its white counterpart.

Among affluent blacks, distortions in employment patterns owing to discrimination are played out in family relationships. The white upper-middle-class family tends to subordinate everything to the career of the father (as we have seen in the previous chapter on social class). Among blacks, however, this is seldom the path to success. The father alone usually cannot attain upper-middle-class status for his family with only their *moral* support. He needs that, but in addition, the whole family works together as an *economic* unit to get ahead. Women's achievements are important. In fact, the pioneering black upper middle class, seeing itself as part of an embattled minority making its way against heavy odds, places special emphasis on self-discipline and achievement by *all* its members. There is great pressure on all the children to go to college and have successful careers. Willie (1981, 176–79) has argued that, as a result, the affluent black family has held together more firmly than the comparable

Members of St. Paul's Baptist Church in Maryland pause for a moment of prayer before breaking ground for their new building. Black working-class families attend church more, belong to more organizations, and hold a higher status ranking within their own community than white workers do in theirs.

white family—perhaps too much so, since, he suggests, both adults and children of such families pay a price in subordinating their individualism and spontaneity to their achievement.

For the black working-class family, discrimination has had a paradoxical effect. It has kept some blacks in the working class who might otherwise have risen higher. But by putting a lid on *all* black achievement, discrimination has in effect transformed the black working class into a middle class. Compared to their chief reference group, other blacks, their achievements *are* in the middle of the spectrum. It was already noticeable at the time of the Second World War that black working-class families were much more like middle-class families than their working-class counterparts in the white community (Drake and Cayton 1945). They belonged to more organizations, attended church more, and held a higher status ranking within their own community than white workers did in theirs. Willie (1981; 1985) indicates that this may still be so. Whereas the latter, as we have seen, are alienated and withdrawn from a world run by affluent whites, the working-class blacks are more of the respected middle stratum in their own world. Perhaps for this reason, many studies have shown that blacks at the working-class level actually have lower rates of mental illness and higher levels of social participation and self-esteem than comparable whites (McCarthy and Yancey 1971; Farley and Hermalin 1971; Olsen 1970).

The Question of Black Matriarchy

Another argument concerns the deleterious effects of female domination in black families. Above, we have argued that this actually does not occur as much as some writers would have us believe: many black families are in fact rather like the conventional two-parent white family. At the same time, it is admitted that a substantial number of black families, usually in the urban underclass, are headed by women. It is also believed that black women play an unusually strong role within middle-class families (Willie 1981).

The question is: Are the effects of this female power good or bad? The assumption has been made by Moynihan and many others that the psychological effects of this situation have led to delinquency and nonachievement in the black lower class. Others have also focused on "restoring the dignity" of the black male, giving him a dominant place in the family, and implicitly advocating that black women should be in a subordinate position (see Ransford and Miller 1983). But the evidence does not support the contention that equality for women is responsible for psychological problems and low achievement in children.

What *is* true is that black women have more family power than their white counterparts. This is especially true in the lower class, where most families are headed by women, but it is also true in the intact families of the working class and middle class. Because of economic discrimination, black men have had

FEATURE 7.1
Sharing Networks among Lower-Class Black Women

Lower-class black women have evolved special structures of their own for dealing with economic insecurity. The anthropologist Carol Stack (1974) found that, instead of conventional marriages, they use a network of mostly female relatives and friends who help each other out regularly in emergencies. They provide each other with places to stay, take care of each other's children, share the use of a refrigerator, and use whatever money is available to pay the most pressing bills.

This "artificial kinship" network makes it possible for these women to survive at the very depths of the class structure, but it also tends to keep individuals from breaking out of this position. Stack found that one young woman wanted to get

married, but the other members of her network discouraged her because they needed her economic contributions. An older couple in a network received a windfall inheritance of $1,500, which they decided to use as a down payment on a house. But other members of the network put in their requests too: one needed $25 to keep her phone, which they all used, from being cut off; another was about to be evicted if she couldn't come up with the rent; still others asked for train fare to attend their mother's funeral; then the couple's own children were dropped from the welfare rolls. Within a few weeks the money was gone. The network was still intact, and so were the bonds of poverty.

---

unusual difficulties in getting steady or well-paying jobs, and black families have tended to rely upon women's incomes (as well as children's incomes). Black women rate themselves as having more power than white women, and black husbands attribute more power to their wives than white husbands do. Other researchers conclude that black families are more egalitarian than white families at the same class level (Scanzoni 1977, 334–37; Taylor et al. 1991; Broman 1988).

From the point of view of gender equality, this is a positive development. However, there are signs that it may also be a source of continuing, or even increasing, conflict. For example, much African-American rap music in the 1990s has been criticized for using hostile and degrading language toward women, treating black females as sex objects and antagonists. Even among the black middle class, men tend to hold fairly traditional notions of appropriate gender roles, perhaps because black men have been denied access to the jobs and earnings that have enabled their white counterparts to enjoy dominance over women. Ransford and Miller (1983) found that black men were much more likely than whites to agree with statements such as "Women should take care of running their homes and leave running the country to men," and "Most men are better suited emotionally to politics than are most women." Furthermore, black middle-class wives often tend to agree with their husbands, expressing much less support for gender equality than white middle-class wives. This fits Willie's (1981) picture of the black middle-class family, with its strong internal pressures subordinating everything to the task of getting ahead economically.

Willie even refers to such families as "affluent conformists." The result is something of a paradox. We see the black middle-class family fighting its way to some success against racial discrimination but losing some of the female equality it had previously held. Even though middle-class black men tend to believe in male domination, they have a realistic attitude about the importance of their wives' incomes, which have always been so crucial to the success of black families.

At the same time, there are countertrends and signs of impending strain. Working-class and lower-class black women do not share this conservatism about gender equality (Ransford and Miller 1983). On the contrary, many of them are rather militant in their attitudes toward men, not in the larger political sense of taking a feminist position on issues in the larger society, but in the immediate personal realm. Black women have been found to be much more likely than white women to regard men as untrustworthy and economically unreliable (Ladner 1971; Turner and Turner 1974). Black women are much less likely to feel they need a husband to have children, contributing to more black babies being born outside of marriage than within it. With black men's economic prospects declining and black women's increasing (though still at the bottom), many African-American women have given up on the white middle-class ideal of the isolated and self-sufficient nuclear family.

The Case of the Black Underclass

The female-headed black family that makes up such a large proportion of the urban ghettos has given most of us our stereotyped image of the black family. Although only a minority of black families fits this description, this group nevertheless is a crucial focus for the problems of poverty, crime, and other social problems of our day.

However, it should be understood that the majority of poor people in the United States are actually *not* black. Their situations no doubt differ from those of blacks in many important ways. By studying black poor people, then, we can by no means learn everything significant about *all* poor people. As it happens, however, sociologists have studied black poor people much more than white poor people: much of what we know about poverty refers to black poverty. Keeping these limitations in mind, we will see what we can learn about major social problems through the study of the black underclass.

One conclusion, we would suggest, is that although this lower-class family is female headed, that is not necessarily the source of its problems. In fact, one might say that black women have been particularly strong here, not only in holding the family together as much as they have, but also in carving out a protected sphere for themselves as women in a world of chronic poverty and under the threat of considerable male violence.

Let us consider the historical trends. It has often been asserted (Moynihan,

Barton, and Broderick 1965; Billingsley 1968) that the black family broke down and became female-headed because of the experience of slavery. Recent historical studies (Gutman 1976) show instead that most slaves grew up in two-parent families, and that the family began to break down only after 1925. This tendency for black families to be headed by a single parent *accelerated* after 1960 (Wilson and Neckerman 1986). About half of black families are married couple households, but those with children are more likely to be headed by women than others. In 1970, over half of all black children lived with two parents (59 percent), but by 1992 only about a third lived with two parents (36 percent) (U.S. Bureau of the Census 1992f). In the early 1990s, a majority of black families with children were single-mother households.

There are other trends, too, that indicate that the female-headed lower-class black family of today represents a *relatively recent type of family structure*. We know, for instance, that whereas blacks used to marry at younger ages than whites, ever since about 1950 blacks have been waiting longer to marry than whites (Cherlin 1992). Blacks have a higher divorce and separation rate than whites at all class levels. They wait shorter intervals after marriage to divorce and take longer to remarry; more blacks never marry (Taylor et al. 1991). Although most blacks have been married, the proportion has dropped significantly since the 1970s. In 1970, one out of five black adults had never been married, but by 1992, one out of three were in this category (see figure 7.1). The trend toward nonmarriage has been especially strong in the black lower class, whose difference in this respect from whites has been accelerating in the last few decades. In the early 1990s, the proportion of never married adults was almost twice as high for blacks as it was for whites (U.S. Census 1992f). Nonmarital birthrates, too, show a sharp divergence between blacks and whites, with blacks many times more likely to have a child outside of marriage. Among black teenagers, birthrates have declined in recent years, but they remain more than double the rates of white teenagers, primarily because of lower rates of contraception and abortion (Taylor et al. 1991).

The recent dramatic increase in the number of black children born out of wedlock parallels a smaller but similar trend for whites. Neither trend is the direct result of increases in the rate at which unmarried women bear children. Unmarried women are *not* significantly more likely to give birth to a child, it is just that there are more unmarried women because both blacks and whites are waiting longer to get married and sometimes not marrying at all. Also, since married women are having fewer babies than they used to, a larger percentage of all babies are born to unmarried women. Both of these factors contribute to a much greater likelihood, especially among blacks, for children to be born out of wedlock. There is now little or no stigma attached to having an illegitimate child among poorer blacks. In 1990, about two-thirds of all black births occurred to women who were not then married (O'Hare 1992).

FIGURE 7.1: Marital Status of Adults by Race, 1970 and 1992

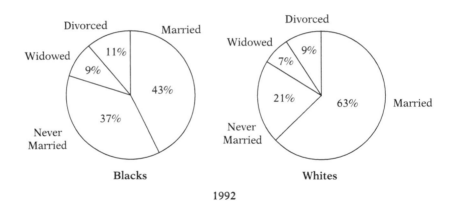

Being single has become more common, especially for blacks.

Source: U.S. Bureau of the Census, 1992. *Current Population Reports,* Series P20, No. 468.

Explanations for Marital Instability

What underlying factors account for the greater instability of marriage and higher rate of out-of-wedlock births for blacks as compared to whites? The most important explanations include: (1) differences in social class or economic position; (2) the impact of family assistance benefits (welfare); (3) the relative economic status of men and women; and (4) a culture of poverty (Jaynes and Williams 1989). We will review these explanations in turn; a fifth explanation, scarcity of black men, is treated in feature 7.2.

Differences in Social Class or Economic Position

Researchers have found that high rates of birth to unmarried women are regularly associated with the following conditions (Goode et al. 1971, 301–5): low socioeconomic status, urban residence, little education, times of economic depression, prior prevalence of divorce and separation, home background of unwed parenthood, weak parental controls over children, and lack of severe

FEATURE 7.2
The Scarcity of "Marriageable" Black Men

Staples (1985) has argued that there are structural causes for the decline in the proportion of traditional two-parent households and the continuing increase in female-headed families. The most important point is: *there are not enough men for black women to marry, especially black men who can make economically viable marriages* (see also Bennett, Bloom, and Craig 1989; Guttentag and Secord 1983; Spanier and Glick 1986). Of the black women who do marry, 98 percent marry black men (see figure 8.2 on interracial marriages in chapter 8). But there are almost 1.5 million more black women than men over the age of fourteen.

Since women tend to marry men who are two to three years older than themselves (see chapter 8), when the population is growing, there are more women than men in the same "marriage pool." For instance, when the black population growth rate is 2 percent per year, as it was for over a decade after World War II, the number of women born in any year will be 6 percent greater than the number of men born three years earlier. What this means is that black women born during the post-World War II baby boom must "compete" for relatively few men. The sex ratio for young unmarried whites in 1985 was 102 men per 100 women, but only 85 men per 100 women for blacks (Jaynes and Williams 1989, 537–38). And this does not take into account other factors, such as the exceptionally high mortality rates among black males or their lower earning potentials.

While there are slightly more males than females born, there is a much higher mortality rate among young black males. One of every 21 black males is murdered before age twenty-four, whereas the white ratio is 1 in 131 (*Science* 1985, 30: 1257). Rate of suicide, accidental death, and drug overdose are also much higher among young black males than females. Blacks make up a disproportionate number of the armed forces and are especially likely to be assigned front-line duty and to be killed in combat. A substantial number of black men are in prison. Recent studies estimate that one in four black men between the ages of twenty and twenty-nine is either in jail or otherwise under the control of criminal courts through parole or probation (Savage 1990). Others, with little chance of employment, have dropped out of the labor force and live a transient existence.

Marital "availability ratios" take into account the fact that women looking for marriage partners may exclude those too old or too young, those who are incarcerated, those with educational attainment deemed too little or too much, or those with too little income (Goldman, Westoff, and Hammerslough 1984). Using these techniques, Jaynes and Williams (1989, 539) calculated that black unmarried women above the age of twenty-two are in marriage pools with relatively few men. For example, at age twenty-six, black women with less than a high school education are in a marriage pool that has 651 men per 1,000 women. For twenty-six-year-old black women with some college education, the ratio is 772 men per 1,000 women. Conversely, black men can select from a large pool of unmarried women. Black men who

(Continued next page)

FEATURE 7.2 *(Continued)*

are twenty-five years of age or older are in a marriage market containing between 1,100-1,200 unmarried women per 1,000 unmarried men.

Of the black males who are actually available for marriage, those who are in the proportion who do well economically tend 'to have quite conventional family lives. Approximately 90 percent of black male college graduates are married and living with their spouses. But this group of successful black males is not enough to provide marriages for most black females. For one thing, twice as many black men as women marry outside their race. For another, black women have higher average levels of education than black men. They are more likely to finish high school, and they make up 57 percent of black college students and an even larger proportion of those earning degrees. In the past, black women tended to marry down in educational level. But now there is increasing resistance to doing so, and hence about one-third of college-educated black women are remaining unmarried into their thirties.

censure or social sanctions for premarital sexual relations and pregnancy. Since blacks lag far behind whites on most indicators of economic and educational status, they are much more likely to be exposed to the conditions that produce out-of-wedlock births. In a major five-year study on the status of black Americans, Jaynes and Williams (1989, 529) note that the above background conditions are typically found in clusters, so that families with few economic and social resources are least able to control youths or to reward them for confining childbirth to marriage.

One persuasive argument is that black men at the lower-class level are reluctant to get married because they do not feel confident that they could support children. And those marriages that do exist tend to fall apart because of economic pressure. There is plenty of evidence that people are dissatisfied with their marriages when economic pressures are severe (Voydanoff 1991), and these pressures hit especially hard in the black lower class.

Studies conducted since the 1970s show that black marriages are more likely to end in disruption than white marriages. Whereas white women are more likely to be divorced than separated from their husbands, black women are more likely to be separated than divorced (Wilson and Neckerman 1986, 237). These differences in marital disruption and marital status for blacks versus whites primarily stem from the underlying effects of social class and racial segregation. During the 1970s and 1980s, when rates of marriage for blacks were declining, the black underclass was growing. Economic restructuring and inflation drove up poverty rates for blacks, but underclass communities were mainly created where there was also a high degree of spatial segregation. Patterns of isolation and segregation in housing encouraged physical decay, crime, drug use, joblessness, welfare dependency, and nonmarital birth among the black underclass living in urban ghettos. According to Massey and Denton (1993), spatial segregation of blacks in this country is a relatively recent phenomenon. "American Apartheid," as they call it, is encouraged by discrimination and prejudice, but once in place, such racial segregation tends to produce even more economic and social disparities between blacks and whites, leading to further discrimination and more segregation. In this view, black marriage patterns are a major outcome of segregation and economic inequities, rather than their cause.

Family Assistance Benefits

Some conservative observers have suggested that it is primarily welfare laws that have caused illegitimacy among blacks. Because mothers have received aid according to how many dependents they have and whether the father was absent, antigovernment politicians blamed welfare for out-of-wedlock births. Some have argued that AFDC (Aid to Families with Dependent Children) encourages unmarried pregnant women to have more children and set up their own households, and for married mothers to divorce their husbands. The assumption is that in the absence of such payments, women and men would be forced to rely on other options, such as avoiding pregnancy, having an abortion, living together, getting married, sharing a residence with parents, or staying in an unhappy marriage. If this was the case, then those areas with higher levels of benefit payments might be expected to have higher rates of births to unmarried women and higher divorce rates. A number of studies have tested this hypothesis with statistically detectable but modest associations between these factors (Ellwood and Bane 1985; Garfinkel and McLanahan 1986; Moore and Caldwell 1976; Duncan et al. 1988; Hayes 1987). Ellwood and Summers (1986) found no significant relationship between the percentage of black children in single-parent households and the level of AFDC benefits in different states. In addition, the number of black children living in female-headed households rose sharply during the 1970s, at the same time that AFDC declined (Jaynes and Williams 1989). Cherlin (1992) notes that even if assistance payments accounted for some part of recent illegitimacy, this does not explain why divorce rates and births to unwed mothers grew to be so high in the period between 1925 and 1960, before significant welfare benefits were available. In sum, the evidence seems to suggest that family assistance benefits play a relatively small role in the overall trend toward greater divorce and illegitimacy among blacks.

Research does seem to suggest, however, that public assistance payments may reduce the likelihood that an unmarried pregnant woman marries in haste or obtains an abortion (Moore and Burt 1982). Higher family assistance benefits are also associated with mothers marrying less rapidly after divorce (Hutchens 1979). Perhaps the strongest and most consistent finding is that the availability of AFDC encourages single mothers to form their own households rather than living with their parents (Ellwood and Bane 1985).

The Relative Economic Status of Men and Women

Many researchers suggest that high rates of illegitimacy and marital disruption among blacks result from the relative economic status of black men and women. Wilson (1987) claims that low rates of employment for black men make marriage less attractive for both men and women. In 1985, a black male was about 2.5 times as likely as a white male to be unemployed or out of the labor force (Jaynes and Williams 1989). In general, black marriage rates have declined most in those regions of the country where the ratio of employed black men to employed white men has declined most. Wilson and Neckerman (1986) show that the ratio of employed black men to women in younger age groups has fallen rapidly in the recent past, and that these changes parallel changes in marriage rates. While joblessness is an important concern, even earnings of fully employed black men have declined since 1973, with blacks in all income categories earning much less, on average, than their white counterparts.

As the employment and earnings of black men have decreased, employment and relative earnings of black women have risen. In the 1940s, the wage and salary earnings of black women were about a third of those of black men; by the 1950s black women's earnings were over half of black men's; in the 1960s they reached two-thirds; and by the 1970s, black women's income was three-fourths of black men's income. By the 1980s, black women were earning over 80 percent of what black men did. For whites, a different pattern holds: in the 1940s, women's median earnings were about half of men's, and by the late 1980s, white women's earnings had only increased to 64 percent of white men's earnings (Jaynes and Williams 1989; England and Browne 1992). To be sure, black women are disadvantaged relative to both white women and black men, but the relative gap between their earnings and those of black men has been declining. The question thus arises, Does increasing income independence for black women make marriage less attractive to them? In general, researchers find that increased economic independence for women leads to increased divorce and separation and that men's joblessness is associated with family stress and marital disruption (Garfinkel and McLanahan 1986). With black women earning more, and with the scarcity of black men who are able to support families, many black mothers are likely to remain unmarried. But the economic interpretation may not be the only one, for it is apparent that black women, *as women*, have been fighting for power and independence vis-à-vis men longer and more successfully than white women. The circumstances of racial discrimination and poverty are largely responsible for this, but black women have taken the opportunity and created a type of family system that gives them greater power and economic leverage, while reducing their subordination to men.

Poverty and Black Family Form

Although changes in family form have often been thought of as *causing* poverty among blacks, it is important to remember that family arrangements are more typically the *result* of economic conditions. Recent research indicates that for whites, the formation of female-headed households from divorce or separation can be seen as a cause of poverty. For blacks, however, poverty is often the result of an already poor two-parent household splitting up (Bane 1986). Hill (1983) discovered that the long-term income of children in two-parent black families throughout the decade was lower than the income of children in female-headed nonblack households. We can thus conclude that it is not family structure per se that causes poverty among blacks.

The important trends in the marital status and living arrangements of blacks closely parallel those of whites. Since 1960 these trends include:

- lower marriages rates
- a delayed age at first marriage
- lower birth rates
- higher divorce rates
- earlier and increased sexual activity among adolescents
- a higher proportion of births to unmarried mothers
- a higher percentage of children living in female-headed households
- a higher proportion of women working outside the home
- a higher percentage of children living in poverty

With so many similarities between trends for the two groups, it is misleading to place sole reliance on race-specific differences to account for recent trends. Black men, like white men, are much more likely to get married, stay married, and be involved fathers when they are employed in secure jobs that earn decent wages (Taylor et al. 1991).

Once-popular explanations that relied on slavery or African cultural heritage can tell us little about contemporary family patterns. While all of the factors discussed above undoubtedly contribute something to higher rates of out-of-wedlock birth and marital disruption for blacks, the most convincing explanation is economic. A much larger proportion of blacks than whites live in poverty.

The crucial group of black females that accounts for the growing proportion of female-headed households and illegitimate childbirths is the poverty class. Not all female-headed households are of this type; a majority of them consist of women who have been married, but who are now divorced or separated from their husbands. It is unmarried teenage mothers who are producing the largest number of children. Staples (1985) points out that these women have particularly poor prospects of marriage. Not only are there relatively few men available, but the ones with whom they are in contact have poor economic prospects. Unemployment rates are over 50 percent for black teenagers in inner-city neighborhoods. If we count dropping out of the labor force entirely—that is, giving up on seeking a job and registering for employment—the figure rises to 75 percent. It is estimated that one-third of these black inner-city teenagers

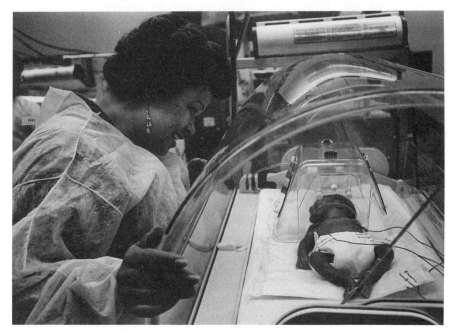

Low birth weight is a significant factor in infant mortality. Black infants are more than three times as likely as white infants to be born at less than 3.25 pounds in the United States.

have serious drug problems and others are alcohol abusers. Given these conditions, desirable marriages for young black women are not very available (see feature 7.2). And when marriages do take place, the chances of divorce are high (two-thirds of all black marriages end in divorce), due especially to the volatile economic positions of the men.

Added to this picture is the fact that poverty-level black women tend to have traditional attitudes about motherhood. They are much less likely than white teenagers to use contraceptives, and only a small minority (4 percent) terminate premarital pregnancies by abortion. Staples (1985) suggests that for these young women, bearing a child is the most important thing they can do with their lives. Their own career prospects are low; their chances of a traditional, stable marriage to a man with a respectable job are even lower. Rather than live without having children and a husband, they choose to have the children and rear them themselves.

The effects of extreme poverty conditions on many young black mothers and their children are reflected in recent data on infant health. Overall infant mortality rates are significantly higher in the United States than in most other industrialized countries, but the rates for U.S. blacks are more than double the rates for whites. Nearly 39,000 of the 3.8 million babies born each year in the United States die before they reach the age of one year. The United States' infant death rate of 10.1 per 1,000 live births is higher than that of Ireland, Spain, Italy, and twenty other countries. For blacks the picture is even worse. The infant mortality rates in cities with large black populations, such as Washington, D.C., and Philadelphia, are higher than those in some Third World countries, such as Jamaica and Costa Rica. In the United States, not only are black infants more than twice as likely to die in their first year as white infants, but black babies are more than three times as likely to be born at less than 3.25 pounds. Low-birth-weight babies are forty times more likely to die during the first months of life and two to three times as likely to be disabled by conditions such as blindness, deafness, and mental retardation (Hale 1990). These trends are directly related to poverty and the attendant lack of adequate nutrition and medical care. For this group of black families, things appear to be getting worse.

HISPANIC FAMILIES

One of the fastest growing segments of the American population is composed of Hispanics, members of ethnic groups that have migrated to the United States from Spanish-speaking areas. There were over 24 million Hispanic-Americans in 1992. Hispanic-Americans comprised less than 3 percent of the total U.S. population in 1950, but by the early 1990s, they were almost 10 percent of U.S. residents (O'Hare 1992).

Hispanic Subgroups

"Hispanic origin" is considered an ethnic identity rather than a race, which presents problems for census takers who must classify people into categories. People of any race can be classified as Hispanic, on the basis of tracing their ancestry to countries where Spanish is the primary language. There are several rather different Hispanic groups in the United States: Mexican-Americans (or Chicanos) make up 60 percent of the Hispanic population; Puerto Ricans make up 12 percent; Cubans, about 5 percent; and persons from the many different Central and South American countries, 23 percent (O'Hare 1992). Mexican-Americans are largely located in the Southwest, from Texas to California. Puerto Ricans are located primarily in New York and surrounding states; Cubans are mainly located in Florida. Most Cubans migrated as refugees from the Castro regime between 1960 and 1973, though there was one last, dramatic boatlift of 125,000 persons in 1980. The fastest growing group from 1980 to 1990 was the catchall "other Hispanics" composed of immigrants from Central and South America. Most Cuban-Americans came from the upper and

TABLE 7.1 Profile of Hispanic Subgroups in 1991

	Mexican-American	Puerto Rican	Cuban	Central/South American
Under 15 years old	30%	30%	14%	25%
65 years and older	6	7	25	5
Median age	24	27	40	28
Education				
Less than 5 years of school	15	7	6	8
4 or more years of college	6	8	18	16
Families below poverty level	30	39	18	25
Female-headed households	21	41	22	25
Family Size				
2–3 persons	43	55	67	50
4–5 persons	40	37	29	40
6 or more persons	17	8	4	10
Birthrate per 1,000 persons	23	20	10	na

Source: U.S. Bureau of the Census, 1992, p-20., No. 465; U.S. Department of Health, Education and Welfare, 1989, Vol. 38, No. 3.

middle classes, and hence they are quite different from the other two large Hispanic-American groups. Cubans on the whole tend to be much older than the average of the American non-Hispanic population, whereas Mexican-Americans and Puerto Ricans tend to be much younger. Cubans in most respects are much like the majority white population: relatively high levels of education, relatively few families below the poverty line, and small families and low birthrates (see table 7.1).

The other two large Hispanic groups are quite different from the American mainstream. Not only are there more young persons and far fewer old persons among Mexican-Americans and Puerto Ricans, but both groups have relatively low levels of education and lower incomes. Puerto Rican families, in particular, are likely to be below the poverty line and to be female-headed with no husband present. Mexican-American families have a poverty rate that is higher than that of the mainstream, but lower than that of Puerto Ricans. Mexican-American families have a rather low level of female-headed households but the percentage has been increasing in recent years. An overall impression is that Puerto Rican families are somewhat like the black families in the poverty sector, where there is a predominance of female-headed families.

The proportion of adults who divorce is also higher for Puerto Ricans than whites. Whereas the Puerto Rican familiy reflects a more female-centered pattern, the Mexican-American family tends to be viewed as a conservative, father-centered structure (Mirandé 1977, 81). Mexican-Americans are less likely to be divorced than the majority population, and much less likely to divorce than blacks or Puerto Ricans. When separation is included with divorce, Mexican-Americans look much more like non-Hispanic white households in their rates of marital break-up (Bean and Tienda 1987). This means, incidentally, that we must look elsewhere for causes of poverty among Mexican-American families. Although there are no conclusive studies of this issue, which has received much less attention than the issue of black poverty, it is quite likely that a combination of low educational credentials and employment discrimination is responsible.

Mexican-American Families

Recent research on Mexican-American families parallels the larger body of research on black families. As a corrective to earlier studies that held minority family structure accountable for poverty and other problems, recent scholarship has focused on the strengths and adaptability of Chicano families. While some of this research tends to idealize or romanticize certain aspects of Chicano family life, it avoids the previous tendency to blame the victim (Vega 1991). Social scientists, many of them Chicanos or Chicanas, have tended to focus on three general topics related to chicano family form and functions: (1) historical trends, (2) family solidarity, and (3) gender relations. We will briefly review some of the major findings in each of these areas.

Historical Trends

In 1848, after the Mexican-American War, the United States annexed territory in what is now the American Southwest. In general, land ownership was transferred from Mexican to American hands, in spite of treaty provisions that guaranteed the original Chicanos continued ownership of their lands. In the late 1800s, indigenous Chicanos who had worked for themselves as subsistence farmers began to be employed as subordinate wage laborers in agriculture, ranching, railroads, and mining (Barrera 1979; Zinn and Eitzen 1987). The population of the American Southwest grew rapidly, supplemented by large-scale immigration from Mexico. People tended to move from Mexico to the United States in "immigration chains," whereby family units once living in Mexico were gradually relocated over time in the United States.

Divisions of labor in traditional Mexican families tended to be rigidly defined by gender. This was also true of Mexican-American families living in the southwestern United States in the mid-nineteenth century. Men performed virtually all of the productive work outside the household and women did virtually all the housework, child care, and—in rural areas—tended gardens. After about 1870, Chicanas and their children were increasingly recruited into the labor force as domestics, as well as in the textile industry, in canning and packing houses, and in other agricultural endeavors, especially seasonal harvesting. Wages tended to be extremely low in all of these jobs (Zinn and Eitzen 1987).

Hispanic women tend to be the guardians of their families' cultural traditions.

One of the most distinctive historical features of Mexican-American family life was the reliance on extended networks of kinfolk for emotional and material support. Not only was household size larger for Mexican-American families throughout the Southwest, but the family also consisted of an extended network of relatives living outside the household. "La familia" included grandparents, aunts, uncles, cousins, married brothers and sisters, and their children. A unique *compadrazgo* system also linked godparents and children in a system of mutual exchange and support. *Padrinos*, or godfathers, and *madrinos*, or godmothers, were nonbiologically related individuals who became members of the extended family and participated in the major religious ceremonies of the children's lives, including baptism, confirmation, first communion, and marriage. Godparents acted as *compadres*, or coparents, disciplining children, offering companionship to parents and children, providing emotional support, and offering financial aid (Griswold del Castillo 1984).

Because of high fertility, Chicano household size tended to be large. Many households were made even larger with the addition of compadres, boarders, adopted and visiting friends, and servants. While almost all Mexican-American families had kin who lived in the same town during the nineteenth century, recent research shows that most did not share their houses with relatives.

Griswold del Castillo (1984) notes that extended family households were a temporary and impermanent creation of circumstance arising out of old age, sickness, death, or economic misfortune.

Maxine Baca Zinn (Zinn and Eitzen 1987, 92) describes Mexican-American women of the late nineteenth and early twentieth century as doubly oppressed: they held the most subordinate jobs outside the home and in their private lives were subject to "a distinctive system of Mexican patriarchy." Mexican-American women were instructed to be obedient and subservient to their husbands and parents and were obliged to shoulder full responsibility for all housework and child care. While Chicanas were viewed primarily as wives and mothers, they often performed wage work to make ends meet, though such work was considered much less important than that of the men. On the whole, the women guarded Mexican cultural traditions through daily family life and with customs and rituals including folklore, songs, birthday celebrations, saints' days, baptisms, weddings, and funerals (Garcia 1980; Zinn and Eitzen 1987).

Family Solidarity

The concept of "familism" has often been invoked to describe both traditional and contemporary Mexican-American families with their higher than average birthrates and household size. Familism refers to "a constellation of values which give overriding importance to the family and the needs of the collective as opposed to individual and personal needs" (Bean, Curtis, and Marcum 1977, 760). Embeddedness in close-knit extended kin networks and valuing family solidarity over individuality or personal achievement have typically been thought to be Mexican cultural patterns handed down through the generations. Recent research contends, however, that reliance on actual and fictive kin is common among lower-class blacks and whites (Stack 1974; Rubin 1976) as well as Mexican-Americans (Bean et al. 1977; Zavella 1987). For this reason, some have suggested that dependence on close-knit kin networks might more appropriately be viewed as an adaptation to a marginal existence (Zinn and Eitzen 1987). Nevertheless, Mexican-Americans are typically found to have the highest levels of extended familism when compared to blacks and Anglos of various class levels (Mindel 1980). One of the benefits of strong patterns of kin interaction among Chicano families appears to be enhanced mental health (Ramirez and Arce 1981).

Acculturation theories based on a melting-pot ideal would predict that Mexican-Americans' patterns of extended family interaction might become more like Anglo-American family patterns over time. Recent research reveals that acculturation is not a simple linear process, however, particularly when applied to Mexican-Americans. Keefe and Padilla (1987) found that long-term residents of Mexican descent did not rely on *compadres* as much as earlier generations had. Nevertheless, they did rely heavily on primary and secondary kin for emotional and material support. Counter to the acculturation theories, local extended families were *more typical* of long-term residents than new immigrants, and the local extended family network grew stronger in conjunction with many processes assumed to promote individualism and sole reliance on the nuclear family—including English language acquisition, urbanization, and social and economic assimilation. Similarly, Zinn (1980) reports that many Chicano families are simultaneously modern *and* traditional. In light of findings like these, many contemporary scholars conclude that there is no *one* type of ethnic family structure and that Mexican-Americans construct unique family patterns based on cultural inheritance, economic necessity, religious beliefs, and a host of other factors (Keefe and Padilla 1987; Zavella 1987; Williams 1988).

Gender Relations

Mexican-American families were characterized as rigidly patriarchal in most early social scientific research (e.g., Rubel 1966; Heller 1966; Clark 1959). The father was seen as having full authority over the mother and children, and wives were described as passive, submissive, and dependent (Zinn 1980). William Madsen (1973, 22) focuses on the destructive aspects of Mexican-American *machismo* by comparing the men to roosters: "The better man is the one who can drink more, defend himself best, have more sex relations, and have more sons borne by his wife." According to such depictions, authoritarian child rearing and wife beating are assumed to be common in Mexican-American families.

Some researchers acknowledge male dominance as a persistent feature of Mexican-American families, but see it as benevolent rather than malevolent. In a review of literature on Mexican-American fathers, Mirandé (1988, 102) suggests that genuine *machismo* is characterized by bravery or valor, courage, generosity, and ferocity. Sociologically, what should be recognized here is that we are describing a family structure imported into the United States from a largely traditional and rural society. Thus, Mexican-American and many other Hispanic families come from a background which is much like the agrarian societies described in chapter 3. We saw there that these societies are highly stratified and organized around patriarchal households, with the upper class dominating the population of small land-holding peasants and farm workers. This structure has changed rather slowly in Latin America, and many of the Hispanic migrants to the United States have come to escape from the unfavorable conditions of this form of stratification.

The culture of "*machismo*" thus looks very much like the culture of an agrarian society of patriarchal households. Where coercion and even violent threat are prominent features of class relations, men value courage as well as loyalty to their group. Moreover, as we should recall from the theory of gender stratification (chapter 3), male dominance over females is maximal where there is a high level of violent conflict and where there are weapons in the household. As we recall, gender stratification is at its height in agrarian societies. Male and female roles are sharply distinct, especially in regard to sexual behavior; males act both to protect and control the females of their own families.

The ethos of the original agrarian society is slowly adapting to new conditions in the largely urban and industrial United States. The conditions that supported a patriarchal family structure are gradually eroding. The "machismo" style is apparently weakening, and Chicano spokespersons such as those quoted above are attempting to redefine it to fit a less coercive modern setting. While it may be true that youth gangs in working-class Hispanic neighborhoods maintain some aspects of the traditional conflicts and the "machismo" style, the realities of adult life are now quite different than in a traditional community dominated by the power of *caudillos* and rural landlords. The status of Hispanic families in the United States now depends upon their individual economic success rather than on the power coalitions to which they belong. The power resources of Hispanic men are declining as well. Simultaneously, the movement of women into the labor force is creating a new balance of gender resources and reshaping the Hispanic family. It appears that gender role expectations in Hispanic families will change as social and economic conditions require (Vega 1991).

Some recent research on marital interaction seems to support the notion that gender relations in Mexican-American families are more egalitarian than the traditional model assumes. Studies of marital decision making have found that Chicano couples tend to regard their decision making (but not necessarily their division of household tasks) as relatively shared and equal (Hawkes and Taylor 1975; Cromwell and Cromwell 1978; Ybarra 1982). Mexican-American women, like their Anglo-American counterparts, exercise more marital decision-making power if they are employed outside the home than if they are housewives (Ybarra 1982; Zinn 1980; Coltrane and Valdez 1993).

Mirandé (1988) suggests that contemporary Chicano fathers take more responsibility for child care than in the past and Zavella (1987) notes that Chicano husbands sometimes respond to demands from employed wives to do more around the house. Nevertheless, most researchers reject the notion that marital roles are truly egalitarian in Mexican-American households (Williams 1988; Hartzler and Franco 1985; Zavella 1987). (Neither are marital roles truly equal in Anglo-American couples; see chapter 11). Because Mexican cultural ideals require the male to be honored and respected as the head of the family, Zinn (1982) contends that Chicano families maintain a facade of patriarchy. At the same time, Chicano families can be characterized as mother-centered because women are responsible for child rearing and domestic chores and thus assume authority over day-do-day household activities. It may be that many minority families grant symbolic power to the male head of household as a gesture of appreciation for trying to earn a family wage in a discriminatory labor market. A similar pattern of symbolic deference often occurs in white working-class families. Power in such families, in terms of everyday interaction, may in fact reside more securely in the women's domain.

Segura (1984) documents how Mexican-American women, as a group, have been concentrated in the lowest-paying jobs and have made little progress over the past decade in terms of educational attainment, occupational status, or income. Limited employment opportunities for Chicanas coincide with a strong sense of obligation to the family and primary self-esteem being derived from being wives and mothers. Garcia-Bahne (1977) notes that adopting the individualism common to many white middle-class families runs counter to the collectivity of the Chicano family, but that idolizing familism can be limiting for Chicanas. She describes how various "myths" about Chicano families serve both positive and negative functions, with family loyalty, self-sacrifice, and modesty often contributing to Chicanas' social and economic oppression.

Pressures for Change in Chicano Families

In a study of Chicana cannery workers in central California, Zavella (1987) reports on how some cultural ideals about the Chicano family directly affect women's lives. Tensions and conflicts in the families she studied tended to revolve around the uses of women's time and labor. "Because women were considered primarily as homemakers who happened to work, husbands (and even children) expected the women to continue deferring to them and maintaining the needs of family members. When women contested these assumptions, conflicts emerged" (Zavella 1987, 159). As the Chicana cannery workers began to view their jobs as essential to the family's well-being, they challenged taken-for-granted assumptions about Chicano family structure. One of the cannery workers talked about how she automatically enacted traditional gender expectations, but then began to question obligatory deference to her husband:

> I found myself doing things that I didn't agree with. I'd tell the kids: "Be quiet because your father's asleep, and he's tired and he needs his rest." I'd make a meal and leave something for him to warm up. I always had this feeling, "Well he's the daddy, so I'd better make the kids be quiet, and I'd better cook his meals, or whatever." He's a grown man; he could do things for himself. But you're taught to do these things without thinking, because he's the man of the house. It's inbred in you. (Zavella 1987, 87)

Researchers have noted how life in both working-class and middle-class Chicano families begins to change when employed wives demand more participation from husbands and children (Coltrane and Valdez 1993, Zinn 1980). One Mexican-American woman interviewed by Baca Zinn commented:

> Since I've gone to work, the kids have more responsibilities. Sometimes I have to tell them what to do, but you know, they're more independent now. They

don't rely on me so much. Believe it, even Juan [husband] helps, but this has been a long time in coming. At first, he reacted by going against me; he wouldn't get after the kids, but things are different now. And it's good for the whole family. Just this morning when I was getting ready for work, there was Christopher sewing his pants. (Zinn 1980, 53)

Zinn also reports that wives' employment outside the home and attempts to share some of the household labor did not weaken the families' efforts to maintain and develop their Mexican heritage. All of the families were Catholic and many of their activities with kin tended to revolve around religious celebrations. While many were "much like non-Chicano families in day-to-day activities" (p. 57), they spoke Spanish at home, preferred Mexican and Spanish

FIGURE 7.2: Educational Attainment Differs among Racial/Ethnic Groups

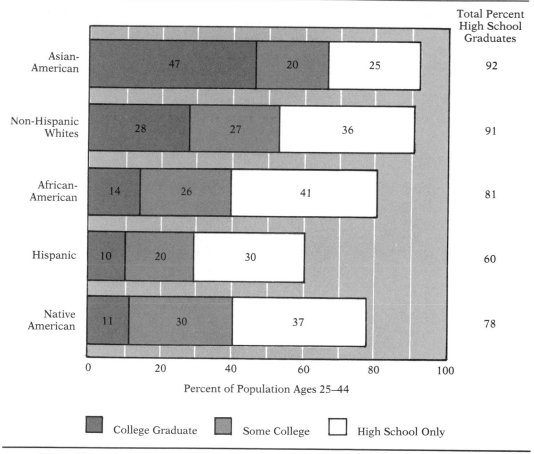

Source: O'Hare, 1992.

food, and listened to Spanish or Mexican music. In these ways, many Chicano families combined some modern family patterns with more traditional ethnic practices.

As with African-American families, there is no one typical form of Hispanic family. Even within Hispanic subgroups there is great variation depending on social class position, recency of immigration, place of residence, education, and other factors. Although Hispanics are less segregated than blacks, where urban barrios exist, the Chicanos or Puerto Ricans who live within them are faced with a large number of problems that shape the nature of family life.

Among young adults, Hispanics have the lowest educational levels of any race or ethnic group. This is partly due to the large number of Hispanic immigrants who had little schooling in their home countries, but this is also true for American-born Hispanics (see figure 7.2). Educational levels for all racial ethnic groups have been improving, but only for some members of each group, so that there is increasing educational disparity within minority groups. Because education is often the key to a good job and a promising future, the lower levels of achievement among many Hispanics is of major concern.

Hispanic Families in the Future

Hispanic families have become increasingly important in the United States and will become even more so into the beginning of the next century. While the rate of population growth of the non-Hispanic white majority has slowed, the Hispanic populations constitute a major source of what growth there is. Almost half of all immigrants to the United States between 1960 and 1990 were Hispanics. Like other immigrants, some come to escape harsh conditions at home and many are drawn by the prospects of economic prosperity. Many also come to join family members already here, and while most come legally, many slip across the border without proper documentation, or remain long after their student or work visas have expired (O'Hare 1992, 13). It is hard to know the exact numbers, since there has been a large amount of illegal migration; hence, a sizable population, especially of Mexican-Americans, is not reported by the Census.

We do know that the Hispanic population in the United States has been growing because of its high birthrate. Even if there were no more immigration (whether legal or illegal) from now on, the Hispanic population would still reach a sizable proportion of the U.S. population by the early twenty-first century. Hispanic-Americans have a birthrate that is 50 percent higher than the average of non-Hispanic women. Leaving aside Cubans, who are similar to the majority population, we can see (table 7.1) that Mexican-Americans have a birthrate of almost 23 per 1,000, and Puerto Ricans of 20, which compares to the majority white birthrate of 17 per 1,000. Hispanic women tend to bear their children earlier, too. When we add to this the fact that the Mexican-American and Puerto Rican populations are a good deal younger than the majority population, we can expect that proportionately more Hispanic persons are going to reach child-bearing age in the coming decades. These are some of the reasons for the present Hispanic population bulge, which will continue for some years to come. By the year 2020, projections indicate that Hispanics will comprise over 15 percent of the total U.S. population and will be slightly more numerous than blacks, whose percentage is also increasing (O'Hare 1992).

What difference will it make to the United States as the family structure becomes proportionately more Hispanic? In part, this depends on how much these families are assimilated into the majority culture (and which culture they assimilate to). The question has been little investigated, but there is some interesting information about the use of the Spanish versus the English language (National Commission on Employment Policy 1982, 29). Most Mexican-Americans (61 percent) either speak nothing but English or predominantly English. Among Puerto Ricans, however, only 41 percent speak nothing but English (14 percent) or predominantly English (27 percent). To put it another way: English is the dominant language of Mexican-Americans. Despite their sizable poverty sector and low educational levels, they are by and large an assimilation-oriented group, at least in terms of language. Even though most Mexican-Americans first came to the United States as temporary workers, the typical process has been one of making social ties, bringing more family members, and gradually attaching themselves to permanent settlement (Massey 1986). Puerto Ricans, on the other hand, are predominantly Spanish-speaking (59 percent). Cubans overwhelmingly emphasize speaking Spanish (80 percent). That is probably why, despite their high educational and occupational levels and their conventional family structure, Cubans remain a highly distinctive ethnic group within the United States.

Use of English depends, in part, upon whether Latinos are foreign- or native-born, and if native-born, how long they have lived here. Preference for English or Spanish depends on one's job, neighborhood, and family life-style. In urban centers with large Latino populations (New York, Los Angeles, Miami), primary Spanish speakers have access to newspapers, radio, and television stations, and many businesses cater to Spanish speakers. Elsewhere, pressures to adopt English are greater. Still, about three of four Hispanic immigrants speak English on a daily basis by the time they have been in the country for fifteen years. Younger immigrants and the children of Hispanic immigrants usually become fluent English speakers, even though many retain some knowledge of Spanish. Most observers predict that bilingualism will persist longer for Hispanics than for other immigrant groups, though it is likely to decline if immigration drops significantly (Valdivieso and Davis 1988).

ASIAN-AMERICAN FAMILIES

Asians are the fastest growing minority in the United States. In 1990, there were more than 7 million Asian-Pacific Americans in the United States, double the number that were here just ten years before (O'Hare 1992). Population projections estimate that by the year 2000, 10 million Asian-Pacific Americans will constitute about 4 percent of the total U.S. population (Gardner, Robey, and Smith 1985). The Census Bureau enumerates and identifies a total of seventeen Asian-American groups and nine groups of Pacific Islanders, the most numerous being Chinese (23 percent), Filipino (19 percent), Japanese (12 percent), Korean (11 percent), Asian Indian (11 percent), Vietnamese (9 percent), Hawaiian (3 percent), Laotian (2 percent), and Cambodian (2 percent) (O'Hare 1992). While these groups are often lumped together in reporting demographic statistics, the groups can be quite different from one another. Many Chinese and Japanese families have lived in the United States for several generations, while groups like the Vietnamese and Cambodians tend to be recent immigrants.

Historical Precedents

In the mid-nineteenth century, many Chinese men came to western United States to work on the railroads and in the mines. Most were unable to

The long-standing tradition of Chinese-American self-reliance is captured in this 1946 photograph of the Board of Chairmen of the Chinese Six Companies Benevolent Association, which was founded in San Francisco. This organization settled differences between Chinese groups, assisted the needy, and represented the Chinese in difficulties with local and national authorities.

establish families because of limits on immigration and laws that prohibited Asians from becoming citizens. Prejudice, violence, and legal and extralegal discrimination against the Chinese in the late 1800s and early 1900s kept them segregated, poor, and undereducated. As a result, Chinese-Americans tended to rely on their own resources and developed their own institutions. They created their own businesses, ran their own hospitals, formed their own insurance companies, and established their own social clubs. This self-reliance led in turn to accusations that Asians were separatists who were not assimilable into the dominant community (Kitano and Daniels 1988).

The stereotypes applied to Asians in this country have changed dramatically in the latter half of the twentieth century. Asian-Americans, especially Japanese and Chinese, have increasingly come to be seen as "model minorities" who work hard and easily conform to American ways of life. Whereas Asian men were previously stereotyped as wily and devious and Asian women as exotic and mysterious, these images have given way to the stereotype of Asian families as cohesive, conformist, and highly successful (Sue and Kitano 1973). As we will see, however, the stereotype does not fit all Asian-American families.

Modern Demographics

The majority of Asian-Americans live in western United States, including Hawaii, with only Asian-Indians, and to a lesser extent Koreans, spread relatively evenly across the country. More Asians live in California than in any other state, with twenty-five California counties each counting over 5,000 Asian-American residents and the four largest cities collectively holding almost a million Asian-Americans. In San Jose, nearly one out of five people is of Asian descent, and in San Francisco more than one in four is Asian (O'Hare 1992).

Most Asian-Americans live in large metropolitan areas, though their neighborhoods of residence are about evenly divided between central cities and the suburbs. About half of all Asian-Americans live in just five large urban centers: Honolulu, Los Angeles, New York, Chicago, and the San Francisco Bay Area. This pattern contrasts with general residence patterns among whites, two-thirds of whom live in the suburbs, and a quarter of whom live in rural areas (O'Hare and Felt 1991).

Population gains among Asian-Americans have resulted primarily from changes in U.S. immigration policy. Beginning in the 1960s, a new wave of Asian-American immigrants began coming to the country in record numbers. This trend continued and even accelerated during the 1970s and 1980s. Although the number of Asians coming to this country is less than the number of Hispanics, the rate of growth for Asians is higher than for any other group, including whites. This phenomenal growth is likely to continue because recent immigration policies favor family reunification. This means that more Asians will be eligible to apply for entry. Newer immigration policies also favor admitting more highly educated and well-trained workers. Because Asians are more likely than others to have advanced degrees and job skills, continued high levels of Asian immigration are expected in the future (O'Hare and Felt 1991).

Although birthrates for Asian-American women have been below those of whites in the past, they now exceed those of whites. In 1992, the average white woman had 1.9 children, compared to 2.3 for Asian-American women, 2.5 for African-American women, and 2.7 for Hispanic women (U.S. Bureau of the Census 1992a). Whereas Hispanic and black women tend to have their first child at a younger age than do whites, Asian-American women tend to have first births later than white women. Because Asian-American women are more likely than others to stay married, however, they tend to have slightly more children than their white counterparts. Newly immigrating Asian groups tend to have the highest birthrates. For example, Vietnamese-Americans have birthrates approaching those of Hispanics.

Stresses and Strains

While research on Asian families is rare, it typically focused on the importance of unique cultural influences. For instance, in a decade's review of literature on minority families, Staples and Mirandé (1980, 495) comment that

Asian-Americans set a premium on their children's educational achievement and obedience to authority.

"culture seems to be the key element in Asian family life." The cultural approach to studying Asian family life focuses on traditional child socialization practices and shows how individual needs are sacrificed to the overall welfare of the family unit. Asian child-rearing patterns are characterized as instilling obedience, loyalty, self-control, and a desire for educational achievement. Traditional Asian family relations also include rigid task and role divisions, with wives expected to stay home and assume major responsibility for household tasks and child care, while husbands fill the role of breadwinner. In the traditional Asian family model, the father has final authority and wields power sufficient to enforce his decisions. Children and wives are expected to be obedient and submissive, and children are supposed to defer to both parents' wishes (Kitano and Daniels 1988, 115).

Autobiographical accounts reveal some of the stresses and strains that result from traditional Asian family patterns. For instance, Jean Wakatsuki Houston (1985) discusses how her mother, a first-generation immigrant from Japan, subordinated herself to her husband and sought ways to elevate his position in the family:

> He was always served first at meals. She cooked special things for him and sat next to him at the table, vigilantly aware of his needs, handing him the condiments and pouring his tea before he could ask. She drew his bath and massaged him and laid his clothes out when he dressed up. As I was growing up I accepted these rituals to be the natural expressions of a wife's love for her husband. . . . This attitude, that to serve meant to love, became an integral part of my psychological make-up and a source for confusion when I later began to relate to men. (1985, 13)

Houston describes an issue common to many Japanese-American families: the tension between first-generation immigrants *(Issei)* and their children *(Nisei)* that stems from the younger generation adopting the dominant culture's language and customs:

> Whenever I succeeded in the *hakujin* [Caucasian] world, my brothers were supportive, whereas Papa would be disdainful, undermined by my obvious capitulation to the ways of the West. I wanted to be like my Caucasian friends. Not only did I want to look like them, I wanted to act like them. I tried hard to be outgoing and socially aggressive, and to act confidently, like my girl friends. At home I was careful not to show these traits to my father. For him it was bad enough that I did not even look very Japanese; I was too big, and I walked too assertively. My breasts were large, and besides that I showed them off with those sweaters the *hakujin* girls wore! My behavior at home was never calm and serene, but around my father I still tried to be as Japanese as I could. (Houston 1985, 17)

Like many *Nisei*, Jean Wakatsuki eventually married a Caucasian (the novelist John Houston). Younger Japanese- and Chinese-Americans have extremely high rates of out-marriage, that is, marriage to people of a different race (see table 8.1). For example, a majority of third-generation Japanese-Americans marry non-Japanese, with Japanese women much more likely than men to out-marry. A primary reason appears to be the more acculturated Asian women's dissatisfaction with more traditional Asian men's limited attitude toward women (Kikumura and Kitano 1973; Staples and Mirandé 1980).

The stereotype of Asian-Americans as model minorities can be harmful in that it masks some of the unique problems and strains they face. For instance, the common perception that Asian families are extremely cohesive is supported by the relatively low number of divorces among Asians, especially first-generation immigrants. Nevertheless, divorce rates are increasing for Asians, as they are for other races. Acculturation and generational changes have produced different motives and expectations for marriage among Asians, and these differences have contributed to increases in rates of marital disruption. Because most Asians come from societies in which women rarely work for wages outside the home, the modern necessity of having two wage earners can lead to special conflicts and tensions in Asian-American families. As noted above, intergenerational conflicts between "old fashioned" parents and their Americanized children have been common for most immigrant groups. Since Asian-American families tend to observe strict privacy norms, they rarely seek counseling and thereby reinforce the notion that they are "problem-free" (Kitano and Daniels 1988, 166).

Another area in which Asian-Americans are assumed to be an ideal minority is educational attainment. For instance, in 1992, almost half of Asian-Americans between 25 and 44 years old had graduated from college (see figure 7.2). All Asian groups except the Vietnamese have higher rates of completion for high school, college, and doctoral degrees than other minorities or majority whites. Selective migration of better educated people from Asian countries partially accounts for the higher levels of education, but native born Asian-Americans also have extremely high levels of educational attainment. The stereotype of Asians doing well in school is also supported by higher-than-average Scholastic Achievement Test (SAT) scores in mathematics (Kitano and Daniels 1988).

Nevertheless, not all Asians do equally well in school and one out of five Asian-American adults has less than a high school education, about the same percentage as for white adults. The stereotype of Asians as overachievers can have deleterious effects on good students as well as on below-average students. Some researchers report a "pushy parent" syndrome, in which children are forced to study continually and are channelled into focusing only on the natural sciences. Instead of praise for a report card with all As and one B, the pushy parent criticizes the student for receiving the one B grade. Young Asian-Americans can also receive mixed messages from their parents. On the one hand, they are expected to stay in close contact with their parents and remain dependent on them. On the other hand, they are expected to work very hard, be independent, and earn money while in school. The result can be high levels of stress (Kitano and Daniels 1988, 165).

Once out of school, job discrimination can add to the pressures inspired by the stereotype of Asians as overachievers. In all age groups and at all educational levels, white Americans earn more than Asian-Americans (O'Hare and Felt 1991).

Kitano (1976, 93) suggests that Asian men have lower earnings because they are excluded from high-earning occupations, are prevented from advancing in particular occupations, and experience higher levels of unemployment. Thus, while Asian-Americans tend to be well-educated, they receive lower returns on their education than European-Americans.

Unlike other minority groups, Asian-Americans tend to have higher than average household incomes. Similar to blacks and Hispanics, however, the differences among Asian families are substantial, and recent trends indicate ever-widening gaps between families in this large and diverse group. For example, the poverty of Asian-Americans has increased over the past decade and is nearly twice as high as that of whites. At the same time, the average household income of Asian-Americans is higher than that of whites (O'Hare and Felt 1991).

What might account for these differences? For one thing, Asian-Americans tend to live in households with more workers than any other group. Although official government figures do not even include many Asian-Americans who work in family-owned businesses, almost one of every five Asian-American families includes at least three wage-earning workers. The relatively high number of workers is related to typical living arrangements. Three out of four Asian-Americans live in married-couple families, slightly more than the percentage for whites. But Asian-Americans are only half as likely to live alone, and they are twice as likely to live in extended family situations or to double-up in one household. More workers means more earnings, and the average income of Asian-Americans reflects this. In 1990, a greater percentage of Asian-Americans than white Americans lived in households with incomes of over $50,000 per year (O'Hare and Felt 1991).

Another reason for the wide variation among Asian-American families is the composition of recent immigrant pools. Less advantaged migrants from Southeast Asia have swelled the ranks of the poor at the same time that more highly educated people with special skills have come here with immediate high earning potential. It makes a big difference whether one considers the social and economic situation of recent Laotian or Cambodian immigrants—who tend to be poor and to utilize government welfare programs—or the situation of Japanese or Taiwanese immigrants—who tend to have higher earnings than virtually any other foreign-born group in the United States today. Second- and third-generation Asian-Americans also move into the ranks of the middle and upper-middle class at higher than average rates.

While researchers who study black and Hispanic families have moved away from an emphasis on culture to an emphasis on institutional and economic constraints, those studying Asian families have relied almost exclusively on cultural explanations (Staples and Mirandé 1980). Evelyn Nakano Glenn (1983) criticizes the cultural approach for portraying Asian-American families as essentially static. By tracing historical shifts in the economic, legal, and political constraints faced by Chinese-American families, she illustrates how Chinese family structure has changed drastically: from split-households, to small-producers, to dual-wage earners. Rather than seeing traditional Chinese culture as determining Chinese-American family form, she shows that individual families devised strategies that varied according to the structural conditions that prevailed during a given historical period. Split-households, prevalent until about 1920, adopted the strategy of sending married men to the United States to make money, while the wife and children maintained a separate household in China. The small-producer household, prevalent from about 1920 to the 1960s, relied on the labor of all family members in small family-run businesses such as laundries or restaurants. The dual-wage pattern, common among immigrants arriving after 1965, is based on a strategy of both husbands and wives working in low-wage jobs outside the home. Glenn's (1983) analysis reminds us that culture alone does not determine family form, and that families continually adapt to changing economic and institutional circumstances. This is as true for Asian families as it is for black, Hispanic, and white families.

NATIVE AMERICAN FAMILIES

The term *Native American* (sometimes called "American Indian") refers to a diverse assortment of tribes and ethnic groups living in North America before Europeans and others settled the continent. Although estimates vary, there were probably between 10 and 18 million Native Americans living north of Mexico when Europeans arrived in the late fifteenth century (John 1988). Native American family and kinship structures varied enormously among different cultural groups, as did social customs, lifestyle, subsistence practices, religion, and political organization. North American Indians, including Eskimos and Aleuts, spoke over 300 different languages and represented over 100 different major cultural groups before contact with white settlers. With such diversity, it is impossible to identify one dominant historical type of Native American family.

European settlement almost wiped out the native population through disease, destruction, and displacement. By the end of the nineteenth century there were only about 250,000 Native Americans in the entire United States (Martin 1978; Dobyns 1983; Taylor 1994). In spite of high levels of infant mortality and a relatively short life span, the North American Indian population began to grow in the twentieth century. The 1990 Census counted 1.9 million American Indians, including 81,000 Eskimos. Part of this dramatic increase is due to the fact that Native Americans have the highest birthrate of any ethnic group (2.9 children per woman). Part of the increase is also due to better census reporting, though there is still probably substantial undercounting, especially on reservations. The remarkable increase in the number of people who identify as American Indian is also likely the result of a resurgence of ethnic pride among Native Americans (O'Hare 1992). In 1990, American Indians claimed membership in over 500 tribes. Most, however, identified with one of the eight largest tribes: Cherokee, Navajo, Chippewa, Sioux, Choctaw, Apache, Iroquois, and Pueblo (includes Hopi, Zuni, Acoma, and Laguna).

Nearly half of all Native Americans live in the western part of the United States. Very few tribes have been able to retain their ancestral homelands, though some native peoples were never forced to leave their native geographical areas (mostly in Alaska, Arizona, and New Mexico). About half of Native Americans lived in rural areas in 1990, with about half of these living on one of the 314 reservations left in the United States. Still, over a fourth of Native Americans now live in suburban areas, and another fourth live in larger cities such as Los Angeles, Phoenix, Tulsa, and Milwaukee (O'Hare 1992; Snipp 1989; Taylor 1994). Not surprisingly, Native American families look different depending on tribal affiliation, geographic location, and socioeconomic status.

Diversity among Native American Families

Research on American Indian families is still relatively rare compared with the large number of studies on other racial ethnic groups in the United States. We are beginning to identify some of the important features of American Indian families, including extended family structure, respect for elders, and distinctive socialization practices. Nevertheless, diversity among different tribal groups and differences between rural and urban residence makes generalization about American Indian families difficult.

In the past, nearly all the principal variants of marriage and family found among nonindustrial peoples were commonly practiced by Native Americans. Looking across societies, one can find arranged marriages; bride price and dowry; adoptive and interfamily exchange marriages; trial marriages, monogamy, polygamy, and temporary polyandry; premarital and extramarital sexual relations; divorce; patrilocal, matrilocal, and neolocal residence patterns; patrilineal, matrilineal, and bilineal descent systems; and other kinship variations (Driver 1969; John 1988; Taylor 1994). Differences among indigenous North American families were undoubtedly as numerous as the similarities among them.

This family belongs to the Yazzie tribe. Most Native Americans traditionally have derived their identity from their family and local kindred group.

Historically, most Native Americans derived their identity from the family and local kindred group, so that the externally imposed label of "Indian" made little sense to one who lived daily life as a Navajo or Cherokee. Even today, there is a strong sense of tribalism and pride among specific groups. Native American families continue to speak different languages, observe different customs, and maintain different kinship, political, and social practices. For example, most Navajos tend to socialize with and marry other Navajos and most Sioux tend to socialize with and marry other Sioux, in spite of residence on reservations or in cities. Some Native American groups, such as the Pueblo peoples of the American Southwest, maintain especially strong tribal and family identities. The Hopis, Zunis, and other descendants of the ancient Anasazi remain loyal to their matrilineal clan systems, native tongues, and religious ceremonies, emphasizing sobriety, self control, and inoffensiveness (Benokraitis 1993; Olson and Wilson 1984).

In spite of tribal differences, a common history of oppression and the recent upsurge of mobilization of protest movements have helped to foster a shared sense of American Indian culture among diverse peoples in the United States and Canada, and to promote a modern "pan-Indian identity" (Jarvenpa 1988). Continuing high levels of spatial segregation on reservations have also promoted a sense of unique American Indian culture and have allowed many traditional family practices to be retained (Red Horse 1980; Taylor 1994). Over the past several decades, however, the range and diversity of Native American family patterns have been reduced, as more Indians take up urban residence, attend majority schools, and marry non-Indians.

Common Features of Native American Families

Some researchers claim that "the extended family network" is a universal cultural feature among Native Americans of both the past and the present. Indian extended families or family networks have been described as structurally open, including several households of significant relatives and nonkin as family members. Extended family members engage in obligatory mutual aid and actively participate in important ceremonial events (Red Horse 1980). Some scholars suggest that the extended family form was a fair description of Native American households in the past, but that today it is more of an ideal than a reality (John 1988).

It is hard to determine whether extended family structures and practices are widespread because we have few representative studies of Native American families. Another problem is that modern survey measures may not be sensitive enough to capture unique Native American family structures. Yellowbird and Snipp (1994, 180) point out that the U.S. Census Bureau defines a family as "two or more persons, including the householder, who are related by birth, marriage, or adoption, and who live together as one household." They suggest that this definition tends to obscure the existence of extended family relationships among Native Americans, for it treats all related people under a single roof as just one family and it treats parents or grandparents living in a mobile home near children or grandchildren as entirely separate households and families. A related problem is that Census terminology does not reflect Indian cultural conceptions of family membership. For example, an Indian "grandmother" may actually be a child's aunt or grand-aunt in the Anglo-Saxon use of the term, and "cousin" may have variable meaning not necessarily based on birth and marriage (Yellowbird and Snipp 1994). The picture is further complicated by the fact that terms, meanings, and family relations can vary significantly from tribe to tribe.

According to 1990 data, Native Americans (like Hispanics and Asian-Americans) are more likely to live in married couple households than either African-Americans or European-Americans. Census data also show that Native Americans (like Hispanics and African-Americans) are more likely than whites to have more and younger children, to have teenage pregnancies, and to be single mothers (O'Hare 1992). As noted above, however, these demographic data don't tell us much about what goes on within and between families. Most researchers have concluded that American Indian families are more firmly based on interdependence than white American families. Although joint residence among Indians may be decreasing, a strong norm of inter- and intra-family exchange is still common, especially on the reservations. For many, family and tribal identity are made salient through patterns of reciprocal exchange, ancient cultural customs, and public participation in important life course rituals (Taylor 1994; Yellowbird and Snipp 1994).

Using an acculturation framework, some researchers have identified different Native American family types. For example a three-part longitudinal study of 120 American Indian families living in the Oakland, California, area found four distinctive styles of adaptation to city living. A *Traditional* group endorsed and retained Indian values and behaviors, living similarly to ways they had on the reservation. A *Transitional* group left traditional Indian values behind and tried to adopt white values and behaviors. A *Bicultural* group held onto Indian values and behaviors at the same time as they adapted to white expectations for behavior. A *Marginal* group abandoned Indian ways but did not fit into the white culture either, feeling alienated from both worlds. In this study, the author found that the Bicultural group, by retaining some native language and customs, was the most successful in "social and psychological adaptation." (Miller 1979, 479).

American Indian families have also been characterized as affording a special role to elders. Through elders' meetings and councils, older tribal members have directed the spiritual, social, and cultural needs of the family and community. Children are taught to respect their elders and elders expect family members to offer assistance without being asked. By virtue of their age, older Native Americans are assumed to have pleased their creator, and move into a stage of life that Red Horse (1980, 464) calls "assuming care for," wherein they are seen to be important stabilizing forces in the extended family and in the tribe. Grandfathers take grandsons and nephews for long walks to talk about life, nature, and tribal values. Grandmothers are often the center of family life and assume substantial responsibility for child care and for passing on tribal language and family traditions.

Some of the "core values" of Indian families are different from those of the dominant culture (Snipp 1989). Besides specific tribal identity, scholars have pointed to Native American conceptions of "time," "leadership," "sharing," "cooperation," and "harmony with nature," as departing from western ideals. For example, the Pueblo Indians blend past and present in their conception of time and say "time is always with us." Western cultures, in contrast, view time as linear and are often preoccupied with "watching the clock." Leadership among American Indians entails a kind of servitude, and the good leader is one who is more sacred, person-oriented, honest, and intuitive. For the dominant culture, in contrast, leadership is often characterized in terms of ambition, aggressiveness, and the ability to dominate situations (Yellowbird and Snipp 1994, 196). These cultural differences help shape family life and lead to some unique socialization practices in Native American families.

Responsibility for children tends to be shared by a wider range of adults than in the dominant culture. In general, there is little stigma attached to having children outside of marriage, reflected in rates of nonmarital childbirth for Native Americans that are second only to those of African-Americans. Most Indian cultures encourage children to become self-reliant at earlier ages than European-American families, and place more emphasis on traditional ceremonies and spiritual values. Rather than direct intervention, many Indian families teach by example and attempt to foster good listening skills and respect for authority (Yellowbird and Snipp 1994). These values sometimes conflict with the emphasis on competition and accumulation common to many American families.

Poverty rates are about as high among Native Americans as they are among African-Americans, leading to a variety of social problems. High levels of infant mortality, suicide, alcoholism, psychological distress, poor health, and shorter life expectancy are strongly associated with poverty and are higher among Native Americans than among most other groups. Although poverty rates are highest for Indians living in rural areas, the extended family structure that is more common in those areas helps to mitigate some of the adverse personal impacts of impoverishment. Transition to urban areas often brings increased economic advantages for Native American families, but also carries risks associated with social isolation and loss of supportive extended family networks.

Rates of out-marriage are higher for Native Americans than for any other racial ethnic group in the United States. Over half of all marriages of Native Americans are now to people of other races (O'Hare 1992). These high rates of out-marriage are probably due to the small number of Native Americans in the population, the increased migration of Indians to urban areas, expanding opportunities for education and employment in nonreservation settings, and a shift toward more favorable attitudes about Indian culture. Unlike among Asian-Americans, out-marriage is not strongly associated with gender. Whereas intermarriage rates for Asian-American women are much higher than for Asian-American men, rates for Native Americans are over 50 percent for both men and women.

Interracial marriages are also leading to more multiracial children. Since 1970 the total number of multiracial children in the United States has increased more than fourfold, led by high rates for Native Americans and Asians. In over half the births to a Native American parent in the early 1990s, the other parent was of another race, usually white. In addition, American Indians have the highest total fertility rate of any ethnic group, with almost three children born in a lifetime to every Native American woman (O'Hare 1992). Although the dominant culture usually assumes that assimilation is a desirable objective for any minority group, there is some concern among American Indians that they will "marry themselves out of existence" (Yellowbird and Snipp 1994, 239).

As the structural and social conditions affecting Native Americans change,

Rates of out-marriage are higher for Native Americans than for any other racial or ethnic group in the United States. The "black Indians" of the Pequot tribe trace their lineage to slavery days.

we can expect that their family structures will also change. For example, some American Indian scholars observe that rigid gender roles in the family are loosening as more women are employed outside the home and as more Indian families move to urban or suburban areas. Others point to a history of matrilineal and matrilocal family systems that promoted "loose" marriage ties and afforded women substantial authority in many Indian families to begin with. As with other ethnic minority groups, there is a growing split among Native Americans between the haves and the have-nots. Many of those in rural areas and on reservations live in substandard housing and rely on government subsidies just to get by. Native Americans living in cities are also more likely to be poor than their white counterparts. Urban residence and high rates of single motherhood in the cities tend to compound the problems of some Native Americans because they cannot rely on extended family support networks more typical of rural areas. Nevertheless, increasing numbers of Native Americans are finishing high school and attending college, and the percentage of American Indians in professional and managerial jobs is increasing (O'Hare 1992). Combined with a greater likelihood of intermarriage and increasing numbers of dual-earner couples among Indians, we can predict that the American Indian middle class will continue to grow, and that these families will be most likely to assimilate into the dominant culture. Of course, even modern middle-class families can identify with their Native American ancestry and maintain important aspects of their cultural identity, but without being embedded in dense social networks and participating in routine ritual practices, we can expect their ethnic identification to become more diffuse. In the face of changing social and economic circumstances, we can expect American Indian families to flexibly adapt, but because the life circumstances of Native American families will continue to differ, we can also expect to see continuing diversity among Native American family forms and practices.

SUMMARY

1. Blacks, Hispanics, and Asians are the most numerous minorities in the United States. Because of higher than average fertility rates for Hispanics and blacks, and because of high rates of immigration for Asians and Hispanics, the proportion of minority children and families is increasing.

2. Over one-third of black families in the United States are not "broken" or matriarchal but have two parents present, with the father employed. Another third of the black population, however, constitutes an underclass with very low levels of income and education and with families typically headed by unemployed women. Economic pressures seem to be primarily responsible for the failure of marriages in this group.

3. As a result of racial discrimination, black workers earn less than their white counterparts at all class levels. Because of economic discrimination experienced by black men, black families have relied heavily upon women's incomes. Black women's real earnings have been increasing, while black men's real earnings have been declining. One result has been that black marriages are more likely than white marriages to end by separation or divorce. Another result is that black women have had unusually strong positions within their families compared to white women.

4. There is no evidence that black families with strong female influence have negative effects upon their children's achievement or psychological health. Working-class blacks tend to have higher levels of social participation, better mental health, and greater self-esteem than do working-class whites.

5. Since about 1950, there has been an accelerating tendency for women in the black lower class to remain unmarried and bear children out of wedlock. In fact, two out of three black children are now born to unmarried mothers. One interpretation for this phenomenon is that these women have become increasingly unwilling to enter into a relationship of subordination to men that gives them little economic advantage. Another structural cause is that there are relatively few marriageable men available in relation to the number of women. The availability of welfare payments to single mothers has sometimes been blamed for encouraging divorce and illegitimate births among blacks, but these payments have contributed

relatively little to the overall trends in marital disruption and out-of-wedlock birth.

6. Hispanic families include Mexican-Americans, Puerto Ricans, Cubans, and Central and South Americans. The groups vary in terms of family demographics and social class standing, with Cuban families most likely to be middle class, Mexican-American families most likely to have large numbers of children, and Puerto Rican families most likely to be female-headed and poor.

7. Mexican-American family ideology, with its emphasis on family cohesion and male dominance, derives from its agrarian and patriarchal roots. This ideology is undergoing change as most Mexican-American families now live in urban areas and most Mexican-American wives and mothers are now employed outside the home.

8. Asian-American families, including those of Chinese, Japanese, Korean, Filipino, Vietnamese, and East Indian descent, tend to be concentrated in the western United States. While these groups differ from one another, most are characterized by high levels of educational attainment and high rates of out-marriage (particularly for women). Traditional Asian families value family well-being over individual needs and grant respect to an authoritarian father. As economic conditions shift, and as native-born Asians become acculturated, Asian-American family patterns are coming to resemble those of majority whites.

9. Native American families were so diverse historically that it is impossible to identify one typical American Indian family form. Nevertheless, researchers have suggested that Native Americans tend to have open and extended family networks that provide mutual support, respect elders, collectively raise children, and reinforce tribal identity. The American Indian population is growing because of high birthrates, but more Native Americans are moving to cities, getting an education, and marrying non-Indians than ever before. American Indian families, like African-American families, have higher rates of poverty and nonmarital childbirth than other racial/ethnic groups.

10. Variations in family form and life-style are most often the result of economic and institutional constraints, and minority family patterns quickly respond to changes in economic conditions. For instance, heavy reliance on extended kinship networks for child care, emotional support, and financial assistance is typical of poor white families as well as of poor black, Hispanic, Asian-American, and Native American families.

11. Rates of marriage, divorce, and birth differ between ethnic groups, but the overall trends for all groups parallel those of whites. Since 1960, these include lower marriage rates, later marriage and birth, higher divorce rates, more out-of-wedlock birth, more employed mothers, more single-parent households, and more children living in poverty.

PART THREE

LOVE, MARRIAGE,
AND CHILDBEARING

8

LOVE AND THE MARRIAGE MARKET

Love and the marriage market: this seems an incongruous pair of topics. The first has the idealized ring of one of our highest values; the other sounds cold and cynical. Nevertheless, there is good reason to treat them together. The relationships between men and women before marriage involve both. A few years ago a standard text would simply have called this courtship.

If that term sounds a little archaic now, it is because the relationships of unmarried men and women are becoming rather different. Many people are putting off marriage until they are older, though they are still planning for it eventually. Others talk of alternatives to marriage. The old term *courtship* carried with it the idea of old-fashioned boys and girls sitting on the porch swing under the watchful eyes of their relatives in the parlor, with wedding bells on everyone's mind. Instead, we now have more of an image of "swinging singles," who certainly do not feel that they are courting; they are interested in having fun in the here-and-now. Even the fun-oriented dating of a few decades ago seems to be turning into something else.

Beneath the surface, however, the relations of unmarried males and females may not have changed so much. They still involve two old themes, love and the kind of bargaining that leads to marriage, even if we now analyze them in a somewhat different light. Love is highly idealized, but we are becoming more realistic about what it really involves, even to the point of seeing how it relates to something as practical as the economic pressures of a marriage market. Recent feminist theories point out how the traditional ideal of love contributed to a system of gender domination. Love may be a beautiful interpersonal bond, but the family structure built around it can also be a part of a system of female—and male—bondage.

We like to regard love as something ultimate that transcends worldly considerations. If there are material interests involved in a relationship, or even only sexual ones, we tend to regard it as not really love. But in fact our feelings about love are solidly anchored in the social, material world. For one thing, our very idealization of love is a construct of our culture. To think about romantic love as an ideal is unusual in the history of the world. Most traditional societies, such as ancient or medieval China, the Middle East, and Europe, regarded it as rather dangerous and aberrant. We live in a historical era that is almost unique in making romantic attachment the proper, and indeed the only real, basis for marriage. We don't live up to this ideal so very often, but the fact that almost all of us believe in it shows that cultural forces are at work here.

Moreover, *when* people fall in love, and *with whom*, is strangely predictable. "Love matches" are distributed across social classes, races, and age groups in very standard ways—within, of course, a certain range of statistical variation. Romance typically affects males and females differently. Men, for instance, tend to fall in love more quickly than women do. This can be explained by the use of sociological theories, especially those informed by a feminist viewpoint.

Love, in short, is tied into the market system through which people choose the partners with whom they enter the sexual, economic, and emotional relations called marriage. As we shall see, even if those relations don't turn out to be marriages in the conventional legal sense, they still are shaped by the social structures that we inhabit.

In this chapter we begin with a discussion of love and its ambiguities. We examine feminist theories as well as more traditional social-psychological theories of love, and introduce a theory of love as a ritual symbolizing emotional possession. Then, we turn to some sociological findings on the distribution of love: how often it happens, when and to whom and with whom, and how long it lasts. To find out why these patterns exist, we examine the second component of our analysis: the marriage market. We look at who marries whom, and at explanations of how various personal attributes of men and women are matched up or traded off on the marriage market. Romantic love, we argue, is a feeling that appears at certain moments in this bargaining and matching process.

While some form of love has probably been present in all societies, modern Western culture (European/American society) places unique emphasis on romantic love. As we pointed out in earlier chapters, in past times and in other cultures, marriages were arranged by parents or other community members, not by the prospective bride and groom. Western culture, and especially contemporary American society, is unusual in considering romantic love to be the only proper basis for marriage. Because we are so preoccupied with being "in love" before we marry, we often overlook the fact that selecting a marriage partner is influenced by social and market forces outside of ourselves.

Call it courtship, a marriage market, or a modern sexual friendship: the

FEATURE 8.1
Three Love Poems

I thought it was snowing
Flowers. But, no. It was this young lady
Coming towards me.
    From the Japanese of Yori-Kito,
    Nineteenth Century

Will he be true to me?
That I do not know.
But since the dawn
I have had as much disorder in my thoughts
As in my black hair.
    From the Japanese of Hori-Kawa

Dew on the bamboos,
Cooler than dew on the bamboos
Is putting my cheek against your breasts.
The pit of green and black snakes,
I would rather be in the pit of green and black
    snakes
Than be in love with you.
    From the Sanskrit, Fifth Century

basic theory is the same, even if we get somewhat different outcomes now than we did in the past. And if the market seems a rather harsh environment in which to situate something as tender and beautiful as love, that is largely due to our society. It is what constrains us now and puts the ambivalence into modern love relationships. If we are to push toward something better, it is best to examine where we are right now with our eyes open wide.

WHAT IS LOVE?

Feminist Theories of Love

Feminist writers have argued that the scenario of romantic/nonromantic love in our culture and the social structures built around it constitute one of the main forces keeping women tied into traditional gender roles and subordinate to men.

It is because women are tied by bonds of love and dependence to their husbands (and subsequently to their children) that they have fitted so easily and without protest into the traditional gender-segregated division of labor (Greer 1970; Firestone 1970). Traditionally, love meant wedding bells and a happy ending. After the wedding cake was divided up and the white dress put away, there came the reality of "forever and ever, until death do us part"—a young woman retiring to the interior of her new home to become a housekeeper, wife, and mother. Romantic love was a glorious episode, but it exacted a price; it meant the end of whatever independent career she might have started. "Let me take you away from all this," said the prince in the fairy tale. The unfortunate reality is that he does, and that is the end of the story. And if perchance the woman wants to get out again on her own, the well-known answer from her husband invokes the sentimental tie: "Don't you love me anymore?"

In the traditional scenario, then, love serves to recruit women into subservient positions in a system of exchanging goods and services (Rapp 1982). Love also acts as a bond that keeps women in traditional gender roles and prevents them from attaining the economic freedom that could provide a basis for equality. Nevertheless, one might say, if the love is real and true, then at least she has that—and after all, what could be worth more in one's life? But the feminist critique goes further. What is commonly called love is often permeated with uneasiness. It is the grounds on which is fought what Constantina Safilios-Rothschild (1977, 3) calls "the battle of the sexes." In this battle, men and women use "love" as a subtle weapon. Both attempt to win rewards from the other, but without giving up too much of themselves. Love itself—along with male and female personalities—becomes distorted in the effort. Feminist theorists have thus become interested in Freudian theory in an effort to explain how people can fall into these destructive traps.

A traditional view of the battle between men and women emphasizes the difference between what men and women want from the opposite sex (Safilios-Rothschild 1977). In this view, men approach relationships primarily with sex in mind, while women downplay the element of sex and emphasize emotional attachment and intimacy. A battle then ensues over who has to give what. Men may become cynical about emotional attachment and regard it as a feminine trick. When they do fall in love, other men regard them as having lost the game—and they may come to feel that way themselves. On the other hand, men may feign love in order to win sex. Then there is a battle over the level of commitment in the relationship: Will there be a brief encounter or a lifetime marriage?

This is not to say that men and women don't *both* experience sexual desires and feelings of love, but they experience them in different ways. Feminist writers have looked at this issue from various perspectives in an effort to understand the complex relationships between love, sex, and power. For instance, Chodorow (1978), Gilligan (1982), Rubin (1983), and others draw on Freudian theory to suggest that child-rearing arrangements produce men and women with basically different personalities. Briefly, they suggest that exclusive maternal care encourages women to value and depend on close emotional relationships, while it produces men who fear close emotional connection and value independence. Cancian (1985), Sattel (1976), and others suggest that

A traditional view holds that men approach relationships with women primarily with sex in mind.

men and women define love differently because of differences in power. Men equate love with sex or providing help, whereas women define it as verbal exchange and emotional rapport. By focusing on conflict and exchange theories, these approaches show how men's style of loving reaffirms and perpetuates men's dominance over women. Germaine Greer (1970) places great stress on the point that women are more likely to repress sexual feelings and that they become passive and dependent as a result. According to Greer, if women's sexual energies were not made inaccessible through repression, they could help to fuel female creativity and success in the larger world. Shulamith Firestone (1970) points out that females tend to repress their own sexuality because both men and women have regarded sex as primarily the male province. In this view, only "bad girls," the *femmes fatales*, are openly sexual. Hence women tend to repress each other's sexuality as well as their bids for independence.

When men fall in love, they tend to overromanticize and idealize their women. She becomes an idol, to be protected from the world. She is someone whose favors have to be striven for mightily; once she is won, she should be safely stored away. Although these attitudes seem to glorify women, they end up imprisoning them. The protected woman on the pedestal is as confined and dependent as a child. And when the man is finally sure of possessing her, he begins to lose interest. She is no longer a mysterious and lofty goal to be striven for; she turns out to be an ordinary person, and the ideal vanishes into somber reality. What was once idealization can turn into subtle contempt.

Ironically, Safilios-Rothschild (1977, 3) points out, men tend to be allowed to indulge themselves in love as well as in sex. Men are more likely to marry for what they believe is love (even though it may be mainly sexual desire with an idealized screen thrown over it); women have to be more prudent, choosing their marriage partners on the basis of their need for economic security. This is not to say that women don't feel some emotions for their chosen partners, but their feelings of love develop much less spontaneously. Women become "love experts," who analyze and manipulate their own feelings to give themselves, and their partners, the illusion of love. All too often it turns out to be only temporary (see feature 8.2).

Is the feminist position too cynical? It seems to regard love as a counterforce in the war over male/female domination, merely a weapon of the sexist society. But this is not necessarily so. Feminist theorists, like other sociologists, distinguish among various types of love. Love can involve sexual passion, romantic idealization, affection and companionship, or altruism (extending oneself for the well-being of the other). There are many gradations and combinations. In general, when feminists critique the bondage of love relations in our society, they have love of the passionate/romantic type in mind. Their criticism is stated from the point of view of a higher ideal: love that is more truly spontaneous and unmanipulative, love that allows equality. It is what Safilios-Rothschild (1977, 10–11) calls "mature love"—genuine caring for the

FEATURE 8.2
Love and Marriage

If someone had all the other qualities you desired, would you marry them if you were not in love with them?

—— yes —— no —— undecided

In 1967, William Kephart asked over 1,000 college students this question. Almost three-quarters of the men (65 percent), but just under one quarter of the women (24 percent) answered no, that they would not marry someone if they were not "in love" with them. Only 12 percent of the men and 4 percent of the women said yes, they would marry the person. Less than one quarter (24

percent) of the men were undecided, as compared to 72 percent of the women. One woman commented "I'm undecided. It's rather hard to give a 'yes' or 'no' answer to this question. If a boy had all the other qualities I desired, and I was not in love with him—well, I think I could talk myself into falling in love!" Kephart suggests that a young woman has "a greater measure of control over her romantic inclinations" than a young man.

Source: Adapted from William M. Kephart "Some Correlates of Romantic Love," *Journal of Marriage and the Family* 29 (1967): 470–74.

other person while preserving one's own honesty and respecting one's own rights. A two-way relationship of this sort can only take place between mature persons (of whatever ages), both of whom give support and affection to the other while allowing a breathing space for their own independence.

Equity Theory

Elaine Walster (Hatfield) and G. William Walster (1978) have proposed a model of falling in love that ties the psychological level of interaction to a larger sociological tradition of exchange theory. Relationships work best when both individuals feel they are getting a "fair exchange" for what they have to offer. The man and woman do not have to offer the same things, but the total "worth" of what the two are offering must somehow feel approximately equal. This worth may include attractiveness, social status, personality traits, and admiration by other people around them. A man who felt he could easily get other partners, while his girlfriend could not, would tend to move out of the relationship; a woman who had more opportunities than her boyfriend because of her attractiveness would similarly be motivated to leave. In the Walsters' research, falling in love could be predicted by matching up the traits of the two individuals and finding whether the combination gave them both a sense of an equitable situation.

The Walsters' *equity theory* (Walster, Walster, and Berschied 1978; Berschied 1985) is a comprehensive framework within which the findings of other social psychologists can be included. Individuals may pair off somewhat randomly at first, but they gradually gravitate toward those with whom they feel most appropriately matched. This process seems to go through various stages, in which more superficial traits are matched first, then other similarities or complements are worked out (Backman 1981). Individuals need not be exactly alike, although similarities do help to establish a sense of a "fair trade"; the two can also offer different traits, which must compensate for each other so that the whole package is balanced. This is essentially a social-psychological view of the market process that we will examine below. It also is a process that can be rather callous and put considerable strain on the individuals involved. And one of its typical outcomes, especially in many of those couples studied a decade or more ago, was a trade-off in which traditional gender roles of domination and subordination were negotiated.

Love as High-Intensity Ritual

The foregoing theories may seem somewhat remote and abstract. They bracket love inside a large-scale frame, as if viewed through a telescope. Love, after all, is something dramatic and emotional. Can sociology focus in on the immediate reality of love at all?

It can, but we must shift our level of attention down to the microstructure of interaction and consider love as a ritual. We can make use of the theory of interaction ritual inspired by the classic sociologist Émile Durkheim (1893/1947), and developed by Erving Goffman (1967) and Randall Collins (1975, 97–102, 153–54). A ritual has the following ingredients:

1. It brings people together face to face;
2. focuses their attention on some common object or activity;
3. promotes a shared emotional tone among the participants, which grows as the ritual proceeds; and
4. produces an emotionally charged symbol, which represents the partial sense of membership in the group.

Romantic love can be seen as a ritual in precisely this sense. In fact, it is an extremely high-intensity ritual. Let's look at the list of ritual "ingredients" again.

Face-to-Face Interaction

The face-to-face quality of love is famous. Lovers want to be in each other's sight; when they are together, they tend to ignore everyone else, "on a cloud by themselves." Lovers prefer to be alone, with no one else around. They constitute a group of two, cut off from the rest of society: so much so that sociologists have sometimes referred to love as "dyadic withdrawal" (Slater 1963). Sociologists who have actually measured the process find that the

The more infatuated
two people become with
one another, the longer
they sustain their
face-to-face interaction.

intensity of love is indeed correlated with the length of time lovers spend staring at each other (Rubin 1973).

Focus of Attention

Not only are lovers together, but they have a common object of attention: themselves. Passionate love means being obsessed with the person one loves, forgetting everything else. Lovers talk about each other and about their love; mundane subjects apparently do not exist. Or, instead of talking, they are looking, touching, making love: activities that absorb all one's attention. Love, in fact, probably involves a more intense focusing of attention than any other type of interaction. Love may well claim to be the most high-intensity ritual.

Shared Emotion

Love, of course, is an emotion. It is other things too; without the face-to-face encounters and the strong focus of attention on each other, the emotion of love could not reach its full power. What kind of emotion is it? Initially, it is desire and admiration, perhaps mixed with apprehension; or it may even be a much weaker emotion at first, a vague kind of attraction or interest. But *the process of going through the high-intensity interaction strengthens the emotion and turns it into passion,* for the fact that the lovers isolate themselves from the rest of the world, and focus intensely on each other, means that whatever emotions each has are then shared by the other. They settle into a common mood—one that is intensified by their exclusive attention to each other.

The physical acts of making love have exactly this structure. Kissing, stroking, intercourse itself: all these involve a narrowing in of one's attention to just oneself and one other person. Sexual excitement builds up between one partner and the other; each one's arousal makes the other more aroused. We are not saying here that love is simply equivalent to sexual arousal, but the two are connected, above all by the fact that both of them are highly intense forms of ritual interaction between two persons. Sexual arousal, precisely because it is contagious and shared, can generate the sense of oneness that is the hallmark of love.

Symbols and Sacred Objects

Love is known for its symbols: hearts, rings, presents given by one lover to another. If love is a high-intensity ritual, then the ritual must work to attach the shared emotions to some emblem that can represent them. This symbol serves as a reminder of the emotions and also as a touchstone for setting the boundary between insiders and outsiders. Bear in mind, the theory of rituals was originally developed as an explanation of how religious symbols acquired their sacred status: the idol of a pagan cult, the Bible or cross of Christianity (Durkeim 1912/1954). The ritual of love, then, creates a little private cult with its own object of veneration—the loved person. That is why lovers idealize their loved one to such a high degree: the intense ritual absorption that they are involved in automatically elevates its object into something that is more than human.

Lovers idealize each other most strongly early in a love relationship (Kerckhoff and Davis 1962). Later, they acquire a more realistic view of one another, probably because the stage of highly ritualized interaction cannot last forever. The lovers find they can't spend all their time together, staring at one another and ignoring the rest of the world. Eventually, the claims of ordinary life reassert themselves. When that happens, the idealization diminishes, and so does the intense emotion of love. It is for this reason that lovers, after they are married, inevitably feel some letdown from their emotional peak. They stop carrying out the social conditions that made up a high-intensity ritual, and the effects of the ritual diminish as well.

The ritual-charged symbols, however, can retain their power for some time after the rituals are over. Such symbols represent membership in a group of two, much as respect for the symbol of the cross represents membership in the religious group of Christianity. That is why lovers are so touchy about the emblems of their relationship. The man who carelessly loses a trinket given to him by his lover may precipitate an emotional crisis in their relationship, because the gift has become charged with symbolism. In losing it, he appears to be reflecting a loss of devotion. Actions, too—kissing, holding hands, not to mention sexual intercourse—symbolize the emotional bond. Forgetting to perform them—or performing them with the "wrong" person— opens the way to the emotional strains and jealousies that make love affairs so dramatic.

Love and Hate

It is because of love's inherent riskiness and the lovers' vulnerability that love and hate are so closely tied together. Passionate love and hate might seem to be polar opposites, but they actually have a great deal in common. Both of them are high-intensity emotions; both of them tie individuals together in a passionate relationship. The opposite of love is not hate but indifference; hate has a different emotional quality, but it is of the same structural nature as love. Hate arises when the emotions of love are threatened or negated; its intensity is related to the intensity of the love it supplants.

Lovers are intensely connected; the whole ritual process works toward that result. Their emotional bond amounts to a form of emotional property, for the group that is formed by the love ritual consists of just two persons, and there is no room for outsiders. Moreover, the ritual has charged up various symbols representing this exclusive relationship, and these symbols are vulnerable to abuse. For example, holding hands is a symbol of their relationship; hence, if the woman so much as touches another man, her lover may become jealous. Since staring at one another is part of their love ritual, if the man glances at another woman a split second too long, his lover may go into a rage.

These types of jealousies and angers belong especially to the most romanticized part of a love affair—a first love or, for more experienced loves, the earliest phases of intense feeling. But long-term commitment certainly does not preclude jealousy. When the emotional bond incorporates a strong sexual tie, the lovers tend to demand a kind of property right over their partner's body. Any act of physical sex comes to symbolize the whole emotional relationship. Since the emotional relationship is exclusive, so must the sexual relationship be. If this sexual exclusiveness is violated, the excluded partner often feels an uncontrollable rage, perhaps out of all proportion to any consequences. The emotions are parallel to how a worshiper would feel if his religious symbols were desecrated by an unbeliever—for example, if someone spit on a Bible or tore down the cross. The aggrieved lover feels a righteous anger: he has no doubt of being in the right and may not feel inhibited from punishing the offender, even violently.

This, then, is the negative side of the intense and exclusive love bond. As long as all goes well, there is strong devotion and idealization between the lovers. But if some symbol of their relationship is violated, the result can be jealousy and uncontrollable and seemingly irrational anger. The outcome, as we will see in later chapters, may be quarrels, violence, or even murder.

Where and When Does Love Happen?

There are no surveys that will tell us exactly how many people are in love at any given time. In our society, love is supposed to be the basis of marriage, and to marry someone whom one does not love is considered a somewhat abnormal and pitiable situation. Since 95 percent of the population marries, we might expect that most people have some experience of love.

How significant this experience is, though, can vary a great deal. Constantina Safilios-Rothschild (1977, 19) estimates that 50 percent or more of the people in the United States marry for reasons other than love: because of loneliness, or the desire for children, financial security, or the status of being married. She quotes a woman who said:

> I wasn't madly in love. I was about 31 and I wasn't getting any younger. I was getting tired of being by myself and I like him quite a bit. He wouldn't leave me alone. In fact, we decided to get married because he spent so much on phone bills. (P. 17)

Or a man:

> I was ready to settle down. I felt I had played around long enough. (P. 18)

Nevertheless, when people marry, they usually do go through at least a temporary period during which they feel they are in love with each other, or at least put up the pretense of being in love.

Who Loves More, Men or Women?

We ordinarily think of love as being a predominantly female province. Men are regarded as more worldly and cynical, women as more romantic. But the evidence seems to be that men are much more likely than women to fall in love spontaneously and to base their marriages upon it (see figure 8.1).

Men fall in love earlier in their relationships than women do (Kanin,

FIGURE 8.1: Women Are More Wary of Love at First Sight

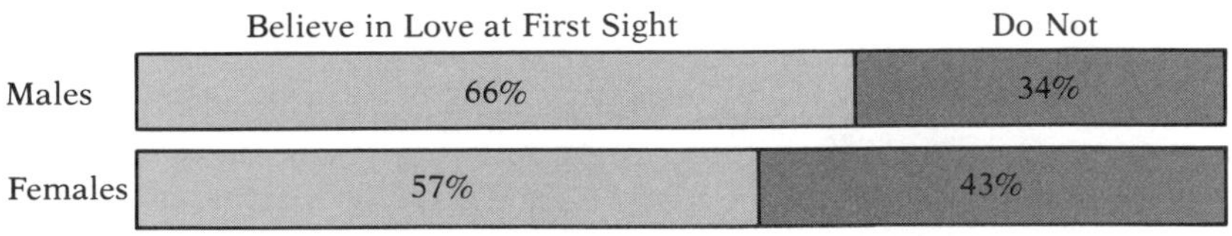

	Believe in Love at First Sight	Do Not
Males	66%	34%
Females	57%	43%

Source: *Public Opinion*, vol. 7 (Aug./Sept. 1984), p. 23.

This woman is analyzing the man's feelings for her rather than focusing on her own.

Davidson, and Scheck 1970). Moreover, in research following couples over a period of two years, the men not only fell in love more quickly but also stayed in love longer and tried harder to resist breaking up a relationship (Hill, Rubin, and Peplau 1976). Men were also more likely to end up marrying someone they loved. There are twice as many marriages in which both partners perceive the husband as loving more as marriages in which the wife is the one who loves the most (Safilios-Rothschild 1977, 72).

How do we explain these patterns? For one thing, men are more likely to choose their partner on the basis of physical attractiveness (Combs and Kenkel 1966; Kephart 1967). Hence, a man may fall in love almost, as the saying goes, at first sight. For a woman, however, the emphasis has been upon other qualities in a man. As long as a woman's social status was determined by her husband's, economical and social considerations had to be higher in her mind. In a sense, marrying for love was a luxury that only a man could afford.

This is not to say that women have had no interest in love. They have, but with special emphasis upon *the man's love*. Women have given much more weight to the intensity of the man's feelings for them than to their feelings for the man (Safilios-Rothschild 1977, 27). This makes sense: if a woman is going to be heavily dependent upon her husband, it is important for her to be sure that he is strongly attached to her. Thus, men have not only been able to marry for love more than women have, they have also been more able to focus on purely personal traits of a woman—not only her attractiveness, but her personality—since they do not have to consider her as a means of support.

In a traditional society, the economic relationship would have been considered openly, and a woman would not have bothered to raise the question of love. In our society, however, love is viewed as the ideal. Hence, women who are considering marriage make some efforts to fall in love, if they are not in love already. Arlie Hochschild (1975, 1983) calls this *emotion work*. In her research, she finds that women, much more than men, consciously analyze and discuss their feelings about their partners. This process shapes and manages their emotions, so that they end up feeling about a man the way they choose to feel. If a man is a good match for them, they allow themselves to be overwhelmed by their emotions, provided that they spontaneously feel a liking for him; if the spontaneous feeling is not there, they work on developing it. On the other hand, if the man for whom a woman has strong feelings is not a socially acceptable person for her to marry, she may rather deliberately work at falling out of love.

This sounds somewhat cynical and manipulative, but it is not necessarily quite that conscious. A woman is usually much more deliberate than a man about getting into the intimate situations in which the rituals of love will take over and generate strong emotional attachments. And this seems to be backed up by the evidence, cited above, that women not only take longer to fall in love than men but also are faster to break up a relationship.

This imbalance in love between males and females is specific to our own type of society. In earlier historical periods, romantic love may not have occurred in most people's lives at all, and it was not expected to. Modern women have performed "emotion work" on their feelings of love because their social status has depended upon how they negotiated a marriage for themselves. If a different kind of marriage market emerges in the future, in which a woman's status does not depend so heavily upon her husband's, we can expect that the nature of love may change too. Whether women will move toward a male pattern of falling in love, or men will move toward the female pattern, or some new pattern will emerge remains to be seen.

THE MARRIAGE MARKET

Contrary to the old saying, love is *not* blind. It does not generally appear as a bolt from the blue or a tidal wave sweeping away all in its path. Rather, there is considerable predictability in who will marry whom.

Who Marries Whom?

Another old saying, at odds with the one we just referred to, goes: "Don't marry for money. Go where the money is and marry for love." Consciously or not, this is the way it tends to work. In general, people tend to marry individuals who resemble themselves in social class background, race, religion, and education, as well as in various personal traits. Sociologists label the tendency to marry those with similar social characteristics "homogamy."

Class Background

Most people marry within their own social class (Winch 1958; Carter and Glick 1976). This is true whether one measures both the husband's and wife's class background by that of their parents or by their own occupations (using the wife's father's occupation as a measure of her social class if she does not work). Not everyone marries into exactly the same class, of course. But when people marry outside their social class, they usually do not move very far; they are most likely to marry into the adjacent social group, up or down. In the past, women have been more likely than men to use marriage as a means of status enhancement, marrying men just above them in social class. The farther apart social classes are, the fewer people climb or descend that distance through marriage.

Race

Most marriages occur within the same racial group. Some of this, of course, may not simply be due to the marriage market. Seventeen states still had laws prohibiting miscegenation (interracial marriage) when the U.S. Supreme Court declared such laws unconstitutional in 1967. Even today 98 percent of all marriages take place within the same race. Nevertheless, the number of interracial and interethnic marriages has increased significantly since 1970 (see figure 8.2), and attitudes toward intermarriage have become more favorable.

Contrary to the common image, marriages between blacks and whites are not the most frequent type of interracial marriage. Only 0.4 percent of all marriages are black-white intermarriages (Surra 1991). The most common types of interracial marriages, in fact, are not very highly publicized. Native American women are the most likely to intermarry (54 percent of their marriages are interracial). Next come Japanese-American women (41 percent) and Filipino women (32 percent). The great majority of these interracial marriages are with white men (Kitano and Daniels 1988).

When intermarriages do occur, there are very distinctive patterns by sex of the out-marrying partners. Sixty-six percent of the black/white marriages in existence in 1992 were between black men and white women (U.S. Bureau of the Census 1992f). In Japanese/white and Filipino/white marriages, the situation is almost the reverse, with mostly white men marrying nonwhite women. For Chinese/white and Native American/white marriages, the ratios are about even (U.S. Bureau of the Census 1980). How can these patterns be explained? Some may be due to the experience of American military men abroad, who brought home Japanese and Filipino brides. The question remains why they would be particularly likely to marry Asians. It is sometimes suggested that Asian women are traditionally more passive and subservient to males; hence, American men

FIGURE 8.2: The Number of Interracial and Interethnic Marriages Is on the Rise

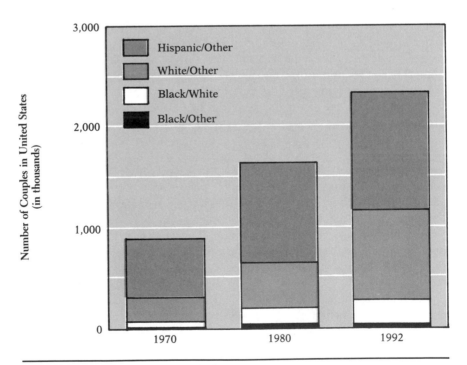

Source: U.S. Bureau of the Census, 1992f, table F

The majority of black-white intermarriages are between black men and white women, unlike this couple.

who marry them are looking for more dominance in their marriages than they could get from other brides. Some have also suggested that out-marriage rates are higher for Asian-American women than for Asian-American men because the women can exercise more independence if married to non-Asians (Kitano and Daniels 1988).

The black/white pattern is more mysterious. We do know that these marriages are most likely to occur among highly educated persons. Black men who have attended graduate school are the most likely to have interracial marriages (Carter and Glick 1976, 414). But why is there no such tendency for highly educated black females, or white males, to marry interracially? Moreover, it is not just a matter of availability. There is a fairly severe imbalance of numbers in the black population, with twelve females for every ten males. Hence the intermarriage ratio is just the opposite of what one would expect. The higher tendency of black males to marry out reduces even further the chances of black females marrying.

One explanation may be that our cultural ideals of beauty have been based on white standards. Images of thin young Caucasian models predominate in the media, and while this may be changing somewhat, the stereotype of female beauty tends to be defined in terms of white features and styles.

Another possible explanation may be that there is a particularly strong distinction between male and female cultures in the black community, and a particularly vehement battle for gender domination. African-American women, as we have seen (chapter 7), have a tradition of independence; they are not very subservient to men. This may contribute to their own lower rate of intermarriage and to the tendency of black males to seek wives elsewhere.

Religion

There used to be a strong tendency to marry within the same religious faith. Seventy-nine percent of Catholics, 91 percent of Protestants, and 92 percent of Jews married someone of the same religion in the 1960s (Carter and Glick 1976, 140–41). This sounds like Catholics were most likely to marry out, and Protestants and Jews were both equally endogamous. But we must take into account the fact that these groups are of different size, and that the sheer chances of marrying someone from a large group are much greater than from a small group. Protestants are by far the largest religious group in the United States (70 percent of the population), while Catholics constitute 26 percent and Jews approximately 4 percent. Thus, Protestants marry other Protestants slightly more than we would expect just by chance, by a ratio of about 1.7 to 1. Catholics marry within their religion at an even higher rate, about 5 to 1 over what would happen purely randomly; and Jews marry in at a ratio of about 52 to 1.

Today, religious intermarriage rates are much higher. For example, Jews now marry non-Jews about 40 percent of the time (Krantowitz and Witherspoon 1987). Some researchers have found that religious intermarriages are less stable (i.e., more likely to end in divorce) (Bumpass and Sweet 1972; Glenn 1982). This is probably because those who marry someone outside of their own faith are also less bound by a variety of traditional norms. But inter-religious marriages are not less happy than others. Recent research indicates that Catholics who marry someone of a different religion are just as likely to report being happily married as Catholics who marry Catholics (Shehan et al. 1990).

Personal Traits

People tend to marry others who resemble themselves in quite a number of ways. The most prominent feature is the degree of physical attractiveness (see feature 8.3). Married partners tend to be relatively similar physically, correlating in weight (given, of course, the fact that men tend to be heavier), height, hair color, condition of health, and even basal metabolism (Winch 1958; Burgess and Wallin 1953; Murstein 1972). Some of these features are probably artifacts of the factors mentioned above. Hair color, for instance, is partly an index of race and ethnicity; and height, weight, and health are correlated with social class. Homogeneity even extends to psychological characteristics. Husbands and wives tend to correlate in IQ scores at a level of about .50, which is approximately the same as the correlation between the IQs of siblings (Taylor 1980, 23–30; Alstrom 1961). Since there is evidence that a good deal of IQ is cultural (see chapter 12), this probably indicates, more than anything else, that people are attracted to each other because of similarity in what we would call their "cultural capital."

Market Resources and Opportunities

There is a way of explaining the various patterns of marriage just mentioned. Meeting and being attracted to someone takes place in a large-scale system of exchange. It is like a market, only what is being exchanged are not economic goods and services but people's own social and personal traits. This characterization seems harsh, as if people were cattle to be bought and sold. In fact, the process has a rather harsh side, particularly when it takes place in situations, like our society, where there is a good deal of status stratification, as well as inequalities between the sexes.

For us, love is supposed to be a highly personal experience. Hence, we do not wish to be conscious of the market forces at work around us that mold our opportunities for falling in love or marrying. In traditional agrarian societies, where the ideal of love was not important, people were much more open about marriage as a market. In Middle Eastern societies even today, or places such as rural Greece or Sicily, men speak quite plainly of the value of women they might acquire, or of their own daughters and sisters as property that they might use to make an advantageous deal with some prospective suitor (Safilios-Rothschild 1977, 26–27). Traditionally, a man's value was his social status and wealth. A woman's value was measured by her beauty and youth and especially by her virginity. A woman who had lost her virginity before marriage was no longer of value on the marriage market. Her father or brothers felt dishonored and might even take her life; such murders tended not to be punished since they were regarded as crimes of honor. In Bangladesh, women who were raped by Pakistani soldiers during a war, through no fault of their own, were nevertheless devalued; men refused to marry them, and husbands often would not take back their own wives (Safilios-Rothschild 1977, 27).

Such societies show a market system at its most brutal. Even though some of the characteristics of the market have changed, we are still subject to similar pressures. One of the major differences is that families no longer control the marriage market for their children, and men no longer barter away women. Instead, all individuals, male and female, have to enter the marriage market on their own and strike the best deal they can for themselves. A further difference is that the emphasis upon virginity as a key attribute of a woman's value has declined precipitously, especially in recent years.

But the market still operates in other respects. Social status, earning capacity, and attractiveness are among the resources that each person has to offer. A person's ability to attract others is heavily influenced by the bundle of resources at his or her disposal. The market consists of all the people who might come into contact with each other and then "compare resources." Eventually, they find out not only who is available and whom they want but

FEATURE 8.3
The Influence of Physical Attractiveness

We like to believe that we are above merely physical considerations, that we choose our friends and loved ones for their "selves," not their bodies. But there is a lot of evidence to the contrary, summarized in Gordon L. Patzer's book, *The Physical Attractiveness Phenomena*, 1985. (See also Hatfield and Sprecher 1986.)

Generally, there is consensus about how attractive someone is among both males and females. The major exceptions are:

1. Individuals tend to be inaccurate in judging their own attractiveness; they think they are better looking than the way in which they are perceived by other people. Interjudge reliabilities are about .87–.89; whereas self- and other ratings correlate .17–.22 (Patzer 1985, 24–26). Better-looking persons tend to be more accurate about their own appearance, especially if their attractiveness is actually quite high. Low- and average-attractiveness persons tend to overestimate themselves. Men are especially likely to overestimate their own looks, whereas unattractive females are more accurate about other's estimations (though still overestimating themselves somewhat). This is because women get more feedback about their attractiveness and are made aware of it. In general, the contrast of self-concept between females of high and low attractiveness is greater than between their male counterparts.

2. Spouses, as well as engaged couples, overestimate the attractiveness of their husband or wife (Patzer 1985: 25–26; Murstein 1972; Murstein and Christy 1976). Wives overestimate their husbands' attractiveness even more than husbands overestimate their wives'. On a 10-point scale,

with 1 very unattractive and 10 very attractive, the average values in one study found husbands rated 3.6 by judges, 5.4 by self, and 7.2 by wife; wives were rated 3.6 by judges, 5.1 by self and 6.5 by husband.

This says something about the process of marriage. Not only is there a market in which people select someone of about the same attractiveness level as themselves, but their own egos become involved in their beliefs about the attractiveness of their partners. Spouses who overestimate their partners' attractiveness the most have greater marital satisfaction.

Patzer (1985: 140–63) reviews evidence on what traits people judge are the basis of attractiveness. In men, the face is most important (explaining 50 percent of the variance; the next most important, explaining 10 percent, is weight and its distribution); in women, by far the strongest factor is not being fat (50 percent of variance; this goes up to 75 percent when females judge females); face is second at 10 percent. A woman's bust size makes little difference; medium-sized breasts were judged as more attractive than large ones. People tend to believe a large bust creates an air of sexiness, but the large bust also calls up a stereotype of immorality, lack of intelligence, and immodesty. Fat stomachs are generally considered the worst trait in men.

Females are concerned with their own attractiveness more than males are concerned with their own attractiveness. But males consider their partner's attractiveness more important than females do. For this reason there is higher correlation between dating popularity and attractiveness for also whom they can get; and the latter depends upon how valuable one is compared with the other people on that market.

To put it very simply: everybody has a market value. Whom a person will be able to "trade" with depends on matching up with someone of about the same market value. Thus, when any two people come together, each has the following market position:

females than for males (Berscheid et al. 1971). For example, college females' attractiveness correlated .61 with the amount of dating, males' attractiveness correlated .25. Better-looking women date more, have more friends, are in love more, and have more sexual experiences than average or less-attractive women.

Before meeting, both males and females choose dating partners largely on the basis of attractiveness. For actual dates, though, males select the most attractive females only if they feel assured of being accepted by them; otherwise they pick partners of moderate attractiveness, assuming more-attractive girls would be less accepting of them as dates. Females do the same. Females also are less likely to pick the better-looking male as a date if their own self-rating of attractiveness is low. In bars, men approach women more often whose attractiveness matches their own. In other words, although physical attractiveness is highly valued, people are usually aware of the realities of competing on the sexual marketplace and adjust their aims accordingly.

Once dating takes place, personality plays a role in the couple's involvement, but attractiveness has a continuing and undiminishing effect on romantic attraction through at least five encounters (Patzer 1985: 87). Persons who considered their partners to be more attractive tended to have greater love for their partners; they were also more submissive to better-looking partners (Critelli and Waid 1980). Better-looking persons worried less about their partners becoming involved with other persons. Hill, Rubin, and Peplau (1976) found that breaking up is seldom mutual; the female partner is more likely to perceive problems and initiate a breakup. Differences in physical attractiveness are recognized as a significant factor in ending relationships.

The process of forging a romantic tie is structured around *similarity* of the partners' attractiveness. This happens because the marriage market brings people together according to their own levels of resources. Feingold (1982) found that the couples who fall in love or become more intimate are those who are more closely matched in looks, and these well-matched individuals progress more rapidly in their romance. Dating couples, like married couples, tend to be similar in attractiveness (Patzer 1985, 92). A marriage is less stable when the attractiveness of husband and wife is not similar. Matching for attractiveness is correlated with marital adjustment even among elderly couples.

The general pattern is that women are more conscious than men of attractiveness ratings, that they compare themselves more to other women than men do to other men, and that women suffer more anxiety over these ratings. Women are particularly severe in judging other women, and themselves, by their weight. This is probably the most important cause of anorexia nervosa, an eating disorder. All these are strains resulting from the competitive marriage market, in which men's main resource is their economic prospects and women's main resource is their personal appearance. This pattern reflects the sexual stratification of the larger society. As women break through into better-paying, independent careers of their own, attractiveness would become somewhat less important, since it would no longer be women's major resource on the marriage market.

1. Your own resources include your social status (class, race, ethnicity; your family background; your current occupation), your wealth and your prospects for making more in the future, your personal attractiveness and health, your culture. Even personality traits such as your "magnetism" or charisma are social resources (moreover, as we shall see, they are produced by one's social position, including one's position on the market itself). Your market opportunities consist of the people you know who are of the opposite sex, who are heterosexual, and who are not already involved in a relationship that prevents their becoming involved with you. The people in your "opportunity pool," of course, have their own degrees of attractiveness depending on their resources.

2. The other person's resources are also social, economic, and physical. His or her opportunity pool also consists of persons of the opposite sex who are known and available, ranked by the attractiveness of their resources.

In every encounter, then, an implicit comparison takes place. You weigh how attracted you are to a person by comparing that person to others whom you know *and* by determining how confident you feel about being able to "strike a deal" with him. Most people, it should be stressed, are not consciously very calculating about this. But they do make such comparisons, *whether they consciously want to or not*, partly because people have been conditioned to do so on the basis of previous success or failure in both dating and other social situations.

Suppose that a woman meets someone she finds extremely attractive. He must have traits that she values: appearance, self-confidence and good humor, style, interesting conversation, politeness, and sympathy. But her evaluations themselves, if we analyze them further, turn out to be influenced by her previous experiences in the social marketplace. If one finds someone physically attractive, that always means that one finds him more attractive than some other people. That, in turn, is affected by how many other attractive, or unattractive, people one has known. A woman who knows a lot of handsome men is going to be affected differently by meeting a moderately handsome fellow than another woman who has mostly associated with a pack of scruffy and unattractive guys.

The same holds true for other traits. One's style comes to a considerable degree from one's social class background; the things one has to talk about, too, depend upon one's social experiences, most of which are acquired as a result of social status (education, travel, acquaintances in important positions, etc.). Even one's emotional qualities are affected by social experience. People who appear relaxed, charming, and good-humored are that way to a large extent because they have had many favorable experiences in previous social encounters. That is to say, they have had a balance of social resources that has made them attractive to most of the people they have met. In the social world, and especially in male/female encounters, nothing breeds success like previous success.

Even how polite and attentive the other person is to you has been influenced by the market, for a high degree of attentiveness means that you are a relatively attractive person to him; it implies that the resources you present rank relatively high compared to what he has been used to. There is evidence that men tend to prefer women who are hard for others to get (i.e., who have a high market value) but who are easy *for themselves* to get, instead of uniformly hard-to-get or easy-to-get women (Walster 1974).

The marriage market, then, is a vast sorting process. People move from one encounter to the next, attracted or unattracted by the people they meet depending on whom else they have met or expect to meet. Sometimes one is attracted to another person but finds that the attraction is not reciprocated; one's resources do not match up with what other people have been showing to that person. Sometimes the shoe is on the other foot, and you find someone is much more attracted to you than you are to him. Both kinds of mismatches tend to make those encounters relatively brief, as one or the other person will move away from it. Someone may be very persistent, of course, in pursuing a consistently rejecting partner. When this happens, it may not be so much because of that particular person's personality traits (stubbornness, insensitivity, passion) but simply because of the way the market works for him. A man may chase a woman who rejects him simply because the other women he knows are much less attractive, or because he does not know very many other women.

Eventually, though, people discover that they get the fewest rebuffs, and find the most rewarding person they can, by sticking to someone whose resources most closely match what they themselves have to offer. This is why marriage partners tend so often to be similar in social class background, education, and many other social traits. There is even evidence that couples remain together most often when they are similarly matched at levels of attractiveness (Stroebe et al. 1971; Hill et al. 1976; Murstein and Christy 1976).

The Winepress of Love

We are beginning to see some of the reasons, then, why love can be an emotionally draining, or even brutal, process. As long as a market is operating, people are being shunted around by forces beyond their own personal control. One's attractiveness to other people changes as rivals come on the scene, as new opportunities open up or others close off. And these things can go on without our conscious awareness. New resources elsewhere in the network of people who make up a community can send rippling effects all down the line. When an attractive man or woman moves into a singles apartment complex, his or her mere presence may be enough to change the character of several existing romantic liaisons.

A formal party is an opportunity for people of the higher social classes to ritually enact their social status.

Even in the normal course of events, a market of this sort puts a great deal of strain on the people who go through it. One's status goes up and down without one's knowing precisely why or being able to control it, depending upon whom one happens to encounter. There is a lot of manipulation, especially when people of different social classes put their unequal resources into the market. Most Americans are not very aware of social-class distinctions, but the resources operate nevertheless. People from higher social-class backgrounds have resources that enable them to be more socially attractive and popular than people from lower ranks. Invitations to attend parties or to join fraternities, sororities, and clubs are a normal part of many youthful settings, and whether one is invited or not can have have a powerful effect on one's self-esteem. Hidden in the background is the fact that all of these are affected by social class. It is the higher social classes who are much more likely to give parties: they have more money to spend on entertainment, and their lifestyle is much more oriented around acquiring the cultural skills for this sort of leisure mingling. Even though no overt class discrimination may be in anyone's mind, nevertheless the socially popular youths tend to be those with the higher class backgrounds. All of this makes up a major part of the setting for the marketplace of sexual relationships. It is no wonder that young people often feel a good deal of strain in their lives.

Ironically, romantic love is itself produced by the experiences of being on this market. The market is the major reason that love is so dramatic. For many people, it is the single most exciting thing that happens in their lives. And it does seem to happen, at least once, to a large proportion of people today. Why should this be so? Because virtually everyone is on the sexual marketplace at some time in his or her life; in modern societies, one has to find one's own partner and negotiate one's own marriage. Thus, everyone experiences the ups and downs of the market.

There is the uncertainty about whom a person will meet, and about whether a given pair will like each other. There is no guarantee that one will hit on a compatible person right away. Even if one does, the pushes and pulls of rivals and rejections in the larger market will usually make for a lot of momentarily raised hopes followed by disappointments. When one finally finds "the right person," there are still uncertainties. Will he call? What does she really think of me? Quarrels. Misunderstandings. Reconciliations. Rivals, and rivals' rivals, spreading out into a network of men and women whom one has not even seen.

FEATURE 8.4
Love Advice from the *Kama Sutra*

Now, a girl always shows her love by outward signs and actions such as the following: She never looks the man in the face, and becomes abashed when she is looked at by him; under some pretext or other she shows her limbs to him; she looks secretly at him, though he has gone away from her side; hangs down her head when she is asked some question by him, and answers in indistinct words and unfinished sentences, delights to be in his company for a long time, speaks to her attendants in a peculiar tone with the hope of attracting his attention toward her when she is at a distance from him, and does not wish to go from the place where he is; under some pretext or other she makes him look at different things, narrates to him tales and stories very slowly so that she may continue conversing with him for a long time; kisses and embraces before him a child sitting in her lap; draws ornamental marks on the foreheads of her female servants, performs sportive and graceful movements when her attendants speak jestingly to her in the presence of her lover;

confides in her lover's friends, and respects and obeys them; shows kindness to his servants, converses with them and engages them to do her work as if she were their mistress, and listens attentively to them when they tell stories about her love to somebody else; enters his house when induced to do so by the daughter of her nurse, and by her assistance manages to converse and play with him; avoids being seen by her lover when she is not dressed and decorated; gives him by the hand of her female friend her ear ornament, ring, or garland of flowers that he may have asked to see; always wears anything that he may have presented to her, becomes dejected when any other bride-groom is mentioned by her parents, and does not mix with those who may be of his party, or who may support his claims.

We shall now speak of love quarrels.

A woman who is very much in love with a man cannot bear to hear the name of her rival

(Continued next page)

FEATURE 8.4 *(Continued)*

mentioned, or to have any conversation regarding her, or to be addressed by her name through mistake. If such takes place, a great quarrel arises, and the woman cries, becomes angry, tosses her hair about, strikes her lover, falls from her bed or seat, and, casting aside her garlands and ornaments, throws herself down on the ground.

At this time the lover should attempt to reconcile her with conciliatory works, and should take her up carefully and place her on her bed. But she, not replying to his questions, and with increased anger, should bend down his head by pulling his hair, and having kicked him once, twice, or thrice on his arms, head, bosom, or back, should then proceed to the door of the room. Dattaka says that she should then sit angrily near the door and shed tears, but should not go out, because she would be found fault with for going away. After a time, when she thinks that the conciliatory words and actions of her lover have reached their utmost, she should then embrace him, talking to him with harsh and reproachful words, but at the same time showing a loving desire for congress.

When the woman is in her own house, and has quarreled with her lover, she should go to him and show him how angry she is, and leave him. Afterward the citizen having sent the Vita, the Vidushaka, or the Pithamarda to pacify her, she should accompany them back to the house, and spend the night with her lover.

Thus ends the discourse of the love quarrels.

The means of getting rid of a lover are as follows:

1. Describing the habits and vices of the lover as disagreeable and censurable, with a sneer of the lip and a stamp of the foot.
2. Speaking on a subject with which he is not acquainted.
3. Showing no admiration for his learning, and passing a censure upon it.
4. Putting down his pride.
5. Seeking the company of men who are superior to him in learning and wisdom.
6. Showing a disregard for him on all occasions.
7. Censuring men possessed of the same faults as her lover.
8. Expressing dissatisfaction at the ways and means of enjoyment used by him.
9. Not giving him her mouth to kiss.
10. Refusing access to her jaghana, that is, the part of the body between the navel and the thighs.
11. Showing a dislike for the wounds made by his nails and teeth.
12. Not pressing close up against him at the time when he embraces her.
13. Keeping her limbs without movement at the time of congress.
14. Desiring him to enjoy her when he is fatigued.
15. Laughing at his attachment to her.
16. Not responding to his embraces.
17. Turning away from him when he begins to embrace her.
18. Pretending to be sleepy.
19. Going out visiting, or into company, when she perceives his desire to enjoy her during the daytime.
20. Misconstruing his words.
21. Laughing without any joke; or, at the time of any joke made by him, laughing under some other pretense.
22. Looking with side glances at her own attendants, and clapping her hands when he says anything.
23. Interrupting him in the middle of his stories, and beginning to tell other stories herself.
24. Reciting his faults and vices, and declaring hem to be incurable.
25. Saying words to her female attendants calculated to cut the heart of her lover to the quick.
26. Taking care not to look at him when he comes to her.
27. Asking from him what cannot be granted.
28. And, after all, finally dismissing him.

Attributed to Vatsyayana, India, *ca.* A.D. 400.

All these dramatic shifts and uncertainties make the experience an emotional one. Negative emotions—anxieties, anger, and the like—are set against the positive emotions that arise when a person finds someone compatible, someone who seems to reciprocate one's positive evaluation. Love, we have argued in the foregoing model, is produced by a type of social ritual. Its ingredients include, besides the privacy that focuses the energies of a small group of just two, the sharing of a common emotion. Where does that emotion come from? It comes from the experience of the market itself. The special feeling about the person whom one loves, which is so strong right at the first, freshest point in the relationship, is the joy that comes from having successfully negotiated a passage through the uncertainties of the market. After the inevitable buffeting about that comes from encounters with persons whose market resources are either too great or too small, a match is generally found. Not only are the two compatible, because having similar resources tends to make them socially similar, but also, the matchup of market resources means that both have the same motivation to stay together. The market bargaining, with its disappointments and blows to the ego, is over. A haven from the market has been found, and the feeling of joy and relief a couple experiences provides the first impetus for the ritual that generates the symbolic attachment of love. Love is a sentiment that arises upon escaping from the market into a safe harbor.

SUMMARY

1. Feminist theories regard love as a bond that has tied women into domestic roles and kept them from economic independence. Safilios-Rothschild stresses that women have had to appear sexually attractive but control their emotions in order to acquire a man of the right social status. This bargaining has produced a good deal of dishonesty in romantic relationships.

2. The Walsters' equity theory predicts that people fall in love when they feel they are making a "fair exchange" of resources and traits.

3. Ritual theory analyzes love as based upon: face-to-face interaction, with an intense focus of attention on each other, excluding all outsiders; this intensifies the lovers' shared emotions and results in symbols representing common membership in the dyad. Love is thus a kind of private cult of mutual worship and idealization.

4. Because love produces strong bonds of attachment, amounting to emotional property, small symbolic violations are likely to provoke righteous anger. For this reason, love easily turns to hate.

5. At least brief experiences of love are very widespread in our society. Nevertheless it has been estimated that 50 percent of Americans marry for reasons other than love, such as the desire for financial security or the status of being married.

6. Men are more likely to act upon spontaneous feelings of love. Women are more concerned about the intensity of a man's love for them than about their own feelings. Because more of their social and economic status hinges upon a marriage, women do more of what Hochschild calls "emotion work" in order to shape their emotions toward the appropriate man.

7. Most marriages match persons who are similar in social class background, race, religion, and many personal traits.

8. These similarities in marriage partners can be explained by the workings of the marriage market, which enables people to exchange mainly with others who have social and personal resources similar to their own, in competition with the "opportunity pool" of persons whom they know.

9. The experience of finding the right person on the marriage market can involve much uncertainty and many bruises to one's ego. The emotion of joy and relief at finally finding a match is a prime ingredient that goes into the ritual producing love.

THE EROTIC TIE: MARITAL AND NONMARITAL SEX

O western wind, when wilt thou blow,
That the small rain down can rain?
Christ, if my love were in my arms
And I in my bed again!
　　　Anonymous, medieval England

What is sex? The answer seems obvious, but once we begin thinking about the subject, the possibilities seem limitless. Sigmund Freud (1924) once commented in a lecture that sex is so hard to define that he finally concluded it is whatever we consider to be shameful or dirty. He was joking, but only partly. For the rather Victorian audiences of the early twentieth century, sex was a uniquely shameful, unmentionable subject. It is no longer unmentionable, but it is still surrounded by a tension of its own that makes it different from anything else in our lives.

Why did Freud say that sex is so hard to define? From a biological viewpoint, sexual behavior is that which leads to procreation. But what about kissing, or masturbating? These are obviously sexual, so Freud pointed to the need for a wider definition. What about anything having to do with the genitals or erogenous zones? But the hand is not an erogenous zone, although holding hands, especially the first time a couple touch one another, can have tremendous erotic significance at the moment. So can stroking someone's hair. Freud pointed out, moreover, that a common theme in many of his neurotic patients was that they had eroticized certain objects in their environment while repressing and displacing their erotic drives for the genitals themselves. For some patients the foot was the prime object of sexual arousal. Others had made a fetish out of an object of clothing, or had a neurotic phobia of snakes or small dogs because they unconsciously symbolized a penis. These examples come from the realm of psychiatric symptoms, but Freud held that the borderline

Affectionate gestures and embraces can be erotic for intimate couples.

between the normal and abnormal in matters of sex was not very wide. Whether or not we accept the details of Freud's theories, it is not hard to think of examples of normally accepted sexual symbols: a woman's spiked-heeled shoes are considered sexy in our culture, though, technically speaking, one might call this a shoe fetish.

In short, we must distinguish between the biological and the cultural. Recall the discussion in chapter 1 regarding the terms *sex* and *gender*. *Sex* refers to the biologically given genitals and secondary characteristics (facial and body hair, pelvic size, musculature, etc.) that make one male or female. Most persons have a distinct sex, although there are occasional hormonal mixtures and even cases where a person's genitals are partly of one sex, partly of the other (hermaphroditism). *Sex* also refers to behavior related to eroticism and procreation. *Gender*, on the other hand, is a social role, a learned way of performing erotic or sexually distinctive behavior.

SEX AS AN INTERACTION RITUAL

Is there any sociological significance in research on what happens during intercourse? One pattern that stands out is the underlying similarity of the sexual response in men and women. The sex researchers Masters and Johnson (1966) found that in more intense sexual experiences, there is the same sexual flush over the body and the same postorgasmic reaction. Timing of rhythmic contractions up to and during the orgasm itself is remarkably parallel in men and women. All this suggests that males and females are fundamentally similar beneath the surface, although the similarity can be covered over by the way their lives have been socially organized.

We are reminded that embryonically, both sexes start out the same, and only in the third month does the undifferentiated genital tissue shape into male or female organs. Thus, the penis is shaped from the same embryonic tissue as the clitoris, the scrotum is analogous to the labia minora and labia majora, and there are vestigial breast nipples in the male. Intense sexual arousal strips off the social differences, so to speak, and brings a man and woman closer together in a common pattern they share as human beings.

We can understand sexual intercourse as an intense kind of interaction ritual. We have met this line of sociological theory, deriving from Durkheim and Goffman, at various points earlier in the book. It was used in chapter 6 to examine the emotional and moral consequences of social class differences in the "ritual density" of interaction. Even closer to our concerns here, the theory was spelled out in chapter 8 to analyze love as a powerful interaction ritual creating a symbolic bond between two persons. Sexual intercourse, on a more physical level, also has the ingredients of a naturally occurring interaction ritual.

Let us recall the components that go into generating a ritual. A group is assembled; they focus their attention on the same activity; and they share a common emotional mood, which builds up as they become increasingly aware of each other's focus on the same thing. If all this happens, the result is a feeling of group membership, attached to symbols that are respected because they represent the relationship. Sexual interaction certainly has the elements of this: it brings together a little group of two persons; they focus attention on each other's bodies; and there is a common emotion of erotic arousal.

It is well known, of course, that both partners are not always equally aroused. One may be dominating the lovemaking while the other is passively following along, or even just lying there letting the partner "do their own thing." A couple are not always in the same mood or even on good terms with each other at the moment. Though the basic ingredients are there, sexual intercourse does not always build up the intensity that makes it a successful interaction ritual. But this is true of other interactional situations: sometimes an intense ritual focus is achieved and sometimes not. The key is whether a cumulative process is set in motion, so that the mutual focus gets stronger as the interaction goes along.

Sexual intercourse has a somewhat unique quality, though, as compared to other kinds of potential interaction rituals: if the physical bodies can start getting attuned, there is a kind of physiological escalator that will pull the partners together more and more intensely as they go along. The right emotional mood helps to get it started; so does the right physical arousal—of the genitals or other erotic zones. Often a couple has their own private arousing gestures—not necessarily overtly erotic, but just a touch of the hair, or the face, perhaps—that have come to symbolize their own personal connection to each other. (That is to say, they carry over symbolic gestures charged with meanings from previous interaction rituals.) On the other hand, though the lovemaking couple may be psychologically comparative strangers to one another, they may turn each other on enough physically so that the physiological escalator starts taking off of its own accord.

On the sheer physical level, sexual intercourse has a built-in rhythmic mechanism. If it can get started, the rhythm can take over on its own. Furthermore, this rhythmic sexual arousal is contagious from one partner to the other. The intensity of one partner building up to an orgasm spills over to the other partner and builds up his or her intensity in turn. That is why lovemaking couples are usually concerned about each other's sexual arousal; even if they are not in love, and the sex is engaged in fairly casually or even selfishly, each person's own arousal is usually more intense if the other person is aroused. Sophisticated lovers try to turn each other on because there is a kind of self-interested, as well as altruistic, interest in the other's orgasm.

When and if this happens, intercourse truly becomes an interaction ritual, on an unintentional, physical level. The contagious sexual arousal is like the emotional contagion that happens in more ordinary rituals, pulling in the participants' attention and making them more and more sharply focused on their common experience. It is no wonder, then, that sexual intercourse itself becomes a powerful symbol of the social bond and even gets invested with a "sacred" quality. It becomes more than a matter of individual pleasure (though it is that, too); it becomes an emblem of the relationship between those two persons. Sex comes to symbolize love. It comes to symbolize the possessive aspects of a marital (or quasi-marital) relationship, too. Lovemaking is an exclusive possession, and its violation—or even jealous thoughts about it—becomes the most serious of all transgressions.

To be sure, not every act of sexual intercourse between a couple actually builds up to this level of intensity. Much of the time, probably, one or the other partner is not so aroused. Their attention may be wandering from their bodies, and there is little mutual build-up of contagious physical or emotional effects. But total absorption occurs at least occasionally, and people orient themselves toward those high points. However infrequently this level is achieved, sexual intercourse nevertheless is the most potent symbol of an intense social tie. Even comparatively unsuccessful acts of lovemaking become significant representatives of the couple's relationship. Whether they are only maintaining their marital routine or changing their relationship for better or worse, it is likely to be felt symbolically in their lovemaking.

PREMARITAL SEX: A MODERN EROTIC REVOLUTION?

It is often claimed that there has been a sexual revolution in the last twenty years—during the lifetime of many students reading this book. Undoubtedly many things have changed, particularly in the area of premarital sex. College dormitories housing both males and females were introduced only in the 1970s; before that there were numerous regulations concerning hours during which members of the opposite sex could be in one's room, whether the door had to be kept open or others had to be present, or the like. In the 1950s and before, even regulated intersex visiting was prohibited entirely on many campuses. Despite all that, the change has not necessarily been so drastic as it appears. There had been a great deal of change already throughout the earlier part of the century, and our grandparents and great-grandparents may not have been completely restrictive in their sexual behavior, although they were not very open about it. On the other hand, premarital sex is not the *official* standard even today. Sweden is the only country in the world whose laws do not officially confine sex to within the bonds of marriage (Safilios-Rothschild 1977, 66–67). Most states prohibit sex below age eighteen except for married partners, as "contributing to the delinquency of a minor." These laws are not enforced much, although numerous people believe they are morally correct, and teenage girls (but not boys) are sometimes arrested on charges of delinquency if they are caught committing sexual intercourse.

It would be more accurate to say that the United States today has a variety of different sexual standards and practices and that different groups are frequently in conflict. Surveys of the American public in recent years show that 27 percent believe that premarital sex is always wrong, 39 percent believe it is always permissible, and the remaining 34 percent think it is sometimes right or wrong, depending on the circumstances (National Opinion Research Center 1986, 236). In other words, the country is split, with about a third at each extreme and the rest in the middle but with an edge for the libertarian side. Moreover, this has been a relatively recent shift. In 1963, more than 75 percent of the populace stated that premarital intercourse was always wrong (Reiss 1967), although, as we shall see, most of them did not live up to their own statements. Today, the division is sharply along age lines: 81 percent of persons below the age of 30 would not prohibit premarital sex, as opposed to 37 percent of persons over age 50 (*Gallup Report*, June 1985, p. 28). The trend toward permissive attitudes that began in the 1960s has continued fairly steadily (Thornton 1989).

Words versus Actions

Perhaps the clearest difference between today and earlier times is that sex is talked about a good deal more openly. In the Victorian period, anything involving sex was unmentionable, although a surprising amount went on behind the backs of polite society. This was a hundred years ago, and the situation lasted nearly up to the present in the United States, which has been in many ways an especially puritanical society. Farther back in history, however, there were societies in which sexual matters were much more open: for instance, public rituals of some tribal societies featured sexual interaction (see chapters 3 and 4 for more historical information).

A notable change that has occurred in the last few years is a major increase in public talk about sex. Popular magazines are full of discussions of it; men's magazines are openly pornographic; movies and some television channels have explicit eroticism; advertisements place tremendous emphasis upon the sexy; and formerly taboo four-letter words are commonly used by both men and women (although at the same time some persons strenuously object to this). But all of these references are impersonal. People may sometimes allude to their own sexual experience, but it is still not quite an ordinary topic of conversation. One does not often hear people talking about exactly what happens when they make love, even with their own partners; in fact, this lack of explicit communication has been cited by sex therapists as a common cause of sexual difficulties.

Following are some questions worth keeping in mind as we examine the data. Is there more sex today, or only more talk? And has the kind of sex changed? Is there still a dual standard for males and females, so that what is allowed to be experienced by one is prohibited for the other? We shall see some trends, but at the same time we should be aware of the possibility that trends are not all moving one way, that there are strong differences among social groups, and that there are currently ideological battles over conflicting standards for sexual behavior.

The Accuracy of Statistics about Sex

Getting reliable information on sexual behavior is a problem that has still not been completely solved. Standards of greater openness in talking about sex make it easier to do surveys today, but the fact that these standards have changed recently means that older information may not be strictly comparable; previously people may have been more inclined to hide what they were doing. Even today, there is considerable disagreement about the appropriateness of discussing sex, so that one's sample may be badly biased toward those who are willing to talk about it. In 1974, *Playboy* magazine carried out a national survey of sexual behavior; 80 percent of those contacted refused to cooperate (Hunt 1974). Another *Playboy* poll (Petersen et al. 1983) was done simply by having readers who felt like it mail in questionnaires torn out of the magazine. This resulted in 80,000 responses, which is by far the largest number of people ever surveyed on sex, but they are obviously among the segment of the populace most interested in it. The fact that 60 percent of the *Playboy* respondents almost always engaged in oral-genital sex, for instance, or that 35 percent had had sex with more than one person at the same time probably does not mean that the rest of the population was doing the same.

This problem of biased sampling can be overcome by selecting respondents randomly from the population. Such surveys are attempted most often of women of child-bearing age in connection with attempts to estimate the birthrate. But this not only leaves out men's sexual practices but also tends to yield relatively superficial data. A stranger asking persons a few questions about their sex life is likely to get perfunctory answers, skewed toward what is the conventionally respectable thing to say in a particular social group. For this reason, really good, in-depth surveys of the sex lives of the general population are comparatively rare. For many of the more intimate issues, the best source of data is still the Kinsey reports (Kinsey et al. 1948; 1953), carried out by the Institute for Sex Research at Indiana University.

Alfred Kinsey was not a sociologist but a zoologist, who in 1937 realized that he knew more about the sexual behavior of insects than anyone did about the sexual behavior of humans. Between 1938 and 1950 Kinsey and his colleagues interviewed 16,000 persons, making strenuous efforts to ensure reliable responses—for instance, by comparing the answers of husbands and wives. Kinsey attempted to bring in the full range of individuals by contacting groups such as churches and trying to interview 100 percent of their members. By questioning a wide selection of groups, the effort was made to represent the entire population of the United States. Kinsey did not succeed in this, but different social classes, educational levels, and other subgroups were sampled, and it was possible to compare their patterns of sexual behavior. Probably the weakest part of his data are those on the working class; for this group, Kinsey relied heavily on interviews with imprisoned criminals, although their sex lives had probably been more active than the rest of the working-class population.

Since the late 1980s, due to social awareness of the AIDS epidemic, there have been new efforts to carry out large-scale surveys that accurately represent the sexual behavior of the whole population. These surveys have generally found that sexual behavior is more conservative than the impression given by previous research, especially by the more popular surveys. Even these modern surveys are not entirely ideal; often they have a narrow interest only in the question of how a disease like AIDS is transmitted and do not provide information that is relevant to how sexual relations operate to tie people together, and how they affect the family in this era of increasing gender equality.

For more intimate data, we often must continue to rely on some of the older studies, such as Kinsey, or on studies of more limited groups. Especially when we are trying to estimate long-term changes in sexual behavior, we have no choice but to put together estimates drawn from all the sources that are available. By comparing and cross-checking the overall pattern, we try to get the most reliable picture of what has been going on.

Recent Trends in Premarital Sex

Virginity

It is clear that the percentage of persons who are virgins at marriage has declined considerably (see figure 5.1). Among women coming to maturity in the 1970s, some 80 to 85 percent had had sexual intercourse before marriage; among men, premarital experience was nearly universal—95 percent. But no sudden revolution took place in the 1960s or at any other point. Nonvirginity appears to have been fairly widespread all the way back to the beginning of the century, and it has gradually increased. Among men in the 1920s, about two-thirds (67 percent) had had premarital intercourse; in the 1940s, Kinsey found the figure was already up to about 85 percent. There has been no sexual revolution for men as far as nonvirginity is concerned.

For women the shift has been much sharper. Around 1900, about three-fourths (73 percent) of women had no premarital intercourse, and of those who did, it was generally with their fiancé (Reiss 1960, 230). The percentage of virgin brides declined in the 1920s and '30s to around 45 to 50 percent; most of the increase in nonvirginity, though, came about because more women were having intercourse with their fiancés. As late as the period 1960–64, a slight majority (52 percent) of women marrying were nonvirgins; this number rose sharply to 72 percent in 1970–74 and 79 percent in 1975–79 (*Family Planning Perspectives*, 1985). The trend leveled off in the 1980s. By 1988, about 75 percent of unmarried women had had intercourse by age 19 (as compared to 80 percent of unmarried teenage males) and 82 percent by age 29 (*Family Planning Perspectives*, 1985; Sonenstein et al. 1991; Mosher 1990). In the long run, there has been a continuous decline in virginity for women throughout the twentieth century, although there was a big leap in the late 1960s and the 1970s. This period seemed even more revolutionary at the time, because the nonvirgins became a large majority, and the public finally realized that the traditional position enunciated in public all along was now practiced only by a small minority.

Prostitution

By comparing the sets of figures for men and women, we see that consistently, a larger share of men have been sexually active. In Kinsey's day, about 50 percent of American women had premarital intercourse, compared to 85 percent of American men. Not only that, but most of the women's premarital sex was with their prospective husbands, whereas at least half the men had intercourse with women other than their fiancées. How was this possible? Obviously a small proportion of women were having intercourse with a large number of men. For the most part, this occurred through prostitution.

Back in the 1940s, Kinsey, Pomeroy, and Martin (1948, 286–88, 597–603) found that about 70 percent of American men had visited a prostitute at least once. (This activity was not entirely confined to unmarried men; 10 to 20 percent of married men at various ages also had intercourse with prostitutes, although this constituted only about 1 percent of their total sexual outlet.) Intercourse with prostitutes was most common among men born before 1900 and has declined for each generation since then. Thus, the *Playboy* survey of 1974 (Hunt 1974) found that males were much less likely to use prostitutes than were Kinsey's subjects, even though this group was probably above average in sexual activity. Prostitution still exists today, but it is no longer as visible as it was previously, when organized brothels flourished in "red-light" districts of major cities. In its place is a smaller-scale form of prostitution consisting of streetwalkers and brothels disguised as massage parlors, often catering to older rather than young unmarried men (Winick and Kinsie 1971).

The decline in male use of prostitutes is probably due to the trends in premarital sex. Prostitution was to a considerable extent the result of a large imbalance between male and female sexual behavior. As female behavior has become more like the male pattern, prostitutes have become less important for unmarried men. At the same time, female sexual behavior continues to place more emphasis upon affectionate contacts rather than sheer promiscuity; in that sense, the shift of male attention from prostitutes to girlfriends probably indicates that in some ways the male pattern is getting closer to the female one as well.

Who Does What How Often?

Figures on nonvirginity are a little misleading in some respects. The loss of virginity requires only a single act of intercourse; how frequently premarital sex takes place is another question. The sharp rise in female nonvirginity, or for that matter the long-standing, near-universal male nonvirginity, does not mean that either women or men are having a great deal of nonmarital sex.

Figures on frequency of intercourse indicate that the nonvirginity issue does indeed overdramatize what is actually going on. Whether one is a virgin or not has been a major symbolic issue in the past, and hence such figures still make headlines. But in fact the rate is rather low. Among unmarried teenage women who had intercourse at all, the average was about 2.5 times per month (Zelnik, Kantner, and Ford 1981, 86–87). In 1988, a little less than half of these women were having sex at least once a week (Sonenstein et al. 1991). If we go back to Kinsey's data for the years before 1950, we find that frequencies then were lower. Kinsey et al. (1953, 289) found premaritally experienced women had intercourse once every one or two months if they were under age twenty, and about 1.3 times per month if they were older. It is true that Kinsey's sample of women is mostly confined to the middle and upper classes, whereas Zelnik had a better sample of all social classes; so it is possible that the working-class women in Kinsey's day might have brought the total frequency up to nearer the contemporary figures.

In either case, the figures are fairly low. Kinsey and his associates found that premarital intercourse happened more often for men than for women, but even for men it did not happen very often (1948, 283). Below the age of twenty, sexually experienced males were having intercourse about eight times per month (twice a week). Although for women, premarital sexual activity increased as they got older, for men it *decreased*, falling slowly to about four times a month (once a week) by the age of forty. There is, of course, a lot of variation among individuals. In the 1983 *Playboy* sample about 50 percent of the men said they had intercourse at least two or three times a week, and 20 percent said almost every day (Petersen et al. 1983). The women in that sample were even more active than the men: about 63 percent of them reported that they had intercourse at least two or three times a week, and 28 percent almost every day. But this probably represents far and away the most sexually active group in the population.

Teenagers Compared with Young Adults

The public image is that the teens are the wild years, the stage when all the premarital sex occurs. But in fact this is a myth. Teenagers tend to be a

There is a considerable amount of change going on in all aspects of teenagers' lives, and this affects their sexual relationships. Typically, these relationships are monogamous but short-lived, lasting only a few months.

good deal more conservative than the adults just ahead of them, in sexual behavior as in most other respects—in religious, political, and social attitudes. It is the young unmarried adults, especially those in their twenties, who are most sexually active (Kinsey et al. 1948, 1953; Reiss 1960; DeLamater and MacCorquodale 1979; *Family Planning Perspectives* 1985). What this means, though, is that teenagers do not engage in intercourse as *often* as young adults; it is true that the rate of nonvirginity has fallen among teenagers.

This is clearly a social pattern, not a biological one. Male youths in their late teens have the maximal amount of sexual drive (Kinsey et al. 1948, 219–21, 232–33). A surprising proportion of very young males are capable of as many as five ejaculations within an hour or less, though capacity for multiple orgasms (at the rate of two or more in short succession) falls off quite rapidly with age. Kinsey (1948, 231) even found evidence that the angle of erection declines with age. Among females an equivalent figure is rather hard to estimate. Kinsey did find (1953, 718–19) that a woman's tendency to have multiple orgasms becomes higher than a male's by about age twenty-five and remains higher throughout the rest of their lives. But it makes sense biologically that in the human species, as in all others, peak reproductive ability is attained soon after sexual maturity and falls off gradually thereafter. We see this in the declining fertility of women after their twenties, until the age of menopause around fifty.

The lower incidence of sex among teenagers is yet another instance of how sexual behavior is affected by social processes. Sexual relationships among teenagers are more unstable than those among adults; there is a considerable amount of change going on in all aspects of teenagers' lives, and this affects sex, too. Most sex takes place during an affair that is monogamous during the time when it is going on. For instance, about four-fifths of male teenagers had some experience of sexual intercourse; but their frequency of intercourse was only 2.7 times per month, i.e., less than once a week (Sonenstein et al. 1991). One reason the average was so low was that these sexually experienced youths typically spent six months out of the year without any sexual partner; only 21 percent of them had more than one partner in any one month during the year. There are not many teenagers who are having roving sexual adventures day after day; much more typical is two sexual affairs per year, each lasting a few months, separated by several months without a sexual partner. It is the same thing for teenage females; in the years since the early 1970s, as the number of sexual partners has gone up, the frequency of intercourse actually declined slightly (Zelnik, Kantner, and Ford 1981). It is just because of this switching between partners that the amount of sex is restricted; so much time is taken up in breaking up and finding a new partner.

How Many Partners?

Of all the different aspects of sex, one of the most difficult to get reliable information on is how many sexual partners people have. Zelnik, Kantner, and Ford (1981, 79–85) found that among unmarried female teenagers, the majority (54 percent) of those who had sexual relations at all had them with only one person. At the upper extreme, 18 percent had had intercourse with four or more different partners. White women had an average of more partners (2.8) than black women (2.2). Interestingly enough these figures do not seem to have changed much since Kinsey's day (1953); he found that 53 percent had had a single premarital partner only, and 13 percent had had six or more partners. By 1988 there had been a little more widening of sexual experience, and 59 percent of experienced teenage women reported more than one partner (Sonenstein et al. 1991).

Somewhat surprisingly, the number of partners reported by men has declined since the time of the Kinsey surveys (Weinberg and Williams 1980). Kinsey found that high-school educated men (i.e., the lower social classes) had about nineteen premarital partners, and college-educated men (the higher social classes) had about ten. Thirty years later, these figures had *dropped* to below thirteen and eight, respectively. It is true that in some recent surveys, we get very high figures: the 1983 *Playboy* sample gave an average number of twenty partners for men, sixteen for women (Petersen et al. 1983, 244), and about 10 percent of each sex said they had had more than fifty. But this group is probably rather out of the ordinary. In the overall population in the late 1980s, adult men were saying they had had, on the average, twelve sexual partners since age eighteen, and adult women were averaging a little more than three (Smith 1991). Probably the men were exaggerating upwards and the women may have been minimizing somewhat. In general, men seem to be dropping closer to the female pattern and in some ways women are behaving more like men.

Is Gay and Lesbian Sex Replacing Heterosexuality?

The period since the late 1960s has not only been a time of heterosexual liberalism. It is also the time of gay liberation—more accurately, the movement of gay men and lesbian women to overcome discrimination and gain equal rights in society. Here again we need to distinguish between changes in what

Spectators at a gay rodeo in California. Greater tolerance of homosexuality does not necessarily indicate a widespread shift in sexual behavior.

people are willing to say in public, and changes in what they are actually doing. There has been a huge change in the degree of openness in talking about homosexuality, and some changes toward greater tolerance in the attitudes that people express about allowing homosexual preferences. Through all the controversies and conflicts, it is not clear if there is a real change in sexual behavior, or only greater attention to the topic. Is a shift taking place from heterosexual to homosexual preference? Does homosexuality coming into the open, along with the greater tolerance of premarital sex in general, indicate that the family is gradually declining as the main center of sexual life?

Surveys of sexual behavior made in the late 1980s have begun to give a picture of what has been happening. Somewhat surprisingly, these figures turn up less homosexual behavior than was suggested by the earlier Kinsey studies, which may have been biased toward sampling the more sexually active respondents. A large-scale national survey (Smith 1991) asked people about their sex lives since the age of eighteen; respondents indicated whether they were sexually active, and the types of sexual activity in which they had engaged:

heterosexual activity	90.9%
bisexual activity	5.6%
homosexual activity	0.7%
no activity	2.9%

In other words, about 97 percent of the adult population were sexually active at some time in their lives (about 3 percent never had sex); less than 1 percent were exclusively homosexual; slightly more than 5 percent were bisexual; and the rest were heterosexual. Although Kinsey estimated that there were over twice as many homosexual men as women, in modern surveys the figures for men and women were approximately the same. It also appears that gay and lesbian sex is concentrated in particular periods of a person's life, rather than going on constantly throughout. When the survey asked about sexual activities that happened during the past year, between 1 and 2 percent of the respondents said they were either bisexual or exclusively homosexual during that time.

If one can believe these most recent surveys, gay men and lesbian women are a rather small minority. This of course does not justify any kind of social discrimination against persons on the basis of their sexual preference. In terms of the sociology of sex and the family, there probably never was a threat to the institution of the family from this direction. There are still many questions unanswered. We lack good trend data on the relative amounts of homosexuality and heterosexuality in the past; and we do not know whether the AIDS epidemic that began in the 1980s was responsible for reducing the amount of homosexual sex. It seems an accurate generalization, though, that the great majority of sexual relationships have been heterosexual in the past and continue to be so in the present.

Causes of the Sexual Revolution

Although changes in sexual behavior in recent years have received a great deal of publicity, they have not always been accurately understood. There has been a revolution, perhaps, in the way we talk publicly about sex. And there is less insistence that sex be tied to marriage, although the change partly means less hypocrisy about the realities of premarital sex. Changes in the overall frequency of sexual intercourse and other forms of sexual behavior are not so great. The importance of commercialized sex in the form of prostitution has actually declined, and premarital sex for both men and women is now more likely to take the form of affectionate and quasi-stable, if not permanent, pairing. *Cohabitation* of unmarried couples has gained wide acceptance as part of this trend to relationships that fall somewhere between traditional marriage and wide-open promiscuity; in fact, living together is much like a very flexible form of marriage.

FEATURE 9.1
A Famous Literary Love Scene

Premarital sex has a power in real life that is scarcely hinted at by bare sociological statistics. It is a matter of passion and adventure, for many people perhaps the most dramatic experience of their lives. It is no wonder then, that it is the subject of some of our most famous literary works. The following is a notable literary love scene from James Joyce's *Ulysses*. An Irish woman lies in bed next to her sleeping husband, as thoughts of their courtship go through her head:

> so there you are they might as well try to stop the sun from rising tomorrow the sun shines for you he said the day we were lying among the rhododendrons on Howth head in the grey tweed suit and his straw hat the day I got him to propose to me yes first I gave him the bit of seedcake out of my mouth and it was leap-year like now yes 16 years ago my God after that long kiss I near lost my breath yet he said I was a flower of the mountain yes so we are flowers all a womans body yes that was one true thing he said in his life and the sun shines

for you today yes that was why I liked him because I saw he understood or felt what a woman is and I knew I could always get round him and I gave him all the pleasure I could leading him on till he asked me to say yes and I wouldnt answer first only looked out over the sea and the sky I was thinking of so many things he didnt know of Mulvey and Mr Stanhope and Hester and father and old captain Groves and the sailors playing all birds fly and I say stoop and washing up dishes they called it on the pier and the sentry in front of the governors house with the thing round his white helmet poor devil half roasted and the Spanish girls laughing in their shawls and their tall combs and the auctions in the morning the Greeks and the jews and the Arabs and the devil knows who else from all the ends of Europe and Duke street and the fowl market all clucking outside Larby Sharons and the poor donkeys slipping half asleep and

(Continued next page)

FEATURE 9.1 *(Continued)*

the vague fellows in the cloaks asleep in the shade on the steps and the big wheels of the carts of the bulls and the old castle thousands of years old yes and those handsome Moors all in white and turbans like kings asking you to sit down in their little bit of a shop and Ronda with the old windows of the posadas glancing eyes a lattice hid for her lover to kiss the iron and the wineshops half open at night and the casanets and the night we missed the boat at Algeciras the watchman going about serene with his lamp and O that awful deepdown torrent O and the sea the sea crimson sometimes like fire and the glorious sunsets and the figtrees in the Alameda gardens yes and all the queer little streets and pink and blue and yellow houses and the rosegardens and the jessámine and geraniums and cactuses and Gibraltar as a girl where I was a Flower of the mountain yes when I put the rose in my hair like the Andalusian girls used or shall I wear a red yes and how he kissed me under the Moorish wall and I thought well as well as him as another and then I asked him with my eyes to ask again yes and then he asked me would I say yes to say yes my mountain flower and first I put my arms around him yes and drew him down to me so he could feel my breasts all perfume yes and his heart was going like mad and yes I said yes I will Yes.

From *Ulysses* by James Joyce. Copyright © 1934 and renewed 1962 by Lucia and George Joyce. Reprinted by permission of Random House, Inc.

Two major factors appear to be involved in this change. One that is frequently mentioned is the "contraceptive revolution" that took place with the development of the birth-control pill in the 1960s. Other forms of birth control were available before that time, such as rubber condoms, vaginal foams, and diaphragms. But the birth-control pill made contraception easy, unobtrusive, and reasonably reliable, and this may have sharply reduced women's fears of pregnancy and hence increased their premarital sexual behavior. (See table 9.1 for further information on the effectiveness of various forms of contraception.)

The revolution in contraceptives, however, is not likely to be the whole story. Changing social attitudes were necessary in order for the pill and IUDs to become widely and even openly available. Moreover, it is clear that many women do not use them, as indicated by the rising rate of illegitimacy (discussed further below). Contraceptive styles shift from time to time, as when health hazards reduce the popularity of a method (witness the decline in the use of birth-control pills in recent years). It is likely that contraceptives of various kinds, although contributing to sexual behavior, are only part of a larger story that is basically social rather than a matter of medical technology.

The other major shift that coincides with the sexual revolution is the feminist movement. For some people this may seem incongruous, in that feminists have been described as antimale or antisexual. But in fact the structural pattern makes a good deal of sense. The feminist movement is part of a wholesale shift in the way women organize their lives. It involves the

TABLE 9.1   Effectiveness of Contraceptive Methods

Percentage failing to prevent conception in 1 year of use:

Male sterilization	0.1%
Female sterilization	0.4
Pill	3
IUD	6
Condom	12
Sponge	18
Rhythm	20
Diaphragm	2–23
Foam/cream/jelly	21

Source: "Developing New Contraceptives," National Academy of Science, 1990.

tendency for women to finish college and professional training, to pursue full-time careers, and to put off marriage while they are doing such things. All these are trends that became very strong in the 1960s and '70s. Moreover, they fit together with the sexual revolution, for *that* revolution has consisted mainly of closing the gap between men and women in sexual behavior.

Thus, there has been more premarital intercourse on the part of women, and it is now reaching almost the same level as for men. There has been a slower increase in the number of sexual partners women have, although part of the pattern here may be that men's premarital sex has become less promiscuous and closer to the female pattern. To some extent the dual standard, which had prescribed one pattern of sex for men and another for women, has been disappearing. The shift in the way we now publicly talk about sex is part of this change, since the old taboo on using sexual language in the presence of women was part of the dual standard that allowed men to do what they wished as long as they kept it among themselves (and with a small female minority of prostitutes).

More important structurally has been the fact that as women pursue careers and put off marriage, they are putting off marital sex as well. Hence, the rise in women's premarital sex is partly a way of finding the sexual gratification that they would otherwise find in marriage. Also, as women emphasize having their own incomes, it becomes less important for them to confine sexuality to marriage as a way of acquiring a husband to support them. For all these reasons, the women's movement has probably contributed to the changing patterns of sexual behavior.

Has the AIDS epidemic, which began in the early 1980s, had an effect on sexual behavior? There has been a good deal of public speculation that the sexual revolution has come to an end, and that people are behaving more conservatively because of the deadly disease. Survey data show that a relatively small proportion of people (a little more than 10 percent) have changed to less risky sexual behavior because of the danger of AIDS; that is, they have shifted

Safe-sex campaign materials are targeted toward Native Americans, a population that is considered to be at risk for AIDS.

to a lower number of sexual partners and avoided casual sex with persons they did not know well (Smith 1991). The same surveys, though, show that overall sexual behavior has not changed much in recent years; there is still the same general level in the number of sexual partners and in the amount of marital fidelity. On the whole, AIDS does not seem to have had much impact. In part this is because many people ignore dangers; but it is also due to the fact that the majority of people are fairly conservative in their sex lives. As we have seen above, they may have a series of different sexual partners during their lives, but usually they have only one partner at a time; and many unmarried persons go through periods without any sexual partner. About 25 percent of both unmarried and divorced persons are completely abstinent in any one year (Smith 1991). For these reasons, most people don't feel much need to change their sexual behavior.

The Influence of Social Class

The sexual revolution has not occurred uniformly throughout society. In fact, there seem to be opposing trends in different groups, and a polarization of conflicting attitudes has built up. In the early part of this century, there appears to have been more premarital sex on the part of the lower classes. In Kinsey's sample (for people interviewed in the 1940s, including older persons whose sexual experience went back to the beginning of the century), college-educated males were most likely to be virgins at marriage, had the lowest rates of premarital intercourse, and were involved with fewer premarital partners (1948, 374–50; Weinberg and Williams 1980). By the 1960s, the college-educated group had caught up in some ways with the less-educated groups (which, generally speaking, meant the lower social classes), and both groups now have the same rate of premarital nonvirginity. But the higher classes still are less likely to begin intercourse as early in their teen years, and they still have fewer different sexual partners (Weinberg and Williams 1980).

For females, moreover, the trend has been for social class differences to become *greater*. In Kinsey's 1940s sample, women of *all* social classes tended to be equally restrictive (although lower-class women did begin intercourse earlier, and upper-class women were more likely to have masturbated). By the 1970s, social class differences had emerged, so that lower-class women were having premarital intercourse (as well as various other forms of genital contact) in their teens at a much higher rate than higher-class females (Weinberg and Williams 1980). Women of the higher social classes have remained more sexually traditional, at least before marriage.

The sexual liberalization of the upper classes seems to hold only for the white population. As we have already seen in chapter 6, the middle- and upper-middle-class black population is actually quite conservative in social matters (though not necessarily in politics). This holds for sexual behavior as well. Thus, black men and women not only have more traditional attitudes about gender roles (Ransford and Miller 1983), but also blacks of the higher classes are less likely to have premarital sex than blacks from the lower social classes (Reiss 1967).

Parents' Influence

There is one important factor that cuts across social class: young people's sexual behavior is strongly influenced by their parents. Not that teenagers simply follow what their parents tell them regarding how to behave sexually, as there is little correlation between how sexually permissive parents say they are and their children's own attitudes (Walsh 1970). The crucial factor, as John DeLamater and Patricia MacCorquodale (1979) found, is how close teenagers are to their parents. The more they see of each other, and the closer the terms on which they get along with one another, the more restrained the children are sexually. This is a good example of the general sociological principle that conformity is stronger the more regularly a group interacts.

Similarly, adolescents who are not living with both of their natural parents are especially likely to have early intercourse (Newcomer and Udry 1986). This seems to imply that the family acts as a socially conservatizing influence on sex, and perhaps even as an emotional substitute for it, for the family is the location of a high density of interaction rituals of its own. These interaction rituals of the family group are quite different from the two-person interaction ritual that takes place in sexual intercourse. Inside the family, only the parents have this sexual ritual. The children are very pointedly excluded from it. Any violation, though it may sometimes occur, is regarded as a shocking event, *incest*. The interaction rituals that do take place among parents and children are explicitly desexualized, and this desexualization seems to carry over into inhibiting sexual behavior in general.

The Realities of Sex

In reality, statistics do little to communicate the emotions associated with sex itself. In real life, sex may be excitement, a game or a lure that keeps life going when times are otherwise dull. For some men it is a challenge, a kind of ego-building sport to see how many women they can possess, or how many virgins. Some boast of having had intercourse with over one thousand women (Safilios-Rothschild 1977, 29), which obviously means little time or emotion invested in any one individual. For other people, sex is sin, temptation, or just plain hassle—a superficiality that boils down to realities that are ugly, coercive, and disillusioning. For still others, sex is a kind of narcissism, an admiring of oneself in the mirror of admiration one gets from others. This style is especially common among very beautiful women, although there is a male counterpart as well. Freud (1914/1957) noted that a narcissistic preoccupation with oneself often is alluring to others, as if they could recapture the self-centered bliss that everyone once felt as an infant, by sexually possessing someone else who still gives off that kind of narcissism. Constantina Safilios-Rothschild (1977, 42–47) also comments that sexual teasing is a weapon women can use to get back at men for the power they hold.

The fact of the matter is that we each negotiate our own path through the various alternatives offered by our social environment. Sex means all sorts of things to different people, and we are constantly encountering different combinations that open certain possibilities and shut off others. Each individual embarks on a sexual "career," just as he or she chooses and develops an occupational one, and each person builds up a kind of investment in one or the other that determines how much effort he or she will expend. This choice depends upon how much ego success one builds up in one sphere or another.

Sex is also a way of negotiating oneself into a marriage, although this is of primary importance for some people and secondary or irrelevant for others. Sex shades over into love too. There is a complex continuum with extremely idealized love at one end and sheer horny lust at the other, and infatuation and many other emotions somewhere in between. Our language is full of expressions for where we think we fall on that continuum, from "pure" and "eternal devotion" and so on in one form of rhetoric, to the many examples of derogatory sexual slang at the opposite extreme. Perhaps there is some reality to these verbal differences, as there is evidence that a belief in romantic love is lower among people who have more premarital partners (Israel and Eliasson 1972, 186). But basically these are just forms of talk, and both the people who use very idealized language *and* those who speak most cynically about sex tend not to live up to what they imply. The social circumstances out of which love emerges (as we've seen in chapter 7) are too similar to those in which prolonged sexual contact occurs for there not to be a connection between love and sex. The fact that the connection may be getting stronger is indicated by the popularity of an expression that means both of these: "making love."

MARITAL SEX AS A SOCIAL TIE

As we have seen in chapter 2, a main ingredient of the family is exclusive sexual possession between husbands and wives. If lovemaking is an interaction ritual establishing relationships, marital sex is the interaction ritual that most centrally symbolizes the ongoing bond. During the marriage, this is supplemented by other ties—especially shared property and children—and the amount of sexual activity goes down. But sex continues for most married couples at least occasionally, even into old age, as a token reminder of what their tie is about.

Premarital sex is now quite widespread; but it is not generally as frequent or as regular as during marriage. Moreover, premarital sex is often a pathway to marriage, since a considerable percentage of people who engage in premarital sex end up marrying each other.

Marital sex is less of a negotiation and an adventure and more of a routine. But what kind of routine is it, and how do married people feel about it? Is it the essence of wedded bliss or a source of domestic frustration? Are we dealing here with couples making love forever and ever until death do them part, or with horny husbands and unsatisfied wives? Marriage in general is a situation of power and conflict in which happiness and unhappiness depend on many conditions. These conditions carry over into a couple's sexual life as well. How a couple get along sexually is an indication of how their marriage is going in general.

TABLE 9.2   Frequency of Marital Intercourse Over One Month

9.2%	None
16.3	Once or twice during the month
26.7	About once a week
30.2	About twice a week
10.7	About every other day
6.9	Nearly every day

Source: National Opinion Research Center, General Social Survey, 1991, 1992.

Differences in Intercourse by Age and Social Class

How often married couples have intercourse varies a great deal, depending on how old they are, how long they have been married, their social class, and other factors. Overall, the rate is about once or twice a week, but different couples may range all the way from no intercourse at all to every day.

The rate has gone up since 1965 (Westoff 1974, 136). This happened among all age groups; so it was not merely a phenomenon of the youth culture of the late 1960s. It seems to be correlated with the increasing use of contraceptive pills, which came into popularity at that time. Also women who are employed, especially at careers rather than only to supplement the family

TABLE 9.3 Frequency of Marital Intercourse by Age

Age	Average frequency per week
16–20	3.7
21–25	3.0
26–30	2.6
31–35	2.3
36–40	2.0
41–45	1.7
46–50	1.4
51–55	1.2
56–60	0.8

Source: Kinsey et al., 1953, 394.

TABLE 9.4 Maximum Frequency of Marital Intercourse Ever in Any Single Week

Frequency of intercourse	Percent of women	Cumulative percent who have experienced at least this frequency
1–3	10%	100%
4–6	32	90
7 (once per day)	23	58
8–14 (up to twice per day)	25	35
15–21 (up to three times per day)	8	10
22–30 (up to four times per day)	2	2

Although the average rate of marital intercourse is about twice a week, most people at some time in their marriage have a period in which sex is a good deal more frequent. The above table shows that about one-third of all married women had a peak period in which they were having intercourse more than once a day.

Source: Kinsey et al., 1953, 395.

income, have higher rates of intercourse; so the increase in marital sex may also be due to the greater career liberation of women in general during that period. Again, as we've seen in regard to premarital intercourse, the general trend of growing feminism has not been antisexual, as some critics contend, but quite the opposite.

Frequency of intercourse generally decreases with age. Teenage married couples have intercourse the most often. The rate falls off gradually thereafter until it is less than once a week for couples in their late fifties (see table 9.3).

These figures are from Kinsey (1953, 394) and refer to the 1940s; more recent data (Hunt 1974, 190; Blumstein and Schwartz 1983, 196–98; Smith 1991) show the same pattern, with slightly higher averages for all ages. These are, of course, averages over lengthy periods of time. Most couples have at least occasional periods of more intense sexual activity (see table 9.4).

There is an interesting pattern by social class. Contrary to the image that the lower social classes are less inhibited and more direct about sex, it turned out that earlier researchers found that higher proportions of middle- and upper-middle-class persons have intercourse in total nudity, or with the lights on, whereas the working class tends to be more prudish about these matters (Kinsey 1948, 363–74; 1953, 399–400; Hunt 1974; Westoff 1974). The higher social classes are also more likely to use a variety of positions in intercourse and to practice oral and other forms of sex. More recent data (Weinberg and Williams 1980) indicate that these social-class differences have now narrowed, although the traditional differences remain in some respects.

There seem to be several reasons for these class differences. One is that working-class homes are smaller and more crowded. Especially at the poorest levels, numerous people are likely to be sleeping in the same room, perhaps including children as well as adults. Under the circumstances, intercourse is more difficult to arrange and has to be more surreptitious. It simply isn't possible to have a leisurely, totally nude erotic experience with the lights on if the couple are trying not to wake everyone else up.

But there is another important factor. Generally, the lower the social class, the stronger is the division between traditional male and female roles. As Lee Rainwater (1964) pointed out in a study of poverty-level people in Latin America, England, and the United States, males in these groups regard sex "as a man's pleasure and a woman's duty." "Good girls" are not supposed to know anything about sex, whereas men take pride in whatever sexual pleasure they can get from prostitutes or "loose women." There is a strong male belief in the importance of possessing a virgin at marriage. Given these differences about sex, the elaborateness of erotic behavior tends to be kept to a minimum. The honeymoon experience is often traumatic for the woman, and it sets the tone for unpleasant sexual relationships during the subsequent marriage.

Rainwater's data, concerning the most traditional classes, are over thirty years old. They probably draw a somewhat exaggerated picture of how negative sexual relationships are in the lower classes. (In fact, Rainwater's own data from Puerto Rico show that only half the wives found marital sex to be a completely unpleasant experience.) But Masters and Johnson (1966, 202) also found that in their sample only 14 percent of the males whose education was below college level expressed any concern with whether their sexual partners achieved satisfaction, as compared to 82 percent of the college-educated group. More recent data show that traditional sexual attitudes have certainly not disappeared from working-class male culture (LeMasters 1975).

There is some evidence that social class differences have decreased in recent years. We have already seen this in terms of premarital sex. In Kinsey's day, the lower social classes were much more likely to have premarital sex than the higher classes, whereas today that is no longer true. Kinsey drew a picture in which the college-educated groups did not have much intercourse before marriage, either with prostitutes or with others. But this was not a matter of sheer sexual inhibition in the higher classes, since they also masturbated before marriage much more than the lower classes, and once they were married they had a higher rate of intercourse. Kinsey's picture is still by and large true, aside from the shift in premarital virginity (Hunt 1974; Weinberg and Williams 1980). As Lillian Rubin (1976) points out, working-class sex is still not as elaborate as in the middle classes. The sexual "revolution" of recent years has in many respects affected the higher social classes more than the lower.

EXPLAINING BY COMPARISON: EXTRAMARITAL SEX

We turn now to examine extramarital sex—sexual activity that breaks the marriage bond. This has always been a favorite dramatic topic, but there is a sociological reason for looking at extramarital sex that is more significant than the fact that it entertains us with its scandal. We have seen that marriage involves an element of possessiveness, which can be analyzed as a form of sexual and emotional "property." In traditional societies, this sense of marital property was strong. It was part of one's public honor (especially of a husband and father) to maintain control of one's sexual property. That proprietary ideal is not so strong today, and the very notion of someone being someone else's "property," especially sexually, is generally considered distasteful. Nevertheless, as we have seen, the ritual aspects of love (in chapter 8) and of intercourse (earlier in this chapter) tend to make one's spouse's body into a symbol of the couple's special bond. Even among persons who consciously disapprove of jealousy and favor individual freedom and openness, there are powerful social mechanisms in the forms of intimate interaction itself that tend to produce an emotional, symbolically loaded sense of possessiveness about sex.

It is worthwhile, then, to look at extramarital sex precisely because it is such a useful comparison. If marriage is a strong possessive tie, what kinds of conditions could break this tie apart? How do people deal with conditions that are *non*possessive sexually? This is part of the research strategy of explaining things by contrasts; we learn about what holds a sexual tie together by seeking what pulls it apart. We will compare various kinds of arrangements that break down sexual exclusiveness: conventional adultery, as well as deliberate arrangements of sexual sharing such as "open marriages" and swingers' groups, and the way that sex has been arranged in communes. Do these various sexual arrangements work? If so, what social conditions are involved in them? These conditions, we will see, are revealing in explaining not only the variety of sexual arrangements, but also the features that hold a conventional marriage together.

The prohibition on extramarital sex is a very strong one in our cultural history. Officially, sex is supposed to be confined to marriage. Unofficially, there has been a double standard, in that infidelity has been much more strongly condemned when it is the woman who is unfaithful. Traditional English has the term *cuckold* for the husband whose wife has sex "behind his back," but no equivalent term for the wife whose husband does the same. The implication is that it is much more of a disgrace for a man to fail to keep control of his sexual "property" than it is for the wife. In fact, the opposite is much more likely to be the case, since it is men who traditionally have committed adultery and condoned it in other men.

Although some groups now have shifted their standards somewhat, there is still a great deal of strong belief that extramarital sex is morally wrong. Recent polls of the entire American population find that about 74 percent believe that it is always wrong, and another 14 percent believe it is almost always wrong (National Opinion Research Center 1986). This level of disapproval has been fairly constant since the early 1970s after a small surge of liberalization in the late 1960s. Only 1 percent of women and 3 percent of men say there is nothing wrong with extramarital sex (Thornton 1989).

In the years of the great controversy over sexual liberation (the late 1960s and early 1970s), some radical proposals were made to change this situation. Some authors (O'Neill and O'Neill 1972) advocated what was called *open marriage*. This is a marriage arrangement in which either spouse has the right to seek fulfillment of various personality needs—including sexual needs—with whomever he or she chooses. It is agreed that there will be no jealousy, no claims of exclusive sexual or emotional property. The marriage is supposed to be open, with everything out front and discussed. This approach eliminates the need to sneak out for a secret affair, with all its guilt, anxiety, and worries about being caught, not to mention jealousy, suspicion, and angry confrontations. The relationship is egalitarian because it holds equally well for both husband and wife; neither partner needs to feel wronged, because both have equal rights to have sex with whomever they choose. Two West Coast lawyers (Ziskin and Ziskin 1973) advocated that the couple draw up an explicit extramarital contract, specifying exactly who can and cannot do what. From the experience of several hundred clients, they argued that such arrangements work well as long as certain rules are followed. The main rule is that extramarital relationships should not be a threat to the married partners; hence, they must be predominantly sexual and defined as transient and secondary to the marriage. Controlled in this way, it was claimed, sexual variety tends to revive and improve the love relationship of the couple themselves.

Nevertheless, it must be said that open marriages have not worked out very well. One of their earliest advocates, Nena O'Neill, was writing two years later to explain why she could no longer endorse them—writing, incidentally, no longer with her former collaborator and husband (O'Neill 1974). Open marriages themselves seem to be unusually transient. Either the couple break apart or they decide to end the agreement for extramarital affairs. What is the source of strain? Interestingly enough, it is predominantly the *man* who finds the relationship frustrating. What tends to happen is that once the couple become actively engaged in their "open" sex lives, the woman finds extramarital affairs much easier to come by than the man does. People who are married to each other tend to be approximately equal in attractiveness; there is no advantage to either one on this count. But compared to a man of about the same degree of good looks, a woman who is openly willing to have casual sex finds a lot more receptive partners. This means that the wife is usually much busier with her affairs than her husband is with his. And since there is open discussion, they can't help but be aware of this, not to mention the fact that she is not around so often and available for sex when he wants it. In short, the open marriage seems to turn rather quickly into a sexually competitive marriage, and the woman usually wins the sexual popularity contest hands down. This is structurally caused, and not personal, so the man can't help feeling that there is some unfair advantage stacked against him, without understanding what it is. Despite their agreement that there will be no jealousy, jealousy emerges. The result is that the open marriage gets renegotiated back into being a regular marriage or into a divorce.

The "regular" marriage that they are going back to, however, still has its infidelities. For the most part they are hidden, and they are also relatively frequent. There is some evidence that extramarital sex is on the increase and that the dual standard is disappearing, since most of the increase in extramarital sex is on the part of women. At the same time, there are other forms of openly extramarital sex, including both "swinging" and sexual communes. We will examine these in turn, starting with the traditional clandestine affair.

Adulterous Affairs: Emotional and Erotic

Kinsey (1953, 437) found that about half of all men and a quarter of all women had at least one extramarital affair. About half the women who had any extramarital sex had only one affair, while about 20 percent of them had more than five extramarital partners. (Kinsey did not collect systematic data on the number of the husbands' adulterous partners.) That extramarital affairs happen at all is probably more significant than how much intercourse happens in that fashion; Kinsey found the largest proportion of extramarital sex involves men and women in their forties, when it makes up about 10 percent of their total sex lives. The incidence of extramarital experience generally increased from the beginning to the middle of the century (Kinsey 1953, 422; 1948, 414), although it should be noted that this increase was largely on the part of women rather than men. More recent surveys show that this trend has continued (Hunt 1974; Tavris and Sadd 1977; Blumstein and Schwartz 1983, 273–76). A 1989 survey (Smith 1991) found the level was up to 70 percent of men and 35 percent of women who had at least one adulterous affair during their lives. But only 2 percent of husbands and 1 percent of wives admitted they were currently having an affair.

What kind of people are most likely to be involved in these affairs? For one thing there is a social class difference, at least for males. Men of both higher and lower social classes have adulterous affairs, but for the lower classes extramarital sex begins very soon after marriage, while for the higher classes it happens later. Kinsey (1948, 586–88, 600) found that working-class women were most likely to have extramarital sex when they were in their early twenties; the middle- and upper-class men were beginning it past the age of thirty or later. As we have seen in general, men of the higher social classes tend to be more sexually active than those of the lower classes, but *only after they are married*. They start later, but once they start they have more sex in general, both marital and extramarital. Prostitutes provide about 10 percent of the extramarital sex of the youngest married men, and this rises to over 20 percent for married men past the age of fifty.

Kinsey (1953, 427) pointed out that one of the major factors in a woman's likelihood of having an affair is whether she had premarital sexual experience. Women who did so were more likely to have an extramarital liaison as well, whereas women who were virgins when they married were less likely to commit adultery. Women's extramarital affairs have not been very far-flung. About half the wives who had any extramarital affairs at all had only one extramarital partner, both in the Kinsey data and in more recent surveys (Tavris and Sadd 1977). There is also evidence that career women have higher rates of extramarital sex than traditional housewives (Tavris and Sadd 1977). No doubt this is due partly to greater opportunity as well as to changing attitudes that women should be able to choose to do what men have traditionally done. Since both the degree of premarital experience and the proportion of women who work have gone up, the rate of extramarital sex on the part of women has also increased, and probably will increase in the future (Atwater 1979). Interestingly enough, there has been little apparent change in the rate of extramarital sex by men, which is understandable enough in that no major change in their premarital experience or occupational status has occurred.

A certain amount of what is technically "extramarital" occurs during the period when a couple have separated but not yet legally divorced (Hunt 1974). Since the rate of divorce has gone up considerably in recent decades, the increase in extramarital sex may be partly due to this factor as well. We might imagine, conversely, that extramarital sex is a prime cause of divorce: either because somebody gets caught, or because the affair leads a partner to fall in love with a third party or otherwise desire to break up the marriage. But this does not seem to be the case. Morton Hunt (1969) extensively studied some

FEATURE 9.2
Husbands, Wives, Gays, and Lesbians: The Influence of Gender Roles on Sex

Blumstein and Schwartz's study, *American Couples* (1983), provides an unusual opportunity to examine the effects of gender—the cultural expectations and definitions of what it is to be male or female—upon sexual behavior. Their study consisted of ordinary married couples and cohabiting couples who were not married. In addition to these heterosexual couples, Blumstein and Schwartz also included two types of homosexual couples: gay males who lived together in a permanent arrangement and lesbians in similar arrangements. This sample gives us the opportunity to see what effect cultural gender is having, because it separates out the effects of one's sexual partner from one's own sexual behavior.

In heterosexual relationships between a man and woman, how much of their behavior is determined by the male or by the female? What we observe is the interaction between the two of them, and this can be affected by both. One side may be more influential than the other, but it is difficult to separate out how much. In *American Couples*, though, we can also see what happens when a relationship has two males in it, or two females. By comparing these patterns with the one male/one female pattern, we can see in what direction the cultural role of being male or female is exerting its influence.

For example, consider the following figures on the frequency of sexual activity for couples who have been together for two years or less:

	Married Couples	Gay Males	Lesbians
Sex 3 or more times a week	45%	67%	33%
1 to 3 times a week	38	27	43
Once a week or less	17	6	24

Source: Calculated from Blumstein and Schwartz, 1983, 196.

It is apparent that gay males are much more sexually active than heterosexual couples and that lesbians are the least sexually active. In other words, when there is a male and a female in a relationship, the amount of sex is moderate. When there are two males, the amount of sex goes up. When there are two females, the amount of sex goes down. Similar figures apply to couples who have been together for longer periods. In all cases, the frequency of sex drops after they have been together for two years, and drops still further after ten years, but the pattern between gay males, heterosexuals, and lesbians remains the same.

It looks as if sexual activity is culturally defined as a male province, and this turns out to be true whether the activity is heterosexual or homosexual! Furthermore, males are more attached to their sex lives. Married males, as well as gay men, tend eighty affairs and found that marital breakup was in fact very low. Perhaps the very fact that the affair was kept clandestine meant that the person involved did not want to break up his or her marriage, for economic or emotional reasons. Thus, even if the affair was discovered and an angry dispute ensued with a spouse (this seems to happen about 30 percent of the time; Kinsey 1953, 434), the culprit might well be motivated to try to make amends. If one really valued the extramarital sex more than the marriage, it is likely that one might make

to be dissatisfied with their whole relationship when there isn't much sex in it, while lesbians (like female cohabitors) are no less satisfied with little sex (Blumstein and Schwartz 1983, 201).

We see a similar pattern in regard to whether persons from various groups feel it is important that their partner be sexy looking (Blumstein and Schwartz 1983, 249):

Gay males	59%
Husbands	57%
Wives	31%
Lesbians	35%

Again, the pattern is that males are quite similar, whether they are homosexual or straight; females, too, generally are similar, in a much lower emphasis on the sexiness of their partner.

If we ask what percentage of each group is possessive of their partner, we find another interesting pattern (Blumstein and Schwartz 1983, 254):

Wives	84%
Husbands	79%
Lesbians	74%
Gay males	35%

Sexual possessiveness and jealousy over breach of possession is widespread among wives and almost as widespread among lesbians. Married men are

also quite possessive, though less so than their wives. Gay males are generally rather unpossessive. What this pattern seems to mean is that in our society, possessiveness is an especially strong part of the female cultural role when it comes to sex. In the ordinary marriage consisting of a heterosexual couple, the male tends to be quite possessive as well. The possessive bond goes both ways, although it seems to be particularly oriented around the female. In a couple in which there is no female— that is, gay males—most of the possessiveness drops out, perhaps because there is no female to be possessive or to be possessed. This isn't the whole story, of course, since about a third of the gay males are possessive. There is at least some influence in any kind of sexual relationship that works in the direction of possessiveness.

The same pattern is apparent in the extent to which people think it is important that they themselves be monogamous (Blumstein and Schwartz 1983, 272):

Wives	84%
Husbands	75%
Lesbians	71%
Gay males	36%

In general, then, it appears that sexual activity remains more a male than a female province in

(Continued next page)

FEATURE 9.2 *(Continued)*

our society. Left to themselves, men have more sex, women less. Each group has corresponding levels of concern about sex as a basis for relationships. Females are more concerned about emotional ties, possessiveness, and their own fidelity, although when men are married to women, the possessiveness of that relationship affects them too.

Since Blumstein and Schwartz collected their data in the early 1980s, does this analysis still tell us anything sociologically significant? For one thing, the AIDS epidemic has dramatically grown in the public consciousness. In the intervening period, one of the original authors, Philip Blumstein, has died of AIDS. In response to the AIDS epidemic, gay men have to some extent reduced their number of sexual partners (Kimmel and Levine 1992; Weinberg et al. 1993). But the relative differences remain, including lower levels of sexual activity in general as we go down the table from gay males, through heterosexuals, to lesbian couples; this is what we would expect theoretically, if there are socially constructed roles for sexual behavior (Kleinberg 1992; Kinsman 1992). Gay men still share in the aspect of our culture that makes sexual initiative and adventure more of a male province than a female province. Conversely, the same pattern is found in the opposite direction, in studies of lesbian women, and discussions by lesbian activists about the relatively lower level of lesbian sexual activity (Vance 1989).

Another theme that has developed in recent years is the argument that homosexuality is genetically determined and not social at all. A modest but growing number of studies by geneticists lends support to the argument that at least some proportion of homosexuality is genetically transmitted. But an important point should be borne in mind. Even if there is a genetic mechanism, that does not make social factors irrelevant. For instance, homosexuality in ancient Greece was predominantly between older men and young boys, whereas modern homosexuality is overwhelmingly between adults. In our own society, the number of persons who are bisexual exceeds those who are exclusively homosexual by a ratio of about 8 to 1; and bisexuals switch back and forth at different times in their lives between an emphasis on same-sex and heterosexual partners (Smith 1991; Weinberg et al. 1993). There is much to be understood about how biology interacts with society and culture; but it is clear that how genetic predispositions are expressed, and whether they get expressed at all, is affected by social factors. For this reason, it is important for us to be able to examine the key comparisons, which show us how various social factors operate, while keeping biological factors constant. It is this type of comparison that makes Blumstein and Schwartz's data so significant and more useful than statistics that merely count how many people are doing what at some point in time. His ongoing contribution to this understanding makes this research a fitting memorial to the life of Phil Blumstein.

little effort to hide it and would openly abandon the spouse. In support of this interpretation, Hunt (1969) noted that most affairs were transient episodes, a moment of erotic pleasure rather than any emotional commitment. Spouses seem to know this and may tend to condone affairs for that reason. Kinsey (1953, 434) found almost half of the women who had affairs said that their husbands were aware of it, but major difficulties occurred in only 40 percent of these cases. Blumstein and Schwartz (1983, 268–71, 313) found that over two-thirds of adulterous spouses said their husband or wife was aware of their affair. There was some tendency for the discovery to lead to divorce, but a majority of these marriages nevertheless survived.

Not all extramarital sex involves male-female intercourse. Both Kinsey (1953, 438) and Hunt (1974) found that there was quite a lot of extramarital kissing and petting, sometimes but not necessarily to the point of orgasm. Since many survey respondents may think of "extramarital sex" only as intercourse itself, we have probably underestimated how frequently this goes on. In our society, the definition of sexual "fidelity" is so sharply concentrated on intercourse that many people—perhaps even a majority—indulge in these other forms of extramarital sex without thinking of themselves as truly unfaithful.

It is also a mistake to think of extramarital sex as necessarily heterosexual. A considerable proportion of persons who are homosexual—at least part of the time—are married, and their homosexual affairs are thus necessarily adulterous. There is little information about how spouses react to such affairs, and we have no folklore or traditional terms (equivalent to *cuckold*) to describe them. Does the spouse get more or less outraged to discover he or she is being "cheated upon" for a homosexual partner? (How often does anyone say, "I might have expected I would be jealous of another woman, but of another man?") Perhaps the number of people affected in this way is rather small, or homosexual affairs may be particularly easy to conceal. There is some evidence that women who are struggling with a traditionally sexist marriage, though, may turn to other women for both emotional and sexual intimacy, and sometimes this turns into an exclusive love relationship (Blumstein and Schwartz 1976).

In general, extramarital affairs are more likely to happen in unhappy marriages than in happy ones, but they seem to be relatively frequent in both. The *Redbook* sample of women found that half the unhappily married women had had an affair, but so had a quarter of the happily married ones (Tavris and Sadd 1977, 122). Approximately the same figures were found in another survey (Bell, Turner, and Rosen 1975). There is also some evidence (Glass and Wright 1977) that after couples have been married longer than about ten years, the rate of extramarital sex is the same whether the marriage is happy or not. Marriages that have held together that long tend not to break up. They also involve people in their thirties and forties, whose frequency of marital sex has usually dropped. For these reasons, and because marriages at this stage have usually become rather habituated emotionally, it may make little difference whether extramarital sex happens or not.

In general, the pattern seems to be that adultery is least disruptive to the marriage if it is purely sexual rather than emotional. It is falling in love with the wrong person that is most likely to pull a marriage apart. To be realistic, though, this is not so removed from sexual affairs, since falling in love usually brings sex along with it. But it is a particular form of sex: the claim to have sex as an exclusive relationship, a kind of token of one's emotional property rights over someone else. There is one curious exception to this generalization. Many high-ranking executives develop a kind of pseudomarital relationship with their personal secretary (Cuber and Harroff 1965, 146–51).

Do Swinging Marriages Survive?

Another form of extramarital sex that was popular for a while was *swinging*. This refers to groups that meet for "mate swapping" or group sex. They are, in other words, planned and regularly scheduled orgies. This sounds very free-flowing and uninhibited, but in fact there are certain constraints (Bartell 1971; Gilmartin 1978). First of all admission to the group is generally by couple only. An attractive unattached female would of course be welcome, although she does not happen along very often in these groups, but the ironclad rule is that all males must bring a female with them. Males are generally the ones who organize swingers' groups, and they do it on the basis of strict exchange: every male provides a female to share in the group, preferably an attractive one. It has been suggested that in some business circles a man will use an attractive wife in this manner as a way of gaining "brownie points" with superiors (who may be older and have less attractive wives) (Safilios-Rothschild 1977).

Another important rule is that the group makes a sharp distinction between sex and love. The purpose of the group is sex. Members are not supposed to become involved with one another personally or to see each other individually outside of the group context. This keeps the sexual infidelities "safe" for the people involved, since each spouse knows exactly what is going on and how it is limited. Unlike open marriages, swingers' groups seem to feel that the experience is both good for their sex lives and an enhancement of their marriage; after all, it is something that they enjoy doing together! Swinging does seem to be a very absorbing activity, and swinging couples spend a great deal of time in talking about it and planning their next encounter. Gilmartin (1978) matched his sample of California swingers with other couples in the area according to social class and other variables, and found that there was no difference in the average level of marital happiness of swingers and nonswingers. Swinging does not seem to be an escape from bad marriages, nor does it particularly raise marital happiness above the average. But Gilmartin (1978) did find that swingers were happier with their marriages than the 20 percent of his nonswingers' sample who were having conventional, clandestine extramarital affairs.

It has been estimated that 1 to 2 percent of married couples have tried swinging (Hunt 1974, 271; Gilmartin 1978, 472), although these guesses may be too high. Usually it is arranged among strangers, by using ads in swingers' newspapers or by contacts in bars known to specialize in this. The first time the two (or more) couples meet, they usually do no more than talk, testing their mutual compatibility for sexual adventures. At subsequent meetings, the sexual activities evolve. After about two years (or sometimes less), most swingers' groups break up, and members who stay tend to attend less frequently. Partly they just get tired of it. But interestingly enough, it is usually the men who want to pull out first.

The reason is that women enjoy the intense sexual stimulation more than the men. Although the women tend to be more hesitant about swinging at first, they discover that they have the capacity for half a dozen or more orgasms in one evening, while their husbands are usually not capable of more than a few. Just as in open marriages, the women end up being more sexually active than the men. Since these groups always have at least as many women as men, there is a tendency for most of the wives to become involved in homosexual sex with each other while their husbands are recovering—and also encouraging them, perhaps to keep everyone from being bored. (No male homosexuality, however, is allowed.) The result is that the men start becoming jealous, not of other men, but of their own wives. Finally, the men who talked their wives into swinging in the first place end up pressuring them to quit.

What kind of people take up swinging? The studies show a predominantly white-collar, college-educated group, for the most part in their thirties. It is definitely not the equivalent of the "swinging singles" or other youth cultures. Many of the swingers' groups consist of people who are otherwise quite conventional, politically conservative, who have children and even send them to Sunday school (Bartell 1971). But there are no good systematic samples that deal with this question. Probably some proportions of other social groups have engaged in swinging, but they have been less visible to researchers.

The Fragility of Sexual Communes

Yet another form of experiment with regularized extramarital sex occurred in certain types of communes. These were group living arrangements, which became quite famous, although not necessarily very widespread, a number of years ago. In contrast to the swingers' groups, which tend to be otherwise rather conventional, the communes were notably radical. Whereas the swingers were usually people in their thirties, employed full-time in standard middle-class occupations, communes were organized primarily as part of an explicit political and religious program. Their members tended to be young and militantly unconventional, dropouts or rebels against existing society and its corruptions.

One conventional practice that some communes rebelled against was the "normal" standard of sexual possessiveness. Some regarded this as a manifestation of capitalist society or of pathological individualism or the result of a lack of sufficient love and trust to share oneself physically and emotionally with those around one. The conventional form of marriage was seen as a form of exclusive sexual property rights over another human being, a kind of dehumanization that communes set out to replace with a better and more open way of personal life. The communes organized group households to combat selfishness fostered by holding private property; many also tried to overcome the concept of "private property" in the realm of sex. This was a revolutionary aim. In fact, however, communes usually substituted one form of sexual controls for another.

All kinds of different sexual arrangements were found in communes. Contrary to the popular image, however, more communes were on the highly restrictive side sexually than on the libertarian side. Thus, out of 120 communes (including many of the larger ones in the United States) studied by Benjamin Zablocki (1980, 339): 1 percent had *group marriage*, in which everyone was included; 12 percent had partial group marriage (some individuals not participating); 19 percent had shifting sexual relationships, not organized under any particular rule; 47 percent had monogamous couples, in some cases officially approved and enforced by the commune; and 13 percent had *celibacy*, either as a compulsory policy or as a voluntary religious observance. What is surprising here is that the most common form of sexual organization (observed in almost half the communes) was monogamy. This was particularly the case in communes under religious auspices, some of which insisted quite strongly upon the sanctity of marriage. (These usually involved Christian rather than Eastern religions.) But quite a few of the other types of communes (political, psychotherapeutic, hippie counterculture) also advocated monogamy, apparently as a fairly obvious practical solution to the need for regulating sexual life. In the close quarters of a group, the emotions and strains involved with sex are a major cause of problems.

One way to handle sex, then, is for the communes to try to keep traditional couples together. Another method is to prohibit sex entirely. In fact this was done quite frequently—not so much by prescribed celibacy for everyone (found in only 13 percent in Zablocki's sample), but by the individual practice of celibacy (chosen by about 30 percent of all commune members). (See table 9.5) Most sexual practices tended to become less frequent after people joined the commune. Beforehand, commune members were more likely to have been involved in group marriage, open marriage, homosexuality, public nudity, and sexual orgies than after they joined. This was even true of falling in love.

The communes seemed to know what they were doing, in that sex is rather dangerous to the group structure. The primary aim of regulating sexual life is to avoid disruptive attachments. Thus, among the minority of communes that did not have monogamy or celibacy, there was an effort to put sex on the basis of a regular schedule, so that everyone slept with everyone else and individual possessiveness and jealousy were avoided. In the group marriages studied by Larry and Joan Constantine (1973), about half had a rule of fixed rotation among sexual partners. Sex still was mostly private in all these kinds of communes, and free-for-all orgies were very rare. (We can see from table 9.5, though, that about 14 percent of the commune members did at least occasionally have sex with more than one partner at a time.)

All of the communal ways of dealing with sex turned out to have their own

TABLE 9.5 Sexual Activities Declined after Joining a Commune

Percent who have ever participated in:	Before		After	
	Males	Females	Males	Females
Sex with more than one person at a time	23%	24%	14%	14%
Homosexuality	15	11	8	11
Open marriage	16	16	6	7
Group marriage	3	1	1	1
Public or group nudity	44	35	32	31
Falling in love	85	83	47	50
Celibacy	23	24	31	30

Source: Zablocki, 1980, 115–16.

strains. Communes tended to disintegrate rather frequently, and this happened most often when there were intense emotional relationships among most of their members. Zablocki found that it made no difference whether these relationships were sexual or not; even nonsexual ties, if they were intense and loving, tended to create crisscrosses of strain and jealousy that pulled the groups apart. Besides, sex tended to be a problem even when it was tightly regulated. Even those communes that enforced celibacy as an official rule often had illicit affairs—rather like a conventional marriage, one might say, except that one is committing adultery not against a particular individual but against a whole group! Some groups attempted to provide an outlet by enforcing celibacy only on the premises, while visits to outside lovers were allowed. The result, though, was that individuals began to leave more and more often, until they finally stayed away entirely (Zablocki 1980, 340–41). On the whole, Zablocki found that in at least 60 percent of the communes that tried to restrict sex, by either celibacy or strict monogamy, violations had come to the attention of the group. Moreover, these often involved the leader of the commune, the charismatic individual who laid down the moral rules in the first place.

Given the social structure of the commune, this is not surprising. The commune is an intentionally self-enclosed group, isolated from the outside world. It constantly focuses everyone's attention on one another, constituting what we have referred to (in chapter 6) as *high ritual density*. The leader, if there is one, is especially likely to be at the center of the group's emotions. Sex is an emotional process, hard to separate entirely from other emotions, and hence it is not surprising that a leader—even a celibate or highly moralistic leader—should be a kind of sexual magnet in the group. The upshot, of course, is to put a good deal of strain on the ideological beliefs of the groups, and scandals or disagreements involving the leader are one of the main ways communes have come apart. For ordinary married couples, too, the commune pressure chamber tends to be fatal. There was a high rate of marital breakup in the communes Zablocki studied, and in the great majority of cases the cause was a dispute over an extramarital affair (1980, 120).

CONCLUDING REFLECTIONS ON THE SIGNIFICANCE OF SEX

Aside from being a real-life version of X-rated entertainment, does sexual behavior have any theoretical significance for the family in general? It does, if we recall that the family consists of economic, sexual, and emotional possession. There is much more sex in marriage than outside it. Young married couples have intercourse over four times a week, while for unmarried persons, even the highest regular frequency is about half that—for males—and even lower for females. Prostitution, as commercialized sex, might seem to be readily available outside marriage; but Kinsey (1948, 601–3) found that the average male visited a prostitute no more than five times a year. Unmarried couples who are cohabiting have a higher rate of intercourse, but they are also involved in a kind of exclusive sexual possession (which is to say, cohabitation is really an informal variation of marriage).

Of course, sex isn't love, and love isn't sex, as people say over and over. But in fact, there is a tendency for *permanent and mutually possessive* sex to become the same thing as the bonds of love (and vice versa). In real life, a great deal of what is implied in "love" is exactly this claim for permanent possession of another, sexually and emotionally. Marriage is not merely sex, but it is a claim to *exclusive* sexual possession between two partners. Extramarital sex in its various forms is interesting theoretically, quite apart from its soap-opera aspects, because it tends to prove that marriage is an institution of sexual possessiveness. Affairs are usually clandestine because people think they are wrong; even those who don't think so usually try to hide them anyway because of pressures from their partners, who want to maintain exclusive possession.

The various modern liberalized sexual experiments, in which the partners try to allow open extramarital sex, tend to confirm this. These arrangements allow only *some* sexual nonpossessiveness, as long as it doesn't threaten one's basic rights as the first possessor. These experiments maintain the ritual and symbolic part of possession, which is not the sexual behavior itself but the avowals of attachment. Hence the rules clearly recognized by swingers' groups: no love affairs, only short-term and immediate sex with a clear and equal trade of partners. Open marriages try to maintain this same primacy for the first sexual partner. They break down precisely because they cannot carefully arrange limited and balanced sexual trades; instead they put both husbands and wives back on the sexual market. This is devastating to the marriage because marriage is a way of getting *away from* the sexual marketplace, with its constant negotiations, the need to make efforts to find attractive partners whom one is able to attract, and the need to constantly check how one is doing in the eyes of others. Worse yet, open marriages make husbands and wives compete against *each other* over how extramaritally popular they are. Before long, whichever spouse feels most left out or abandoned wants to end the arrangement. This, of course, is a likely reaction to most cases of conventional, clandestine adultery.

Communes teach a similar lesson. They try to avoid jealousy and possessiveness, surprisingly often by banishing sex (through celibacy) or enforcing traditional monogamy. (Not that they do this very successfully, since the opportunities for sexual relations in a living group of this sort are continually tempting.) Or else communes try to enforce nonpossessiveness by having universal and equal sex (e.g., by rotation), as well as mutual group love. But these sexual arrangements are extremely hard to maintain, and only a tiny percentage of communes ever put them formally into effect. Most others that try it only manage a loose version of sexual sharing, in which some people are always left out and some are more sexually popular than others. In other words, the market for sexual relationships rears its ugly head once again. Along with these market negotiations, of course, comes the dramatic and sometimes pleasant side—love—but also attachment, possessiveness, jealousy, strains, and—the equivalent of extramarital scandals for a commune—blowups. Communes are only the most spectacular example of how sexual relationships turn into *emotional* property.

The conclusion seems to be that sexually possessive marriages are here to stay. This isn't to say that monogamous marriages work so smoothly either (or that monogamy is the only way in which sexual possession can be organized; see chapter 3, especially feature 3.1). Conventional marriages do tend to strain over the issue of possessiveness, and there is a constant undercurrent of extramarital affairs. Nevertheless, by far the greatest amount of married persons' sex (over 99 percent by Kinsey's, as well as more recent, figures) is marital rather than extramarital. At most, extramarital sex accounts for 10 percent of a married person's sexual activity in any given period of the marriage, and that high point occurs during middle age when marital sex has slowed down.

Marriages are overstepped because people chafe over limits on sexual partners. But extramarital sex means going back on the sexual market, which never solves sexual desires permanently. People come back from their experiments, and even from swinging and from adulterous affairs. They either retreat into their conventional marriage—or else break it up, in most cases to form another.

SUMMARY

1. Sexual intercourse at its most intense is a naturally occurring interaction ritual, in which the mutual physical arousal of the partners makes them strongly feel their personal tie. Though intercourse may not often reach high levels of intensity, it tends to become a symbol of the couple's social relationship.

2. The greatest change in sexual practices during the twentieth century is that they have recently become publicly acknowledged. Traditionally, the populace was always more sexually active than official standards admitted.

3. During the 1920s, about two-thirds of men had at least some premarital intercourse; this figure has risen to around 95 percent in recent years. For women, the percentage of nonvirgins at marriage has risen more sharply, from about 45 percent to about 85 percent. The majority of these women were having intercourse with only one partner, however, usually with their future husband.

4. Nonvirginity, however, does not mean that premarital intercourse happens with great frequency; the average rate is about once or twice a week for men, three times per month for women. Earlier in the century, men resorted to prostitutes a good deal more than in recent generations. This shift is probably due to the growing equalization in sexual behavior between the majority of men and women.

5. Earlier in this century, college-educated males had much lower rates of premarital intercourse than men from lower-status groups. In recent years, this class difference in premarital virginity has disappeared, although the lower social classes still begin sex earlier and have more sexual partners. Among women, all social classes used to be almost equally restrictive sexually. In recent years, though, women of the lower social class have become more sexually active than other women. The higher social classes have now split into two groups, one more sexually liberal than the average of the population, another more sexually conservative.

6. Marital intercourse takes place much more frequently than nonmarital intercourse. The average rate in the total married population is about twice a week, although this varies from considerably higher in young married couples to less than once a week for couples beyond age fifty-five.

7. The higher social classes tend to have more elaborate erotic practices than couples of lower social-class levels. This is connected with a sharper distinction between male and female sexual cultures in the lower classes. However, these class differences have been diminishing in some respects in recent years.

8. A large majority of Americans believe that extramarital affairs are morally wrong, although some have practiced open marriages, swinging, or group marriage. Even among those who say they do not approve of extramarital sex, however, clandestine affairs are fairly widespread. About two-thirds of all married men and a third of all married women have had at least one extramarital affair. Working-class men tend to have adulterous sex early in their marriages and less as they get older, whereas middle-and upper-class men tend to participate in more adultery when they are older. Women are most likely to have extramarital sex if they have had premarital intercourse, and career women are more likely to have extramarital sex than traditional

housewives. The rate of extramarital sex has gone up in recent years for women but apparently not for men.

9. Although extramarital affairs lead to angry confrontations for about 30 percent of the people who have them, there is also a fairly widespread tendency for spouses to condone such affairs. This is especially likely if the affair is transient and purely sexually oriented and does not involve the possessive bonds of love.

10. *Swinging* is a term for couples who meet periodically for mate-swapping orgies. In order to control jealousy, they always have strong rules that relationships are purely sexual, do not involve love, and are confined to the group encounters. Though males usually initiate these groups, the females (because of their greater capacity for multiple orgasm) tend to become the ones who enjoy them the most, and males are the first to demand that the couple withdraw from swinging. Similarly, open marriages (in which extramarital sex is openly allowed to either partner) tend to break down because males find they are not as sexually popular as their wives.

11. Although communes were regarded as places where group sex was practiced, in fact the majority of communes attempted to control the disruptive effects of sex by instituting celibacy or strict monogamy. A minority of communes had formal or informal rotation of multiple sex partners. In *all* these types of communes, however, sexual jealousies frequently created strains, and the intense interpersonal atmosphere has been a cause of the breakup both of individual marriages and of entire communes. In short, there seems to be no perfect solution to problems of sexual possessiveness, either in conventional marriages or in utopian experiments.

CONTRACEPTION, CHILDBEARING, AND THE BIRTH EXPERIENCE

FERTILITY CONTROL IN CROSS-CULTURAL AND HISTORICAL CONTEXTS

Before exploring the ways that our own society organizes contraception and birth, it is useful to look at how other societies have regulated birth and population growth. All human populations must strike a delicate balance with nature if they are to survive (see chapter 3). Overpopulation has often been avoided because many people die from disease or warfare. Societies have also regulated their birthrates in order to ensure a sufficient food supply for their members. Intentional population control has typically been accomplished through infanticide (the killing of infants) and contraception (techniques to prevent pregnancy). The most common forms of contraceptive practices have been celibacy, delayed marriage, abortion, and minimizing sexual intercourse through taboos or prolonged breastfeeding. Late weaning of young children and low percentage of body fat also seem to have been effective means of birth control among hunting and gathering societies (Blumberg 1984). Even without scientific knowledge of human reproduction, many societies discovered and used various contraceptive barrier devices (e.g., roots, pods, animal intestines, grasses) and contraceptive or abortive oral concoctions (mostly herbs, roots, or metals). It is unlikely that these contraceptive devices, or the ritual practices that accompanied their use, were very effective, but they probably did reduce the chance of pregnancy to some degree (Green 1971).

High infant-mortality rates from disease or malnutrition have naturally limited population growth, but the most common form of *intentional* population control has been infanticide. Although it is comfortable to attribute the killing of newborn babies to some distant and primordial tribes, it may have been most common among those we like to think of as most civilized. For example, Plato and Aristotle advocated infanticide as a means of population and disease control. William Langer (1972) reports that infanticide was especially common in Europe and Britain during the eighteenth and nineteenth centuries, and that women were almost never punished for taking the life of their infants:

> In the eighteenth century it was not an uncommon spectacle to see the corpses of infants lying in the streets or in the dunghills of London and other large cities. . . . In 1862 one of the coroners for Middlesex county stated infanticide had become so commonplace "that the police seemed to think no more of finding a dead child than they did of finding a dead cat or a dead dog.". . . The *Morning Post* (September, 1863) termed it "this commonest of crimes." (Langer 1972, 96–97, cited in Skolnick 1987)

Abandoning children, or putting them in foundling homes, was also quite common in eighteenth century Europe and amounted to almost the same thing as infanticide. While foundling homes were started by reformers, such as St.

Vincent de Paul and Napoleon, who were shocked at infanticide, the conditions were so bad in these homes that between 80 and 90 percent of infants placed there died (Skolnick 1987, 317). In some traditional peasant societies, such as China or India, the killing of girl babies has also been common for many centuries.

These historical practices suggest that there is no universal "maternal instinct" that causes mothers to treat babies a particular way, and that belief in the sacredness of infant life is the product of specific cultural and historical influences. Cross-cultural studies show us that when women are given free choice—that is, when children are not needed for old-age security, and religious, cultural, and economic pressures are minimal—they choose to have only a few children (Steinmetz, Clavan, and Stein 1990, 221).

The Birthrate Today

Women in a given society tend to give birth at similar ages and to have a typical number of children. Theoretically, it is possible for women to have thirty or more births, but even ten is rare. Partly this is a matter of health, since many women used to die in childbirth before the development of modern standards of medical practice. But especially it implies that social factors are more important than biological ones in determining how many children are born.

Our own birthrate has been declining and has reached the point of *zero population growth*. There is an economic reason for this decline. In our society, children are increasingly expensive to have and raise, and they bring in no income (see chapter 12).

In many traditional societies this was not the case (Blumberg 1984). In the agrarian, or medieval, system of peasants and rural landlords, for example, child labor was useful for the parents. Children could perform many tasks, such as caring for livestock or helping in the largely self-subsistent household economy. Moreover, in these societies the landlords were usually pressuring the peasants hard and taking most of their surplus crops, and hence there was even greater pressure for peasants to have as many children as possible to squeeze out even more agricultural productivity. On top of this, as Rae Blumberg points out, women in these societies tended to have little power, so that economic pressures plus male dominance turned them into breeding machines—despite the difficulty of childbirth under poverty conditions and the high maternal death rate.

Conversely, in many traditional societies where women had greater power, they took measures to keep the birthrate more comfortably under their own control. In tribal hunting-and-gathering societies (see more on these in chapter 3), women usually spaced their children quite far apart, typically more than four years; so they had fewer small children to deal with. Today, in the agricultural countries of the Third World, women also tend to reduce their own birthrates when they acquire more economic resources of their own. This is sometimes done in opposition to their husbands, who cling to the more traditional, machismo ideal in which a man's status is measured by the number of his children (which ensures that his wife is too busy to challenge his authority). But even in traditional parts of Latin America, women have increasingly resorted to illegal abortions in order to gain some control of their own lives (Kinzer 1973). When women in Puerto Rico entered the labor market on their own, for instance, the first thing they did was to initiate contraception (Weller 1968). And because their husbands often depend upon their income, they tend to get their way regarding family size.

These same causal conditions seem to operate in our own society. We are at an advanced stage of this process, because the movement of women into full-time work has proceeded quite far here. Thus, all the general factors point toward a decline in our birthrate. First, children are not an economic asset but a sheer expense; hence the economic motivation for having children has disappeared. Second, women have entered the labor force in large numbers

FIGURE 10.1: Changing Public Opinion Now Favors a Smaller Family

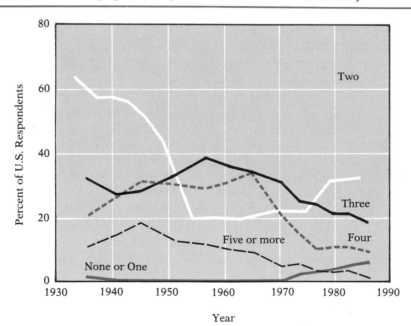

"What do you think is the *ideal number of children* for a family to have?"

Source: Chadwick and Heaton, 1992, table D1-7 (adjusted for mission data and no opinion).

and are now committed to independent careers. This both reduces their motivation for having children (especially early in their lives) and increases their power to make their own decisions about childbearing. The result is a long-term slowing of the population growth.

We are currently very close to zero population growth, with a birthrate of about two births per woman. Our population is not quite stable yet for two reasons. One is that immigrants continue to come in, primarily from Latin America and from Asia. The other reason is that the actual numbers of children born depend on how many women are of childbearing age. Since there was a "baby boom" in the 1950s (see chapter 15), there are more women now who are young adults and are now bearing their own children. This is sometimes called the demographic "echo boom." But in the 1990s and 2000s, the number of mothers will be rather low, reflecting the low birthrate of the 1960s and 1970s, and the number of children being born will decline (Westoff 1986).

There are a number of factors that enter into this low birthrate. On the whole, people are older when they marry. Although many births occur out of wedlock, it is marital births that account for the large majority of births. More and more women are waiting longer to have their children, many delaying into their thirties. The result is that as women reach childbearing age, they do not add their children to the population so quickly, which also reduces the rate of population growth.

But most people still want to have children. They just do not, for the most part, want to have large families any more. As figure 10.1 shows, a majority of Americans (56 percent) regard two children as the ideal family size, a proportion that has been growing ever since the early 1960s. In the 1950s, as many as 45 percent of Americans thought that four children or more were the ideal number for a family. Today, it is the trend-setting sections of the population who are most strongly in favor of having only two children: the college educated more than the less educated; whites more than blacks. And women of childbearing age—the persons who are most directly affected by having to care for children—are especially likely to favor two children (61 percent), although a majority of men (53 percent) also favor that number. All this suggests that the two-child family will become even more strongly entrenched as time goes on.

NONMARITAL CHILDBIRTH

"There was something mysterious about that house," related a woman interviewed in a midwestern town, speaking about a time forty years ago. "There was somebody in there who was never seen, never spoken about. Everybody knew she was there, but nobody would ever let on that she existed. It must have been terrible for her. She used to come out at night sometimes, just to drive around

Most indicators suggest that the two-child family will become increasingly entrenched as the standard of family size.

in a car, but never stopping anywhere, just to get out." She was a woman who had had an illegitimate child. Abortions were not easily available; she had given birth, put the child up for adoption, and gone back to live with her parents. It was all done in great secrecy, but everyone in the town knew about it, speaking in hushed tones with accents of gravest scandal. Her parents took her in, and she was never heard from again.

This kind of story, with all its variants, must have happened millions of times throughout the last centuries. A woman's life was circumscribed between social pressures: men pressing her toward sexual adventure (perhaps along with her own rebelliousness, independence, or sexual desires), while social condemnation awaited her if she did not avoid any appearance of sex outside of marriage. In the 1500s and 1600s in England, when most of the wealthier households had female servants, it often happened that a young woman was made pregnant by her employer, then dismissed from service with no chance of getting another job; the result was that she would often drift into poverty or prostitution (Stone 1977, 382, 392). Though men were often the exploiters in these everyday dramas, the worst condemnation frequently came from other women, who scorned unmercifully any woman who fell from the narrow path of socially defined virtue.

Times have changed, but only recently, and we have hardly yet emerged into a modern-day utopia, if one exists at all. Nonmarital childbirth and various ways of dealing with it are no longer hidden in the shadows or subject to traditional persecutions. Nevertheless, there are major problems and controversies, and various movements are pressing either for new solutions or for ways to turn back the clock. We now examine some of these trends.

Pregnant Brides

Pregnancy can be, among other things, a route to marriage. In colonial America as well as in England during the late 1700s, for example, almost half of all brides were pregnant at the time of the wedding ceremony (Stone 1977, 386–88). It has been suggested that a "shotgun marriage" was one way that a woman could maneuver a less-than-willing boyfriend into getting married—a rather dangerous game, in view of the dire social sanctions if she failed, but for that very reason pregnancy was likely to put tremendous moral pressure on the man "to make an honest woman out of her." There may actually have been a fair amount of complicity on the part of the man, however. Such couples, especially in the less "respectable" lower classes, seem to have often deliberately begun intercourse about the time it was economically possible for them to marry and only delayed the actual marriage until a pregnancy occurred (Laslett 1971, 135–40).

This kind of "normal" premarital pregnancy appears to have become less common in the 1800s, and by the early 1900s middle-class standards had become much tighter throughout the whole society (Smith and Hindus 1975). But it was never completely eliminated, and in the last few decades it has made a decided comeback. Today again, one-quarter of all new brides either are pregnant or have already given birth. This is particularly true of teenage brides: among those fourteen to seventeen years old, about one third have their first child within eight months of the wedding. Among black women below the age of twenty-four, more than half are already pregnant or have given birth before they are married. Or, to put it another way, over one-third of all American women, black and white, are pregnant by age nineteen, 80 percent of these premaritally. Not all female teenagers are sexually active, of course; but among those who are, one out of three becomes pregnant. The chances that teenagers will become pregnant are five times greater in the United States than in most other modern industrialized countries (see figure 10.2).

So the old pattern of pregnant brides is very much with us again. But today there is a difference. The results of premarital pregnancy are not as dire as they once were, and it is possible that when marriage follows, it is less of a forced choice and more of a preferred decision. Couples living together, for instance, are most likely to become officially married when they decide to have children or find one on the way. There are also more alternatives, although these tend to differ sharply by social class. Women who are college graduates are much less likely to marry when they become pregnant than are high school dropouts.

FIGURE 10.2: U.S. Teens are Most Likely to Have Babies

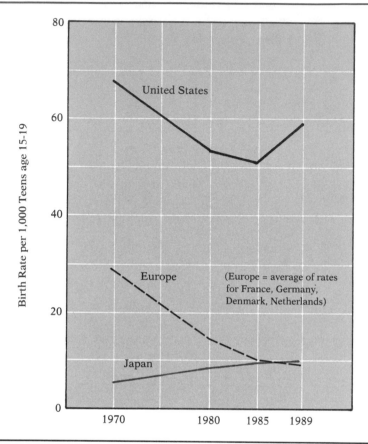

Source: Eurostat, Demographic Statistics, 1991; Japan Ministry of Health and Welfare; U.S. National Center for Health Statistics, Final Natality Statistics; Ahlburg and DeVita, 1992; Population Bulletin, Vol. 47, No. 2, Table 4.

Instead, these highly educated women are more likely to have an abortion rather than marrying or bearing a child out of wedlock. Of course, being pregnant is one reason why the women at the other end of the spectrum dropped out of school; but the pattern also shows that the women of the higher social classes are much less likely now to allow a pregnancy to disrupt their lives.

The Nonmarital Childbirth Explosion

The proportion of births to unmarried women of all ages has gone up continuously during the last half of the twentieth century, so that now there are over 1 million nonmarital births each year in the United States (see figure 10.3). In 1960, nonmarital births accounted for 5 percent of all births, but by 1989, the share had increased to 27 percent, more than one out of every four children born (Ahlburg and DeVita 1992). It used to be that most nonmarital births happened to teenagers, but the recent dramatic increase in the birth rates outside marriage is driven primarily by women age twenty and older, who now account for about two-thirds of all nonmarital births.

The trend has been especially sharp among minority groups. Two of every three black infants are now born to unmarried mothers, as are one of every three Hispanic infants, compared to one of every five white infants. Although most nonmarital births to whites, blacks, and Hispanics occur to persons who are below the poverty line, an increasing number are to older women who are better off. Various explanations have been offered for these trends, including some claims that the upsurge of nonmarital childbirth shows a breakdown in morals or in the structure of the family. The actual causes are more likely

FIGURE 10.3: The Nonmarital Childbirth Explosion

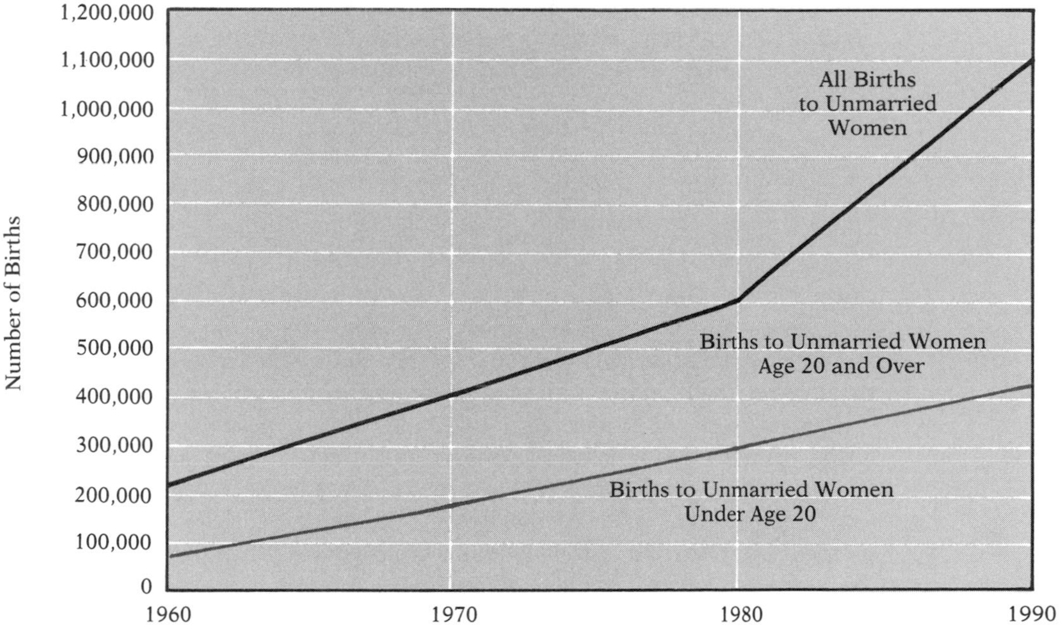

The increase in nonmarital childbirth is primarily the result of older unwed mothers having babies.

Source: Ahlburg and DeVita, 1992, figure 8, p. 22.

real-life, practical matters. One possible reason is the reforms in welfare laws, which make it easier for poor women to support their children alone than with a nonearning husband. But as we have already seen in chapter 7, the welfare law changes in the late 1960s came too late to account for the upsurge of nonmarital childbirth that had already started earlier—in the case of the black population, as far back as 1940. Another factor may be that some of these women, especially the blacks, are evolving a family system that gives greater power to women by excluding exploitative men. This too has been discussed in chapter 7, where it was also pointed out, though, that the black population is splitting sharply in this regard: the lower classes are going in one direction, while the higher classes foster a type of family that is *more* conservative than the average white family. But more black and white women in recent years have been choosing to remain single, and hence the total population in which nonmarital childbirth can occur has gone up.

Some observers are alarmed by the high percentage of births to unwed mothers, but trends in the United States are roughly similar to those of many modern industrial societies. The share of births occurring outside of marriage in our country is similar to the share found in Canada, France, and the United Kingdom, but only half the rate found in Sweden. In contrast, rates in Japan are many times lower than in the United States (Ahlburg and DeVita 1992). A combination of factors account for the differences in rates between countries, but the most obvious cause of teenage pregnancy has to do with low levels of contraceptive use.

CONTRACEPTIVE USE

Contraceptive methods of various types have been known for a long time, although they have not always been very reliable. In 1844 the vulcanization of rubber was invented, and by the late 1800s rubber condoms and diaphragms were in use (Himes 1963; Reed 1978). In the rather puritanical public atmosphere of the late nineteenth and early twentieth centuries, however, contraceptives were available only covertly. Margaret Sanger coined the term *birth control* and launched a campaign to bring information about it to the public, after visiting the deathbed of a young woman who died in 1909 while undergoing an illegal abortion. But contraceptives remained illicit in many places, with some states passing laws influenced by Catholic doctrine that held (since 1869) that any form of contraception was murder. Federal customs agents used to confiscate women's diaphragms at the docks when they were returning from a trip abroad. Only in 1965 did the U.S. Supreme Court finally strike down laws prohibiting birth control, on the grounds that these laws unconstitutionally infringed upon the individual's right to privacy.

By the 1960s, the birth-control pill and the intrauterine device (*IUD*) had

Margaret Sanger, who coined the term *birth control* and promoted its practice, is confronted by a mother of seven in 1917.

appeared. These constituted easier forms of birth control, since they could be controlled by the woman (unlike the condom) and did not require preparing for every act of intercourse (unlike the diaphragm). As these became available and widely publicized, it was expected that unwanted pregnancies would virtually disappear. As it turned out, however, this was not the case, as we know from the increase in both nonmarital births and abortion (see table 9.1 for a summary of the effectiveness of various contraceptive methods).

In the last few decades there has been a revolution in the acceptance of birth control in most countries, including the United States. This is true even though the official doctrine of the Roman Catholic Church—America's largest religious denomination—condemns all forms of birth control as murder, except for abstinence or the rhythm method. Today, contraceptive use by Catholics is virtually identical to that of Protestants (Westoff 1986).

Since levels of sexual activity in the United States are comparable to levels in most Western countries, the high rates of teenage pregnancy in our country can be attributed to low levels of contraceptive use. Research has consistently shown that U.S. teens are slow to adopt contraceptive techniques. Few teenagers report wanting to have a baby when they begin having sex, but 10 percent of whites and 20 percent of blacks report that they didn't think about pregnancy or didn't care about whether a pregnancy occurred when they first had sexual intercourse. (Miller and Moore 1991). About two-thirds of unmarried female teens who are having sex report using some form of birth control (primarily the pill). Nevertheless, only about half of females ages fifteen to nineteen report having used some method of contraception at first intercourse. Recent data also suggest that condom use has increased substantially in the last decade primarily as a result of awareness (Miller and Moore 1991). Over half of male teens surveyed in the late 1980s were using condoms (Sonenstein et al. 1989). Nevertheless, about four of ten male teens said they did not use a reliable form of birth control at first intercourse and two of ten said they did not use birth control the last time they had sex.

The foregoing refers to teenagers. However, it is women in their twenties who are most sexually active. Nearly half of women aged twenty to twenty-nine have never married, and a large majority of them have had intercourse at least once during the previous month. Nearly all these women have used a contraceptive at some time; they are not ignorant of contraceptive methods. Still, over one in ten had not used a contraceptive at the time of their most recent intercourse. This is a good deal higher than among teenagers, but it still represents much unprotected intercourse. Furthermore, many women do not start using birth control until some time after they first have intercourse.

The question then becomes, Why don't unmarried women use contraceptives more regularly? One response is that they do use them, especially if they have intercourse with more than one partner, have intercourse relatively frequently and over a long period of time, and discuss the contraception explicitly with their boyfriends in advance. But notice what this means: the women who are the most sexually active and sophisticated are the ones most likely to use contraceptives; the women who don't use them are the ones who have only occasional intercourse, with perhaps only one partner—and these are the women most likely to become pregnant. In short, the "good girls," who stay closest to the traditional pattern, are least apt to control unwanted pregnancies. Kristin Luker, in a book appropriately titled *Taking Chances* (1975), discovered why. To deliberately use contraceptives, and especially to talk about them in advance, is a way that a woman announces to herself (and perhaps to others) that she is sexually sophisticated. The young woman who regards herself as "virtuous" in the traditional sense is the one least likely to use contraceptives—not because she is ignorant of them but because doing so would contradict her moral self-image. For example, religious young women are more likely to postpone sex, but also less likely to use contraceptives when they do have sex. As teens become more sexually experienced, they become more consistent contraceptive users (Hofferth and Hayes 1992). Availability of family planning services and parental communication about sex also seem to increase contraceptive use (Miller and Moore 1991).

Even though the use of contraceptives would significantly reduce the level of unintended pregnancy, sex education programs have not been very effective in promoting their use. The mass media rarely mention contraceptive use, family planning services are not readily available, many boys continue to view pregnancy as the young woman's problem, and teenage girls are reluctant to admit that they are sexually active.

Use of birth control increases dramatically after about age twenty, and remains relatively constant from age twenty-five to the end of the reproductive years (Westoff 1986). The pill is the most common contraceptive method among women ages fifteen to forty-four, used by almost a third of all women. The second most common form of contraception reported by women of childbearing age is female surgical sterilization (28 percent), followed by male sterilization (12 percent), condoms (15 percent), diaphragm (6 percent), and IUD (2 percent) (Ahlburg and DeVita 1992).

ABORTION

The incidence of *abortion* has risen sharply since abortion became legal in various states around 1970 (see feature 10.1). Prior to that time, illegal abortions were occurring at a rate estimated at between 200,000 and 1.2 million per year. Currently, illegal abortions have decreased to almost zero, while the number of legal abortions has risen to 1.5 million per year (see figure 10.4). This is an extremely large number when one considers that the total

FIGURE 10.4: Legal and Illegal Abortions in the United States

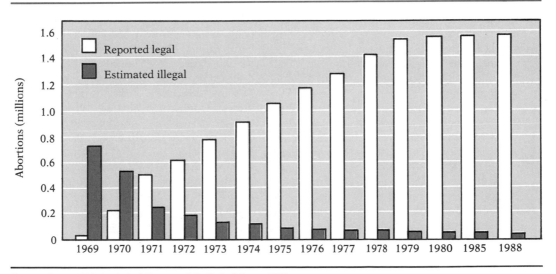

Source: Cates, 1982; *Statistical Abstract of the United States,* 1993.

FEATURE 10.1
Two Centuries of Abortion Battles

Before the early 1800s, the law in England and the United States allowed abortion in the first half of pregnancy. The crucial time was considered to be the "quickening": the point at which the fetus first began to move in the womb. This usually occurred late in the fourth month or in the fifth month, after which the fetus was considered to have begun a separate existence. After the 1840s, public controversy over abortion began to emerge in America. This was the height of the Victorian era, and abortion as evidence of sexual behavior was morally condemned. But apparently large numbers of women, including married women, were resorting to abortion, mostly by medically unlicensed practitioners. Medical techniques in general were not very advanced, even in the regular medical profession, and the operation involved considerable pain and danger. By the 1860s, the American medical profession was attempting to reform itself, to get beyond its own harmful practices such as bleeding patients with leeches; antiseptics, for example, were not invented until 1865. But as the best-educated doctors strove to upgrade their own practice, they also began to organize politically to drive out unlicensed practitioners. Among those whom they attacked were the abortionists, although the arguments that they chose were not so much ones of medical safety as moral arguments that abortion was murder.

The historian James Mohr (1978) has called this "the physician's crusade against abortion," and he argued that the political rhetoric that was most effective in getting antiabortion laws passed was an appeal to ethnic division. It was claimed that nativeborn white Protestant women were the ones who were having abortions, while the new European immigrants were breeding at a dangerous rate. Hence outlawing abortion was claimed to be a way of keeping "pure" native American stock in the majority. The same sentiments were to lead, early in the twentieth century, to laws largely shutting off immigration. By 1900, virtually every state had passed a law making abortion of any kind a crime.

Throughout the first two-thirds of the twentieth century, abortion was illegal everywhere in the United States. But it did not go away; it only went underground. Kinsey (1953) found that 90 percent of the premarital pregnancies among the women in his sample had been terminated by abortion, and that 22 percent of the women had an abortion during the time they were married. These illegal abortions were expensive, difficult to get, and sometimes dangerous, but they happened. In the 1960s, public opinion began to switch. The catalyst was the case of Mrs. Sherri Finkbine, a mother of four who had taken the drug thalidomide during the early months of her pregnancy. Thalidomide was a tranquilizer but its effect upon pregnancy was discovered to be that the child was born deformed, with "flippers" instead of arms and legs. Mrs. Finkbine applied for an abortion to her local medical board in Phoenix, Arizona, which denied her request after considerable national publicity and pressure from religious and political organizations. Mrs. Finkbine finally was able to have an abortion in a Swedish hospital, where she delivered a deformed fetus. As Mrs. and Mr. Finkbine desperately applied to one American medical board after another, the publicity dramatized the abortion issue, and 87 percent of American doctors favored a liberalization of abortion policy by 1967 (Mohr 1978, 256). In the same year, Colorado passed a law legalizing abortion, and seventeen other states had followed suit by 1972. In 1973, the U.S. Supreme Court, reviewing the case of *Roe v. Wade*, struck down all remaining antiabortion laws, dating from the late 1800s.

In the *Roe* decision, the Court held that three

rights were involved: the constitutional right to privacy, the right to protect maternal health, and the right of the state to protect developing life. In reconciling these rights, the Court declared that in the first trimester (three months), pregnancy was a private matter involving a woman's right to decide her own future, and struck down all laws interfering with that right. During the second trimester, as abortion became medically more dangerous to the mother, the state acquired an overriding right to regulate abortion by ensuring proper standards of medical procedure. During the last trimester of pregnancy, the Court held that the fetus had become "viable," in that it could likely survive outside the womb with artificial aids. During this time, the right of the state to protect potential life became paramount, and abortion could be legally prohibited except under extreme conditions to preserve the mother's health or life.

Since that time, there has been considerable legal and political controversy over abortion. An antiabortion, Right-to-Life movement sprang up in the 1970s in response to the *Roe* decision, In 1981, a bill to prohibit abortions was introduced in the U.S. Senate but was rejected by a majority vote. The populace remains split on the issue. The *Roe* decision was hailed by the women's movement, while opposition was led by the Catholic Church and fundamentalist Protestant denominations. There is evidence, though, that abortion is widely practiced in Catholic countries as well as elsewhere (David et al. 1978) and that a large proportion of American Catholic women favor leaving abortion a matter of individual conscience rather than public law. Opinion polls have fluctuated somewhat since abortion became a matter of controversy in the early 1970s. From the mid-1970s until the present, most people agree on allowing abortion in at least certain cases. About 90 percent

of Americans believe that abortions should be allowed if a woman's health is endangered, 80 percent if she became pregnant as the result of rape or if there is a strong chance of the baby being born with serious birth defects. There is much more controversy over the availability of abortion for other reasons: as of 1992, a majority of Americans favored the *Roe v. Wade* decision, but a majority also favored putting some limits on the availability of abortions, especially for those under the age of eighteen. Catholics are about 6 percent less approving than Protestants, while Jews and persons with no religious affiliation are especially likely to approve abortion for any reason. Young people below the age of thirty were most strongly in favor of allowing abortion, while persons over age sixty-five were most conservative.

In 1979, the Supreme Court ruled that state laws may not prohibit women below age eighteen from obtaining an abortion even if they do not have their parents' consent. Since then, court decisions and state laws have challenged that right. In 1992, thirty-five states had laws requiring minors to inform their parents or obtain their consent to get an abortion. Access to abortion, even for older married women, varies widely depending on where one lives. In 1989, 51 percent of metropolitan counties and 93 percent of nonmetropolitan counties in the United States had no abortion provider. In June of 1992, a Supreme Court decision gave states even more latitude to legislate restrictions on abortion, stipulating that such restrictions not place "undue burden" on women seeking abortion. This decision is likely to create even greater differences between states in laws governing access to abortion and in the availability of abortion services (Ahlburg and DeVita 1992). In 1993, the murder of a Florida abortion doctor raised the acrimony in the abortion debates to a new high.

number of live births is about 4 million annually. Since 1980, the abortion rate leveled off, and even declined a little after 1983, reflecting a decrease in the pregnancy rate (Westoff 1986). But not all pregnancies end in either birth or abortion; miscarriages and stillbirths (when the child is born dead) are almost as numerous as abortions. Thus, of 100 pregnancies there are 56 live births, 23 abortions, and 21 miscarriages and stillbirths.

Back in the 1950s, the evidence seemed to indicate that most abortions were performed on married women who did not want more children (Calderone 1958, 60). By the mid-1970s, however, abortions were overwhelmingly being sought by unmarried women. It has been estimated that 9 percent of all marital pregnancies were terminated by abortion, as compared to 65 percent of premarital pregnancies (Forrest et al. 1978, 275). The latter figure is quite astounding when one considers the huge numbers of nonmarital births that do occur. Abortions are especially concentrated among teenagers and women in their early twenties. Teenagers have about one-quarter of all abortions, and another third are performed on women from twenty to twenty-four years of age; most of the rest involve women in their late twenties (Ahlburg and DeVita 1992); *Statistical Abstract of the United States* 1993). Although many people assume that abortions are associated with not using contraceptives, nearly half of all abortions are for women who are using some form of contraception during the month in which they conceive (Henshaw and Silverman 1988).

The abortion rate in the United States, although quite high compared to the number of pregnancies, is only average compared to those elsewhere. Some countries, such as the Netherlands, Britain, and France, have abortion rates that are half or even less than half of the American rate. At the other extreme, countries such as Hungary and Cuba have abortion rates that are two or even three times higher than ours. The United States is in the middle, along with Norway and Sweden, Germany, Denmark, and some Far Eastern countries (Forrest et al. 1978, 273; Westoff 1986).

The Abortion Pill—RU-486

French scientists developed a pill called RU-486 that can be taken within a few days of a missed menstrual period and causes the uterus to expel the fertilized egg. Testing in Europe has found that RU-486 is 96 percent effective in terminating a pregnancy and causes fewer side effects and complications than the traditional surgical abortion. In 1988, RU-486 was legalized for use as an abortifacient in France. The so-called abortion pill, commonly taken in conjunction with a second hormone, is now also used legally in four other countries.

If the abortion pill became widely available in the United States, the postconception control of pregnancy could be taken out of the realm of medical intervention and public surveillance. Precisely because RU-486 is medically less risky than surgical abortion, as well as quicker and more private, anti-abortion activists have blocked its testing and introduction into the United States. Groups opposing abortion have pressured the U.S. Food and Drug Administration (FDA) and politicians to refrain from beginning testing and licensing procedures so that the abortion pill could be used here. The French pharmaceutical manufacturer has been threatened with consumer boycotts of its other products and has been reluctant to push for introduction into the American market.

Britain, one of the countries that allows use of RU-486, agreed in 1994 to allow Americans and other foreigners to obtain the drug there for a $500 fee. Also in 1994, the French pharmaceutical company, with the support of the Clinton Administration, entered into negotiations with nonprofit groups to provide the drug for formal testing by the FDA and eventual distribution in this country (*Los Angeles Times*, March 8, 1994). This is likely to be a hotly contested debate for many years, but if the abortion pill were introduced, it would change the politics of abortion battles. Instead of targeting medical doctors and abortion clinics, anti-abortion activists would have to devise alternate strategies to promote their cause.

EARLY VERSUS LATE MOTHERHOOD

The most significant fertility trend in the last twenty years has been the timing of parenthood, with two general patterns emerging—one early and the other late. An increasing number of first births reflect an early pattern, involving either a premarital birth or a pregnancy legitimized by marriage. In the 1970s and early 1980s, the teenage birthrate declined and the number of births to women under twenty fell steadily. In the late 1980s, however, the trend reversed itself and there was a sharp upswing in teenage pregnancy (Ahlburg and DeVita 1992). Birthrates among younger teens (ages fifteen to seventeen) rose the most, but rates for older teens (eighteen to nineteen) also increased. While teen birthrates are still below what they were in 1970, the recent shift reverses the earlier trend away from early parenting. Minority teens are most at risk for teen birth, but whites still account for a full two-thirds of all births to teenage mothers.

Early childbearing is associated with lower educational attainment, and since education is closely linked to labor force opportunities, lower attainment often translates into lower earnings later in life. Early childbearers also tend to have larger families (Bumpass, Rindfuss, and Janosik 1978) and a higher incidence of poverty than women who bear children later in life (Hofferth and Moore 1979). The trend toward childbearing during the teen years may have peaked again, but the number of out-of-wedlock births continues to rise. The major change seems to be that women who conceive premaritally are less likely to get married in order to "legitimate" the birth. About half of women under twenty-five who conceived a child and were unmarried in the early 1970s got married before the child was born. By the late 1980s, in contrast, only one in four got married before they had the baby. The big jump in out-of-wedlock births is even more dramatic because abortion services were legalized and made accessible to a large number of women during this time period. As noted previously, the highest rates of nonmarital childbirth are for African-American women, but the rate of increase in nonmarital births has actually been greater for whites than for any other group.

Even more dramatic than the increase in early birth is the trend toward delaying childbearing. Six out of seven women in the United States eventually become mothers, but many are putting off childbearing longer than was once the case. Many women with professional and business careers wish to establish their careers and postpone having children until well into their thirties or even later. Between 1972 and 1982, the percentages of women who waited until they were past thirty to have their first child more than doubled. The birthrate is highest among women who are in their late twenties. Ten years or more earlier it was women in their early twenties who were having the most children (U.S. Bureau of the Census 1989). Four of ten American women under the age of thirty had not yet had children in 1990, although most expected to have children eventually.

Figure 10.5 shows that the percentage of childless women is increasing in all age groups. Three out of four women aged eighteen to twenty-four were childless in 1990, but the biggest jump in childlessness has been for women in their late twenties or older. While somewhat more women are remaining permanently childless, most women are just delaying having children longer than earlier cohorts. As a consequence, the birthrate for women over thirty increased dramatically in the last decade. The most highly educated women are leading this trend, but women of lower educational levels (and lower occupational attainment) are gradually coming to imitate their pattern. Delayed birth is linked to higher levels of employment for women, and tends to be more common in economic hard times.

According to recent studies, the trend toward delaying childbearing is not fully explained by the tendency of people who wait longer to get married. Not only are people postponing marriage longer than was common during the 1950s and 1960s, but they are also postponing having children once they get married (Teachman, Polonko, and Scanzoni 1987, 15). The recent trend toward postponing having children has precedents from the mid-nineteenth century (Morgan 1991), but another demographic factor makes the recent trend even more consequential. Because of the postwar baby boom, there are significantly more people in their thirties than there were ten or twenty years ago. Since there are more people in the later childbearing ages, and because couples are now likely to delay childbearing, demographers predict that the

FIGURE 10.5: More Women Are Remaining Childless

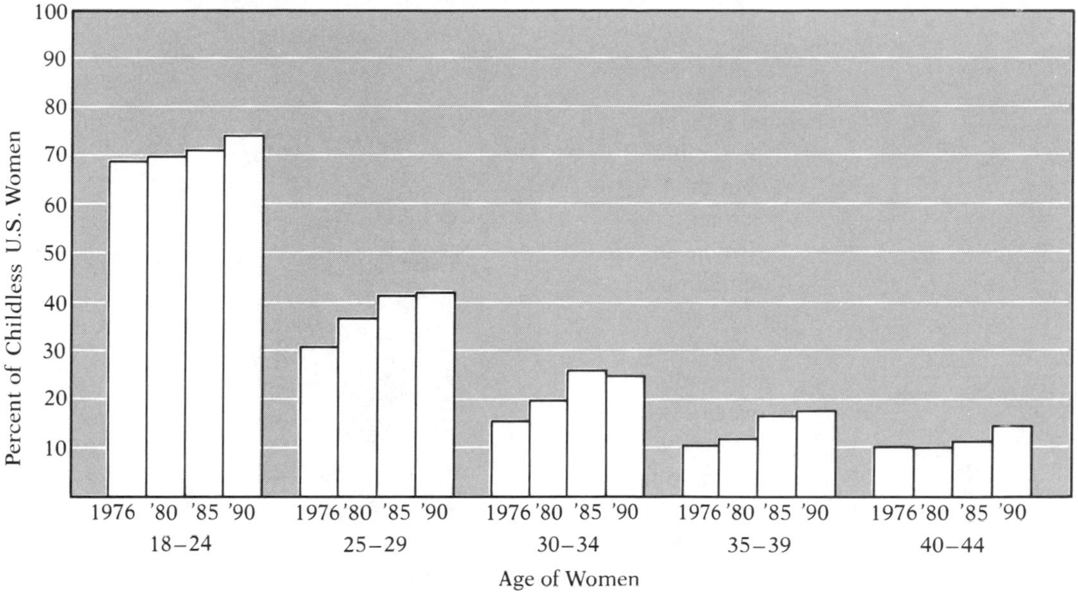

Source: U.S. Bureau of the Census, *Current Population Reports*, 1990, P-20, No. 454.

number of children born to older women will continue to increase, and that more women will end up remaining permanently childless (Rindfuss, Morgan, and Swicegood 1988, 89; Chen and Morgan 1991). In short, the trend toward delaying having children will continue for some time in the future.

Medical Risks of Delayed Birth

How late can one wait to have a baby? Technically, women are capable of reproducing until menopause, in the late forties or early fifties. But fecundity is highest in a woman's twenties and goes down sharply when she reaches her forties (Menken et al. 1986). Ovulation becomes more irregular, especially in women who already have irregular menstrual periods. So a couple who wait until their late thirties or forties may find that pregnancies do not necessarily follow unprotected intercourse. Of course, inability to conceive may occur at younger ages too, and in fact 8 percent of all women who wish to have children do not bear any (Westoff 1986).

The chances of genetic defects do go up somewhat as the parents grow older. *Down's syndrome* (formerly called mongolism), the most common form of mental retardation, happens in 1 out of 1,500 pregnancies with mothers in

A boy with Down's Syndrome displays a proud smile and ribbons after a Special Olympics race. Special Olympics events help physically and mentally challenged individuals to realize their potential.

their twenties; but the odds rise to 1 out of 350 for mothers in their late thirties and 1 in 100 for those in their forties. The chances even for the oldest mothers are still only 1 percent but the risk can be dissuading. However, it is possible to test for Down's syndrome by the technique of *amniocentesis*. This method consists of removing fluid and cells from the womb through a thin needle between the sixteenth and twentieth week of pregnancy. If Down's syndrome is present, an abortion can be performed, although such a late abortion, after fetal movements have been felt, can be emotionally very painful for the parents. Amniocentesis also tests for the sex of the child and for various other (but not all) genetic birth defects. The procedure involves a less than 1 percent chance of causing miscarriage, but it is relatively expensive. Nevertheless, it does provide a way for a woman to greatly increase her chances of bearing healthy children even if she waits until her forties. (Other techniques may increase a woman's chances of bearing a healthy child safely late in life; see feature 10.2.)

Infant Mortality

Infant mortality is often regarded as a measure of a society's quality of life, since low levels of infant death are usually associated with modern technology, effective health services, higher incomes, and proper diets. The paradox is that the United States has a high rate of infant death even though it is one of the wealthiest and most technologically advanced societies on earth, and even though it spends a greater proportion of its gross national product on health care than other developed nations. The U.S. infant mortality rate dropped rapidly and steadily until the mid-1970s, but it has improved only marginally since then (Hale 1990). According to a 1990 national panel on infant health, our progress in reducing the infant mortality rate has ground nearly to a halt. With a rate of over 10 deaths per 1,000 live births, we are now twentieth among developed countries—behind Japan, Canada, Australia, and virtually all of western Europe.

Why is the U.S. infant death rate so persistently high? A large part of the problem seems to be lack of adequate nutrition and prenatal care, for which minorities are especially at risk. Babies who are born with a low birth weight (under 2500 grams or about 5.5 pounds) have a smaller chance of surviving the first month of life, when most infant deaths occur. The United States ranks seventeenth among developed nations in the percent of babies born with low birth weight, but the rate for black Americans is double the rate for whites. The major reason for low birth weights and infant death for both whites and blacks in this country is poverty, with its attendant poor nutrition and lack of routine medical care. In most other industrialized countries, pregnant women receive subsidized prenatal care and home visits after the baby is born. Ironically, our reluctance to spend money on poor mothers actually costs us more in the long run. It has been estimated that $400 worth of prenatal care could make the difference between a healthy baby and one requiring $400,000 of medical care and services throughout its life (Hale 1990). It remains to be seen whether recent attempts to reform the health care system and provide universal health insurance will help alleviate this problem.

Childless Families

Some married couples deliberately decide they do not want children. About 5 percent of all marriages, or one in twenty, are of this sort. The rate is even higher among college graduates, where as many as 14 percent of all couples are childless. The proportion of childless marriages is slowly rising. They are most prevalent among people with career commitments in high-paying occupations. The apparent reason is that children are less appealing than the careers, and child-care responsibilities are a burden to be avoided (Campbell 1985).

This sounds like a masculine motivation, but it isn't necessarily so. The woman, in fact, is more likely to be the member of the couple who does not want to have children. Moreover, if the couple disagree over whether or not to have children, the woman who doesn't want them tends to be much more adamant about her viewpoint. Some husbands said they had changed their minds—initially not wanting children and later wanting them—but their wives were the ones who held firm (Silka and Kiesler 1977).

Successful careers are a motivation for women as well as men to remain childless, but noncareer motivations seem to be even more important. Childless couples seem to be somewhat different than others: they are more independent, live farther away from their own parents, and enjoy being alone more (Nason and Poloma 1976; Houseknecht 1987). The most common reason they gave for not wanting to have children was *to give more time to their spouse*. In other words, their motivation might be broadly called *erotic* reasons. They wanted to spend their love on their adult partner, not on their children.

Being a childless family is not a once-and-for-all decision. Couples do not usually make up their minds to avoid childbearing until after they are married, rather than when deciding to marry. It is not usually part of an originally thought-out plan. Also, many couples deliberately decide to *delay* childbearing. These couples seem to go through the same processes as the ones who forego it entirely. In both cases, it is the woman who has the stronger opinion; and women also give as their main reason for delaying childbirth, not career considerations, but the desire to spend time with their husbands.

America is typically characterized as a strongly pronatalist society: that is, we tend to believe that all married couples should want children and that they should act on that desire (Veevers 1980). When cultural norms and values strongly encourage people to become parents, those who remain childless—through infertility or by choice—tend to be labeled as deviant. Those who are "childless by choice" are the most likely to be negatively stereotyped, and women who choose to forego motherhood are much more likely than men to be considered "unnatural" (Houseknecht 1987; Veevers 1980). But even women who are involuntarily childless often feel discredited. For instance, one woman felt that admitting to possible problems with reproduction was "an admission that you're not a whole person. . . . The ability to reproduce strikes at the very essence of one's being" (Miall 1987, 392). Another woman who tried unsuccessfully to have children commented that she and her husband felt like failures:

> We expected things to work out. We led the golden lives until this happened to us. It's the worst thing that ever happened to us. My husband and I are high achievers. We work together and it works out. Therefore, it's even harder for us to accept this because we don't have control over something that's so easy for other people. (Miall 1987, 391)

In order to avoid the negative reactions and judgments that followed when they told others of their inability to have children, many of the couples in Miall's study (1987) felt compelled to conceal or deceive others about their condition. These examples show how social pressure to have children can influence people's self-concepts as well as their willingness to attempt to reproduce. Even though norms concerning childbearing are changing, most people, and especially women, feel that they should have children in order to feel complete.

The idea that women are required to have children and are fundamentally and essentially defined by being mothers is sometimes referred to as "compulsory motherhood." While the average American woman spends only about one-seventh of her life either pregnant, nursing, or caring for preschool children, being a mother is seen as her ultimate reason for being. Motherhood is thus considered sacred, but only if it happens to a woman who is married. The stigma attached to women who give birth out of wedlock may not be as great as it once was, but it is still quite intense. Labels of "unwed mother" and "illegitimate child" are used to castigate women and children who are not connected to a man. Media images and political rhetoric also reveal an interesting double message about motherhood that is sent to white women and women of other races. Poor black women are sometimes threatened with sterilization or reduced welfare benefits in an effort to discourage their getting pregnant, but educated white women, in contrast, are bombarded with propregnancy propaganda that one commentator has called "maternity chic." The message seems to be that "Real women *should* have babies, successful women *are* having babies, and even feminists *want* to have babies" (Pogrebin 1983, 180).

What seems to be happening is a no-win situation for many women. On one hand, we still have the cultural ideal that fulfillment and true personhood come only with being a mother. On the other hand, we now have the "total woman" dictum that says being a mother is not enough, you must have a career as well. If women want only to be housewives and mothers, they are now made to feel guilty because that is not enough. If they try to "have it all" as suggested by countless articles in women's magazines, they end up perpetually exhausted (see chapter 11). In the face of these conflicting pressures, it is no wonder that increasing numbers of women are delaying childbirth but maintaining their commitment to have children sometime in the future.

INFERTILITY

Although more couples are postponing childbirth or remaining childless, about 10 to 15 percent of couples want children but are unable to have them. Infertility is the label used to describe those who are unable to conceive a child. As a general rule, people are defined as infertile if they have not conceived after having had regular sexual relations for over a year without using any contraceptives. Infertility is also sometimes used to refer to what doctors call "impaired fecundity"—the inability to carry a pregnancy to live birth.

Infertility is about equally likely to be the result of problems attributed to the man as to the woman. Common causes of infertility in men include low sperm production; poor semen motility; effects of sexually transmitted diseases such as Chlamydia, gonorrhea, and syphilis; and enlargement of the prostate limiting the passage of sperm. Common causes of infertility in women include blocked fallopian tubes, endocrine imbalances that prevent ovulation, dysfunctional ovaries, imbalances in cervical mucus, and effects of sexually transmitted diseases (Knox and Schact 1994).

Although more people than ever before are seeking treatment for infertility, it is not the case that the actual incidence of infertility has increased. In fact, the National Center for Health Statistics estimates that the infertility rate has declined since the 1960s (Mosher and Pratt 1990). The infertility rate is higher for those who delay childbearing, and since people are waiting longer to have children, and older parents have more financial resources to treat the problem, more people are seeking medical help for infertility. As discussed below, the increase in help-seeking behavior is also linked to the development and availability of modern reproductive technologies.

Doctors prescribe a variety of drugs to women with infertility problems. For example, fertility drugs can be used to treat hormonal imbalances, induce ovulation, and correct irregularities in the menstrual cycle. Two of the drugs that are used most frequently are Clomid and Pergonal. Taking fertility drugs drastically increases a woman's chances of giving birth to multiple children. In the 1980s, the rate of higher-order multiple births almost doubled, primarily because of the increase in the use of ovulation-inducing drugs for treating infertility (Kiely et al. 1992).

ADOPTION

The most common solution to the problem of infertility is adoption. Almost 120,000 adoptions take place each year in the United States (Flango and Flango 1993). About half of these are by relatives and half by nonrelatives. For nonrelatives, adoption by infertile couples are most common. Over a third of adoptions in the United States are through public agencies, about a third are through independent or private channels, and just under a third are through private agencies (*Statistical Abstract of the United States* 1992).

State agency adoptions are the least expensive because they require only legal fees, but the waiting times are long. It is not unusual to wait up to seven years to adopt a child through a state agency. State agency requirements for placement are also usually the most stringent. People over fifty are usually considered "too old," those under twenty-five are sometimes considered "too young," and those who violate traditional norms are given low priority. Unmarried heterosexual couples may sometimes adopt, but homosexual couples or individuals usually have great difficulty doing so. Private agency adoptions are usually faster, but they are more expensive, often costing $10,000 to $30,000 (Knox and Schact 1994). Some private agencies facilitate foreign adoptions, usually involving children from Third World countries. The waiting period for foreign children is usually less, often only one to two years. About one out of ten adoptions in the United States involve foreign-born children (Benokraitis 1993).

The third type of adoption—direct—entails contracts between a couple who want a baby and a pregnant woman. It is estimated that 80 percent of adoptions in California are of this type. Usually a pregnant teenager realizes that she does not have the resources to care for her baby and her doctor or someone else helps arrange for the couple to legally adopt the child at birth (Knox and Schact 1994). The problems associated with direct adoptions can be similar to those encountered in surrogacy arrangements discussed below.

THE NEW REPRODUCTIVE TECHNOLOGIES

The newest forms of reproductive technology involve procedures that bypass sexual intercourse. For instance, artificial insemination involves the introduction of sperm into the female reproductive system by means other than intercourse. (Children conceived in this fashion have been referred to by some as "turkey baster" babies.) Commercial and nonprofit "sperm banks" collect sperm from male donors (who are usually screened for genetic defects) and freeze it for future use. Artificial insemination is currently offered by the majority of infertility facilities in university clinics and is practiced by most of the larger medical centers in the country. The success rate for artificial insemination is lower than for natural methods, but ranges from 40 to almost 90 percent depending on the woman's fertility potential, the condition of the sperm (fresh or frozen), and the timing and frequency of the insemination (Moghissi 1989, 124–25). It is estimated that between 200,000 and 1 million children have been conceived through artificial insemination in the United States.

Some states have enacted new laws to handle this special kind of procreation. Such laws typically specify that a child conceived by means of artificial insemination carried out with the consent of the husband is the legal offspring of the couple. Some states further specify that the man who furnishes the sperm for the artificial insemination of someone other than his wife is not the child's legal father. In twenty-two states the child conceived through artificial insemination is still in legal limbo (Moghissi 1989, 127). While the legal and ethical issues have yet to be resolved, some people suggest that we should freeze the sperm of especially talented men to be used for the artificial insemination

FEATURE 10.2
Test-Tube Babies

About twenty years ago, scientists in England, Australia, and the United States developed a technique for fertilizing human eggs outside the mother's body called in vitro fertilization. The first baby born by this method was in Cambridge, England, in 1978. Since then, over 300 clinics in the United States have duplicated this procedure, resulting in about 20,000 live births.

The method was developed to counteract sterility due to blocked or damaged fallopian tubes, a condition shared by about 2 million American women. Blocked tubes can sometimes be opened with laser beam surgery or by inflating a tiny balloon in the constricted passage, but when this is not possible, test tube fertilization is an alternative. In this procedure, the ovaries are stimulated to produce multiple eggs, which are taken from the ovary through a small abdominal incision just before ovulation. In a glass petri dish in the laboratory, the eggs are mixed with sperm from the male donor—usually the woman's husband—and placed in a chemical incubator. Finally, one or more eggs are transferred back into the mother's uterus.

Success rates for in vitro fertilization vary, but it is estimated that only about one in ten couples give birth as a result of the procedure (U.S. Congress 1990). About a third of those end up with two or more births because multiple eggs are placed in the uterus. In vitro fertilization is typically quite safe for the mother and there is no evidence linking abnormalities in children with the method. The main drawback is that the method is expensive. Depending on the number of inseminations or special actions that are necessary, the cost of a pregnancy using in vitro techniques is typically over $10,000, and is frequently two or three times that much. In addition, the procedure entails hormone injections, blood samples, ultrasounds, physical exams, and periods of inactivity. Most couples thus report that in vitro techniques are time-consuming and emotionally exhausting (Elmer-Dewitt 1991).

Several new techniques have been developed that are variations on the basic in vitro procedure. These include:

Gemete Intrafallopian Transfer (GIFT)

Because only about one in five fertilized eggs implant on the wall of the uterus during in vitro procedures, doctors developed intrafallopian transfer. In this process, both egg and sperm are placed directly into the fallopian tube where they meet, fertilize, migrate down into the uterus, and hopefully successfully implant on the wall of the uterus. The

of specially selected women. Others suggest that we should also freeze the ova from especially talented women. Proponents of eugenics (the science of controlled breeding) claim that we could produce superior children with these techniques and improve the human race. There is already a sperm bank for Nobel Prize winners with plenty of female applicants (Steinmetz, Clavan, and Stein 1990, 247).

Test-tube babies, or external fertilization, is another technological possibility that is quickly gaining widespread acceptance. *In vitro* fertilization (literally "in a glass") is much more complicated than artificial insemination, but with rapid progress in medical technology, is becoming a viable alternative. Whether it is widely practiced will most likely depend on important legal, political, and ethical dilemmas that have yet to be resolved.

GIFT procedure is reported to double or triple the chances of implantation over normal in vitro techniques (Knox and Schact 1994).

Zygote Intrafallopian Transfer (ZIFT)

This procedure combines in vitro techniques with GIFT techniques. The woman's eggs are extracted and fertilized with the man's sperm in a petri dish as in the normal in vitro procedure. The zygote is then transferred directly into the fallopian tube as in the GIFT procedure. In this way, the doctor ensures that fertilization has taken place, but the natural migration of the embryo enhances the chances of uterine implantation. Success rates for ZIFT procedures are similar to those for GIFT procedures (Knox and Schact 1994).

Partial Zona Drilling (PZD)

Eggs with shells that have been punctured have the highest chance of implanting on the walls of the uterus. The PZD procedure thus entails isolating eggs and drilling tiny holes in the protective shell that surrounds them. Although it is too early to determine the effectiveness of this technique, doctors hope that it will raise in vitro fertility rates.

Microinjection

In some cases, infertility results from sperm with low motility. To overcome this problem, doctors obtain sperm from the man and inject it directly into the egg using microinjection techniques.

Embryo Transplants or Ovum Transfer

In this technique, the sperm of the man is placed by a doctor in a surrogate woman. After about a week, her uterus is flushed and the contents are microscopically analyzed to detect the presence of a fertilized ovum. If one is found, it can be inserted into the uterus of an otherwise infertile woman, who may carry the baby to term. The embryo can be frozen and implanted later, but the highest implantation rates have been found for "fresh" zygotes (Levran et al. 1990). Embryo transplants allow the child to be the biological offspring of the father, and afford the "mother" the experience of pregnancy and childbirth. As with other types of surrogacy, however, this technique raises some provocative moral and legal issues, and some suggest that it allows affluent women to exploit the reproductive capacity of less fortunate women.

SURROGATE MOTHERHOOD

One of the most controversial and troublesome outgrowths of the new reproductive technologies is the possibility of surrogate motherhood. Surrogacy has been described as a miracle cure for infertile couples, but has also been called baby selling, womb renting, and reproductive exploitation. Commercial surrogacy involves a contractual agreement, often negotiated by an agency, between a couple and a fertile woman. The fertile woman is paid to be artificially inseminated with the husband's sperm, becomes impregnated, and carries the fetus to term, at which time this surrogate mother relinquishes the infant to its biological father and its social mother (Whiteford 1989). At least, that's how it is supposed to work.

In the much publicized "Baby M" case in New Jersey, the surrogate/biological mother broke the signed contract and refused to surrender the baby to the biological father and his wife. After lengthy court battles, a judge awarded the baby to the father, but the surrogate/biological mother later won visitation rights. This case left many legal questions unanswered.

Two factors seem to make surrogate motherhood especially troublesome. First, the mother grows the fetus inside of her for nine months and during that time may change how she feels about keeping or giving up the baby. This is what happened in the case of Baby M. Second, there is usually an inequality between parties to the agreement. Some surrogate mothers say they do it because they enjoy being pregnant and want to make other, childless couples happy (Martin 1976). Nevertheless, most surrogate mothers are young, poorly educated, have few job skills, and do it because they need the money. The other party to the contract, the infertile couple, is usually well-off financially and willing to pay all medical and nutritional costs as well as to make a relatively large lump sum payment to the woman upon delivery of the baby. This sets up two classes of parents and leads some commentators to describe the relationship as one of exploitation. Dworkin (1987) likens surrogate motherhood to prostitution and suggests that poor women's reproductive capacity, like their sexuality, becomes a commodity that they are forced to sell in order to survive.

THE MEDICALIZATION OF CHILDBIRTH

The reproductive technologies discussed above may seem like they belong in some far-out science fiction version of a brave new world. After all, they transform relatively "natural" processes—conception and childbirth—into medically controlled and legally governed contractual events. Nevertheless, new techniques like *in vitro* fertilization and amniocentesis are logical extensions of a process that has been occurring for several centuries: the medicalization of childbirth.

Before modern times, pregnancy and childbirth were not considered medical problems, and women typically controlled the birthing process themselves, subject to the ritual practices and customs of their community. While birthing practices have varied considerably throughout history and among cultures, the birth itself has almost always been regulated by the birthing woman herself, with emotional support and technical assistance offered by other women and sometimes by her husband. (See feature 10.3).

Our term for those attending women giving birth is *midwife*, which comes from the Anglo-Saxon "mid-wif," meaning literally, "with-woman." Midwives have practiced their ancient art for thousands of years. Besides attending births, some midwives performed other functions: in ancient Rome, midwives often served as marriage brokers; in medieval England, midwives were called upon to baptize babies; and in most societies, midwives have offered herbal remedies for various gynecological disorders and helped the new mother with domestic tasks (Sorel 1985; Sullivan and Weitz 1988).

Until the seventeenth century in Europe and America, midwifery was the exclusive province of women, as it was considered improper for men to be present during the delivery of a child (Merchant 1980). Midwifery represented women's control of a uniquely female process, though this was sometimes equated with witchcraft by powerful men. In seventeenth and eighteenth

FEATURE 10.3
Birth Practices of Some Nonindustrial Peoples

Eskimos

The Danish explorer Peter Freuchen first traveled to Greenland in 1906 and lived among the Hudson Bay Eskimos for many years. In his autobiography, *Book of the Eskimos*, he describes Eskimos' birthing practices:

> Eskimo women used to talk about giving birth as being "inconvenient." This is not to say that it was any fun, but they had a remarkable short period of confinement. The women used to sit on their knees while giving birth. If the woman was in a tent or a house when her time came, she would most often dig a hole in the ground and place a box on either side of it to support her arms and then let the baby drop down into the hole. If she was in an igloo, the baby had to be content with the cold snow for its first resting place. If the birth seemed to take long, the husband would very often place himself behind his wife, thrust his arms around her, and help press the baby out. (Cited in Sorel 1985, 89).

African Bushmen

In the early 1950s, Elizabeth Thomas went on three expeditions to the Kalahari Desert to study the African Bushmen. She describes how the women gave birth in *The Harmless People* (1958):

Day or night, whether or not the bush is dangerous with lions or with spirits of the dead, Bushman women give birth alone, crouching out in the veld somewhere. A woman will not tell anybody when she is going or ask anybody's help because it is the law of Bushmen never to do so unless a girl is bearing her first child, in which case her mother may help her, or unless the birth is extremely difficult, in which case a woman may ask the help of her mother or another woman. . . . When labor starts, the woman does not say what is happening, but lies down quietly in the werf, her face arranged to show nothing, and waits until the pains are very strong and very close together, though not so strong so that she will be unable to walk, and then she goes by herself to the veld, to a place she may have chosen ahead of time and perhaps prepared with a bed of grass. If she has not prepared a place, she gathers what grass she can find and, making a little mound of it, crouches above it so that the baby is born into something soft. Unless the birth is very arduous and someone else is with the woman, the baby is not helped out or pulled, and when it comes the woman saws its cord off with a stick and wipes it clean with grass.

century Europe and America, it was not uncommon for midwives to be persecuted as witches (Merchant 1980; Ehrenreich and English 1973).

With the growth of man-midwifery, or obstetrics as it was called after 1828, childbirth began to be redefined as a pathological event that required monitoring and intervention by medical men. As part of a cultural shift toward scientific understanding of the world, the authority of physicians grew, and medical practices, such as forceps delivery and pain-killing medications, became prevalent during childbirth. During the era of the "Cult of True Womanhood" (Welter 1966, chapter 4), middle- and upper-class women were expected to be idle, even physically weak, as signs of their cultural delicacy and their husbands' wealth. These women entered the birthing chamber with muscles tensed by fear; rib cages and pelvises deformed by corsets; and natural strength weakened by lack of exercise, frequent pregnancies, and tuberculosis. They accepted pain relief, available only from physicians, as part of their lot (Wertz 1983; Wertz and Wertz 1977). In addition, the "true" woman in Victorian society was considered incapable of learning technical skills, and midwives were excluded from medical training (Sullivan and Weitz 1988, 7).

In the late nineteenth and early twentieth centuries, physicians effectively lobbied to better their position by restricting the practice of midwives, along with that of homeopaths, chiropractors, and others (Starr 1982). Medicalized childbirth techniques promised improved sanitary conditions, control of women's pain, and a reduction in the almost 1 percent of women and more than 4 percent of infants who died in childbirth at the turn of the century. As part of the campaign to convince the nation that obstetricians should replace midwives, physicians tended to exaggerate the dangers and difficulty of childbirth. By 1910, physician-attended births were typical among the middle class, but most American women still gave birth in their homes. At this time, the majority of all U.S. births were attended by midwives, primarily because poorer women continued to rely on midwives. In addition, the midwives themselves tended to be poor (large numbers were either immigrants or blacks), and they generally lacked the financial or political resources necessary to counter the physician's lobbying and claims of unsafe practices (Sullivan and Weitz 1988, 7). It did not take long for childbirth to become a medically controlled event. By 1935, the percentage of American infants delivered by midwives had dropped to just 11 percent. This included over half of all nonwhite infants, but only about 5 percent of whites (Jacobsen 1956).

In spite of real medical advances during this time, replacing midwives with physicians as birth attendants actually increased mortality and morbidity rates (Leavitt 1983; Wertz 1983). The White House Conference on Child Health and Protection in 1932 expressed concern that the United States had higher maternal mortality rates than European nations—where at least half of the births were attended by midwives. Studies at that time attributed birth-related deaths to unnecessary and inept medical intervention, such as forceps deliveries, cesarean sections, and manual removal of the placenta, made possible by the new anesthetics and analgesics and frequently performed by general practitioners (Sullivan and Weitz 1988, 17; White House Conference on Child Health and Protection 1933). In spite of the absence of clear evidence indicating superiority of physician-attended maternity care over traditional midwife care, the practice of midwifery was severely restricted and eventually was subsumed under "nurse-midwifery." The outcome is that licensing and training requirements for midwives today are similar to those for nurses, certified nurse-midwives receive additional graduate-level training, and in many instances nurse-midwives assume a subordinate role to the attending physician. While many lay midwives continue to practice illegally, and some states allow home births with licensed midwives attending, all fifty states accept certified nurse-midwives in attendance at hospital births and free-standing birth centers, accounting for almost 4 percent of all U.S. births. About 95 percent of births occur in hospitals with physicians attending.

Questioning the Medical Model

The organization of our medical delivery systems and extensive media coverage of new medical discoveries encourage us to view medical technology as having almost unlimited power to better our lives. After all, since 1900, the average life expectancy of Americans has increased a whopping twenty-seven years. The most common perception is that our increased life expectancy results from new developments in medical science such as vaccinations, X-rays, antibiotics, insulin, chemotherapy, open-heart surgery, and organ transplants. Although modern medicine has undoubtedly saved countless lives and bettered our chances for living longer, the primary reason for our increased life expectancy is improved living conditions (Starr 1982; Sullivan and Weitz 1988).

According to some observers, our belief in science and our faith in the authority of medical practitioners act as mechanisms of social control and can lead to various negative as well as positive consequences (Conrad and Schneider 1980). Our tendency to believe that active medical intervention is superior to more passive or preventative practices has contributed to the premature adoption of some costly and questionable surgical interventions. Examples include the excessive number of tonsillectomies in the postwar years and the more recent proliferation of coronary bypass surgeries, radical mastectomies, and hysterectomies (Millman 1978). Women's consumer groups have consequently questioned whether the medicalization of childbirth has gone too far.

Because childbirth in the United States is typically considered to fall in the medical domain, decisions about the management of the birth are usually made by physicians who, drawing on medical technology, predict the course of labor. A reliance on technological solutions has encouraged birthing practices such as forceps delivery; amniocentesis; antiseptic and aseptic measures (enemas, shaved pubic areas, scrubbing, draping, and isolation); routine episiotomies (pre-delivery surgical incision of the perineum, the flesh between the vagina and anus); pain-control drugs (analgesics and anesthetics); electronic fetal heart monitoring (electrodes attached to the fetus while inside the uterus); and long hospital stays. The increasing use of pain-control drugs during labor to lessen contractions has been instrumental in promoting the use of other drugs (pitocin, oxytocin) to promote stronger uterine contractions. The administration of these artificial labor-inducing medications and the use of mechanical fetal monitoring (which can produce false electronic heartbeat "reads" or varied ultrasound interpretations) have been blamed for increasing the likelihood of physicians deciding that labor is not progressing "normally." Such determinations frequently justify the use of cesarean section surgery, which has increased from just 4 percent of births in 1964 and 7 percent in 1971 to more than 20 percent of all births today (see feature 10.4; Gleicher 1984; Rothman 1983; Sullivan and Weitz 1988).

FEATURE 10.4
The Cesarean Explosion

Most births now take place in hospitals and under intense medical supervision. There is evidence, in fact, that the medical presence has been getting more and more heavy-handed. *Cesarean section* is the surgical procedure of removing the baby through the mother's abdominal wall instead of vaginally. The rate of cesarean deliveries—or "c-sections"—has increased from just 4 percent of births in 1964 to more than 23 percent of all births today (Sullivan and Weitz 1988; National Center for Health Statistics 1993).

Cesareans seem to have become a recent medical fad, despite the fact that the maternal mortality rate is far higher for cesareans than for normal vaginal births. Physicians seem to have changed the definition of "complications"; whereas not long ago well over 90 percent of all births were judged normal, that percentage has fallen drastically. In recent cases in which a hospitalized woman in labor has refused a recommended cesarean delivery on religious or other grounds, the courts have upheld the right of the physician, as medical expert, to perform the operation against the woman's will (Jordan and Irwin 1989).

The threat of lawsuits, along with rising insurance rates, hospital policies, and government regulations, have encouraged increasingly interventionist medical strategies for childbirth, such as cesarean deliveries. There is also some suggestion that increasing rates for c-sections are related to doctors' convenience, insofar as cesareans are more likely to be performed at times of the day or week when physicians have competing professional and personal obligations (cesareans are much less time-consuming than the process of childbirth).

Until recently, a woman who had a c-section had to have all subsequent births by cesarean. Now, however, about one in five births to a woman who previously had a cesarean is a vaginal birth (National Center for Health Statistics 1993).

Most women who have a cesarean do not plan it in advance with their doctors on the basis of some known problems such as a narrow pelvis. Rather, it is something that is sprung upon them in the delivery room. Frank discussion of this with the doctor beforehand may be the best way to avoid an unnecessary cesarean.

New categories of medical specialists, such as perinatalists (obstetricians trained intensively in high-risk maternity care), tend to encourage the use of the latest technological interventions. For example, the few elite intensive-care nurseries that existed in teaching hospitals in the 1950s have become the model for infant services in virtually every large hospital in the United States (Guillemin, Todres, Batten, and Grodin 1989). In addition, the threat of lawsuits, along with rising insurance rates, hospital policies, and government regulations, have encouraged increasingly interventionist medical strategies. As a result, more babies are delivered via surgery, and new-born intensive-care units are populated by premature infants and babies with severe congenital abnormalities whose lives can be saved but whose condition cannot be corrected by medical intervention. Taken together, the conditions surrounding medicalized maternity care raise a number of ethical, technical, legal, and emotional dilemmas that cannot be resolved easily.

NATURAL CHILDBIRTH

In response to what was perceived as "overmedicalized" maternal care in hospitals, a "natural" childbirth movement arose in the 1950s and grew during the 1960s and 1970s. Following the teachings of obstetricians like Grantly Dick-Read and Fernand Lamaze, childbirth education groups formed to prepare women and their "labor coaches," usually their husbands, for a less medically controlled birth. Childbirth preparation classes help expectant couples understand the birthing process and teach women how to respond to uterine contractions with relaxation, structured breathing patterns, and muscular control (rather than with fear, tension, and pain). In general, research suggests that women trained in these techniques have shorter labors, lose less blood, require less anesthesia and analgesia, have fewer operative interventions, produce more alert and healthy babies, and experience greater personal satisfaction (Sullivan and Weitz 1988). In light of the observation that women who had taken childbirth classes and had a support person present were calmer and more compliant than other patients, even physicians who initially opposed "natural" childbirth became less hostile to it during the 1980s.

Another aspect of the natural childbirth movement concerns husband-wife and parent-infant interaction. Following the treatment of birthing as a medical "problem," the standard hospital procedures in the 1950s and 1960s called for exclusion of the father from the delivery, isolation of the infant after birth, elaborate sanitary precautions, rigid feeding schedules, and long hospital stays. In a popular 1970s book on the ills of hospital birthing practices, one mother commented on how uncaring the rules seemed:

> I couldn't touch my baby for three days. He was in isolation in case there'd been an infection in utero. It was crazy! I had my baby and they took him

away and I couldn't touch him. Then we went home. As I was leaving, they put me in a wheelchair, plopped the baby in my lap, and pushed me out of the hospital. (Arms 1975, 21)

Hospital practices were also criticized for discouraging breast-feeding by not allowing sufficient contact or flexibility to establish a mother-infant feeding routine. Breast-feeding, which was used by nine of ten mothers in 1922, was gradually replaced by bottle feeding during the era of "better living through science." By 1972, about eight mothers in ten fed their infants formula in a bottle instead of breast milk. In the 1960s and 1970s, researchers and proponents of breast-feeding began to document the advantages of breast-feeding over formula feeding. They generally found that breast-feeding provides better nutrition (particularly for preterm and low birthweight babies), protects the infant against allergies and some diseases, and offers mothers some protection against possible breast cancer (Sullivan and Weitz 1988, 33). In conjunction with a growing concern for health issues of all sorts, breast-feeding began to regain popularity among the general population, so that in the 1990s, the majority of birthing women advocated a preference for breast-feeding.

Admitting fathers to labor and delivery rooms was also prohibited by most hospitals in the 1950s and 1960s. The typical image is of the father smoking profusely and pacing back and forth in the waiting room until the doctor comes out to inform him "It's a boy!" In 1972, most hospitals still discouraged fathers from entering the delivery room, and only about one father in four was present at the birth of his child (May and Perrin 1985). By the late 1980s, almost all major hospitals had provisions for fathers being present during delivery, most encouraged the father's presence, and many provided special "homelike" birthing rooms or sleeping-in facilities for the father. By the mid 1980s, it was more usual for the father to be present at the birth than to be absent (Klinman and Kohl 1984). In addition, the average hospital stay dropped to three days in 1984, and many hospitals began to offer twenty-four-hour release programs for normal deliveries, reflecting a growing recognition of childbirth as a natural, nonpathological event (Sullivan and Weitz 1988, 38).

THE BIRTH EXPERIENCE

Oh—ah! Glorious! Fantastically excited! We were crying, my husband and I were crying and laughing and yelling, "It's a girl!" and I was kissing him and yelling— the feeling is—I don't think I've ever been happier. It's—you're so proud, you're so excited, you're—it's impossible to—put the emotion in words—none of the words I'm saying come anywhere near—the overwhelming emotion of it.

Pain—the pain just blocked everything. I was so scared.

Kind of excited and kind of unbelieving, you know.
Women's reactions to moment of birth

All beat to hell. I was stunned—it's very *shocking*. It had a big blob of blood on
its head and a bruise—two bruises on each side of its face—and milky-looking
and yucky—it threw me. It really did.

They tell you to expect the worst—and—I was surprised. He wasn't five minutes old
and—um—he was just beautiful. *In my eyes*—he was really pretty. He had a white
color, not reddish or anything.

    Husband's reaction to birth [Entwisle and Doering 1981, 102, 107]

The experience of childbirth can be a profound event, but parents vary a
great deal in how they respond to it. Some parents find it a beautiful or
ecstatic event. Others are neutral or rather negative about it. What causes
these differences? In general, what is the course of pregnancy and childbearing
like, as a social experience?

Doris Entwisle and Susan Doering (1981) followed a sample of 120 women,
and about half of their husbands, through their pregnancies and the first six
months after childbirth. The group included both middle-class and working-
class couples, and for all of them it was their first birth. The patterns they
displayed seem to be typical of many young couples in recent years.

For most women, the nine months of pregnancy are a time of good health.
Those most likely to experience physical complaints were the ones who were
most anxious about the childbirth. Most of them worried rather little, and
their main concern was about how much weight they were gaining. Nevertheless,
pregnancy did not emerge as a particularly pleasant time. The women's average

FEATURE 10.5
A Sensual View of Childbirth

Advocates of natural childbirth suggest that birth
is a normal process that should occur at home,
though most acknowledge that some pain is involved
in giving birth. One unique group of midwives in
Tennessee published a book, *Spiritual Midwifery*
(1977), in which they attempted to redefine the
childbirth process in positive, sensual terms:

> I began having beautiful, rushing contractions
> that started low, built up to a peak, and
> then left me floating about two feet off the
> bed. [My husband] was lying beside me
> and going through the rushes too. I saw
> that I could breathe very deep and fast and
> rush higher with the contraction. The
> contraction would carry me and I would

breathe harder and harder and then we
would peak—it would slip off and leave us
floating. It felt wonderful, and we were
having a beautiful time. As the contractions
got stronger, it felt like I was making love
to the rushes and I could wiggle my body
and push into them and it was really fine.

    Woman giving birth to her first child after
two miscarriages (Gaskin 1977, 53)

Over and over again, I've seen that the best
way to get a baby out is by cuddling and
smooching with your husband. That loving,
sexy vibe is what puts the baby in there,
and it's what gets it out, too.

    Practicing Lay Midwife (Gaskin 1977, 52)

opinion about their pregnancies was somewhat on the negative side although many enjoyed the feeling "of something living inside you." (Entwisle and Doering 1981, 59). The researchers could not find any factor that seemed to predict which women would feel positive or negative about their pregnancy, though there seemed to be some connection to what the researchers called "sex-role ideology" (i.e., women who expected to breastfeed their babies felt better about their pregnancy).

One thing that women did often mention was how they felt about the fetal movements that began sometime around the fifth month. Some were very negative: "I . . . I'm shocked! I just didn't believe it would move that much; I really didn't. . . . It's very uncomfortable." More women, though, tended to like the sensation: "Oh, I love it! It's my favorite part of being pregnant. It feels like it's gonna come out dancing. It's wonderful feeling that there's life inside" (Entwisle and Doering 1981, 59).

Men tended to be more negative about the pregnancy than their wives. Although most of them were quite interested in their wife's condition and in the course of the pregnancy itself, as it went on they ended up with relatively negative feelings about the experience. Only a small minority (23 percent) actually finished the nine months with positive feelings about their wife's state of being pregnant.

One reason may have been sexual. Most couples were continuing to have intercourse until relatively late in the pregnancy; only 14 percent stopped before the eighth month, and half of the rest either said they did not intend to stop at all or would stop during the ninth month. (About half the women felt negative about sexual intercourse during advanced pregnancy, and would have preferred to stop.) Nevertheless, the rate of intercourse for most couples dropped considerably during pregnancy, and both husbands and wives tended to have less sex drive than before. At least less sex drive for each other: almost one-third of the men said they were tempted to be sexually unfaithful during that time (although only one admitted actually doing so). Virtually none of their wives, however, was at all worried about the possibility, and most laughed off the question.

Most of the women wanted to be able to go through their labor without drugs. Many had taken classes in natural childbirth and had practiced relaxation methods. Nevertheless, by the time the birth was over, the majority ended up taking at least some drugs. This was a disappointment: most women did not expect the delivery to be as painful as it was. They were simply not prepared for as much pain as actually occurred. The drugs given were fairly strong ones: over half the women had either Demerol (a narcotic), sedatives, or tranquilizers, while the rest had local anesthetics. The drugs cut off the chance of a "peak experience" for some of those who had been looking forward to it. A typical response was that of the woman who said she felt "very disappointed, just not *caring*; as if I had been dying." Most of those who were *less* drugged described the birth as a positive experience (Entwisle and Doering 1981, 96).

The recent trend has been for husbands to be present during the labor, and sometimes even in the delivery room. As recently as the 1960s, fathers almost never witnessed the actual birth. But in Entwisle and Doering's mid-1970s sample, almost all the husbands were present in the labor room and two-thirds were there during the delivery. By and large, the wives were pleased by this arrangement. The vast majority of them felt that their husband's presence was helpful, even though it was mostly a matter of holding their hands and giving emotional comfort. This was particularly so because the majority of women were left alone at some time during their labor or delivery, some for an hour, some for even as much as eight hours. Doctors were present in the room an average of 28 percent of the time, and nurses were there about half the time. Since these were first births, they tended to take longer than later births: about twelve to sixteen hours for the first stage of labor, and one to one and one-half hours for the delivery stage. It was a long ordeal, and most of the women did not anticipate how physically arduous it would be.

The husbands in the Entwisle and Doering study tended to be anxious and worried during the birth, but their degree of worry did not seem to be associated with how much pain their wives experienced. The men in this study were not that closely attuned to what their wives were actually feeling. The pain tends to peak when the cervix is approaching full dilation, which is to say, just before the baby begins to move into the birth canal. The moment at which the head emerges, on the other hand, is least likely to be the most painful time. Some women, though a small minority (4 percent), did not experience any severe pain at any time. About 12 percent more said that the most painful thing was not the birth itself, but insertion of needles and other medical practices.

A key variable appears to be whether the women had learned to push correctly. Those who could simultaneously push while relaxing the muscles around the birth canal actually had a pleasurable moment of birth, whereas those who pushed while tightening their muscles reported excruciating pain. The required combination of relaxation of some muscles while pushing with others is apparently not easy to learn, and is sometimes aided by interpreting childbirth discomfort as different from other types of pain. For instance, some natural-childbirth advocates even encourage couples to view the birth as a sensual experience (see feature 10.5).

The birth experience can be seen as a high-intensity interaction ritual, at least in the typical case when both parents are present and the mother is awake (if mothers receive medication, it is usually a local or spinal anesthesia [epidural] which leaves the woman conscious). The essential components of a ritual are face-to-face interaction, attention focused on a common object or activity, a shared emotional tone that grows stronger, and an emotionally charged symbol (see chapter 8). Most births have these elements, since the father, particularly if he serves as "labor coach," is in close face-to-face contact with the mother; the couple is very focused on the breathing and everyone there is focused on the birthing process; the intensity builds as the labor progresses, and the emotionally charged symbol is the newborn baby.

Fathers who are involved in childbirth in this fashion typically describe the experience as "incredible," "magical," "moving," "wonderful," and "exciting." One new father's comments highlight the ritual nature of childbirth:

> I really participated in it. When I say we did the breathing together, I mean we really worked together and I really felt like I understood how she felt, in the sense that I was in the rhythm with her and what she was going through. . . . I cut the cord, which was symbolic, you know a kind of special thing to do, and the whole thing was very exciting and incredible, to hold the new baby at first. (Coltrane 1988, 249)

THE TRANSITION TO PARENTHOOD

The month or so following the birth is called the postpartum period. Most parents report that this is a time of significant emotional upheaval and adjustment to the demands of infant care. Although many parents experience profound joy during this period, most struggle to adjust to lack of sleep and the strain of being "on duty" all the time. Some researchers have reported that many mothers experience a period of "postpartum blues" characterized by depression, mood shifts, irritability, and fatigue. While this used to be attributed to physiological factors such as changes in maternal hormones, recent research shows that fathers, as well as mothers, experience these symptoms (Zaslow 1981). What is probably happening is that the emotional intensity of the birth, coupled with lack of preparation for parenting duties and overwhelming time-demands placed on infant caretakers is stressful for both parents. Since the mother is more often the primary infant caretaker, she tends to be the one who is most subject to the blues. While the emotional peaks and valleys of the postpartum period are quite common, not everyone reports that they occur, and each person's postpartum experience is different.

In the Entwisle and Doering study (1981), there tended to be a sort of honeymoon atmosphere right after the mother and baby came home. The father was usually attentive, and the couple were emotionally close. However, after a few weeks some strains and disappointments could be seen setting in. Most couples had been over-optimistic about how easy it would be to take care of an infant, and the pressures on their time and energy were larger than anticipated. The expectant couple had romanticized parenthood, and the reality of it was more work and less pleasure than they were prepared for. Also, in this early period (Entwisle and Doering followed up the new parents only for their first six months) some of the women were disappointed because they were unable to go back to work as they had expected, or in some cases had tried working but stopped. Birth was more of a disruption in people's lives than they had counted on.

Earlier studies on the transition to parenthood (LeMasters 1957; Rossi 1968), other studies conducted about the same time as the Entwisle and Doering research (Hoffman 1978; LaRossa and LaRossa 1981), as well as later research (Cowan et al. 1985; Feldman 1987), confirm that becoming parents is both joyful and extremely stressful. Most studies show that the birth of a first child tends to precipitate a crisis for the couple, with the majority of new parents reporting major strains in their lives during the family-formation period. Perhaps the most consistent finding across studies is that both husbands and wives report declining satisfaction with marriage and life in general, as they adjust to the enormous time-demands of newborn care and struggle to realign their marital roles (Belsky, Spanier, and Rovine 1983; Cowan 1988; LaRossa and LaRossa 1989).

Rossi (1968) first suggested how unprepared most parents are for assuming parental duties. She noted that motherhood has been defined as essential to a woman attaining full adult status in our society, but that unlike most other societies, we isolate new mothers in nuclear family households where they are unlikely to receive the help or training that they need. Unlike most other roles that adults assume in our society, we receive little preparation for being parents. Jobs typically allow us to get some training, to serve as an apprentice, or to gradually assume full duties. Parenthood, on the other hand, comes all at once, is irrevocable, and its training period—pregnancy—teaches us almost nothing about infant care. Not only are we unprepared for all the details of caring for a newborn baby, but after the baby is born, someone has to be "on-duty" every day twenty-four hours a day. It's no wonder that new parents experience stress!

Things have changed somewhat since Rossi studied the transition to parenthood in the 1960s. For one thing, most women had babies at an earlier age then. A more dramatic shift, however, has to do with the new mother's labor force participation. When the Census department first began collecting data on working mothers in 1976, they found that 31 percent of women who had a child in the last year were in the paid labor force. That figure rose each year, so that by 1980, 38 percent of women ages eighteen to forty-four with infants under a year were employed. By 1990, 53 percent of mothers with children under one year old were in the labor force, and the trend shows no sign of abating (see figure 10.6). For college graduates, the rate rose even more sharply, from 44 percent in 1980 to 68 percent in 1990 (U.S. Bureau of the Census 1992). In addition, the majority of these women, regardless of educational level, were working full-time after they became mothers.

The increase in maternal labor force participation has been especially dramatic during pregnancy. For those who gave birth to their first child in the early 1960s, fewer than half were employed at any point during their pregnancy.

FIGURE 10.6: Women with Babies Are Increasingly Likely to Be Employed

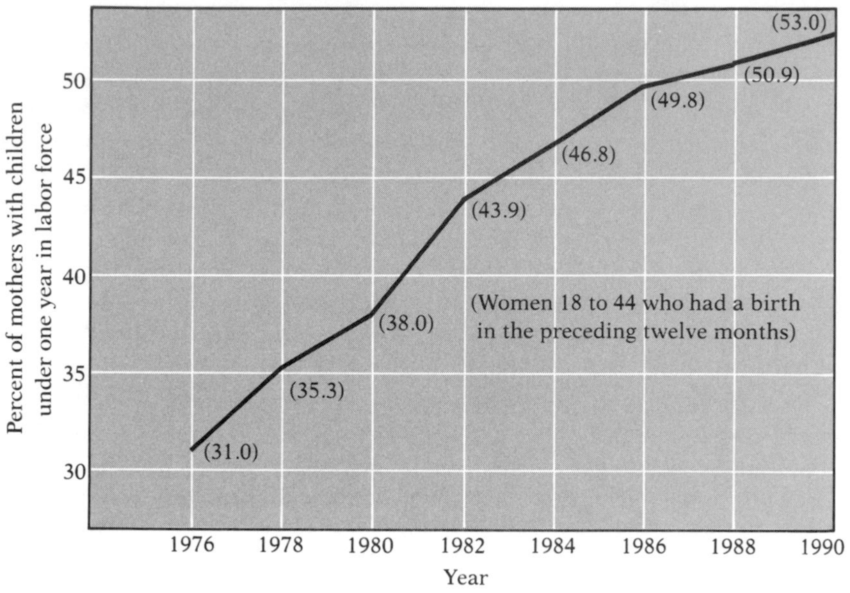

Source: U.S. Bureau of the Census, *Current Population Reports*, 1990, P-20, No. 454.

In contrast, two-thirds of those giving birth for the first time in the 1980s were employed during their pregnancy, and 80 to 90 percent of those were working full time. In the early 1960s, first-time mothers who worked during pregnancy were much more likely to be poor—teenagers, part-time workers, and high school dropouts. Employment patterns had changed substantially by the early 1980s, however. Not only were many more pregnant women working in the 1980s as compared to the 1960s, but the characteristics of those working had also changed. By the 1980s, women employed during pregnancy were more likely to be over age twenty-five and college-educated. These older and more highly educated women were also more likely to work longer into their pregnancies than did their younger and unmarried counterparts from the 1960s (U.S. Bureau of the Census 1990). What we had, then, was a shift among middle-class women toward patterns that had once been common only for mothers at the lower end of the economic spectrum.

The economic necessity of having two workers in every family drove many new mothers to hold onto jobs that they would have abandoned in an earlier era. In the early 1960s, when less than one-half of women worked during their pregnancy, almost two-thirds quit their jobs before their first birth. Things had changed by the 1980s, when only about a quarter of women reported quitting their jobs altogether as the result of having a baby. In the earlier period, less than a third reported some kind of maternity leave, mostly unpaid. In the early 80s, in contrast, over two-thirds of first-time mothers used some form of leave, with two-thirds of those reporting paid leave (mostly for a short time period). Thus, in the space of twenty years, the relative percentages had switched: in the 1960s, women typically quit their jobs when they got pregnant, but by the 1980s, almost three of every four pregnant women kept their jobs and took maternity leave (U.S. Bureau of the Census 1990).

Because most families must rely on two incomes to make ends meet, the transition to parenthood may be even more stressful than it was in the past. While child advocates have called for maternal and paternal leave policies so that either parent can take time off around the birth, most employers do not offer such programs. In the United States, fewer than half of the employed women are entitled to a paid disability leave of at least six weeks at the time of childbirth. The Family Leave Act passed by the U.S. Congress in 1993 will make leave available to more new parents, but because it only mandates unpaid leave and it only applies to larger companies, critics expect it to have little impact on most families (see the discussion of family policy in chapter 12). In contrast, 117 countries have legislated coverage that allows a mother to leave work for at least six weeks at the time of childbirth, guarantees that her job will be there when she returns, and provides a cash benefit to replace all or most of her earnings (Kamerman 1989; Klinman and Kohl 1984, 189). Long-term paid leave for fathers is even more rare in the United States. When short-term paid paternity leave is available, it is only for a few days and is usually deducted from sick leave. Most commonly, birth-related leave for fathers is unpaid and is taken as personal leave or some other category and not designated as official paternity or parental leave (Klinman and Kohl 1984). The absence of paid-leave policies, coupled with lower wages for women than men, encourages a sharp division of family labor based on gender (see chapter 11).

Does Childbirth Promote Traditional Gender Stratification?

One of the major findings of the transition to parenting studies is that household division of labor and marital relations become more traditional after the birth of a child (Entwisle and Doering 1981; Cowan et al. 1985; LaRossa and LaRossa 1981). In other words, women end up doing more of the cleaning, cooking, and child care, with men "helping" the mother, but not assuming major responsibility for parenting or housework (see chapter 11). Competing explanations for this trend focus on four different theoretical perspectives: sociobiology, personality difference, exchange theory, and interaction/conflict.

1. The sociobiological argument suggests that lactation, hormone secretion, and other physiological factors predispose women to be more proficient infant caretakers than men (Rossi 1977). More extreme versions of this approach posit a strong "maternal instinct" in women but not in men.

2. The personality differences argument suggests that the experience of being raised by a mother causes girls to develop different "relational potential" than boys. According to Chodorow (1978), women grow up with fluid ego boundaries and experience the world through relations with others, predisposing them to want to mother their own children. Men, on the other hand, form their identities in opposition to women, form rigid ego boundaries, and consequently have less interest in and capacity for nurturing parental behaviors (see feature 12.3). (It should be noted that Chodorow's theory does suggest that some men are capable of nurturing and that, in fact, their assumption of responsibility for infant care could help promote gender equality by breaking the cycle of "oedipal asymmetries" [Chodorow 1976]).

3. The exchange theory explanation for marital roles becoming more traditional after childbirth rests on the assumption that women's bargaining power decreases after the birth (Nye 1978). New mothers typically make much less money than their husbands, often must cut back on employment after the birth, have few outside resources, and are perceived as lacking the stereotypical prebirth attractiveness that can give them leverage in the relationship. These factors make the woman dependent upon her husband for economic and emotional support and reduce the likelihood that she will be able to bargain for increased domestic labor on his part.

4. The interaction/conflict perspective on postbirth marital roles, like the exchange theory, assumes that spouses have competing needs, but focuses on the ways in which they construct images of themselves as capable of various domestic tasks. In the face of extensive time demands, the new parents "do gender" (West and Zimmerman 1987) as they allocate and perform various child-care and housework chores. Fathers' lack of skill in caring for babies, and mothers' superior nurturing capacities, are created and reaffirmed through interaction with each other and those around them. Men, who get more free time and tend to focus on fun activities, are defined as "not good at" certain tasks or "not aware of" certain details, and their contributions are considered "helping out." Women, in contrast, are expected to and do perform most of the infant care, and thus become "natural" caregivers who almost incidentally assume the position of household manager (Coltrane 1989; LaRossa and LaRossa 1989).

Fathers' Participation in Infant Care

One interesting finding from recent research using the interaction/conflict perspective is that early assumption of infant care by the father seems to encourage later sharing of child care. For instance, fathers of six-month-old babies born by cesarean section have sometimes been found to be more involved in routine infant care than fathers of those born vaginally. This may be because the mother's incapacity requires a more active early caretaking role for the father and allows him to develop competence and confidence in his ability to perform infant care on his own. Interview and observational data also suggest that men develop parenting skills and construct nurturing self-images as a result of performing routine infant care (Coltrane 1990; Field 1978; Pruett 1987). One study found that early involvement of fathers was considered by them to be an important ingredient in their later assumption of responsibility for child care:

Early assumption of infant care by the father seems to encourage later sharing of child-care responsibilities.

> I felt I needed to start from the beginning. Then I learned how to walk them at night and not be totally p.o.'ed at them and not feel that it was an infringement. It was something I *got* to do in some sense, along with changing diapers and all these things. It was certainly not repulsive and in some ways I really liked it a lot. It was not something innate, it was something to be learned. I managed to start at the beginning; if you *don't* start at the beginning, then you're sort of left behind. (Coltrane 1989)

Most of the fathers in this study, all of whom shared significant amounts of child care, talked about having to learn how to nurture and care for infants. Because most of the mothers breast-fed the babies and because everyone else assumed that the mother would be the major caretaker, these couples had to consciously encourage the father's participation. For example, one father commented:

> She nursed both of them completely, for at least five or six months. So, my role was—we agreed on this—my role was the other direct intervention, like changing, and getting them up and walking them, and putting them back to sleep. For instance, she would nurse them but I would bring them to the bed afterward and change them if necessary, and get them back to sleep.... I really initiated those other kinds of care aspects so that I could be involved. I continued that on through infant and toddler and preschool classes that we would go to, even though I would usually be the only father there.

Other fathers talked about "nursing" their babies themselves, bottle feeding them with formula or with mother's milk that their wives had expressed using a simple hand pump. These examples illustrate how prescriptions about gender roles in baby care can be altered by those who try to change them. Sharing infant care, however, does not eliminate the stress that new parents experience.

Coping with the Stress

One of the reasons that the transition to parenthood is so stressful for most parents is that they feel isolated and alone. With all the diapers, sleepless nights, and financial worries of parenthood, most new parents sharply curtail their social lives. The isolation that many parents feel is eventually reduced by forming friendships with other parents, but that process takes time. Those who attend childbirth preparation classes have an opportunity to share hopes and fears with their peers and generally say they feel less isolated than others, report a more positive birth experience, and have a less anxious adjustment to newborn care than others (Entwisle and Doering 1981; Cowan 1988). Nevertheless, "prepared" mothers and fathers leave the childbirth class just before the baby is due, and while they may be ready for the birth itself, they continue to need support and social contact after the baby arrives. In more traditional societies and in some subcultures in modern American society, that support is provided by extended kin networks or by other people in the immediate community.

In order to break the isolation and find needed support during the difficult transition to parenthood, parents have recently turned to parents' support groups. Some are organized and run by mental health professionals, some are organized by schools and led by child development educators, and some are run by the parents themselves. All have become increasingly popular in the United States in the last two decades. These groups provide new parents with an opportunity to compare experiences, share advice, work on problems and give and receive mutual support and encouragement. One of the primary benefits of such groups is that they help new parents make realistic appraisals of their expectations, their experiences, and the prospects for dealing with inevitable stresses and strains. Early evidence suggests that such groups help parents to feel less isolated and lead to higher levels of marital satisfaction and stability (Cowan 1988). Extended family networks also continue to provide similar benefits to some new parents, though kin support tends to encourage traditional gender roles (Bott 1957; Riley 1990).

The best predictor of marital happiness and personal well-being after the birth of a first child is marital happiness and well-being *before* the baby is born (Belsky et al. 1983; Cowan et al. 1985; Feldman 1987). If a couple has a strong relationship with good communication and problem-solving patterns before conceiving a child, they are most likely to continue those patterns after the birth and are likely to report lower levels of stress. If, however, a couple decides to save an already troubled marriage by having a baby, the added stresses and strains of pregnancy and infant care are likely to make matters worse.

SUMMARY

1. In past times, high infant mortality rates served to limit population growth. In addition, most societies practiced some form of birth control, contraception, abortion, or infanticide. Cross-cultural studies show that when given the choice, women tend to have only a few children.

2. Rates of birth are determined especially by the economic value of children's labor for the family and whether the male or the female has the power to decide on the size of the family. The U.S. birthrate has declined to near the rate of zero population growth because children are no longer economically valuable, and the power of employed women to decide on births has increased.

3. Premarital pregnancy has increased sharply since about 1940. Today, over one-quarter of all brides are pregnant or have already given birth. At the same time, the nonmarital childbirth rate has gone up sharply, especially in social groups below the poverty line.

4. Though contraceptives are now widely available, a third of sexually active teenage women do not use them regularly. One explanation is that contraceptives are used most effectively by persons who are sexually quite experienced, whereas young women who consider themselves virtuous in the traditional sense tend to regard the deliberate use of contraceptives as contrary to their self-image. In recent years, the most common forms of birth control have become sterilization, birth-control pills, and condoms.

5. Only slightly more than half of all pregnancies result in birth. Of the rest, almost 25 percent are terminated by abortion and another 21 percent by spontaneous miscarriages or stillbirths. The great majority of abortions are performed on unmarried women below the age of twenty-five.

6. The teenage birthrate in the United States declined during the 1970s and early 1980s, but rose again after that. Primarily because of lower levels of contraceptive use, teen pregnancy is much more common in the United States than in most other industrialized nations.

7. An increasing proportion of women are bearing their first child in their late twenties or in their thirties. It is possible for women to delay childbearing until the late thirties or early forties. The chances of birth defects do increase at this age; however, a test (amniocentesis, which also tells the sex of the child) can reliably predict the presence of certain defects.

8. New reproductive technologies, such as artificial insemination and *in vitro* fertilization, are making it possible for infertile couples to have children. Surrogate motherhood, in which a fertile woman has a baby for an infertile couple, is rare, but increasing. These new forms of reproduction raise complicated legal and ethical issues that have yet to be resolved.

9. Women who are childless tend to be stigmatized in our society. Nevertheless, about 5 percent of American couples decide to remain voluntarily childless. Most decisions to avoid having children result from a series of decisions to delay childbearing, and women tend to be more adamant than their husbands about decisions to avoid or delay childbirth. Rates of childlessness are on the rise.

10. The birth experience involves more discomfort than some women anticipate, and without preparation, women are likely to request pain medication. Women who receive less medication generally consider the birth to be a positive experience. Unlike practices in the 1960s and 1970s, most hospitals now allow fathers into the delivery room.

11. The transition to parenthood differs from how it was in the 1960s because most mothers of infants are now in the paid labor force. In the early 1960s, first-time mothers who worked were poor. Now mothers from all social classes are in the labor force, especially if they are well-educated. Because most families now need two workers, women are less likely to give up their jobs when they have children.

12. Most young couples are unprepared for the transition in their lives that

comes with having a baby. Childbirth preparation classes focus on the birth, but cannot prepare new parents for the round-the-clock duties and interrupted sleep that infant care demands. The transition to parenthood is usually a stressful time during which marital satisfaction declines.

13. After the first birth, household divisions of labor and marital relations tend to become more traditional. This is primarily because women are socialized to become mothers, earn less money in the labor market, and become more dependent on their husbands after giving birth. Women "do gender" by caring for infants and men "do gender" by paying attention to other things. Nevertheless, men who share infant care develop the necessary skills and report interactions and relationships that are similar to those of mothers.

PART FOUR

FAMILY REALITIES

11

HOUSEWORK, POWER, AND MARITAL SATISFACTION

Sociologist Jesse Bernard (1972) suggested that every marital union actually contains two marriages: "his" and "hers." The extent to which "his" is different from "hers" depends on a lot of rather mundane matters, which add up to living in different worlds. Economic realities outside the household shape husbands' and wives' class positions. Inside the home, these same realities are translated into the mundane experiences of doing the housework, leaving it to one's spouse, or hiring someone else to do it. Practical decisions have to be made, and someone has the power to make them. Will it be the wife, the husband, or both equally? The answers will affect all sorts of things, including one's sex life, as well as "his" and "her" marital happiness.

WHO GETS MORE OUT OF MARRIAGE?

Who are more attached to marriage, men or women? Conventional wisdom says that women are. We have the popular image of men being dragged to the altar by women and then chafing under marital restraints and doing everything possible to escape. Like a lot of folklore, this image reveals a bit of truth but also a great deal of falsity. What seems to happen is this: men start out being less attracted to marriage than women, but once they are in it, they end up liking it more than their wives do. Why? The answer will tell us a great deal about the everyday reality of marriage, its advantages and disadvantages for men and women.

First, part of the popular image seems to be true: men seem to be more reluctant to marry than women. One piece of evidence concerns the ages at which they typically marry. Men have always married at an older age than women. In the last few decades, men have waited on the average of two to two-and-one-half years longer than women to marry. Back around 1900, they waited four years longer (U.S. Bureau of the Census 1992f). A more striking piece of evidence comes from surveys that simply ask people about their marital status. For the year 1992 one finds about 56 million married males and about 57 million married females (U.S. Bureau of the Census 1992f). In other words, almost 1 million more women than men *say* they are married. Yet men and women marry each other, so the number of married men and women must be equal. What is going on here? One possibility is that the missing million consists of married men in the armed forces whose wives are not with them (although figures on the armed forces indicate that this does not account for enough of the missing husbands). Another possibility could be bigamy, although it is hard to believe that almost a million men in the United States have two wives. Similarly, males may be harder to interview, especially in the lower classes. Or, a certain portion of the respondents of both sexes may be lying; unwed mothers may count themselves as married, while married men may be more likely to pass themselves off publicly as unmarried. But in general, what

TABLE 11.1   Marital Status and Happiness

	Percentage of men and women who describe themselves as "not too happy"	
	Men	Women
Married	5%	5%
Never married	13	8
Separated	17	28
Divorced	13	16
Widowed	24	14

Source: National Opinion Research Center, 1990, 1991.

appears to be the case is that women are more willing than men to identify themselves as married.

So far, then, we see some evidence for the popular belief that women favor marriage and men avoid it. But there is another side of the coin. Men have a harder time being single or divorced than women do. This shows up in reports of people's happiness (see table 11.1). It should be noted, incidentally, that people tend to put on a favorable public face when they answer surveys; when asked questions about how happy they are, they tend to describe themselves as relatively happier than one would find by examining the intimate details of their everyday lives. Virtually no one will admit to being "unhappy," so researchers have to use the phrase "not too happy" for the lowest level of happiness. In the population as a whole, for example, over 90 percent say that their marriage is "very happy" or "pretty happy". But given this tendency to shift all replies on happiness in an upward direction, there are nevertheless some *relative* differences that are quite revealing.

We notice first that, generally, *married people are happier than everyone else*. Almost half of both married men and married women say that they are "very happy" and only 5 percent admit to being "not too happy." Married men and women seem fairly evenly matched in these assessments.

However, the gender gap in level of happiness increases when people who have never married are polled. Men who have never married are relatively more unhappy (13 percent) than never-married women (8 percent); this time there is a five percentage point difference. In fact, women (but not men) who have remained single are the second happiest group, following married men and women. (Interestingly, men and women who never marry are happier than those who separate or divorce. Apparently, it is much better never to have tried marriage than to have tried and failed.)

Separation and divorce are fairly negative experiences for both sexes. But it appears that women, although they take the separation very hard (28 percent unhappy), rebound after divorce (down to 16 percent unhappy). Men, on the other hand, have moderate levels of unhappiness through both separation (17 percent) and divorce (13 percent). On the whole, then, it looks as if women have a harder time getting used to breaking up a marriage, but they are relatively happy once they are out of it. Widowed women report significantly less unhappiness (14 percent) than widowed men (24 percent).

Overall, although most people say they are fairly happy in their marriages, men are less happy than women when they are not married and when they are widowed. Women publicly report themselves to be quite happy with their marriages, but nevertheless they feel better than men when they stay unmarried or become widowed. Although both men and women are less happy when divorced than married, women who divorce are less likely to remarry than men, and wait longer to do so (see chapter 14).

Not only are married people happier than those who are single, divorced, or widowed, but also they appear to be healthier. Married people have substantially lower mortality rates for almost all causes of death than unmarried people of the same age and sex; they are less likely to die from suicide, tuberculosis, syphilis, pneumonia, strokes, heart disease, cancer, and even automobile accidents (Litwak and Messer 1989).

Although marriage is associated with more happiness and better health among the general population, there is mounting evidence that men benefit much more from being married than do women. We do know, for instance, that marriage has an adverse effect on women's mental health (see feature 11.6). Married women have much higher rates of mental illness than married men (Gove 1972). One out of four women has at least one episode of major depression in her lifetime, as compared to one out of ten men (Holden 1986). This is not because women are inherently or biologically more prone to be neurotic or psychotic, since they tend to have about the same (or sometimes lower) rates of mental illness as men *when they are not married*. Even though married people of both sexes have lower rates of mental illness than unmarried people, it has traditionally been true that marriage is more of a strain for the woman than it is for the man.

In addition, wives are more likely than husbands to report difficulties in the marital relationship, frustration, and contemplation of separation or divorce. Not only are wives more likely than husbands to be upset about finances, religion, sex, in-laws, and a lack of companionship, but they are also more likely to feel anxious and depressed and blame themselves for their troubles. Married women, especially if they are unemployed or have young children, report significantly more troubles than unmarried women. The opposite pattern is true for men. Married men have better physical health, better mental health, and report that they are happier than unmarried men.

Why is it that marriage seems to benefit men more than women? For an answer to that question we first turn to legal images of marriage and then move on to consider the potential influence of unequal distributions of housework and marital power.

MARRIAGE AND THE LAW

Although most people do not usually think of marriage as a legal and financial contract, the law defines what marriage is, how spouses should relate, and what obligations husbands and wives have to each other and their children. Many people become aware of these laws only when they get divorced. While marriage laws are currently undergoing change, it is instructive to review the historical legal bases of the current institution of marriage.

Western law is based on traditions embodied in ancient Roman statutes, including the definition of family as the property of the male head of household (see feature 2.5). The laws of most Western societies have traditionally defined women as inferior beings who must be protected by a man. The ceremonial custom of a father "giving" his daughter away in marriage reflects the legal inferiority of women. In most cases, women have lost legal rights when they marry, and women's obligations in marriage have always been rather severe. As late as 1850, almost all states recognized a husband's right to beat his wife if she did not fulfill her wifely duties.

Traditional marriage law merged the identities of husband and wife. According to the feudal doctrine of coverture, the husband and wife became a unity at the time of marriage, and that unity was the husband. Symbolic loss of identity is still evident today in the custom of a married woman legally adopting her husband's name when she marries: Miss Jane Smith marries Mr. John Jones and becomes Jane Jones or Mrs. John Jones, while the man's legal identity remains exactly the same as it was before marriage. Some alternative customs are more common, but none is as yet standard. A married woman may keep her own name, or she may join her last name with that of her husband with a hyphen (e.g., Smith-Jones). Occasionally, a husband may use a hyphenated name too.

In the past, marriage laws also required a wife to take her husband's legal domicile (place of residence) and obligated the wife to provide various services for her husband. LenoreWeitzman (1981) discusses four essential provisions of the marriage contract that were incorporated into our laws:

1. The husband is the head of the household.
2. The husband is responsible for financial support.
3. The wife is responsible for domestic services.
4. The wife is responsible for child care.

While we might think of these provisions as outdated notions of the past, our courts were still relying on them in the 1970s, and their social ramifications remain with us today. As Weitzman shows in her review of court cases from the 1960s and 1970s, even liberal states tended to treat wives and their earnings as the property of their husbands. For example, the courts refused to enforce contracts in which the husband agreed to pay his wife for housekeeping, entertaining, child care, or other "wifely" duties. Even if the wife performed services considered to be "extra," such as working in the husband's business, the courts voided the contract that obligated the husband to pay her (Weitzman 1981:1189) (see feature 11.1). Today, most states have marriage laws that treat husbands and wives the same, though in some states married women still lose some legal rights to control property or enter into legal contracts on the same basis as men or unmarried women.

HOME AS A WORKPLACE

People usually marry out of an emotional attraction: love, sex, romance, companionship, and other idealized experiences that a couple look forward to. Once into the marriage, though, they experience the hard material realities of family life. The home, after all, turns out to be an economic unit, a place where work has to be done every day just to keep things going. Every home is a combination of hotel, restaurant, laundry, and often a child-care and entertainment center. Each of these activities takes work, work that is often invisible when one is merely the recipient of these services (as children often are, just like customers in a hotel or restaurant). But the work becomes all too real when the young married couple find out they must take care of it themselves, or it will not get done at all. Even the pleasant parts of family life take behind-the-scenes work. Having a party can be fun for the guests but a chore for the wife (or possibly husband) who has to prepare everything, provide service while it is going on, and clean up afterward.

It has often been noted that love and romance tend to decline after the wedding (Safilios-Rothschild 1977, 19). The reason is not simply the disappearance of the ritual conditions that made premarital love an emotional experience, but also the rather rude shock of confronting the plain hard work that makes up an ongoing household. Moreover, the work is not merely mundane and demanding, it is also stratified. It is one of the main arenas in which men and women have different powers and payoffs, and one of the activities around which latent struggle, and sometimes overt conflict, is likely to happen.

Housework: Who Does What?

Housework tends to be divided into men's and women's spheres. Women usually do most of the indoor work, especially cooking, laundry, cleaning house, doing dishes, and caring for small children. Men usually do most of the outdoor work (cutting grass, shoveling snow) as well as indoor and outdoor repairs. Both sexes frequently take part in shopping and caring for older children (Vanek 1974; Berk 1985; Thompson and Walker 1989). Sometimes husbands will help with some of their wives' tasks, although the spheres remain clearly

FEATURE 11.1
Inequities in Past Marriage Laws

In *The Marriage Contract* (1981), Lenore Weitzman presents some examples of the ways that laws regulating marriage gave a husband control over his wife's property. These summaries come from actual legal cases.

Husband's Control of Wife's Earnings

Florence and Samuel Jorgenson, like most people married in the forties, chose a traditional relationship: Samuel brought home the paycheck, while Florence remained in the home to care for their sons, Bill and Ed. Each week, Samuel turned his paycheck over to Florence; she did the family's bookkeeping, paid the bills, and rationed out the money for household and personal expenses.

The Jorgensons didn't have much money. For most of their married life, they lived with Mrs. Jorgenson's mother, not even able to afford a home of their own. But Florence believed that her boys should go to college, and she managed to save a little here and there for their education. She economized in all the little ways in which a housewife can economize. Occasionally, her mother made a donation to the college funds. Slowly but surely, two savings accounts grew.

When Florence Jorgenson died in 1963, she left her teenage sons two savings accounts: $2,000 in trust for Ed and another $3,000 in trust for Bill.

Samuel Jorgenson remarried and moved out of his mother-in-law's house. $5,000 seemed like a lot of money to him. He wanted it, and he went to court to get it. "I earned that money," he argued, "and I never gave it away. I merely entrusted it to Florence to pay the bills. The surplus belongs to me." The court agreed—despite the fact that there would have been no surplus without Florence's careful economizing, despite the fact that Florence's mother had contributed some of the funds. "The general rule in separate property states," said the court, ". . . is that the

excess left after paying the joint expenses of . . . the family remains the property of the husband." The boys' college money went to Samuel and his new wife. Florence's long years of scrimping and saving were an exercise in futility.

Husband as "Head and Master"

Married at eighteen, Helen Tarbel worked double shifts as a nurse to support her husband through four years of college. Her salary paid for all household expenses, and she saved enough to buy a small home. Her husband was unemployed for most of the year after he graduated from college, and Helen continued to support the family. When her husband proposed that they mortgage the house to borrow $5,000, Helen objected. But under the Louisiana "head and master" rule her husband had the full power to mortgage the home she bought—and to control all of their community property—without even asking her. So he ignored Helen's objection, and without her signature or consent, mortgaged the home to take out a loan. When he failed to repay the loan the credit company sought to foreclose the mortgage and take their home. Helen Tarbel objected and challenged the constitutionality of the head and master rule. It was unfair, she said, for her husband to have the power to mortgage the home without her knowledge or permission. And it was unfair that he should control every cent she earned and every item she bought. But the Louisiana Supreme Court denied her appeal, and the U.S. Supreme Court refused to hear the case, thus leaving her creditors to foreclose the mortgage and letting the head and master rule stand.

Husband's Rights after Wife's Death

Fred and Betty Nelson were married in 1950. They lived on a farm in Illinois that Fred had

(Continued next page)

FEATURE 11.1 (*Continued*)

inherited before they were married. At first Betty was a housewife, caring for their children, performing the usual household tasks—gardening, preserving large quantities of food, and cooking five or six daily meals for the hired hands who worked on the farm. Later, when the children were in school, Betty took on an outside job. She continued to do the traditional tasks, and in addition contributed part of her income for family expenses.

After twenty-two years of marriage, Fred divorced Betty. The court awarded her no alimony. Her share of their marital property consisted of only her own clothing and personal effects, a few household items that she owned prior to the marriage, and an automobile which she had purchased in her own name with her own funds. The house and furnishings, the farm with its machinery and livestock, the savings—all went to Fred.

Betty appealed the decision. "Surely, after twenty-two years of hard work, I am entitled to at least a portion of the assets I helped to accumulate," she thought. But the legal system saw things differently. The appellate court upheld the lower court's division of the property.

A spouse seeking part of the other spouse's property, explained the court, must show that she or he made valuable contributions to the property's worth. The court defined a valuable contribution as "money or services other than those normally performed in the marriage relation"; Betty's years of cooking, cleaning, and child rearing did not meet the court's definition of "valuable."

But the money Betty contributed for family expenses from her outside employment was "valuable"; would she not be entitled to some recompense for this financial contribution? No, said the court; she did not keep clear records of what property was acquired through her own effort, and consequently all property was properly awarded to her husband.

identified by gender. Men do not often cook in the home, although they usually take charge of *outdoor* cooking, such as running the barbecue at a picnic. (Similarly, most restaurant cooks are men, especially the head cooks in expensive restaurants, whereas at home it is taken for granted that "women know how to cook and men don't.") Conversely, home repairs are typed as jobs that men know how to do, and hence women are less likely to take up a hammer or screwdriver even for simple repairs.

The few jobs that are related to women's housework but not strongly identified as such are outside tasks that involve a fair amount of discretion (shopping for groceries) or tasks that tend to involve power (looking after older children). Both of these are offshoots of "feminine" spheres (cooking, child care), but are removed far enough from the indoor or the mundane (such as dressing small children, which men rarely do) that men will readily do them. The difference between men's and women's tasks is frequently a matter of status; women's tasks are felt to be especially "feminine," although they also include chores that are dirty and unpleasant, such as cleaning toilets and floors. Some wives explicitly admit that they would feel uneasy about asking their husband to clean the toilets because it would be too degrading for him. The one "dirty" job that is usually considered a man's task is taking out the garbage.

Perhaps this is part of the man's sphere because it involves going outside, which is typically his part of the home, whereas her sphere is inside.

To determine how much men and women contribute to household labor, researchers often include survey questions about who does what around the house. Some surveys ask who is more likely to perform a particular task, others ask who did particular tasks "last week," and others ask people to keep time-budget diaries, recording what they do repeatedly throughout a given day. Depending on the method and sample used, researchers arrive at different estimates of the absolute and relative amounts of time men and women spend cooking, cleaning, shopping, doing home repairs, and so on. Nevertheless, all studies find that wives do two to three times more of the total household labor than husbands do—and, for the most repetitive chores such as cooking and housekeeping, women are found to do four or five times more than their husbands (Berk 1985; Coltrane and Ishii-Kuntz 1992; Pleck 1983; Thompson and Walker 1989).

In spite of the fact that more and more women became employed during the 1970s, the pattern of women doing almost all of the housework persisted. The rigid division of housework based on gender appeared so resistant to change in the early 1980s that one review of research called studies of change in men's housework "much ado about nothing" (Miller and Garrison 1982, 242). More recent time-budget data suggest that change is underway, but that women still tend to do most of the work it takes to run a household.

FEATURE 11.2
The Disappearing Servant

At one time, most households had servants. A large household might have a butler, valets, footmen, and other male servants, but the majority of house servants were usually female. The number of servants has been on the decline in Western countries like Great Britain and the United States since the 1800s. As late as 1900, household servant was still the largest category of workers in Great Britain (Laslett 1977, 35). In 1850, every middle- and upper-class household in the United States had at least one servant (*Historical Statistics of the United States 1965*), usually a maid or housekeeper. Today, there are only 82,000 old-fashioned live-in household servants for the 93 million households in the United States: 1 out of 1,134 households has one. About 1 out of 90 households (about 1 percent) has a cleaning woman or childcare person who comes in regularly (*Statistical Abstract of the United States 1992*). This is an amazingly low figure when one considers that the affluent upper and upper-middle classes comprise about 15 percent of all households. It has been remarked that the upper class has not disappeared but makes less splash than it used to, because the servant class that used to surround it really *has* disappeared. The servants that are left tend to be recent immigrants or black. They are among the most poorly paid of all workers. Being a household servant for someone else is a very low-status job in America today; hence only poverty (and/or migration from a foreign country, especially a very poor one) motivates women to do it.

How Much Has Changed?

In 1985, the Survey Research Center of the University of Maryland collected comprehensive data on how Americans use time. By comparing findings of similar surveys from 1975 and 1965, Robinson (1988) concluded that men are doing more household labor than before. Table 11.2 shows that in 1965, women did 85 percent of total household labor (excluding child care) and 92 percent of housework. By 1985, women were doing 67 percent of total household labor and 80 percent of the housework. From 1965 to 1975, women cut the amount of time they spent on household labor from twenty-seven to twenty-two hours per week, primarily by reducing the time they devoted to housework. During the same decade, men increased the amount of time they spent on household labor from five to seven hours per week, primarily by doing more traditional "manly" activities such as home repairs and lawn care (see table 11.2). Other researchers report similar findings based on data collected in the early to mid-1970s, with women reducing their total amount of housework and men doing about the same amount (Coverman and Sheley 1986; Pleck

TABLE 11.2   Who Does the Housework?

| | Time men and women aged 18 to 65 spent doing household tasks in hours per week | | | |
| | 1965 | | 1985 | |
	Women	Men	Women	Men
Housework				
Cooking Meals	8.4	1.3	6.9	2.0
Meal Clean-up	4.1	0.3	1.9	0.4
Housecleaning	6.7	0.4	5.1	1.4
Laundry, ironing	5.1	0.1	2.2	0.3
Subtotal	24.3	2.1	16.1	4.1
Other Tasks				
Outdoor Chores	0.2	0.5	0.5	1.4
Repairs, etc.	0.2	1.0	0.4	1.8
Garden, animal care	0.5	0.2	0.9	1.0
Bills, other	1.8	0.8	1.6	1.7
Subtotal	2.7	2.5	3.4	5.9
Total Household Tasks	27.0	4.6	19.5	10.0

Women spent 7.5 fewer hours a week doing housework in 1985 than in 1965, while men spent 5.4 hours more.

Source: Adapted from John P. Robinson, "Who's Doing the Housework," *American Demographics*, 10, no. 12 (1988): 26.

1983). Thus men appeared to assume a slightly greater share of the total household labor, but only because their wives were doing less.

Between 1975 and 1985 the pattern began to change as men increased their contributions to cooking, cleaning, and other domestic tasks. During this decade, women continued to reduce the amount of time devoted to household labor, so that by 1985 they were spending an average of about twenty hours per week on all tasks. In 1965 men did about five of thirty-two hours of household labor per week, or 15 percent of the total. By 1985, they were contributing about ten hours per week, or about a third of total household labor. Still, women were doing sixteen hours of housework per week, four times as much as the men (Robinson 1988).

In 1965, getting married meant that women would double the amount of time they spent on household labor, whereas men would do about the same amount as before marriage. Today, married men do about two more hours per week of household labor than unmarried men, though most of that time is spent on traditional men's tasks rather than on housecleaning or cooking. Married women still do over seven more hours of household labor than unmarried women, and most of that difference comes from doing more cooking and cleaning (Robinson 1988). Many comment that the presence of a man in the household contributes more to the need for housework than to its completion (Hartmann 1981). What we see, then, is that while married men are doing a little more around the house, when most women get married, the amount of time they devote to housework goes up, and they still end up doing over 80 percent of the housework.

Although single and childless women do more housework than married men, it is women with children who have traditionally been burdened with the most domestic responsibilities. In 1965, women with children did about ten hours a week more housework (excluding direct child care) than women without children. By 1985, that difference had dropped to about four hours per week. Since 1965, men with children have doubled the amount of time they spend doing housework, from about four to five hours per week to nine to ten hours per week (Robinson 1988). So, while women and mothers continue to do much more than men, there is some evidence that the gender gap in housework is beginning to narrow, especially for parents of young children. A few studies have even found that the total number of hours spent on all paid and unpaid labor (not including child care) is now about equal between husbands and wives (Ferree 1991b; Robinson 1988). In general, American women are likely to spend fewer hours than men on the job, and American men are likely to put in fewer hours than women on household labor, but we may be seeing a convergence in the total number of hours spent doing both.

Not only do women spend more hours performing household labor than men, but the nature of their involvement tends to differ as well. As noted above, women's household work tends to be unrelenting, repetitive, and routine. Tasks such as shopping, cooking, cleaning, laundry, and child care are repeated over and over in a never-ending cycle. The household tasks that men often do, in contrast, are infrequent and irregular. Activities like household repairs, mowing the lawn, or taking out the trash may need to be repeated every so often, but the time between repetitions is much longer than for typical

There is some evidence that the gender gap in housework is beginning to narrow, especially for parents of young children.

"wifely" tasks. Thus some researchers focus on the never-ending aspect of women's household labor, which contrasts sharply with the experience of men's barbecuing on the weekend or occasionally fixing a toaster (Berk 1985; Pleck 1983; Thompson and Walker 1989).

Feelings about Housework

How do women feel about housework? Most women are dissatisfied with it. Oakley (1974, 182) found that 75 percent of her sample of British women thought housework was monotonous and dissatisfying, and they frequently complained of being isolated and lonely working at home.

Both women and men experience boredom, fatigue, and tension when they do household work alone (Baruch and Barnett 1986), but women are much more likely than men to perform household tasks in isolation. Whereas men tend to do much of their family work on weekends, women tend to do housework each day. Women also report doing an average of three household tasks at one time, which may help explain why they find household labor to be less relaxing and more stressful than men do (Hochschild 1989; Shaw 1988; Thompson and Walker 1989).

Which tasks are most liked and disliked? The hierarchy from most favorite to least favorite work runs as follows: cooking, shopping, washing, cleaning house, doing dishes, and ironing. Cooking is considered the most enjoyable task, while ironing is the least enjoyable. When men contribute, they tend to avoid the unpleasant tasks (such as cleaning the toilets). When they pitch in, they are more likely to do child care than housework and more likely to watch or play with the children than to feed them or clean up after them. When they do housework, they are most likely to cook or wash dishes, followed by vacuuming or tidying up. Men are least likely to mop, clean bathrooms, and do laundry.

Women's dislike of housework does not necessarily mean, though, that they reject the role of being a housewife. Although many of the women in Oakley's British study complained that there was little prestige in being "just a housewife," they usually identified themselves strongly with being a mother and a woman in charge of her home.

The same is true of both working-class and middle-class wives in the United States, whether they are employed outside the home or not. Most women experience repetitive domestic chores as boring but important work that is performed for people they love. Even if they do not find the activities themselves enjoyable, most wives and mothers enjoy feeding and taking care of their families and derive considerable self-worth from meeting family members' needs. Since household labor is tied up with what it means to love and care for others, women often have ambivalent and contradictory feelings about it, and report mixed reactions to being responsible for so many of the routine household chores (DeVault 1987; Ferree 1987; Thompson and Walker 1989).

Women as Household Managers

Besides doing more of the actual chores, women also typically act as household managers. That is, they assume responsibility for planning and initiating the vast majority of household chores. Even in families where much of the child care and housework is shared, wives are much more likely than husbands to notice when a chore needs doing and to make sure that someone does it correctly. In most families, husbands notice less about what needs to be done, wait to be asked to do various chores, and require explicit directions if they are to complete the tasks successfully. In line with this division of the responsibility for management of household affairs, most couples characterize husbands' contributions to housework or child care as "helping" their wives (Coltrane 1989).

In general, then, women are more likely to carry the burden of managing the household as well as doing most of the tasks, and tend to worry more about the planning and scheduling of various activities. Women usually plan meals so that the family can eat together, and they typically try to orchestrate the evening meal so that it is a calm and pleasant event (DeVault 1987; Thompson and Walker 1989). Women generally try to keep conversations going and ensure that everyone gets a chance to talk (Fishman 1978). For child-related tasks, women also usually take responsibility for most of the less pleasant activities and attempt to facilitate positive father-child interactions. Women in two-parent families almost invariably take responsibility for organizing, delegating, and scheduling children's activities (Ehrensaft 1987; Berk 1985).

Sharing the Strain

Arlie Hochschild (1989) interviewed and observed fifty-two couples over an eight-year period to see how they managed the complex relationships between paid and unpaid work. She reports that women are "far more deeply torn" between the demands of work and family than their husbands are. She describes how most women work one shift for pay and a "second shift" when they get off work and come home to take care of house and family.

Since they remain responsible for managing the household and performing the more onerous household tasks, Hochschild found that women tended to talk more intensely about being overtired, sick, and emotionally drained. She labeled the common situation of men's favored position in the household division of labor as the "leisure gap" since most men had more free time than their wives. In general, men did not consider the second shift to be "their issue."

Nevertheless, about 20 percent of the men that Hochschild interviewed shared housework equally with their wives. She reports that the men who did half of the domestic chores seemed to be just as pressed for time as their wives (Hochschild 1989). Berk (1985) reports that about 10 percent of the men in her

FEATURE 11.3
The Paradox of Labor-Saving Appliances

One hundred years ago, the work of running a household was much more arduous than it is now. Many houses lacked hot and cold running water; water had to be hauled in from a well or street source and heated on the stove. Carrying water was traditional women's work and had been for thousands of years. To tend stoves and fireplaces, wood or coal needed to be brought in and ashes hauled away. Cooking could be very time consuming under these circumstances, and laundry was a major chore. Even as simple a matter as keeping a house lit at night required filling oil lamps and regularly cleaning off the accumulated smut.

A series of inventions transformed all that. Electricity came early in this century, although many rural homes did not have it, or running water, until the 1940s. Refrigerators meant that food did not have to be bought or prepared every day. Gas and electric stoves, washing machines, dryers, vacuum cleaners, and dishwashers have revolutionized housework. Or so it might seem.

A very surprising finding, though, is that *the total number of hours that American housewives spend on housework has not declined at all since the 1920s.* Joann Vanek (1974) examined studies done on home economics since 1924 and found that full-time housewives were working about fifty-two hours a week in 1926, a figure that *rose* to about fifty-five hours a week in the 1960s (see figure 11.1).

How can this be so? Vanek suggests that women's standards of home comfort have risen as new devices have become available. For instance, before automatic washers existed, doing laundry was

(Continued next page)

FIGURE 11.1: Distribution of Time among Various Kinds of Household Work, 1926–1968

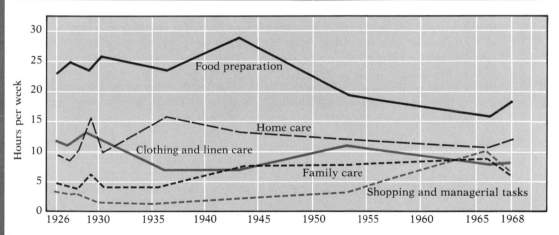

The data relate only to nonemployed women, meaning women who did not have full-time jobs outside the household. Top curve includes cleaning up after meals.

Source: From "Time Spent in Housework" by Joann Vanek, 1974. *Scientific American*, 231 (Nov.), pp. 116–20. Copyright © 1974 by Scientific American, Inc. All rights reserved.

FEATURE 11.3 *(Continued)*

FIGURE 11.2: Time Devoted to Laundry

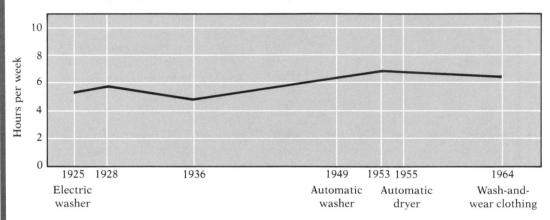

Laundering time has actually increased over the past fifty years, apparently because people have more clothes now and wash them more often. The dates shown for the various appliances and fabrics indicate about when they began to be sold widely.

very arduous. Usually, it was done only once a week. Now laundry is done much more frequently; it takes less time to do each load, but housewives now run a load almost every day (see figure 11.2). A result is that expectations for cleanliness have increased; clothing is changed more frequently.

In addition, women have shifted their housework time from certain tasks to others. Less time now is spent on preparing food and cleaning up after meals, due to all the advantages of labor-saving devices in the kitchen, although these continue to be the most time-consuming housework tasks. Cleaning is easier now, but housewives seem to have escalated their standards of how clean their homes should be. Whatever additional time has been made available has been taken up by other tasks. *Longer* hours are spent on caring for children— even though today's families have fewer of them than families of fifty years ago. And more time is spent on shopping and general household management: in the 1920s women averaged two hours a week going to and from stores, while today they spend a full day shopping every week.

Newer labor-saving devices have appeared in recent years, and still more are touted for the future. One can buy an electronic system that automatically turns on and off the sprinkler system or the lights, and home computers and robots are apparently about to become very common. If the lesson of the past tells us anything, though, it is that we can expect housework hours to decrease very little, if at all. Other studies (Cowan 1983) have confirmed that the amount of housework women do has not fallen since colonial days. There seems to be a status competition over home lifestyles, which will go on raising standards endlessly.

sample also spent as much time as their wives on household chores. Coltrane (1990) selected couples on the basis of self-reported sharing of child care and discovered that highly involved fathers experienced stresses and strains similar to those reported by employed mothers.

Justifying the Division

Since most couples do not share very many household tasks, how do they justify the woman's obligation to perform the second shift? From surveys, interviews, and time-use studies, we know that most couples label asymmetric divisions of household labor, with women doing most of the work, as "fair." Berk (1985) found that only 21 percent of the wives she interviewed felt that they should be doing less housework in relation to how much their husbands were doing. Only 6 percent of their husbands, on the other hand, thought the division of housework was unfair. Pleck (1983) reports that only a minority of American men and women believe that, in general, men should do more housework and child care than they are now doing. However, about half of the men reported that husbands should do more around the house when their wives are employed. Pleck (1985) also reports that most mothers want their husbands to spend more time with their children, although this does not necessarily mean that they want the father to do more of the routine domestic chores associated with child care. In general, surveys show that women have relatively low expectations for assistance from other family members, with only one-fifth to one-third reporting that current divisions of family labor are "unfair" (Barnett and Baruch 1987; Berk 1985; Pleck 1983; Rosen 1987; Thompson and Walker 1989).

We can understand some of these findings on the division of household labor and couples' willingness to accept what look like extremely unequal arrangements in light of the concept of "doing gender" mentioned in the last chapter. West and Zimmerman (1987) point out that to be classified as a man or woman, and to be judged a competent member of society, everyone must "do gender." Doing gender consists of interacting with others in such a way that people will perceive one's actions as expressions of an underlying masculine or feminine "nature." Thus one is not automatically classified as a man or a woman on the basis of biological sex, but on the basis of appearance and behavior in everyday social interaction.

Household labor offers people a prime opportunity to "do gender" because of our cultural prescriptions about the appropriateness of men and women performing certain chores. Doing household chores allows people to reaffirm their gendered relation to the work and to the world (Berk 1985). Thus, women can create and sustain their identities as women through cooking and cleaning house and men can sustain their identities as men by *not* cooking and *not* cleaning house. With peoples' sense of self so tied up with doing separate gender-linked activities, it is no wonder that people perceive their household arrangements as fair.

Social Class Differences

There is also a fairly strong social-class division in women's attitudes toward the housewife role. Both middle-class and working-class women have about the same negative feelings toward house*work*, but working-class women have been much more likely to identify strongly with the housewife *role*. Oakley (1974) found that the women who had held what were then considered relatively high-status jobs before marriage (e.g., computer programmer, fashion model, even manicurist) were most dissatisfied with their current housekeeping tasks. Similarly, Helena Lopata (1971) found a strong difference between highly educated and less-educated housewives in whether they felt their task was primarily to keep the home spick-and-span (the working-class pattern) or to emphasize their social role in relation to their family and the outside community (the middle-class pattern).

Many of the working-class women, in particular, had organized their lives to maximize their own satisfaction in their housework. They did this, interestingly enough, by making their work more extensive. Oakley (1974) found that many housewives had worked out a highly specific routine of work to be done each day (or each week) and took pride in being able to live up to their routine. The more tightly specified the routine, the longer hours these women worked: in the British sample, from ten to fifteen hours every day of the week. They set very high standards of meticulousness in their homes and took satisfaction from being able to meet their own standards. A similar pattern has been found among American working-class housewives (Rainwater, Coleman, and Handel 1962). This shows us, as well, how fully employed women are able to keep up with their housework: having other sources of status and subjective satisfaction, they treat their housekeeping in a more utilitarian manner. One might also say, following the theory of class cultures, that working-class housewives are more likely to make a ritual out of housework, as a way of giving meaning and value to the confined circumstances of their lives.

Class differences in household labor can also be analyzed with reference to the family's embeddedness in social networks. As we saw in chapter 6, working-class families tend to have dense social networks that are segregated by gender. For both work and leisure, women spend time with women, and men hang out with other men. Working-class women often live near other female kin and fictive kin who regularly provide them with companionship, advice, substitute child care, and material support (Bott 1957; Lein 1974; Stack, 1974). Working-class men spend most of their time with male coworkers who use joking and ridicule to discourage each other from performing cooking, cleaning, and other "women's work" (Coltrane 1991; Lein et al. 1974). In addition, the convenient availability of female kin to help with shopping, cooking, and child care, coupled with an ideology of male dominance, reduces the demands on working-class husbands to perform these tasks.

Middle-class families, in contrast, tend to be more mobile and to have more individualistic social networks composed of more friends than kin. The loose-knit networks of middle-class couples are also less likely to be segregated by gender. As a result, some researchers claim that middle-class families are more likely to share housework and child care than are working-class families. Husbands and wives with more years of education are found to share more domestic chores than those with fewer years of education (Eriksen, Yancey, and Eriksen 1979; Huber and Spitze 1983; Ross 1987). Nevertheless, other researchers find that less-educated working-class husbands actually perform as much housework as middle-class husbands (Beer 1983; Hartmann 1981). The difference may boil down to differing vocabularies and ideologies: Better-educated husbands don't do much housework, but accept the idea of sharing and therefore report that they help out. Less-educated husbands reject the idea that housework is their job and report doing little of it, but end up doing domestic tasks out of necessity.

The Struggle over Sharing Housework

What happens when men do share housework and child care with their wives? Most studies show that wives whose husbands share more household labor typically are more satisfied with marriage than other wives (Staines and Libby 1986). Wives who want help from their husbands and receive it are less depressed than other wives (Ross, Mirowsky, and Huber 1983). Husbands' willingness to "help out" with family work has "symbolic meaning" for wives that enhances their psychological well-being (Kessler and McRae 1982). In a study of working-class couples, Rosen (1987) found that wives viewed their husbands' household contributions as expressions of love and concern. Some middle-class couples who share housework and child care also report that sharing the domestic labor puts them on an equal footing and enhances their marital interaction (Coltrane 1989; 1990).

Other studies discover that when women demand help from their husbands, it puts added strains on the marital relationship (Barnett and Baruch 1987; Crouter et al. 1987). Many wives have difficulty pushing for an equitable division of labor if it means fighting with loved ones. Some find that sharing child care and housework is accompanied by increased marital conflict, although those involved do not always define this in negative terms (Russell and Radin 1983; Kimball 1988). Most wives who attempt to share domestic chores complain about their husbands' lower standards for child care and housework (Lamb, Pleck, and Levine 1986), and some report that equitable divisions of household labor are dependent upon establishing congruent standards

(Coltrane 1990; Ferree 1987; 1991b). Wives accuse men of acting inept to avoid housework and complain that husbands must be trained to notice important details. Some suggest that when wives criticize the quality of their husbands' domestic work they are protecting threatened territory and maintaining control over the only domain in which they have power (LaRossa and LaRossa 1981; Polatnick 1973; Thompson and Walker 1989). Blumstein and Schwartz (1983, 118–54) found a widespread belief that the more successful partner should not have to do housework. Interestingly enough, this was found not only among conventional male-female couples, but also among gay male couples. Doing the housework, then, appears to be rooted in power relationships, and not just in male/female roles.

FAMILY POWER

One reason that men seem unhappier being unmarried than women, then, is that they get more out of marriage. For men, being married has generally meant receiving household services, whereas for women it has meant giving services.

There is a good deal of evidence that men usually have more power in marriages than women do. Let us be careful of what this means. The term *power* might seem to imply that there is a struggle over who gets to dominate whom, and that men generally give the orders while their wives meekly obey. That, in fact, is a description of what traditional marriages once looked like, but it does not describe most people's experience today. On the surface, most marriages now seem to have very little to do with power. No one gives orders (except, sometimes, parents to their children). If interviewers ask, "Who has the power in this marriage?" typically, husbands and wives say that it is equally shared.

Nevertheless, this kind of answer is usually superficial. We can tell that because more detailed studies, done when the researcher gets to know the respondents better, find there are real differences in who gets what inside a marriage. For example, power has often been studied in terms of who has how much influence over family decisions.

In one famous study that has been repeated many times, Blood and Wolfe (1960) asked Detroit area housewives about decision making for buying a car, purchasing life insurance, taking a new job, or going on a vacation (see feature 11.4). Based on answers to these questions, they concluded that "the American family has changed its authority pattern from one of patriarchal male dominance to one of equalitarian sharing" (Blood and Wolfe 1960, 47). This optimistic conclusion has been challenged by many family scholars (Cromwell and Olson 1975; Safilios-Rothschild 1970; Scanzoni 1979; Szinovacz 1987). Not only do people give superficial answers to such questions, but the scoring procedure

FEATURE 11.4
Who Has the Power?

Think about a relationship you have been in, or about your parents, or some other couple you know. Who usually makes the final decision about the following items; the husband or the wife?

1. What car to buy?
2. Where to go on vacation?
3. What house or apartment to take?
4. Whether or not to buy some life insurance?
5. What job the husband should take?
6. Whether or not the wife should go to work or quit work?

7. What doctor to have when someone is sick?
8. How much money the family can afford to spend per week on food?

In the 1950s, researchers asked housewives these questions about family decision making. How well do you think these questions measure "Family Power"?

Source: Adapted from Robert O. Blood and Donald M. Wolfe, *Husbands and Wives* (New York: Free Press, 1960).

treats all decisions as equal and assumes that people's answers always express power relations.

For major decisions, such as whether to move to a new house or a new city, or to buy a car or appliance, men typically dominate, making the decision themselves, or at least exercising veto power over a wife's decision (Rubin 1976, 110; Ostrander 1984, 51). The wife of a top executive, for example, recalled: "He wanted to move to the country, and I didn't. So we moved to the country." Another woman in this social class said: "When he's worked hard all day and calls up and says let's get a tennis game together, I do it. He'd get upset if I didn't. . . . If he calls me in the middle of the morning and wants to have a party the next day, I've got to do it" (Ostrander 1984, 37, 45). These women, however, have particularly powerful husbands, and their social lives are an adjunct to business. These couples do not fight over power; but they are maintaining a certain lifestyle in which the husband takes the initiative in what they do.

Life is less formal in the middle and working classes, and correspondingly, power differences may be less blatant. Moreover, there is the strong ideal of love between married partners, which in today's world, at least, has replaced the old-fashioned deference to the family patriarch. Especially in the minor decisions of the daily running of the household, women typically have great autonomy. But there are subtle aspects of power that affect even the events of everyday life. For instance, as we've already seen, women do a good deal more housework than men. Even counting the different tasks that men do around the home (especially outside), women put in many more hours than men; and they do it even when they have jobs. Even when children are given household chores, girls work twice as many hours as boys—as if they were being groomed to fit the same gender pattern of housework in the next generation. Usually, not very much thought is given to these arrangements. They are just taken for granted. But they are a subtle form of power inside the household. Husbands and wives may love and respect each other, but at the same time there is another set of forces operating beneath the surface of their awareness—and these determine how much power they have at home, whether they think about it or not.

Occupation

It is generally true that the higher the husband's occupational level, the more likely he is to dominate at home (Blood and Wolfe 1960; Gillespie 1971). There is an especially sharp increase in power at the white-collar level, with middle-class husbands having much more power than husbands who are skilled manual workers. The pattern is somewhat surprising, since middle-class husbands are more likely to *say* that they believe in equality between husbands and wives. But resources are what count, not sentiments, and these men's occupations give them the resources to get their own way. While new career opportunities for women that have developed out of the feminist movement have challenged this state of affairs, women's occupational levels, and therefore, power, are still generally lower than those of men.

What seems to happen is that middle-class men are less likely to demand that their wives defer to their wishes *as men*, but instead they ask them to defer *because of the importance of their jobs*. The ideology is different, but the outcome is even more biased against women than with an outright sexist claim to male domination. Working-class men, on the other hand, are more likely to speak overtly about male prerogatives, but since their occupation brings them less prestige and other resources, they are less able to translate their demands into realities. This is especially so when their resources do not outweigh their wives'.

In the lower part of the working class, though, the trend reverses again. Unskilled workers have relatively more power over their wives than the skilled workers of the upper working class. Since these men have even fewer resources, their power must have some other basis; there are strong indications that it involves a greater willingness to use force (Rubin 1976; Szinovacz 1987).

Income

Income is even more important for marital power than occupation. The more income a man brings in, the more he gets his way at home. Again, this must be balanced against the wife's income. Ironically, this tends to make the gap between the husband's and wife's powers even wider in the upper-middle-class families than in working-class families. Since working women of all social classes tend to be in clerical and other low-paying jobs, upper-middle-class men tend to earn a much higher multiple of their wives' income than working-class men. This tends to make upper-middle-class men especially powerful at home.

In recent years, the proportion of married women who work outside the home has increased to about 60 percent, while the proportion of men has actually fallen slightly. The traditional barriers to women in high-paying jobs, such as business executives and the more lucrative professions, have come down, at least to the extent of allowing some women into these occupations. These trends, one might expect, would bring a shift in the domestic balance of power. Women's greater incomes should be translating into at least greater equality of power at home. We have no good data available on whether power in the household has really become more egalitarian since the 1950s. But the figures on the incomes of men and women reveal an unpleasant surprise: there has been little change in the average amount that women get paid in relation to men. To be exact, in 1960, women were making 60 percent of what men were being paid in terms of hourly incomes; in 1990, women's hourly pay had risen to 71 percent (see figure 1.3). But these figures are for full-time year-round work and overstate women's gains relative to men, because women are much more likely than men to be unemployed, to work part-time, and to be employed seasonally.

A number of reasons have been advanced for the failure of women to achieve economic parity with men. One prominent argument has been that women are less committed to their careers, because of their family responsibilities, moving to follow their husbands' jobs, or quitting to take care of their children. Nevertheless, these explanations do not hold up well against evidence that women with children do hold down jobs as consistently as men, and that family responsibilities do not affect the pattern of job movement (Reskin 1984). The major explanation of women's lower wages remains that the job market is heavily segregated by gender (Reskin 1984; Michael, Hartmann, and O'Farrell 1989; England and Browne 1992). Even though some women have crossed the barrier into high-paying "men's jobs," they are not enough to offset the general pattern and the lack of promotion to higher level positions—or "glass ceiling" —limits women's upward mobility.

Changing income and earnings potential set the scene for the shifts in the positions of men and women that have taken place in the last few decades—at exactly the time of the rise of the women's movement, the declining birthrate, and the rising divorce rate. Women are still disadvantaged in the marketplace, but they have left the old pattern in which only a minority of women worked for a new pattern in which work is the most typical female lifestyle. Women still have a long way to go to overcome economic discrimination, but now a majority of women are confronting it directly in their own careers, whereas previously it was more typical for them not to make the attempt. In this sense, women have become economically more mobilized. Though they are discriminated against, they have more money simply because of doing more work; this has probably given them at least some leverage in subtle family maneuvers over power. Judging by the figures on how much housework wives and husbands do, women's economic mobilization hasn't had much effect yet on daily life. But it is probably implicated in the fact that the divorce rate has risen so sharply during this time. Being unable to change the power within marriages, women instead have tended to break up their marriages.

Recent data (Blumstein and Schwartz 1983, 53–67) show that income is still a main source of domestic power. This is true both for married couples and for cohabiting couples who are not married. In both cases, whichever partner had more income had relatively more domestic power. And since men generally earned more than their wives or female partners, they generally had more domestic power. Blumstein and Schwartz also compared homosexual couples on this issue. This is a useful comparison, since it enables us to see if income is really having the effect on power, independently of gender. The result is rather striking: income is the source of power, regardless of the couple's gender. That is, in couples composed of gay males, the person with the greater income played the role of the "husband," while the one with less income acted as the "wife," deferring to the other's job demands, yielding in major decisions, and generally taking care of the house. This suggests that the roles of "husband" and "wife" actually have nothing to do with sex. It is only because men have generally had more income (along with occupational power and other resources) that they have dominated their marriages. Traditional expectations built up along the assumption that the major income earner will be a man have made this into a cultural role.

Blumstein and Schwartz found one interesting exception to this pattern of income determining domestic power. Among lesbian couples, there was no relationship between income and power. Neither partner had more domestic power; both shared equally in doing the housework and in making decisions. Why this anomaly? It appears that males, even if they are homosexuals, carry on the dominant practice in which income translates into power. Females, when they are married to or living with a male, also follow this pattern to some extent. Lesbian couples are special. They are a relatively unusual type of household and may be operating with different ideas of what is fair and how it should be determined. For instance, England (1989, 26) suggests that we must consider the extent to which partners embrace a self-interested and competitive stance versus an empathetic and cooperative one. Various feminist theories (Chodorow 1978; Gilligan 1982; Keller 1982) describe how and why men and women develop different personalities, moralities, and ways of thinking. In summary, these theories suggest that women come to experience the world in connection with others while men tend to experience the world as distinctly separate individuals. England (1989) comments that economic, exchange, and rational-choice theories minimize emotional connection and altruism and inadvertently assume that all people feel and act like the stereotypical man in our society. In order to understand how marital power is exercised, experienced, and evaluated, we need to complicate our theoretical models to include consideration of when and why both genders would be empathetic as well as selfish, and how preferences could change over time. Thus, while the spouse who has more income may exercise marital power by making more decisions or

Feminist theories suggest that men tend to experience the world as distinctly separate individuals—not in connection with others.

doing less housework, power may not be the primary concern of the people involved.

Education

Education also has an effect on domestic power. Generally, the spouse with more education tends to dominate (Gillespie 1971; Komarovsky 1962). To some extent, this has helped women, since women have tended to complete slightly more schooling on the average than men. (At least this was the case up until the last decade, when the "mass-production" system of graduating from high school and the proliferation of junior colleges brought men up to and even slightly above the national average of 12.7 years of schooling *(Statistical Abstract of the United States* 1989). Skilled workers, in particular, have been likely to marry high-school graduates, which is one reason why they have tended to have relatively less power compared to their wives. Among the higher social classes, though, men tend to have a sizable advantage in education over their wives—another hidden reason why their egalitarian sentiments do not translate into reality. Where the wife in a professional-level family does have more education than her husband, that sometimes translates into greater outside contacts and prestige and hence into greater domestic power. Given the very recent tendency for women to attend college at a higher rate than men, we may eventually see another shift in domestic power. For the most part it has not happened, despite the spread of feminist beliefs in the higher social classes, primarily because men still control the upper-level professions.

Again, among the lower part of the working class, the resource of relative educational levels does not count for much. Neither husband nor wife has a lot of education, but the lower-working-class husband is likely to be more in touch with the world outside his family. He thus takes on the patriarchal role of "representing the family to the world," while his wife is especially likely to be confined to a domestic routine.

Social Participation versus Isolation

The more one participates in outside organizations, the more power one has at home (Blood and Wolfe 1960; Gillespie 1971). Some women thus acquire power because they are active in civic organizations, clubs, churches, or other areas. This gives them outside prestige that translates into some degree of deference at home. But it is much more likely for men to participate in outside organizations than their wives; thus this resource usually just reinforces the male advantage. Lower-working-class men in particular benefit from this resource; while they usually don't belong to many *formal* organizations, they are much more likely to have networks of acquaintances and to appear in the public arena of bars, street corners, and other "hangouts" than their wives, who are relatively isolated.

It has even been noted that when families move from the city to the suburbs, the domination by the male tends to go *up* (Gillespie 1971). The popular image of the suburbs is a place dominated by domestic and neighborly activities run by women. In fact, the suburbs tend to isolate women, especially working-class women who have neither the cultural resources nor the self-confidence to make widespread social contacts. Hence their husbands, who are less isolated because they commute away from the suburbs to work, end up with even greater relative power.

Children

It might be thought that once a woman bears children her status would go up, since she is fulfilling the important role of mother. But the fact is that *exactly the opposite tends to happen*. A woman's domestic power *declines* when the first child arrives, and it tends to decline further the more children she has (Szinowacz 1987). It reaches a low point during the time she has small children at home, before they go to school. Again, real resources count for far more than ideologies. Women with small children are maximally confined to the home, with the greatest number of pressures upon them. They simply do not have the time to acquire any of the other resources—income, outside sources of prestige—that would give them power.

Women's Outside Sources of Power

When women have more of these resources than their husbands, their power over domestic decisions tends to go up. Women who work at paying jobs have more power than housewives, and their power tends to go up the longer they hold their jobs. This is one reason why upper-working-class women have relatively better power positions vis-à-vis their husbands than wives of middle-class men; they are more likely to be employed. In dual-career families, the wife's influence seems to be proportional to the amount she contributes to the family's total income. As noted, having relatively more outside contacts in formal organizations and being better educated than her husband tend to give her a relatively better power position. Except in the lower working class, it seems; there, if a wife earns more money than her husband, she may provoke jealousy and anger, with resulting quarrels and violence.

Some Domestic Sources of Women's Power

Even if they lack economic leverage, women have sometimes been able to dominate men at home through specific tactics. Constantina Safilios-Rothschild (1977, 50–51) and Mirra Komarovsky (1962, 231–34) have pointed out that some women have gained power by playing the traditional roles in a kind of "front-stage" manner while covertly making their husbands dependent upon them. In public these women are deferential and submissive, and they go along with their men's ambitions and beliefs about themselves. In reality, they treat their husbands as "big children" to be humored and pampered and at the same time protected from real decision making. These women may end up controlling all family finances and even business and career decisions.

Another source of power is the emotional attachment of love and sexual attraction. There is an advantage in a woman's marrying a man who loves her more than she loves him; this gives her power according to the "principle of least interest," and in fact women are more concerned about their prospective husband's love for them than vice versa. Safilios-Rothschild (1977, 47–48) describes how sex can be used as a weapon to get one's way; she argues that women's ability to control their own sex drives is a great advantage to them, since they can more easily use the threat of withdrawal of sex to get their own way. This is a somewhat dangerous tactic, however. Its value probably declines as a woman's attractiveness diminishes, and it is at its height early in a marriage when the novelty and passion are still highest. Furthermore, especially in very traditional marriages, men are likely to procure sex outside of marriage (through prostitutes and affairs), so that a wife's attempt to use sex as a bargaining chip may not give her much advantage and in fact may serve to alienate her from her husband (Safilios-Rothschild 1977).

HAPPINESS AND UNHAPPINESS

Survey researchers studying marital happiness use questionnaires like the Dyadic Adjustment Scale (Spanier 1976) (see feature 11.5). Using such scales to predict which couples will stay together and which will divorce has not been very successful. Some researchers criticize the concept of adjustment (Trost 1985) or suggest that people will provide answers that make them look "too good to be true." For instance, one researcher found that many people would agree with unrealistic statements such as "If my mate has any faults, I am not aware of them," or "Everything I have learned about my mate has pleased me" (Edmonds 1967).

People are influenced by what survey researchers label "social desirability" and will tend to answer questions in a conventional way that makes them appear to be happier than they probably are. Even if we assume that some form of marital happiness or satisfaction is being measured with these questionnaires, there is a bias toward consensus and disapproval of conflict. Spouses who say they have different interests, admit to having disagreements, or acknowledge conflicts end up being rated as less "adjusted." In fact, however, many marriage counselors point out that recognition of differences and the ability to talk about them are signs of emotional maturity and can enhance the marriage. In general, there is no single type or style of marriage that makes the partners

FEATURE 11.5
Measuring Marital Adjustment

How "adjusted" are you and your partner? Even if you are not married, answer these questions about a relationship you are in, or answer about your parents. As you will see, most of the questions focus on whether two people agree about various things.

I. How often do you and your partner agree or disagree about the following subjects? (Always agree, Almost always agree, Occasionally disagree, Frequently disagree, Almost always disagree, Always disagree)

1. Handling family finances
2. Matters of recreation
3. Religious matters
4. Demonstrations of affection
5. Friends
6. Sex relations
7. Conventionality (correct or proper behavior)
8. Philosophy of life
9. Ways of dealing with parents or in-laws
10. Aims, goals, and things believed important
11. Amount of time spent together
12. Making major decisions
13. Household tasks
14. Leisure time interests and activities
15. Career decisions

II. How often do you and your partner do the following? (All of the time, Most of the time, More often than not, Occasionally, Rarely, Never)

16. Discuss or consider divorce, separation, or terminating your relationship?
17. Leave the house after a fight?
18. Regret that you married? (or *lived together*)
19. Quarrel?
20. "Get on each other's nerves?"
21. Think that things between you and your partner are going well?
22. Confide in your mate?

III. How often would you say the following events occur between you and your partner? (Never, Less than once a month, Once or twice a month, Once or twice a week, Once a day, More often)

23. Have a stimulating exchange of ideas
24. Laugh together
25. Calmly discuss something
26. Work together on a project

IV. These are things about which couples sometimes agree and sometimes disagree. Indicate if either item below caused differences of opinions or were problems in your relationship in recent weeks.

27. Being too tired for sex
28. Not showing love

V. All things considered, how would you describe the degree of happiness of your relationship?

| Extremely Unhappy | Fairly Unhappy | A Little Unhappy | Happy | Very Happy | Extremely Happy | Perfect |

Source: Graham Spanier, February 1976. "Measuring Dyadic Adjustment: New Scales for Assessing the Quality of Marriage and Similar Dyads." *Journal of Marriage and the Family* 38:1, pp 15–28.

happy. Researchers have described a wide range of communication patterns and interaction styles that satisfy the partners and contribute to stability in the marriage (Cuber and Haroff 1965; Ryder 1970; Skolnick 1987).

We have seen that married people are generally more likely to be happy than people who have never been married or those who are now separated, divorced, or widowed (table 11.1). Marital happiness is related to how much companionship, sociability, and other mutual enjoyments people derive from the relationship, and negatively correlated with the number of tensions. But the good parts and the bad parts of marriage seem to be independent of each other, and marriages can have a lot or a little of each. It is when there are especially few rewards to balance off the tensions that marriages are most unhappy.

What makes a happy marriage? Some of the answers that people give when asked this question are listed in table 11.3. We should be cautious about interpreting people's responses to this kind of question. For example, only 41 percent said that having enough money was very important for a marriage; but in fact, money is the most common source of arguments (as we will see in the section below, "Economic Success"), and economic strains are often involved in spousal abuse. We don't like to think our happiness depends upon money, but lack of money, at least compared to what we want, is a source of trouble. Probably it has its major effects because of the way it affects the relative power of husbands and wives in the family.

People also seem to believe, in general, that having the same background is not very important in a successful marriage. But if we look at the data on which marriages last and which are most likely to end in divorce (in chapter 14), we see again that "opposites do not attract," and in fact the more similar the backgrounds of the couple, the more stable the marriage. That is what one would expect from the point of view that having a similar culture is what makes it possible for people to communicate well. People recognize that understanding is important (86 percent of the sample), but not the ingredients that make it happen.

TABLE 11.3   What Is Important for a Successful Marriage?

	Percent of People Who Agreed with This Answer
Faithfulness	93%
Understanding	86
A good sex life	75
Children	59
Common interests	52
Sharing household chores	43
Enough money	41
Same background	25

Source: *Public Opinion*, vol. 8, Jan. 1986, p. 25.

FIGURE 11.3: Marital Satisfaction Through the Years

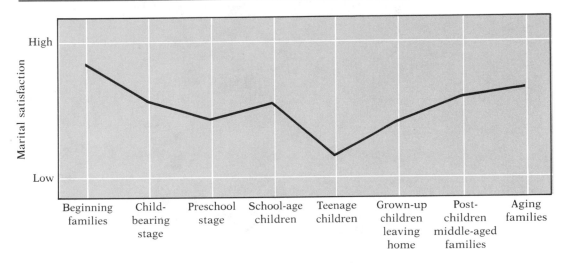

Satisfaction hits a low point when children are teenagers, then rises again.

Source: From "Marital Satisfaction over the Family Life Cycle" by Boyd C. Rollins and Kenneth L. Cannon, *Journal of Marriage and the Family*, 36 (1974): 271–84. Copyright © 1974 by the National Council on Family Relations, 3989 Central Ave. NE, Suite 550, Minneapolis, MN 55421. Reprinted by permission.

A modest majority (59 percent) believes that children are important for the marriage, although as we shall see children are both a source of pleasure and of strain. On the other hand, people are now quite willing to mention sexual matters: 75 percent believe that a good sex life is important, and virtually everyone (93 percent) says that faithfulness is crucial—in fact, this is the most commonly mentioned feature. Sexual fidelity is probably important to avoid serious disruption, but it is doubtful that faithfulness by itself will make a marriage really happy. (And in fact, as we saw in chapter 9, an occasional adultery is surprisingly prevalent.) There seems to be no single factor—a happy marriage is due to the overall blend of various things. We suspect that most people do not have a clear awareness of what is responsible for the degree of happiness or unhappiness of their own situation.

The Pressure of Children

One major factor affecting happiness in marriage is children. The prevalent view of children is a rather sentimental one, and one might expect the presence of children to increase marital happiness. But in fact, just the contrary is the case. Most studies have shown that marital happiness is highest at the beginning of the marriage, before children arrive. With the first birth, the

Tensions between parents and children reach their peak during the teen years, increasing marital strain.

happiness level begins to drop, and it continues to go down, through the children's preschool and school-age years, until it hits a low in the teenage years, just before the children are ready to leave home (see figure 11.3). Finally, as the children leave home, the curve abruptly goes up again (Glenn 1991; Spanier and Lewis 1980).

The conclusion is hard to escape: children are a strain on a marriage (McLanahan and Adams 1987). They tend to disrupt the emotional and sexual relationships of a husband and wife, give them more duties, and cause quarrels and bad moods. The strain seems to be worst in the teen years; perhaps because children are expensive then, perhaps also because the tensions between them and their parents are at their peaks. Children also contribute to marital strain because of the special pressure they put upon the wife to carry out her conventional responsibilities as a mother. This kind of parental "role strain" may be even more important than the life stage of the children (Rollins and Cannon 1974). There is also some evidence that couples with unwanted children are especially likely to have economic frustrations and to experience unhappy marriages.

It should come as no surprise, then, that childless marriages are happier than marriages with children during the years when the children are (or would be) in the household (Campbell 1981). In the later years, after the children have left home, the level of happiness becomes about the same in both types of families.

Children can have both positive and negative effects on happiness. They are a source of companionship, love, and fun. They provide the center of attention around which family rituals are focused (birthdays, Christmas, Halloween, and the like). In this way they help hold families together, creating emotional bonds and the symbols that represent these ties. But taking care of children is also a matter of practical work—work that largely falls to the woman. This is an unacknowledged source of strain in the relationship between husband and wife and probably has an indirect effect on the couple's happiness.

We should note that marital happiness affects one's whole view of life. It is a strong determinant of one's overall degree of happiness or unhappiness. Rarely do people who say that their marriages are "not too happy" describe themselves as personally happy. Women's personal happiness is especially likely to be dependent upon their marital happiness.

Economic Success

People are satisfied with their marriages for different reasons, ranging from the intrinsically satisfying experience of warm, intimate relationships to cold-blooded reasoning that the arrangement is the most advantageous they can get. How many people are there of each?

Purely *utilitarian marriages* seem to be especially common in the higher social classes. John Cuber and Peggy Harroff (1965; see also Blumstein and Schwartz 1983, 345–60) interviewed a sample of very successful Americans: bank presidents, top corporation executives, high government officials, wealthy physicians, college presidents, and the like. They took only those with long, stable marriages who had never considered divorce. Their study produced the following findings. Eighty percent of the marriages were *utilitarian*: the couple stayed together because of economic, sexual, and status advantages, without any important emotional tie. They were either habituated to conflict, or one partner (usually the wife) was completely passive, or the relationship was devitalized and empty. In the last type, the couple went through the motions of a marriage, occasionally conscious that the relationship once had more life and more love. Fifteen percent were *vital*: the couple meshed warmly and closely in at least a few areas of personal interest and emotional need. Five percent were *total*: the marriage satisfied virtually all of its partners' interests and needs, forming a little island against the world.

The great majority of these marriages were empty, yet the group as a whole had never thought of divorce. In fact, the divorce rate in the upper and middle classes generally is lower than in other social classes (see chapter 14). What seems to be happening is that success itself holds a marriage together. Working-

Money management is the main source of conflict in most marriages.

class and middle-class couples have less reason not to separate, and in fact they do separate more often, the less money there is coming in. The affluent upper classes, by definition, are enjoying a higher standard of living, and to a considerable extent, that is what holds them together. The wife of a successful executive, doctor, or businessman has a strong economic incentive to stay married and even to try to keep her husband satisfied by staying attractive, cooking gourmet meals, and the like. Such wives share their husband's prestige and get to travel in his elevated social milieu, which they might have to give up if they lived on their own. The husbands have their careers to consider. Having a wife is not only essential for maintaining their professional images, but for fulfilling their social obligations as well. What is more, their wives provide them with certain basic services (cooking, cleaning, child rearing) so that they are free to pursue their careers aggressively. So, the couples stay together at the cost of personal feelings.

There may also be quite a few empty marriages lower down in the class structure. The subject has never been well explored, although, as we know, there is a lot more overt conflict (as well as separation and divorce) as one proceeds downward into the middle and then the working and lower classes. The reason may be straightforward: money. Having lots of money motivates people to keep their marriages together; having little of it tends to create difficulties and arguments. Blood and Wolfe (1960, 241), for instance, found that money was far and away the most common answer (24 percent) to the question "What do you most often argue about?" Following were children (16 percent), recreation (16 percent), personality conflicts (14 percent), in-laws (6 percent), and gender roles (4 percent). Fifteen percent of respondents reported no arguments.

Fifteen years later, Scanzoni (1975) found that married couples tended to argue about the same things. Major conflict areas were money (33 percent), children (19 percent), recreation (8 percent), and gender roles (3 percent), with 17 percent reporting no conflicts (Scanzoni and Scanzoni 1988, 401). A more recent study by Blumstein and Schwartz (1983, 52, 67–93) confirms this finding; in their sample, the most peaceful relationships were those in which both partners felt satisfied with their income level and had equal control over spending it.

Virtually all families have arguments; only about 15 percent never quarrel (or at least do not admit it). Surprisingly enough, sex is almost never the cause of arguments, nor are remote topics like religion and politics. Children and personality conflicts are both fairly frequent causes of dissension. So is recreation: there is apparently a lot of hassle about where to go when the family goes out, getting the family together, and so forth. And it may be that these samples are not representative of more recent families in the amount of fighting over gender roles.

The point that stands out, though, is that money was far in the lead as the cause of family quarrels. Taking all the evidence together, we can infer that such disputes occur mainly when there isn't enough money to satisfy a family's needs. This fits in with the pattern that couples have higher marital satisfaction when the husband is moving up in his career (Scanzoni 1970). The cost is that the marriage stands a greater chance, perhaps, of becoming an empty marriage, with less fighting but also less happiness. (This is only conjecture, though, because not much is known about empty marriages further down the social scale.)

Single- and Dual-Career Marriages

We come now to a crucial point about the basis of family happiness. We have seen that money is frequently the cause of unhappiness in the family. It is often involved in family quarrels, and in more extreme cases of spousal abuse. It probably has this effect because the power of husbands and wives is connected to their incomes. Quarreling about money may well be a surface manifestation of something more basic. A man angry with his wife over how she spends money is asserting, in a way, his claim to control her behavior because of his superior income. Conversely, a wife who criticizes her husband for his failure to make enough money is attacking his traditional source of dominance. Suppose he has the traditional mentality that symbolizes a man's standing by his economic success and by his dominance in his household—a combination that is particularly likely to be found in working-class families or other localistic, traditional communities. If this is the case, he may likely respond angrily to what he feels is an attack on the basic symbols of his position.

Even without overt quarrels and fights, money affects happiness. Supposing the husband makes all or most of the income, while the wife does most of the housework and defers on family decisions. The woman is, in effect, in a working-class position *inside the family*, although she would probably not define it that way. The idealization of marriage and the sexual and emotional ties of love cover up this practical aspect of the situation. But being in a working-class type of position has its structural effects in the family as it does in conventional jobs. The person who has the power is more attached to the role, while the person who is subordinate tends to be alienated from the role and tries to withdraw from it psychologically. Such a woman is usually cross-pressured: some factors cause her to identify with her role as wife (including love, sex, children, local community status); other factors cause her to be alienated (notably her power disadvantage, due to her economic position). Often this is an uneasy balance. Sometimes the effects are unconscious, coming out in the greater tendency of women to have chronic depression than men (Holden 1986). Sometimes the result surfaces in overt dissatisfaction with marriage, resulting in quarrels or divorce.

A considerable amount of research has focused on the importance of the "housewife role" in explaining why married women have higher rates of psychological distress than men. Bernard (1972) argued that housekeeping has a "pathogenic" effect on wives, and Gove (1972) contended that keeping house is so unstructured but so value-laden for women that housewives often judge themselves harshly. In general, housewives have fewer sources of personal gratification than husbands. While married men have traditionally had multiple roles inside and outside of the home, housewives have had to rely solely on their domestic role to bolster their self-esteem. Gove and others argued that this makes housewives more vulnerable to the problems associated with family life (see feature 11.6).

This is not an irremediable situation. The problem suggests its own cure. If women are being alienated, subtly or overtly, by the power imbalance in a marriage, then correcting the power imbalance should improve the happiness of the marriage. This is in fact what we do find. Where women improve the resources that give them more power in the marriage, there is a tendency for marital happiness to be greater (McLanahan and Adams 1987).

The emotional cost of a traditional, successful marriage seems to be paid especially by the wife. Women try hard to keep the marriage together, at the price of making themselves subservient to their husbands. This seems to happen only when the wife is a full-time housewife (Macke, Bohrnstedt, and Bernstein 1979). When she has no job of her own, her self-esteem has been

FEATURE 11.6
Housewives in the Mental Hospital

Carol Warren (1987) studied the records of a group of housewives committed to mental hospitals in California in the late 1950s and early 1960s. The women were diagnosed as schizophrenics: psychotics who have lost all contact with reality. But their own complaints were those of typical housewives: they felt lonely, depressed, burdened by their housework, cut off from other adults, and overly dependent upon their husbands. They complained of a lack of communication in their marriages and felt that their husbands had little interest in them except for the domestic and sexual services they provided. None of these women worked outside the home except for two who worked for their husbands—a situation providing no relief from the family circle. A number of these women said that they would like to work but their husbands would not let them. This was during the height of the wave of adulation of traditional home life during the 1950s. Some women in the sample who very strongly wanted a career were regarded as crazy for having that ambition; it was taken to be a rejection of femininity and a symptom of mental illness.

It was not their feelings of depression and loneliness that got them in the hospital but the fact that their behavior created troubles for their husbands. Typical troubles involved being sloppy and inadequate at housekeeping or child care, or money—spending too much, frivolously or "crazily" buying things the husband felt he could not afford. Often troubles involved sex as well, although this did not always lead to the husband defining his wife as psychiatrically disturbed. If the wife reacted by extreme withholding of sex, the husband was often angry but tended to regard the behavior as part of the typical female pattern. If, on the other hand, a wife was overly demanding sexually, this more likely raised suspicions of psychiatric problems.

Husbands, however, did not necessarily jump to the conclusion that their wives were mentally ill and ought to be committed, even though under California law at the time either spouse had the power to have the other committed, with the concurrence of one admitting physician. Most men resisted this course of action for some time because they did not want to lose their wife's services as housewife or her availability as a sex object. When troubles with spending, housework, and/or sex made the rewards no longer worth the cost, though, they would finally have their wives committed. This situation could be defined as a family emergency, and husbands usually got female relatives to help them out by taking over the housework.

In the hospital, the women often felt some respite from their housework, but it was hard for them to get away from their husbands and families psychologically. Many felt guilty about "abandoning" them, and several actually developed the delusion that their husbands and children were there in the hospital with them. Within a year, though, most of the women were out of the hospital, many of them having experienced electroshock treatment. There were numerous readmissions over the next few years, but by and large their social adjustment improved. The mental hospital did seem effective in getting the women in line. It was not always used merely for home troubles. One woman decided to leave her husband and had an affair with a fellow patient while on leave from the hospital, whereupon her permanent release was blocked by her irate husband, who procured papers to have her permanently committed.

Most of the women returned to their household responsibilities calmer and wiser. Many of them said in interviews that they were determined to behave and do what they were asked and that they did not want to be sent back to the hospital again.

shown to go *down*, the more her husband's success goes *up*. At the same time, her husband's success makes her feel that the marriage itself is a success. But wives who have professions of their own are more detached from their husband's success; it does not lower their own self-esteem, nor does it have any effect on how they feel about their marriages.

Thus, despite the conservative ideology often expressed about the importance of maintaining the old-fashioned family, the fact seems to be that *dual-career marriages* are far more successful. The traditional employed husband and full-time housewife combination often produces fights when the man is unsuccessful, and empty marriages when he prospers. Working women in general are more likely to report happy marriages than nonemployed women, or at least not to report significantly more unhappiness in the marriage (Spitze 1991). Dual-career marriages often seem to have more energy and excitement as well as more genuine equality and respect; hence they don't detract from love but actually enhance it (Rapoport and Rapoport 1971).

Recently, many couples have moved toward marriages of shared responsibility, in which both partners are employed and involved in housework. As the number of these marriages has increased, the rate of depression among wives has decreased. If an employed wife thinks that men in general should do some housework, and if her husband makes significant contributions around the house, then she experiences better mental health (Ross, Mirowsky, and Huber 1983). If the husband shares the child-care responsibilities with his wife, she has lower-than-average depression levels (Ross and Mirowsky 1987). Such findings bode well for both wives and husbands in the future.

This does not mean that dual-career marriages are idyllic. They create new strains of their own. Both husband and wife may now be subject to pressures for success in the corporate world or in their professions (Hertz 1986). Work pressures can result in a shortage of time for one's partner. Women in these careers do not necessarily give up having children, although they go back to work quickly after childbirth and rely heavily on hiring other women to care for their children. As we have seen in chapter 6, jobs in the upper middle class tend to absorb most of one's identity and one's energies. Whereas this previously applied mainly to men, it now can affect both spouses in a family. Thus there are additional strains on the marriage. But these are compensated for by the fact that the wives are personally happier. Having two incomes also helps the overall family situation. On the whole, the result is probably an improvement in family satisfaction.

SUMMARY

1. Every marital union contains two marriages: his and hers. According to legal tradition, "her" marriage includes providing domestic services and deferring to her husband's wishes, whereas "his" marriage entails supporting

the family and enjoying the benefits of his wife's household labor. While laws have changed, wives still tend to do much more housework and child care than husbands and to pay much more attention to communicating with spouse and relatives.

2. Married women are more likely to worry and be depressed than unmarried women. In contrast, married men, because they enjoy the services of a wife, report being happier than unmarried men and enjoy better physical and mental health than their unmarried counterparts.

3. Household and family labor tends to be divided on the basis of gender, with women doing most of the cooking, cleaning, laundry, and child care and men doing most of the household repairs and outdoor work. Over the last twenty years, husbands have increased the number of hours they spend each week doing household tasks, and wives have reduced their number of hours. Nevertheless, wives still do two or three times more of the total household labor than their husbands and about four or five times as much of the housework. When husbands do more, wives are less depressed.

4. Wives usually act as household managers, planning and allocating various domestic tasks. When husbands care for children or clean house, they often perform the more pleasant tasks, and their contributions are typically described as "helping" the wife. Even though wives do much more than husbands, domestic arrangements are usually considered to be "fair." This is probably because unequal divisions of household labor reaffirm people's self-image as men and women.

5. Most women feel dissatisfied with house*work*, although this does not necessarily spill over into feelings about the larger role of house*wife* and mother. Working-class women are particularly likely to identify with their roles as housewives and to receive reinforcement for this from networks of close-by family and friends. Middle-class housewives identify more with their social roles in their families and in the community.

6. Men tend to have greater power in families, especially based on their usually superior resources in occupational prestige, income, education, and participation in outside organizations. Where women have more of these resources than their husbands, their relative power in the family generally goes up. A woman's domestic power declines when she has small children, because her child care usually deprives her of the resources that bring power.

7. Happiness is generally highest at the beginning of the marriage, begins to go down with the arrival of children, and reaches its low point when children are in the teen years. Happiness goes up again after children leave home. Childless couples are just as happy as couples with children and are happier during the years when the children would be in the home. The negative effect of children on happiness seems to operate especially through the strain it places on the wife, by increasing her responsibilities and reducing her sources of power and independence.

8. Among the higher social classes, there seems to be a high proportion of utilitarian marriages, held together not by personal feelings but by habituation and economic advantage. Lower down in the class structure, scarcity of money is the most common source of family arguments, followed by disputes over children, recreation, and personality differences.

9. Inequality in marriages is a major source of strain, especially for wives. Dual-career marriages, in which power resources are most nearly equal, are those in which women's self-esteem is highest and personal relationships seem to be best, provided this is what both spouses desire.

12

PARENTS AND CHILDREN

Parenting may be the most important job we perform as adults, but as we saw in chapter 10, we are not always well-prepared for it. In this chapter we will review what it means to be a parent or a child in our own society, with reference to how it was in past times. We will look at recent changes in the organization of parenting that have resulted from the rapid influx of mothers into the paid labor force, including fathers' increased participation and trends in the use of outside child care. We will also explore the various ways that children are socialized both inside and outside of families and highlight the ways in which child rearing contributes to stratification along the lines of gender and social class.

CHILDHOOD IN HISTORICAL PERSPECTIVE

The innocence and purity of childhood and the love that parents have for their children exemplify what we think is most noble about being human. Nevertheless, the close emotional ties between parents and children that we take so much for granted are a relatively recent historical invention. Before the seventeenth century, childhood was not considered a special time of life, and children were not sentimentalized as they are today. Very little distinction was made between children and adults, and the main value that children held was in their ability to contribute to the subsistence of the family (Ariès 1962; see feature 12.1).

Although we know little about the day-to-day lives of children in preindustrial times, we can piece together a picture of how adults in Europe and America thought about them from existing historical records. The prevailing Christian view was that children were born in sin. In 1535, John Calvin remarked that children's "whole nature is a certain seed of Sin, therefore it cannot but be hateful and abominable to God" (Muir and Brett 1980, 3). A hundred years later, the American Puritans preached that God-fearing people must break the will of their children if they were to be saved from the devil: "Let a child from a year old be taught to fear the rod and cry softly. . . . Break his will now and his soul will live" (Muir and Brett 1980, 101). Throughout the eighteenth century and into the nineteenth century, Puritan and Methodist views on child rearing continued to stress the "corrupt nature" and "evil dispositions" of children. Parents were admonished to demand strict obedience and to use swift physical punishment to curtail the child's inherent evil (Synnott 1983). Duty and obligation were the hallmarks of parent-child relationships, and love, while probably present, was considered secondary. Much like the husband-wife relationships of the time, parent-child relationships were primarily instrumental, with children expected to do their part to keep the household running. (See chapter 3 for a discussion of the patrimonial household.)

The most significant changes in children's lives from medieval times to the modern era were ushered in by the rise of formal education and by the growth of what Stone (1977) calls "affective individualism." Both of these trends, in turn, were promoted by the growth of capitalism and the gradual shift from home-based production to waged labor, industrialization, and a market economy. Before the twentieth century, children were primarily valued for their economic contributions. They usually worked on farms or labored in their parents' trade, and by maintaining such pursuits, provided a kind of insurance policy for their parents in old age. This does not mean that parents didn't love their children, but they were valued less for their intrinsic worth than for the material contribution they could make to the household economy.

Death was much more visible in those days, which must have had some impact on the emotional bonds between parents and children (Uhlenberg 1980). With fertility and mortality rates both much higher than they are today, love was "spread around" to more children who were more likely to die. Emotional involvement was likely to have been less, out of self-protection,

---

FEATURE 12.1
The Discovery of Childhood

According to many historians, the close ties between parents and children that we take for granted in the modern family were invented comparatively recently (Ariès 1962; for a contrasting view, see Pollock 1984). In the patrimonial household, children were just another part of the household political or business enterprise. They were valued for the fighting they could do, work they could provide, or marriage alliances they could bring; often they were apprenticed out at an early age. Many small children who died did not even have their names recorded. There was very little distinction made between children and adults. Children were dressed like little adults, and they were brought into adult activities as soon as they were big enough to take part. In medieval schools, children of ten might be reciting their lessons alongside young men in their twenties; in the army or navy, a young soldier would begin as an ensign or cabin boy in his early teens. Given the cramped living quarters at home and the lack of privacy, even very small children were exposed to

sexual activities and adult talk, unhampered by the modern attitude that such things ought to be kept from children.

When the private household began to emerge in the seventeenth and eighteenth, and especially in the nineteenth, centuries, the family took on a new shape. Affectionate relationships were supposed to hold, not only between husband and wife, but also between parent and child. Edward Shorter (1975) has referred to this as the "sentimental revolution," which he believed peaked around the years 1780–1830. It was during this period, especially in the United States, that the paramount ideals of the modern family were articulated (Degler 1980; Ryan 1981). Marriage was founded on romantic attraction between husband and wife, each of whom was to take care of his or her own separate sphere. The husband was to be the provider who left the home and returned with its economic support; the wife was to devote herself

(Continued next page)

FEATURE 12.1 (Continued)

With the "sentimental revolution," family ritual began to take over from community ritual. Thanksgiving, for example, began as a community celebration but has developed into a family affair.

to the care of her children. And unlike the medieval family, with its practical, work-centered attitudes, children were now to be cared for, not merely physically, but emotionally and morally. Motherhood, for the first time, became the object of a cult, conceived of as a moral ideal. Family ritual began to take over from community ritual.

The medieval family took part in church ceremonies and community festivals; the modern family may have continued these practices, but it also created its own rituals: celebrating birthdays (part of the cult of childhood) and creating new family festivals like Thanksgiving. The family had become private, personalized, and sentimentalized.

though such things are difficult to measure (Synnott 1983, 29). Ariès (1962, 38) quotes a seventeenth-century woman as saying: "Before they are old enough to bother you, you will have lost half of them, or perhaps all of them." In the elite upper classes, where death was a less constant threat, children had more sentimental value, though still much less than among parents in the twentieth century.

From Economic Asset to Sentimental Object

Viviana Zelizer traces the changing social value of children in *Pricing the Priceless Child* (1985). She documents how the economically useful child of previous times was transformed into an "economically worthless, but emotionally priceless child" in the sixty years between 1870 and 1930. During this time, traditional forms of child labor came to be seen as harmful and inappropriate for those of "tender years." In 1870, if a child died in an accident and the courts concluded that another party was negligent, the parents were compensated for the value of the child's labor. By 1930, however, in cases of "wrongful death," the parents were compensated for incalculable emotional pain (Zelizer 1985). The changing value of children is also reflected in the "market" for babies. As we saw in chapter 10, countless numbers of European infants were sent to foundling homes where most of them died. Between the eighteenth and nineteenth centuries, there was a baby "glut" and infants were seen as relatively expendable. In the 1870s, a woman with an unwanted infant had to pay a "baby farm" to take the child off her hands because legal adoption was rare, and there was no market for babies. Many families were willing, however, to take older children (especially boys) into their homes because they could contribute to the household economy. By the mid-twentieth century, the situation had dramatically reversed itself. By that time, it had become extremely difficult to find homes for older children—especially if they were boys. At the same time, adoption had become relatively common, and the sentimental value of younger children had blossomed. Couples who could not adopt an infant from a licensed agency in the 1950s were willing to pay as much as $10,000 for a baby on the black market (Zelizer 1985; Skolnick 1987, 307). By the 1980s, hopeful parents were willing to pay even larger sums of money to women who would agree to act as surrogate mothers and provide them with a "priceless" baby (see chapter 10).

Individual versus Collective Patterns of Child Care

Our current child-rearing patterns are much more individualized than those of the past. While women have been the primary caretakers of children in all known cultures, the assumption that children should be raised by their biological mother alone is a relatively recent invention. In most preindustrial societies, for instance, young children spent less than half of their waking hours in the care of their mothers. In preliterate tribes and in traditional agrarian societies, fathers, grandparents, siblings, and other women had important roles in taking care of the community's children (Barry and Paxson 1971; Rohner and Rohner 1982) (see feature 12.2). Similarly, in Europe from the Middle Ages to the eighteenth century, the community was very involved in caring for children and shaping the individual's fate. In this older and more collective pattern, parent-child relations were regulated and monitored by relatives and other community members, and what happened inside the family was relatively public. A microcommunity of close by adults and older children acted as surrogate parents, and there were always plenty of people around to offer advice on what to do in specific situations. It was not until the nineteenth century that the industrial revolution encouraged a withdrawal of the family from the outer world and created a relatively private family domain (Ariès 1979). According to the "cult of domesticity" that emerged around this time, women were defined as innocent, pure, and sensitive. Their virtue was reflected in motherhood and they were increasingly confined to the home (see chapter 4).

In the twentieth century, the suburbs multiplied and the family came to be seen more and more as a self-sufficient unit with a monopoly on emotions,

FEATURE 12.2
Child Rearing in Nonindustrial Societies

The anthropologist Ronald Rohner (1975) studied child care in over one hundred nonindustrial societies throughout the world and came up with four central conclusions.

1. Parents who respond to infant distress by nurturing them and trying to reduce their discomfort tend to do the same for older children as well;
2. When children are rejected by their parents (denied love, esteem, and other forms of positive response), they grow up having difficulty managing hostility and aggression, have troubles with dependency, and are often emotionally unresponsive;
3. When children are accepted and given a warm and supportive upbringing, they tend to form less hostile relations; and
4. Mothers who are unable to escape the intensity of continuous interaction with their children are more likely to reject them than mothers who can get away from time to time.

In other studies of nonindustrial societies (those relying on hunting and gathering, horticulture, herding, or agriculture; see chapters 2 and 3). Coltrane (1988, 1992) found that child-rearing patterns are linked to the status of women. If fathers have frequent contact with children, and especially if they participate in routine child care and are physically and emotionally supportive, then women in those societies are more likely to hold influential positions and participate in

community decision making. Fathers' relationships to their children are also found to be related to ritual deference and "hypermasculine" behaviors. In societies with distant father-child relationships, men tend to fear and belittle women and are preoccupied with affirming their manliness through acts of bravery or physical prowess. In these distant-father societies, women are often excluded from ritual gatherings and are more likely to be required to bow to the men, serve them first, and stand in their presence. In contrast, if fathers are more involved with children, women are thought of as more equal and do not have to continually defer to their husbands. The other crucial factor is that in societies where women control property, they are less likely to be subjected to men's derogation, and men tend to be less aggressive and less concerned with proving their manhood. What we see then, is that patterns of child rearing and property control are both important to the relative status of men and women. If men share child care with women, and if women share control of property with men, then there is a fair amount of sexual egalitarianism in the society. If, on the other hand, men control all of the property and women do all of the child rearing, there tends to be a fair amount of fear and antagonism between the sexes. Both patterns tend to be reflected in the mythical symbolism of the culture, with distant-father cultures having male-dominated creation myths and shared-parenting cultures having relatively gender-balanced stories of the way the world began (Sanday 1981).

Mothering can blend the egos of mother and child, creating difficulties for both in establishing separate identities.

raising children, and filling leisure time. By this time, the father's image was one of the "good provider," who "set a good table, provided a decent home, paid the mortgage, bought the shoes, and kept his children warmly clothed" (Bernard 1981b). The woman, on the other hand, was expected to be consumed and fulfilled by her "natural" wifely and motherly duties. Isolated in her suburban home, she had almost sole responsibility for nurturing the children, aided by occasional reference to Dr. Spock or some other child-rearing expert. This was the era of the June Cleaver image of the happy household—suburban housewife, breadwinner husband, station wagon, dog Spot, and 2.4 children. This individualistic model of parenting was never a reality for most working-class women, but the ideal enjoyed unparalleled popularity during the 1950s.

Isolated and exclusive parenting by biological mothers is a sort of "all your eggs in one basket" approach to child rearing that produces unique stresses and strains for both parents and children. Because the mother does all of the child care, her feelings become overwhelmingly important to the children and she has little opportunity for taking a break or deriving a sense of self-worth from alternate activities. This can set up a kind of "hothouse" environment in which mothers and children feel responsible for each others' feelings, and ego boundaries become confused. Exclusive and isolated mothering can create "double-binds" in which love and hostility are merged and children experience difficulty establishing independent identities (Coser 1964). Nancy Chodorow (1978) shows how exclusive mothering contributes to personality differences between men and women and perpetuates itself by encouraging women to seek fulfillment through mothering (see feature 12.3).

Our individualistic model of parenting also sets up a two-party conflict and a certain amount of anxiety that was not present in societies with more collective child-rearing practices. Since modern parents make their own rules, breaking or resisting the rules can easily be perceived as a personal threat and personal battles between parents and children can ensue. In the modern case, parents

FEATURE 12.3
Why Do Women Mother?

Why do women do the mothering in our society rather than men? Mothering is not the same thing as childbearing or breast feeding; it is taking care of a child, physically and emotionally. After a child is born, its own biological mother does not have to be the person who looks after it. Since bottle feeding was invented, there is no longer even the period of nursing that required a female, if only as a wet nurse. Mothering is a social, not a biological, role. And it is as much an emotional activity as a physical one: mothering means caring for one's children in both senses of the word.

The sociologist Nancy Chodorow (1978) has produced a theory to explain why this role is nearly always adopted by women. It is not instinctual, since studies show that both men and women react similarly to infants' cries and smiles. In many animal species, males will care for an infant if left alone with one. There is nothing resembling a "maternal instinct" in women who are separated from their babies (for medical or other reasons) immediately after childbirth. The mothering role is learned.

Freud had pointed out that initially both the male and the female infant were cared for by a female. The mother is the first erotic love object for both sexes. This love is the prototype for all later love relations. It is a merging of the infant's

self with that of the mother; in fact, according to Freud's theory, initially the infant has no sense of an individual self. An autonomous self emerges only as the child separates itself from this state of primal merging.

For the child to mature, the erotic attachment to the mother must be broken and directed outward. For the boy, this takes place during the classic Oedipus complex. Because of fear of his father's jealousy, the boy gives up his mother and instead displaces the erotic energy (libido) into a fantasy; this fantasy takes the form of an imaginary father inside his own head, which now serves as his conscience or superego. The boy's erotic renunciation of his mother, then, is a crucial step toward developing the adult psyche.

But what about little girls? A girl's problem is both to renounce her mother erotically (libidinally), while at the same time creating her own fantasy object of identification, her own superego. Since this would also be her adult sexual (and gender) identity, this identification has to be with her own mother. How this happens was never very clear in Freud's writings, and he never developed an adequate theory on this point.

This is where Chodorow offers her own theory. There is less pressure, she points out, for girls than for boys to break their deep primary attachment have individualistic parental power, but generally lack a larger sense of institutional legitimacy because each parenting decision is an individual choice. In the microcommunities of earlier times, children could disobey or fail to do their assigned tasks, but since it was not up to the parents to establish the rules, the conflict was seen as outside of their control. They were simply enforcing a community standard. Not only that, but if the parent did not enforce the standard, someone else in the microcommunity would, providing validation that the parents were not individually responsible for setting standards. The earlier pattern could also be conflictual, and parents had little capacity to change parental practices, but they had much more institutional legitimacy, and the standards themselves were rarely challenged (Skolnick 1987). One of

to their mothers. As a result, Chodorow says, the break is not so sharp. A girl is given more permission to be close and affectionate with her mother, even extending to physical caresses. This has more than a superficial significance, since the girl is not required as much as the boy to separate herself sharply from her mother. She retains more of the original sense of merging with the world that the infant experienced in the original love-relationship with the mother. This has a powerful effect on the development of the girl's personality.

What is the difference between feminine and masculine personalities? According to Chodorow, the feminine personality is less separated from other people, and less sharply individuated. The boy, whose separation from the mother was made more sharply through the Oedipus complex, develops a sharper separation between himself and the world. Men have firmer ego boundaries; they tend to be more distant, domineering, and instrumental. They prefer the world of objects—action, machinery, science—to the world of people and of the self. Women prefer intimacy and warmth in personal relationships, and their identities come more from how the group receives them than from their accomplishments. Their personalities are what Chodorow calls "relational," with more flexible ego boundaries between one's self and others.

The reason that women do the mothering, then, is because the maternal personality is simply a typically female one. A woman's personality needs are to be close to other people and submerge herself in the group. She surrounds herself with her husband and children because she herself remains underseparated from her own mother. Because she never broke her unconscious erotic ties with her mother, she continues to need this kind of close and nurturant relation with others. Women become mothers because their experience with their own mother has given them the kind of personality that needs to mother. Mothering thus reproduces itself in a chain across generations.

As a feminist, Chodorow asks: How can the chain be broken? If men begin to mother children, assuming more of the emotional care and the physical caressing as well as the physical work of looking after their needs, then the next generation of children will grow up with a different psychic structure. Will both boys and girls come to acquire combination male/female personalities as a result? Will boys become more relational and female-like, and girls more object-oriented and male-like? Chodorow does not say. But her theory remains a thoughtful challenge to our understanding.

the important things about the more individualistic modern pattern of child rearing is that it allows for social change. What types of changes might result, and whether they are considered to be good or bad, are the subjects of many current debates over the future of families and children.

Current Idealized Images of Parents and Children

In the 1990s, we continue to idealize and sentimentalize most aspects of parent-child relationships. In our advertisements and television shows, children are usually cute, clean, good-humored, and playful. Their parents are beautiful, contented adults, happily playing with them, caring for their little needs, amused by their pranks, and proudly watching their accomplishments as they grow up. Unfortunately, reality is not quite like this.

Small children are just as often runny noses and smelly bottoms; when and if they are clean and beautiful, it does not happen automatically but because the parents have put a good deal of effort, every day, into cleaning them up and feeding them wholesome food. Children sometimes do play happily, but one characteristic of young personalities is that they are very volatile and can switch from a happy mood to crying and whining almost instantaneously. Children can be very affectionate, and parents derive great pleasure from holding them in their arms; but children also tend to have few inhibitions about throwing things or hitting people. At certain ages, they are very possessive and selfish about their toys, and indeed about everything in their environment they call "Mine!" In short, children act childishly.

PARENTING TODAY

Most new parents have little idea of the difficulties involved in their new role. They have assimilated the ideals, but the realities often catch them unawares. Thus for many couples, especially those who have their first child while they themselves are still young, its arrival can bring about a marital crisis. This is particularly likely for working-class families (chapter 6), which have fewer economic resources to ease the burden. If the baby arrives soon after the marriage, its parents face the problem of a double transition: having to adjust to living with another adult at the same time that the pressures of child care begin. Middle-class couples, and others who put off marriage and childbearing to a somewhat later age, tend to have certain advantages: more money, a better-settled routine, more maturity for dealing with the children. Nevertheless, even here struggle is inevitable.

Not only that, but the worst may be yet to come. We have already seen in the last chapter that the family happiness of a married couple tends to hit its low point during the years that their children are teenagers. This may be partly due to the fact that parents are now in their forties and hence may be undergoing a midlife crisis of their own. But dealing with children has its own costs from the time they are born; for that reason, family happiness starts creeping downward when the first child arrives, hops back up temporarily when the children are out of the house and beginning school, then keeps falling until they leave home for good, whereupon marital happiness jumps back up again. It is hard to avoid the conclusion that children are a strain on their parents.

Who Takes Care of the Children?

How much of a strain children pose and to whom has very recently become a subject for debate. In our society and in virtually all other Western societies, it has traditionally been taken for granted that women take care of the children. And in fact, the mother today still tends to put in most of the hours required to feed, clean, and dress them. Fathers tend to help more with the older children, especially in outdoor recreational matters: which is to say, in more of the pleasant tasks and fewer of the demanding and dirty ones. However, there has been a shift toward somewhat greater egalitarianism in caring for smaller children, and fathers do join in changing diapers or giving an infant a bottle in the middle of the night. The evidence seems to be that fathers are quite capable of carrying out all the child-care tasks, emotionally as well as physically (Pruett 1987); but traditional gender roles still determine who does *most* of this work.

Maternal Employment and the Need for Child Care

Figure 12.1 shows that in the United States in 1990, almost three out of four married women with school-age children, and more than half of wives with preschool-age children were in the paid labor force. The change since the 1950s has been dramatic. In 1960, fewer than nineteen of every hundred married mothers with children under six were employed. By 1990, that rate had tripled. Since single mothers are even more likely than married mothers to be employed, and because divorced mothers usually receive physical custody (see chapter 14), over two-thirds of all children under eighteen years are now living with employed mothers. What's more, about 70 percent of employed mothers now work full time (Phillips 1989).

Because most mothers must leave home to go to work, other caretakers have to be at least intermittently available. In two-parent families, fathers are a primary source of child care when mothers are working. In 1991, one of every five preschool children were cared for by their fathers while their mothers worked outside the home. This represents a sharp increase since the 1980s, when only about 15 percent of preschoolers were cared for by their fathers. For school-age children, the primary source of care while mothers are at work is the formal school system, but fathers are the next most likely source of care

FIGURE 12.1: Rates of Employment for Mothers Have Increased Rapidly

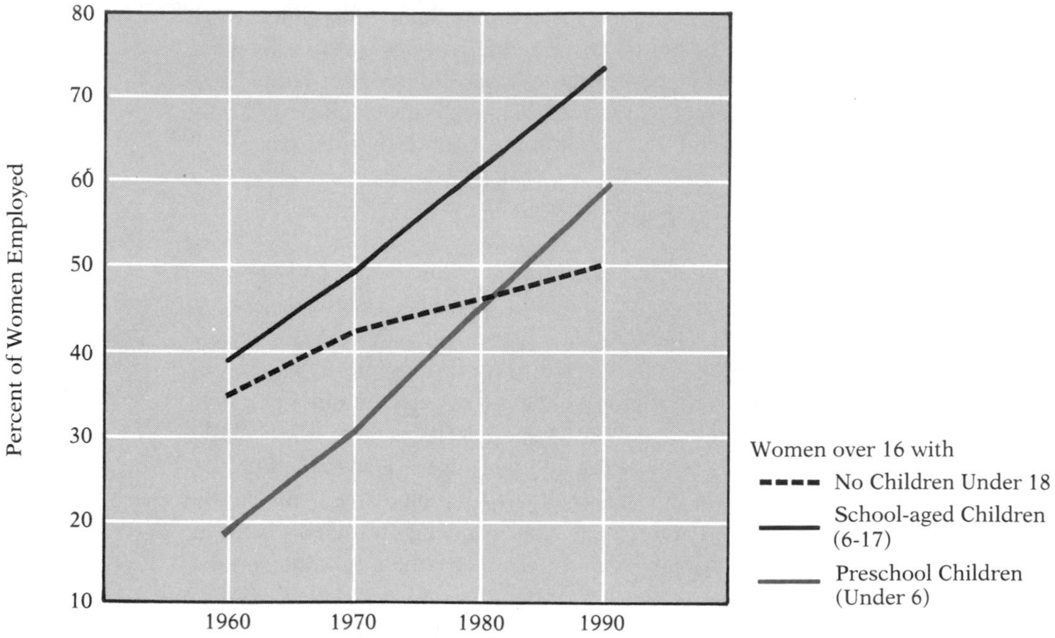

Women over 16 with
- - - - No Children Under 18
——— School-aged Children (6-17)
——— Preschool Children (Under 6)

Rates are for married women 16 years and over in March of each year.

Source: *Statistical Abstract of the United States*, 1992, No. 620; *Current Population Reports*, 1986, P-23, No. 146.

(O'Connell 1993). But fathers are even more likely than mothers to be employed full time, so most families must turn elsewhere for child care help.

In the working class, there is a tendency to leave small children with relatives, such as the mother's own mother. In the middle class, other women may be employed as baby sitters, or preschool children might be sent to family-care homes or child-care centers. Family home-care is usually offered by working-class women who have children of their own and take in additional children during the day. While many of these homes are licensed by local government agencies to ensure adequate health and safety standards, others operate outside of government regulation. Child-care centers, in contrast, are routinely licensed and regulated, and must meet government standards for adult-child ratios, indoor and outdoor space requirements, nutrition programs, health maintenance, and building safety. Some states also have requirements for teacher licensing and have additional curricular requirements for institutions that are designated as "child development centers." A few child-care centers are private for-profit enterprises, but most are run by nonprofit organizations such as schools, churches, or community service agencies. Some businesses have begun to offer on-site child care in response to employee demands. Such centers have been found to increase worker productivity, reduce absenteeism, and minimize turnover and subsequent training costs. Presently, about 5,000 employers out of 6 million businesses nationwide provide some form of child-care assistance (Morgan and Tucker 1991). Child-care allowances and flexible scheduling options are much more prevalent than in the past, but few American businesses provide on-site day care (Phillips 1989). Taken all together, available day-care facilities provide care to only a fraction of the children whose parents need it.

In spite of the fact that most people say they support "quality" day care, the people who care for children in the United States can barely make a living doing it. Ninety-eight percent of child-care workers are women, and on average, they earn less than adults who tend our cars and parking lots, take care of our pets, or haul our garbage. In 1984, nine out of ten child-care providers who worked in private homes earned below poverty-level wages. More than half of the market-based child-care workers also earned wages below the federal poverty line, with a median income for those in educational and social service

Employer-supported child care enables both parents of pre-school children to be part of the paid labor force.

positions of just $9,204 (Phillips 1989, 264). Because of low wages, poor benefits, and stressful working conditions, many child-care workers leave the child-care field for less demanding, less isolating, and better paying employment (Kontos and Stremmel 1987). The average day-care center in the United States has an employee turnover rate of over 40 percent per year (Etzioni 1993). Fifty years ago or more, most middle-class American households had a maid or housekeeper (again almost always a woman) who cared for the children. But the live-in housekeeper has virtually disappeared—a development that is an improvement in the economic status of many minority ethnic women who filled those low-paying jobs, but a problem for the career-oriented middle-class women who must find other daytime child care.

A look at the use of various types of child-care arrangements by working-women since the 1970s shows some clear trends (see figure 12.2). Care by relatives accounts for about half of all the care provided to preschoolers with working mothers. Care by grandparents used to be the most common type of relative care, but it has declined somewhat since the 1970s. Care by other relatives has also declined. Father care has increased the most. For non-relative care, the use of sitters who come into the parent's home to stay with the children has declined a little since 1977, but dropped much more before that (from about 15 percent in 1965 down to 7 percent in 1977). The use of family day-care homes also dropped from 1985 to 1991. Only the use of child-care centers increased significantly among the non-relative care options, nearly doubling between the mid-1970s and mid-1980s (O'Connell 1993; Phillips 1989).

Child-care arrangements change quite frequently because parents work and family arrangements change, because of parental dissatisfaction, because the children grow older, and because of changing availability of caregivers (Leibowitz, Waite, and Witsberger 1988). Shifts from relative care to some form of group care are the most common (Floge 1985). In order to keep working on a continuous basis, most parents use multiple child-care arrangements and make adjustments when one form fails or when they cannot afford the higher quality options (Spitze 1988). Even though most child-care workers do not get paid much, the costs of child care are difficult for many working parents to meet. Among families with children under age fifteen who pay for child-care services, annual costs amounted to $3,300 per year in 1991, or $63 per week. Families with incomes below the poverty line who pay for child care spent more than one-fourth of their monthly incomes on child care in 1991. Families above the poverty line spent about 7 percent of their family incomes on child care (O'Connell 1993).

With the strong emphasis now being placed upon women having their own careers, the problem of child care has given rise to a number of new efforts at a solution. Some women bring their children to work. For instance, about 9 percent of employed women were able to care for their preschool children at their place of work in 1991 (O'Connell 1993). This arrangement tends to work best during the first six months of infancy, when the child is still rather immobile and quiet (and usually even this is possible only in some office environments). Some of these women were also able to work at home, and thereby attempt to combine job and child care without disturbing other workers.

Another fairly typical solution to child-care availability is shift-work (i.e., employment hours other than the typical 8:00 AM to 5:00 PM daytime schedule) or part-time work (less than thirty-five hours per week). A growing number of dual-earner families are juggling their work days and work schedules to meet both employment and child-care needs. In 1991, 42 percent of pre-

FIGURE 12.2: Fathers and Day Care Centers are Caring for More Children

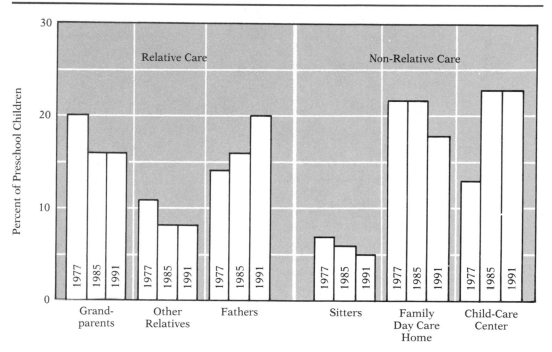

Since 1977, the percentage of preschool children (under age 5) with working mothers who were cared for by their fathers or in child-care centers increased. At the same time, the percent cared for by other relatives, sitters, or family day care providers decreased. (Figures for the share of non-relative care accounted for by grandparents in 1977 were estimated because the population survey did not differentiate them. Child Care Center includes nursery school and preschool.)

Source: U.S. Bureau of the Census, *Surveys of Income and Programs Participation*, 1985–1991; O'Connell, 1993.

schoolers had mothers who worked non-day shifts and 30 percent had fathers who worked non-day shifts (O'Connell 1993). One out of ten dual-earner couples have no overlap between their hours, allowing for alternating child care between them (Presser 1988). Shift work has increased significantly among both mothers and fathers in the past decade. In shift work families, fathers are more likely to be the primary source of child care apart from the mother (O'Connell 1993).

Alternative work scheduling such as "flex-time" is also on the increase. Under such arrangements, workers can vary their schedules depending on their work and family obligations, and in some cases perform duties at home as well as at the job site. Early research suggests that alternative work schedules are associated with more routine child care by fathers (Pleck and Staines 1985). Since the 1980s, part-time work by parents of preschoolers is also on the rise. In 1991, four out of ten mothers and one out of ten fathers worked less than full time. And when fathers lose their jobs altogether, they are likely to take over child care duties. In 1991, about 5 percent of preschoolers with employed mothers had fathers who were unemployed for at least four months, and most were cared for by their fathers (O'Connell 1993).

The Need for Child Care Is Increasing

While women are delaying childbirth longer and having fewer children than in the recent past, one significant demographic factor is driving up the population of young children. Because the baby boom cohort (those persons born during the period 1946–1964) is so large, and because they are now of childbearing age, a "baby boomlet" will increase the number of children born in the United States. As of 1980, the number of preschool-age children began to grow, and in 1990 there were about 23 million children under the age of six. Similarly, the number of school-age children (five to seventeen years) started to increase in 1986 and is now about 45 million. Given the continued growth in the number of employed mothers (see figure 12.1) and considering that most of them work full-time, it is clear that the demand for child care is going to increase in the near future. Maternal employment is not a fad that will reverse itself. Neither is it a phenomenon that is limited to mothers of older children, minority families, or single mothers, as many had previously claimed. These recent trends will undoubtedly put pressure on fathers to take on more of the routine parenting, on businesses to provide flexible scheduling and on-site day care, and on governments and social service agencies to provide decent day-care facilities.

The Family Leave Act of 1993 is the sort of family support reform that might be expected in the United States in the near future. This bill, originally vetoed by President Bush, was re-passed by Congress and signed into law by President Clinton in early 1993. It mandates just twelve weeks of *unpaid* leave to care for a new baby or a seriously ill family member, and only applies to companies with more than fifty employees. That it took a decade to pass this law, with its minimal benefits for parents, does not bode well for future significant reforms in this area.

Most other modern industrial nations make much larger contributions to the welfare of their families and children. In contrast to the minimal unpaid family leave that some U.S. parents now enjoy, Canadian workers receive fifteen weeks of family leave at 60 percent pay (Etzioni 1993, 60). Virtually every European nation provides at least that much support to individual parents, and most provide significantly more. For example, Swedish workers receive 90 percent pay for thirty-six weeks and prorated paid leave for the next eighteen months (Haas 1992). In addition, many European nations provide direct cash payments to parents when a child is born and most also fund the construction and operation of group child care facilities. The difference is that in Europe, child care is viewed as a public responsibility, and social welfare programs have a long history. Some European nations emphasize the benefits of state-supported child care in terms of women's employment and enhanced national productivity, whereas others emphasize the importance of such programs for optimum child development (Kamerman 1989). Either way, these countries invest heavily in the future by subsidizing their children's care. In the United States, we have taken another path:

> The ultimate question is why, despite the far greater investment in child development research in the United States and our knowledge of what is good for children, we continue to dither about who should do what and who should pay. Clearly, we know what to do, but we have lacked the commitment to act. (Kamerman 1989, 107)

Working Mothers: Does a Child Need a Full-Time Mother?

Only two or three decades ago, the question "Who takes care of the children?" would hardly have arisen in a book of this kind. At that time, the "experts" were practically unanimous in the belief that if a mother worked the development of the child would be jeopardized. By the early part of the century Sigmund Freud had already stressed the importance of libidinal/erotic interaction between infant and mother as a normal stage of psychological development. Then during World War II in Britain many infants and young children were separated from their mothers because of bombing raids and other wartime disruptions. These youngsters, brought up in overcrowded and understaffed orphanages, were more likely than usual to succumb to physical illnesses, to lack normal energy, and to be psychologically withdrawn (Spitz 1945). Belief in the negative effects of "maternal deprivation" was further reinforced by the findings of the experimental psychologist Harry Harlow, whose infant rhesus monkeys were raised with a terry-cloth covered wire "mother" and grew up to be very disturbed psychologically.

The strongest advocate of this position was the British psychoanalyst John Bowlby. He was much impressed by the wartime damage to children deprived of their mothers and treated a number of other children for psychotic withdrawal that he attributed to the same cause. In *Child Care and the Growth of Love* (1953), Bowlby argued that a small child needs to have a continuous, unbroken, intimate relationship with the mother or some *one* person who is substituting for her. Even partial deprivation of such contact results in psychological disorders and instability, Bowlby said. He went so far as to declare that a mother should not leave her child for any reason except major emergencies for the first three years of life.

Nevertheless, it has turned out there is reason to doubt these extreme pronouncements. For one thing, the evidence of Spitz, Harlow, and Bowlby all concerns unusual cases. Spitz's orphans were not only deprived of their mothers but went through wartime shock and lived in understaffed institutions. Harlow's monkeys were deprived of *all* living contact. Bowlby examined psychotic children but did not compare them with a group of normal ones to see if the latter also had had mothers who were often away.

More recently, there have been many opportunities to study children whose mothers are away at work for hours every day. The rising rate of female employment has made this a common situation (see figure 12.1). Studies on children with employed mothers have shown no negative effects on their psychological development (Spitz 1991; Hayes and Kamerman 1983; Bianchi and Spain 1986). There is also evidence that women who work provide better role models for their children and are especially likely to have daughters who develop a stronger sense of competence (Hoffman 1979). Daughters of employed mothers are likely to be independent and to plan employment for themselves (Moore, Spain, and Bianchi 1984; Bloom-Feshbach et al. 1982). Both sons and daughters are also likely to hold more accepting attitudes about women having careers and men doing family work. Children of working mothers tend to hold less rigid gender stereotypes than others and view women in general (as well as their own mothers) as more competent (Bloom-Feshbach et al. 1982; Wilkie 1987).

The women who enjoy their employment (i.e., career-oriented women) tend to be at least as good at mothering as full-time housewives. One reason may be that they are less dissatisfied with their situation (a pattern we have seen in chapter 11), and hence they provide a better psychological environment for their children. While employment is generally found to improve women's psychological well-being, employed mothers of young children tend to worry more than employed women without children (McLanahan and Adams 1987). This coincides with the more general tendency of parents with young children to report higher levels of depression and anxiety than nonparents (Gove and Geerken 1977; Pearlin 1974; also see chapter 11). Higher levels of depression can also be the result of money problems. Ross and Huber (1985) found that having children at home leads to economic strain that promotes depression, but that when economic strain is held constant, children actually improve women's psychological well-being. These findings suggest that mental health is more dependent on having enough money to live on than on having children per se.

Mental health is also dependent on having enough time to do everything that one feels is necessary. For instance, some studies have shown that the key factor in whether employment improves married women's mental health is whether husbands contribute to routine child care and housework (Kessler and McRae 1982; Ross, Mirowsky, and Huber 1983; Ross and Mirowsky 1987). Related findings show that employed single mothers experience the most distress and that employed married mothers who receive little help from their husbands are most likely to emphasize the direct costs of children: being tied down, being subject to burdensome demands, and having difficulty organizing one's time (McLanahan and Adams 1987). One study of married women (Ross, Mirowsky, and Huber 1983) found that the psychological benefits of working and having a husband who did some of the domestic tasks depended on a couple's feeling about whether the wife should have a job. Similarly, there is evidence that maternal employment can have adverse effects on children if family members hold unfavorable attitudes toward the mother's dual role (Stuckey, McGhee, and Bell 1982). In other words, children and adults can suffer if they feel that the mother should be performing only the traditional duties of wife and mother. In that case they may end up resenting her for working outside the home, even if she is forced to do so for economic reasons.

Also available are comparative studies of other societies, including some traditional and tribal ones, in which the household arrangement is different from our nuclear pattern. These show that women are best able to fulfill their roles *as mothers* when there are other people around who regularly are able to help with caring for the children (Whiting 1963; Lambert et al. 1980; also see feature 12.2).

There is no evidence, then, that upholds Bowlby's stern warning of the disastrous consequences of not having a mother continuously in contact with a small child. Many societies and social arrangements manage to do otherwise. An infant does need stability in the world immediately around it, and a special bond does tend to grow up with whoever is the child's regular caretaker. But there is no evidence that this person has to be the mother; it can just as well be an affectionate substitute caretaker, such as a relative, regular babysitter or child-care worker, or the child's own father. And having several persons who regularly care for the child seems to be a favorable experience for the child, not a negative one.

School-Age Children, Employed Parents

There has been less debate over psychic harm done to school-age children whose mothers hold jobs. Even these children, however, were called "latchkey kids" because they sometimes appeared at school with a house key hanging from a string around their neck. They were pitied because they returned home to an empty house rather than to cookies and milk with Mom. Because of their economic situations, more and more parents have felt forced to leave school-aged children unattended, or in the care of their siblings, until the adults can get home from work. Today, few researchers or other people think school-age children are generally damaged when their mothers work, but some still worry that leaving children alone will have negative consequences. Primarily because there is no one around to check up on them, unattended children are more at risk for delinquency and drug use. Self-care may have some positive benefits as well, however, and some children report increased independence and maturity as the result of having to take responsibility for themselves.

The Special Strains of Single-Parent Families

Single parents are not much different from parents in general, but their life situations tend to present some unique challenges and some special problems. In 1970, single parents represented just one in ten family groups in the United States, but by 1990, they were three of ten families. With almost 10 million single-parent households some of the stigma of being in a single-parent family has been removed, but many problems remain. For one thing, the relative economic position of such families has not improved. Families headed by women are six times more likely than two-parent families to have incomes below the poverty line, and female-headed households now constitute the majority of poverty households in the United States. One out of three single-mother households were living below the poverty level in 1991 (U.S. Bureau of the Census 1992d). This "feminization of poverty" creates hardship for women and children and increases the likelihood that the household will have to rely on extended kin networks for material and emotional support (McAdoo 1986; Stack 1974; Arendell 1986). Single-father households are actually increasing at a faster rate than single-mother households, but women still maintained 86 percent of the one-parent family groups in 1990 (U.S. Bureau of the Census 1992e).

Many single mothers work at low-status jobs and are subjected to some subtle (and not so subtle) forms of discrimination. Over the past decades, as single-parent households have become more numerous, some myths about single mothers have emerged. One myth is that a low-income, separated, or divorced woman raising children alone is single by choice. In fact many single mothers terminated bad marriages, some were deserted, and most are forced to support their children with little financial help from the absent father. A second myth is that single mothers could pull themselves up out of poverty if only they would get a job and work hard. The fact is that most single mothers are employed, but that the wages they earn are insufficient to keep them out of poverty. A third and related myth is that most single mothers are lazy and unworthy recipients of welfare who have babies just to receive government checks. Again, the fact is that welfare payments—Aid to Families with Dependent Children (AFDC)—are rarely sufficient to make ends meet, let alone generous enough to allow the mother and children to live comfortably. These myths are based on stereotypes that ignore diversity within the single-parent family population and ignore the realities experienced by single-parent families (Mulroy 1988, 4).

Because of the stigma of being in a "broken home," many parents in single-parent families feel pressure to be perfect. Many worry over the children's well-being and some compensate by being overprotective. Most, however, develop especially close relationships with their children. Parent-child relationships in single-parent families tend to be more democratic than in two-parent families. Out of necessity, there tends to be more task sharing and more joint decisions. There is typically more negotiation and bargaining in these families, and the lack of resources and outside help often make it difficult for the parent to be authoritarian. Sometimes single-parent families are characterized as being more "permissive," but what this really means is that the parent tends to give

Most parents in single-parent families develop especially close relationships with their children, which can be mutually supportive.

more weight to the child's wishes and the child is not required to be as deferential to the parent. Children in single-parent families are also sometimes called upon to provide emotional support to single parents, which reverses the traditional image of parent-child dependence (Weiss 1979). With the added responsibilities of being in a single-parent family, children can be said to "grow up a little faster" than children in more affluent two-parent families.

DISCOVERING FATHERS

We didn't just forget fathers by accident; we ignored them on purpose because of our assumption that they were less important than mothers in influencing the developing child. (Parke 1981, 4)

Before the 1960s, social scientists were relatively unconcerned with interactions between fathers and their children. It was generally assumed that if the husband performed the instrumental tasks of his provider role, and the wife her expressive tasks as homemaker and mother, all would be well with the children, the nuclear family, and the larger society (Parsons and Bales 1955). In the 1960s, studies began to include the father, but focused on "father-absent" families, suggesting that children without fathers had considerable difficulty in forming an appropriate gender identity and lagged behind others in academic achievement and moral development. These studies have been criticized for confusing father-absence with other social conditions such as living in a rough neighborhood or being poor. The father-absence studies have also been faulted because they used dubious measures of masculinity that portrayed "normal" boys as traditional unfeeling macho types. One of the assumptions of this research seems to have been that teenage boys needed fathers around to discipline them or else they might turn into juvenile delinquents. Another sexist assumption underlying some of the father-absence research was that boys from single-mother households lacked masculine role models and therefore might turn out to be effeminate "mama's boys."

Since the mid-1970s increasing attention has been paid to fathers as actual caretakers of young children. Research has focused on men's interest in caring for infants and toddlers, their capacity for child care, and on differences between male and female styles of interacting with children. Most of this research has focused on father-child interaction when the children are babies or preschoolers. Studies have documented that many men want to become involved in routine infant care and that men can be competent caretakers of newborns as well as of older children (Lamb 1981). This is not to say that most fathers treat babies the same way that mothers do. Researchers have typically found that fathers are more likely than mothers to engage in rough-and-tumble play, to treat sons and daughters differently (favoring sons), and to be more directive in their interactions with younger, as well as older, children (Lamb 1981). Mothers, in contrast, are typically found to be less directive, to more frequently offer verbal encouragement, and in home settings, to be involved in multiple household tasks while simultaneously tending to the children. The traditional pattern, then, is one in which child care is an ongoing and taken-for-granted task for the mother, but a novel and fun distraction for the father. For example, time-use studies show that many fathers will contribute to household "labor" by playing with the children while the mother cooks or cleans (Berk 1985).

Although some biologically based theories have suggested inherent limitations in men's abilities to nurture children (e.g., Rossi 1977), most researchers have assumed that sex differences in responsiveness to children are socially constructed (Lamb 1981). According to the most current social theories, men and women act differently toward children because: (1) An unequal distribution of power and status requires women to do most of the mundane family labor; (2) different socialization practices for girls and boys create nurturant women and insensitive men; and (3) women exhibit their competence as gendered members of society through mothering and men demonstrate their gendered selves through attention to work and inattention to children.

Although results are sometimes incomplete or contradictory, most child development studies find that fathers have positive effects on their children. For the most part, fathers who are both caring and demanding foster high self-esteem and promote academic achievement in their children. Must of this research, however, focuses on traditional families in which the father assumes few day-to-day caretaking tasks. For instance, one study found that the average American father spent only thirty minutes per week in infant care (Kotelchuck 1976). Other studies have found that fathers of preschool-age children average about two hours per week on direct child-care tasks (Coltrane and Ishii-Kuntz 1992; Russell 1983). With most fathers spending so little time interacting with young children, it is difficult to assess the potential impacts of fathers who play a more significant role in their children's lives. As Lamb (1981, 459) notes, "there is no reason to assume that 'nontraditional' fathers influence their children in the same way that traditional fathers do."

New Patterns of Fathering

If we are to believe prime-time television, fathers are acting more and more like mothers. They use the latest housecleaning products, change the baby's diaper, and even telephone young adult children who have gone off to college. While most fathers do not live up to these media images, it is safe to assume that today's fathers participate more actively in their children's lives than fathers did a few decades ago. What's more, a small but growing number of fathers are assuming major responsibility for raising their children either as single parents (Greif 1985) or more typically as shared-parenting fathers (Ehrensaft 1987; Coltrane 1995). Fathers involved in the routine details of daily child care are found to have even more beneficial impacts on children than kind, but distant, traditional fathers. In general, children with highly involved fathers are characterized by increased cognitive competence, increased empathy, less sex-stereotyped beliefs, and a more internal locus of control (Radin and Russell 1983; Pruett 1983; Radin 1982).

Because it is rare for fathers to share much of the child care in most families, it has been difficult to study the long-term impacts of active fathering. It may be that the positive child outcomes noted above are as much a result of the children being raised by two active caregivers, or in a relatively affluent two-income family, than with anything special about the father's contribution to the child. For instance, Lamb suggests that the positive family context for parents who are committed to sharing child care is probably what is contributing to beneficial child development: "What matters is not so much who is at home, but how that person feels about being at home" (Lamb 1986, 17).

While some researchers focus on the potential benefits of shared parenting for children, some focus more directly on the parents, claiming that further improvement in the status of women is dependent on the increased involvement of men in the family (Lamb and Sagi 1983). Others focus on the gains that might accrue to both men and women if fathers were able to develop their emotional capacities through parenting. The idea is that active assumption of parenting duties could help men shed confining masculine stereotypes and relieve women of the sole burden of providing emotional support to family members (Fein 1978). Because it is primarily men who physically and sexually abuse both women and children, it would be a mistake to assume that increased participation by fathers would be a good thing for all families. However, many families could benefit from fathers being more involved, especially if wives and children desire it.

Recent programs, classes, and research projects have suggested specific ways to encourage fathers' involvement in family life (Klinman and Kohl 1984; Levine, Pleck, and Lamb 1983; Hawkins and Roberts, 1992; McBride 1990). Lack of skill and self-confidence discourage many men from attempting to assume a greater share of child care, but such tentativeness can usually be overcome with encouragement and practice (Lamb 1986). Mothers often act as gatekeepers to men's involvement with their children, and their actions can have substantial influence on a new father's developing sense of competence. If she helps him feel like he is doing a good job (which may not be apparent to either father or child), he is more likely to continue assuming responsibility for child care. If, on the other hand, she continually corrects him and tells him that he is inept, he may give up prematurely.

For the most part fathers, like mothers, learn by doing. The routine experience of caring for infants has been found to facilitate parental responsiveness in men as well as women (Zelazo, Kotelchuck, Barber, and David 1977). Even inexperienced fathers who assume primary caregiving responsibility or share child care with their wives quickly develop the necessary parenting skills (Coltrane 1989; Pruett 1987). If fathers assume major (or equal) responsibility for child care, their interactions with children come to resemble those of traditional mothers more than traditional fathers (Field 1978; Pruett 1983). Unlike traditional fathers who tend to overstimulate infants and engage in rough-and-tumble play, primary-care fathers tend to interact verbally and let infants and toddlers direct the play. Unlike traditional fathers who tend to sex-type their children and give more attention to their sons, shared-parenting fathers appear to treat sons and daughters similarly (Pruett 1987). Thus, shared-parenting families may provide children with the equivalent of two mothers, even though the total amount of time children spend with parents might be about the same as in other families (see feature 12.4).

FEATURE 12.4
Can Men Mother?

In an interview and observational study of dual-earner couples with school-age children, Coltrane (1989, 1990) found that some of the fathers acted like mothers. One woman commented on how most people didn't believe her husband was as involved with their children as she was.

> I think that nobody really understood that Jennifer had two mothers. The burden of proof was always on me that he was literally being a mother. He wasn't nursing, but he was getting up in the night to bring her to me, to change her poop, which is a lot more energy than nursing in the middle of the night. You have to get up and do all that, I mean get awake. So his sleep was interrupted, and yet within a week or two, at his work situation, it was expected that he was back to normal, and he never went back to normal. He was part of the same family that I was.

For those families that shared most of the child care, the distinction between mothering and fathering became blurred. Parents reported that the children were as emotionally close to the father as to the mother and that children would seek comfort from either parent when upset or injured. Selection of one parent by the child typically followed being in the care of that person for the preceding period of time. For instance, if a child had been cared for by the father for the past several days and happened to awaken in the night from a dream, the child would call out for "daddy." If, on the other hand, the mother had more recently been "on-duty," the child would call for "mommy." In addition, many of the children whose parents shared child care roughly equally would confuse parental terms by calling the father "mommy" or the mother "daddy" without realizing that they had done so. In effect, the children were using the gendered form to signify "parent," and in so doing, indicated how mothering and fathering became merged if both men and women share "mothering" duties.

Other studies have come to similar conclusions about the social construction of parenting roles. Risman (1989) studied 141 single fathers and

(Continued next page)

FEATURE 12.4 (*Continued*)

discovered that four out of five had no outside help with housekeeping tasks such as grocery shopping, food preparation, house cleaning, or yard work. Virtually all of the single fathers worked hard to develop and maintain child-centered homes, and all had close relationships with their children. Risman concluded that when men "mother" they are very hard to distinguish from women.

Gender differences in our society are based as much on the differential expectations and role requirements males and females face throughout their lives as upon internalized personality characteristics. . . . When males take full responsibility for child care, when they meet expectations usually confined to females, they develop intimate and affectionate relationships with their children. Despite male sex role training, fathers respond to the nontraditional role of single parent with strategies stereotypically considered feminine. (Risman 1989, 163)

Why Do Some Parents Share Child Care?

The primary motivation for parents to share child care is that they both need to be employed. Whether they are near the bottom or the middle of the income pyramid, couples usually feel that two incomes are needed to maintain their standard of living. This means they need some form of daytime child care. Because full-day child care is costly, and because many believe that children should be with their parents, many are attempting to share the routine care of their children. Most must still use some form of paid child care, but an increasing number are trying to schedule employment so that they can alternate taking care of the children.

In addition, some adult men (and women) express regret that their own fathers were not involved in the emotional life of the family (Osherson 1986). When they contemplate having children, some vow to "do it differently" than their own parents did, and attempt to set up schedules that ensure that both parents will have time to be with the children (Coltrane 1990). Other families, of course, believe that only mothers should take care of children, and if the husband earns enough money the wife is able to stay home with the kids. This traditional pattern is increasingly rare among younger parents, however, who find that they both must work to make ends meet. As we saw in chapter 11, most employed mothers continue to shoulder the responsibility for domestic labor and thus work a "second shift" (Hochschild 1989). The high stress and "burn out" associated with the second shift are encouraging many mothers to demand help from their husbands. As a result, more and more couples are sharing the easier parts of child care—things like looking after children at home or taking them to the park. Whether large numbers of couples will share the more mundane aspects of child rearing—particularly the indirect child-care tasks of feeding and cleaning up after them—remains to be seen.

The handful of studies that have been conducted on shared parenting suggest some of the conditions that might promote such arrangements, but results are tentative and sometimes contradictory. There is some suggestion that sharing of child care (and housework) occurs more frequently when children are younger and the total household labor burden is heaviest (Berk 1980; Lamb 1986; Pleck 1985). In contrast, others find that fathers are more likely to take on or continue family work if child-care demands are low, i.e., when there are fewer or older children (Radin and Russell 1983). Some studies report that fathers' attendance at the birth or involvement in early infancy enhances chances for later shared responsibility (Coltrane 1989; Kimball 1983; Russell 1982; Russell and Radin 1983).

Some also find that couples who wait until their late twenties or early thirties to have children are more likely to share both child care and housework (Coltrane and Ishii-Kuntz 1992; Daniels and Weingarten 1982). This is probably because women are able to receive more education, establish careers, and set up patterns of task-sharing with husbands that persist after they have children. It is probably also the case that women who delay child-bearing are able to establish an identity apart from family roles and may be more willing and able to bargain for participation by the prospective father (Gerson 1985). In addition, men who delay having children are probably somewhat more likely than younger fathers to have established themselves in their occupations and to desire an avenue to emotional fulfillment that active fathering offers (Coltrane 1990; May 1982).

Most researchers also report some differences in family interaction between

In families where child-care responsibilities are shared, children are as emotionally close to the father as to the mother.

shared-parenting families and more traditional ones. For instance, some find more conflict in shared-parenting families, probably because couples cannot rely on standard roles and must negotiate new divisions of labor (Kimball 1983; Russell 1986; Russell and Radin 1983). While negotiations over who does what need not be explicit (Hood 1983; Coltrane 1989), many shared-parenting couples report more talk and more acceptance of voicing anger than in other families (Gilbert 1985; Kimball 1983). In some families, this translates into higher "marital satisfaction" (Gilbert 1985; Scanzoni 1979), whereas in others, it translates into "lower satisfaction" (Kimball 1983). Russell and Radin (1983, 150) suggest that high levels of conflict may tend to be limited to the time when the couple is making the transition to shared parenting, and may decrease as the couple adjusts to the sharing of tasks. Others report that shared-parenting arrangements often revert back to more traditional divisions of labor if things don't work out at first (Radin and Goldsmith 1983).

Reporting on an intensive study of a small number of shared-parenting families, Ehrensaft (1987) found that mothers retained major responsibility for managing child care, and that fathers typically remained in a "helper" role. Even in shared-parenting families, women tend to be responsible for things like buying children's clothes, planning music lessons, and arranging for baby-sitters. Ehrensaft also suggests that there was a "duel of intimacy" in these families. She found that the father's relationship to the children was more threatening to the woman than anticipated, primarily because it was easier for fathers to be intimate with their children than with their wives. Along the same lines, some studies report that when men are more involved with routine child rearing, women are less satisfied with the child-care arrangements (Pleck 1983; Baruch and Barnett 1981). What seems to happen is that issues that are hidden when mothers do all of the child care are suddenly thrust to the foreground when fathers begin to participate. What was taken for granted in the past now has to be negotiated as both parents go about making daily decisions with and for the children.

Mothers' ambivalence about the father's participation in child care is probably also the result of having to give up authority in the one area of life that women have traditionally controlled. For instance, mothers sometimes report that encouraging fathers to participate fully in family work entails accepting his "looser" standards (Coltrane 1989). This can create problems for the mother, who is still judged negatively by others if the house is a mess or if the kids wear something weird. Fathers, too, are sometimes discouraged from spending too much time with their children by comments from their male friends and co-workers (Lein 1979). The presence of women relatives in the neighborhood, especially mothers, also plays a role in discouraging parents from experimenting with new forms of shared child care (Riley 1990). Because the 1950s ideal of exclusive maternal child care is still with us, mothers often feel guilty for not spending more time with their children and fathers feel that they are intruding in the mother's domain. As is true with many forms of social change, actual behaviors often precede shifts in ideology.

GROWING UP

The Struggle over Love and Control

Parent-child interaction is not one-sided. It is not merely a matter, as one might think from reading some guidebooks for parents, of the adults doing certain things to make their children behave and develop in certain ways. Both the parents and the children have wishes and aims of their own, which often may clash or run at cross-purposes. A child, no matter how small, is an active agent. Hence there tends to be a two-way struggle for control: the child is trying to control the parent, while the parent is trying to control the child.

Parents have certain advantages in this struggle: they are bigger and stronger, and hence they can often physically move the child around and make him or her do what they want. Moreover, they were there first; they have set up their local world the way they want it (or at least they have made their own adjustment to the world around them), and the children face a preexisting situation that they must fit into. At least at first, parents have the tremendous hidden power of being able to define reality for their children: to give them their world view, to explain "the way things are," and hence to shape their behavior by shaping their beliefs.

What advantages do children have? Their main advantage is simply their own attractiveness. There is a reason why small children are usually cute and cuddly; the same pattern is typically found in other species of animals, too, in which the young (kittens, puppies, fawns, etc.) are especially appealing. Some thinkers believe in fact, that biological programming accounts for parents' attraction to their offspring, which encourages care for them during the vulnerable early period. Additionally, a special bond can develop between the parent and small child that makes the latter seem especially attractive, even if outsiders don't think so.

This means that the child's main resource is the parents' love. Much of the analysis of parent-child bonds has concentrated on the child's side of the relationship: whether the caretaking is adequate to make the child bonded to the mother or other caretaker. But the other direction is probably even more important: whether the parent becomes bonded *to the child* enough so that the parent will have an emotional need to take care of the child.

Parental bonding does not necessarily happen automatically. It is especially likely in our society (for reasons we will examine shortly). But even here, it is all too often that the parent is not very strongly bonded to the child, with the resulting potential for child neglect or abuse if a sense of duty does not prevail.

Freud believed that human relationships are derived from certain basic drives, especially the erotic one (libido). We need not take this literally, but there does seem to be an important element of truth in the general conception. The parent's bond to the child is a form of love, not unlike adult sexual attachments; similarly, the child's attachment to the parents and desire for parental love are analogous to later sexual demands. There is evidence that women become sexually more responsive after they have borne a child. Nipples are erogenous zones as well as dispensers of milk for breast feeding; hence there is a kind of overlap or mixture of maternal and sexual behavior in the fact that a mother's nipples become erect and she experiences the desire to lactate upon hearing cries of her newborn infant (Rossi 1984). Also, women nursing a baby often feel uterine contractions similar to those resulting from orgasm; these help shrink the uterus back to normal size after childbirth.

The child's major resource, then, is to arouse a feeling of love in the parent (especially the mother, although there appear to be analogous processes in fathers). A good deal of children's cries and behaviors are methods of getting the parents to focus on them, to give them attention. Many of the little struggles that go on between parents and children are of this sort. For example, parents want to talk with visiting friends, while their children run around more and more excitedly making noise. "Why do you have to behave like this just when we have company?" a parent may say in exasperation. But that is just the point: the children act like this at exactly this time because the guests usurp their parents' attention. Moreover, it is one of the characteristics of the "primitive" desire for love that *any* form of attention—even negative—provides some satisfaction of the desire. Children will run around and misbehave if that is the only way they can get attention, even if the attention consists of angry commands or even punishments.

Parents' Control Techniques

Even though a family's resources—time, energy, and emotional investment, as well as money and possessions—do have limits, they can be squandered or multiplied, depending on how they are directed. For example, if parents and children are continually at loggerheads, both sides use up energy with little to show for it. For parents to help children progress through their life stages, resources of time and energy are required, especially in the short run. Over the long term, however, conflict will not consume so many resources, and more will be available for everyone's enjoyment and further growth. This is perhaps the real meaning of the aphorism "Nothing succeeds like success."

As parents seek to guide their children while also preserving their own well-being, various techniques may be employed. Which are chosen depends on many factors—what part of the world they live in, their social class, their personal family customs, the resources available to them, and what they may have learned in their efforts to become proficient parents. Common methods may be categorized as reward, punishment, shame, and love.

Reward. This is one of the most common forms of control, although parents are often not aware of when they are rewarding children's behavior. There are various kinds of **rewards.** *Material* rewards may consist of giving children candy, money, or toys as an incentive for doing what the parent wants them to do: a dollar for a good report card, a cookie if you clean your room, etc. One drawback of this method is that the child comes to expect a reward for every accomplishment and will not perform without one. In addition, children will focus on the reward rather than the action. For instance, they will see no intrinsic value in reading a book but only do it in a perfunctory way in order to get the reward.

Another type of reward is *social:* the parent rewards the child with attention such as play or talk. Here again one has to consider that the child will become more attached to the reward than to the behavior and demand sociability in return for performance. But one might see this as a desirable outcome: the child will like the parent and become sociably oriented.

Control by material rewards does require that the parents have enough wealth. Hence we would expect this technique to be used more in wealthier societies and higher social classes. The social rewards of paying attention to children and devoting time to playing with them are in a sense even more costly, since they require the parent to spend time and energy. These kinds of rewards have rather good outcomes for the children's behavior, as far as parents are concerned; but not all parents are able to use such techniques, because they simply lack the leisure. Hence it is not surprising that social rewards are used most by the affluent middle classes in wealthy societies with plenty of leisure time, such as our own (Whiting and Child 1953; Sears et al. 1957; Maccoby 1968).

Punishment. This type of control consists of either physically spanking or hitting the child or depriving him of something desired. Punishment can consist of threats and angry tones of voice as well as overt actions. The psychologist B. F. Skinner (1969), who experimented largely with animals, argued that punishment is not a very effective means of control. When punished, any creature's first reaction is usually to fight back; if that is impossible because of the opponent's superior strength, then to run away; and finally to comply, but dully and unenthusiastically. If an extreme amount of punishment is used, the subject is too beaten up to be able to comply.

Nevertheless, physical punishment is fairly popular in our society, and sometimes is even escalated to the extremes of causing bodily, and emotional, harm. It is generally used more against boys than against girls. The results tend to be (Goode 1971):

- Boys who fight back against their parents, and who are aggressive toward outsiders.
- Boys who strongly identify with their fathers and acquire very masculine or

"macho" personalities, with authoritarian and ethnocentric (bigoted) attitudes.
- Beliefs about right and wrong that are based, not on internalized moral standards, or conscience, but simply on fear of punishment if one is caught.

Despite all these drawbacks, physical punishment is a common method of discipline. It may often be used, for instance, in hard-pressed working-class families or in rural cultures because it is a relatively cheap form of control, and these people may have no time to spend on more effective, psychological methods.

Shame. Shaming or ridicule is a kind of control in which the child is held up as a negative example to the group. It is widely used in a number of tribal societies, which anthropologists refer to as "shame cultures" (as compared to our own society, which is more typically called a "guilt culture"). But it is used in some families in our own society and elsewhere in the modern world. The anthropologist Lawrence Wylie (1964) gives a vivid description of how this approach was used for school discipline in a small town in rural France: the child who broke some rule was paraded through the town square wearing a placard around his neck, upon which was written the offense ("I threw erasers in the classroom," or whatever), while everyone in the school as well as the town was lined up to watch. Many Americans would consider this a mortifying experience, since our culture has fairly strong feelings against singling out persons for public embarrassment.

Shaming is a form of social punishment that is hard to counterattack or escape. Hence it does not have some of the negative consequences of sheer physical punishment. It tends to produce personalities who strongly emphasize self-control, especially over public demeanor. They typically become very careful of how they express emotions: not that they are necessarily emotionless, but their behavior is calculated to conform to what is expected in a given situation. In other words, shaming leads to a personality type that is most strongly concerned with meeting group expectations. But this is external conformity, not an internalized sense of right and wrong; when the group's demands change, people of this personality type rapidly change their behavior in response.

Control by shaming happens most often in societies in which people live in dense settlements with little privacy. It does not work very well in modern urban societies, which do not provide much surveillance over the individual, and in such societies it is not often used.

Love. Control by love is mostly discussed in the child-development literature as a form of manipulation of the parent's love, as a reward for compliant behavior and a punishment for disobedience. The term *love-*

deprivation is used to refer to such commands as "All right, if you won't behave, I won't love you any more." These kinds of extreme and overt instances are especially common in the psychiatric and psychoanalytic literature, including such variants as "You don't love your mother. If you did love her, you wouldn't have been so bad as to do what you did." But even without saying these kinds of things (which indeed many parents do say), it is possible to convey to children that the rewards they receive from their parents are highly contingent upon how they behave.

Such control by deprivation produces a child who has strongly internalized the parent's point of view (Sears et al. 1957; Maccoby 1968). Deprivation of love poses a devastating threat to a small child, and hence there is little he or she can do but try to comply. Often this is difficult, because of the demanding personalities of the parents who use this kind of control (since they themselves were usually brought up this way). The resulting personality tends to feel that moral standards are absolute obligations, regardless of the consequences in the external world. Children brought up in this way tend to be emotionally inhibited and sexually repressed, though their sexual attitudes may come through in a highly romanticized and unrealistic view of their possible lovers. Often this is combined with strong self-discipline and striving for achievement. Love-deprivation is the only technique that produces strong feelings of guilt for breaking some rule, even if there is no chance of getting caught. In short, this technique produces Freud's classical strong superego and many of the classical Freudian neuroses.

Control by threatening to withdraw love requires that the parent spend a great deal of time and emotional energy on each child. It seems to occur most often in small modern families in which the mother is a full-time housewife with continuous contact with her children. If this arrangement produces a highly moralistic personality in the child, it is not necessarily because we have a guilt culture. Rather, the family pattern that produces this type of personality seems to be tied to the structural situation in which middle-class women derive all their status from their family position as wives and mothers. In such a situation, a great deal of psychological intensity can go into child-rearing. If this method seems to be on the decline, supplanted in recent middle-class families by more emphasis on social rewards, the cause may be in the shift that allows women to derive more status from roles outside the family, especially careers.

There are, however, other ways to love one's children. The *social rewards* discussed above are a use of love as a selective reward for performing certain behaviors, just as love-deprivation uses love as a punishment. As Skinner's principles would predict, the reward method generally has more positive consequences than the punishment method.

Best of all, many psychologists argue, is simply to *love one's children unconditionally.* The parents who spontaneously show affection for their

FEATURE 12.5
Some Nondestructive Ways of Controlling Children

No matter how permissive parents are, there are always times when they have to say no to their children. When kids are small, parents have to prevent them from doing things that are unsafe. When they are larger, parents have to be peacemakers, to check their aggression against each other or even against adults. There are also head-on-collisions between what the adults and children want to do. Typical in almost all families are the times children simply won't go to bed, though the parents don't want to be interrupted or the children need to sleep so they can get up early in the morning. How can this be handled?

The problem with parents' usual methods to get children to stay in bed is that the methods themselves reinforce the behavior that keeps the children awake. They stay awake by demanding attention, and they think of excuses to get it. The request for a glass of water or to go to the bathroom or any of a thousand other "problems" all are successful if they get the parent to come and pay attention. Getting mad does not help, nor does making threats, offering enticements, or anything else the parent says, because the fact that the parent is talking to the child and providing attention is what the child wants. One method that works is to: (1) attend to the child's physical wants before he goes to bed (bathroom, glass of water); (2) tell the child that you are not going to talk to him anymore until morning and the child shouldn't talk either; (3) be firm and stick to it. Every time the child comes in and wants something, the parent should gently, firmly, and *silently* take the child back to bed. This may have to be done repeatedly at first, but by the second or third night, the child will probably go right to sleep.

Another method called *time out* can be used effectively in situations where punishment is necessary. For instance, when a child is attacking other people or destroying things (children do tear things up from time to time) or throwing a tantrum, parents need to intervene. But punishment often does not work, and of course it has many bad side effects. Yelling angrily at the child is even worse, since it creates the hostility and fear that are negative results of punishment and often does not even stop the behavior. The parent who is doing something else and half attentively keeps telling a child "Don't do that!" is actually teaching the child that the words don't mean anything. The child may even keep on performing the annoying act just to provoke the parent into repeating the phrase. Attempting to shame the child or make him feel guilty also has negative effects on personality.

"Time out" avoids these drawbacks. It consists of putting the child *alone* in a room. Often this can be a bathroom or bedroom. The child is told something like: "I'm putting you on two minutes of time out. Please remember while you're in there that you're not supposed to hit your brother with a baseball bat. You can come out when the time is up." For small children, it is useful to use a kitchen timer with a bell, so that they know when their "time out" is over.

Why does the method work? Because it removes the child from the situation in which the misbehavior occurred and provides an opportunity for the child to calm down. It breaks the social situation in which the trouble arose, and it does not substitute a new troubling situation in the form of a fight with a parent, which is what happens in conventional punishment. Two minutes (or even one minute for small children below age five) does not seem very long, but it is effective for children, as their emotions tend to be very volatile and can change completely within a short period. What happens, one may ask, if the child insists on coming out of the bathroom and resuming the forbidden behavior? The parent simply puts the child, gently but firmly, back in "time out" and adds a minute to the time. As in the going-to-sleep-peacefully method, the first time or two this approach is used the parent should be prepared to devote some time to following the routine repeatedly. But within a few days, a child's negative behavior can be very successfully controlled. (Patterson 1976) children during the normal course of the day thereby help maintain a bond with them. This happens apart from whether the child is doing anything good or bad at the moment. It helps maintain a good fundamental relationship, whichever specific control techniques the parent uses in regard to the child's specific behavior (Rohner 1986).

It should be borne in mind that parents are not the only influence on how children develop. Parents may use one or another type of control, or a mixture of them. Nevertheless, the outcome will depend also upon the child's life away from home. Child-care arrangements, and later, schools, add external influences. Playmates and other peer groups become an important social reference point, especially by the preadolescent period. Perhaps even more important is a factor that has received relatively little attention: how much time the child spends alone. We know, for example, that an adult's degree of personal autonomy and creativity is strongly influenced by the amount of time the person spent in solitary pursuits as a child. Not only the controls used by parents, but also how much a child is free of external controls, affects personality. There are also important influences on personality that can happen after one grows up. Since social class settings influence one's style of thinking and acting, moving into different occupations will tend to change one's personality. An individual will change depending on the social reference groups they belong to and the significant others they interact with. Personality is not inalterably set during childhood, and children are not merely passively manipulated by their parents. But the childhood situation is an important beginning for what individuals do later.

The Social Context of Development

Most models of childhood development assume that socialization is a one-way process: as children pass through various developmental stages, parents socialize them to conform to a known set of cultural standards. This premise, sometimes called the "social molding" perspective (Peterson and Rollins 1987) has been criticized by scholars who have come to realize that children socialize parents too, and that what gets learned is rarely what one thinks is being taught. More recent research tries to take account of the fact that socialization is at least bidirectional and attempts to account for influences of social context as well as direct parent-child effects. Things like the parents' marital relationship, sibling relationships, extended kinship ties, the neighborhood, ethnic identification, social class, schools, churches, peer groups, and television are all strong socializing influences on children.

One of the biggest differences between children of today and those of the past has to do with the fact that modern American children spend so much time in school or in front of a television set. Schools, for instance, teach children to obey orders, conform to rules, and to be punctual, along with trying to instill the three Rs. With compulsory education, more children are subjected to similar socialization pressures, which may lead to less regional diversity. Although children spend much of their time in school, the average American child spends a greater amount of time in front of the television. The mass media exposes children to styles, values, and stereotypes that may not be important in any particular family, and again tend to standardize all children's experience. Commercial television also promotes consumerism and teaches children to be passively entertained. All of these influences—some good and some bad—shape how children look at the world and feel about themselves. In order to understand how children develop, we must take these varied influences into account. As our models of child development become more complicated, we begin to get a better sense of what's important, but we also realize that there are no easy answers to the question of what children need.

Psychologist Jerome Kagan (1976) notes that we carry three prejudices regarding the development of the child. The first is that the child is seriously influenced by the actions of others. The second is that children develop in a series of discrete stages that must be mastered in sequential order. The third and most important is that there is an identifiable set of psychological traits that are necessary for a child to develop into a happy, well-adjusted adult. Contrary to this set of assumptions, Kagan says that children do not require any specific actions from adults in order to develop optimally. Why is this so? First of all, assumptions about what is appropriate or healthy behavior for children and adults can vary tremendously from one culture to the next. In order to specify what children need, one must first know what the specific demands of the community are. For instance, middle-class Americans tend to value academic success, autonomy, and independence, and to have permissive attitudes toward hostility and sexuality. These traits reflect an individualistic ego-ideal that tends to favor self-interest over intimacy, competitiveness over cooperation, and narcissism over altruism. When we compare child-rearing patterns in other cultures to our own, we see that these are not traits that all parents want to instill in their offspring. In an effort to identify what all children need, Kagan (1976) specifies the following general psychological requirements.

- Infants: assimilable environmental variety;
                 regularity of experience;
                 predictable human caretaking; and
                 a chance to practice developing skills.
- Preschoolers: exposure to language (talk);
                 affirmation of self-worth;
                 models to identify with; and
                 consistent standards.
- School age: mastery of school's requirements;
                 success in some peer-valued activity; and
                 models to identify with.

The specific actions that parents take with their children may be less important than most of our theories suggest. Contrary to the volumes of material written on the "correct" way to raise children, what children need depends most on their social context. In today's world, that social context is changing faster than ever before.

Some Social-Class Differences in Child-Rearing Styles

Parents' social-class position strongly affects the way they bring up their children. There is a good deal of evidence (Kohn 1977; Kohn and Schooler 1978) that middle-class and working-class parents hold different ideals of children's behavior. Middle-class parents, both mothers and fathers, tend to want their children to grow up happy, curious, interested in the world, and (especially girls) considerate of others. Working-class parents, on the other hand, want their children to be obedient (especially boys), neat and clean. Interestingly enough, the working-class parents stress more than middle-class parents that their children should get good grades in school (even though, in fact, middle-class children tend to get the higher grades). Similar class differences have been found in Canada, Europe, and Japan (Lambert, Hamers, and Frasure-Smith 1980); in all of these places the working-class parents were more likely to censure insolence and temper and to insist on good manners while restricting children's autonomy and comfort.

What is going on here? Melvin Kohn's interpretation is that the working-class parents are stressing behaviors that are vital to at least maintaining the status they have. They want their kids to be obedient, neat, polite, and so forth because failing to meet these standards might lead to trouble with the law and living like "low-class," "not-respectable" people. Their position on the social ladder is often tenuous enough that these feel like real threats. Success in school and a good appearance, on the other hand, are perceived as steps to upward social mobility.

However, working-class parents also stress obedience and punish insolence because polite, obedient, conforming behavior is seen as necessary to job survival, at least in their own lines of work. Middle-class parents, in contrast, tend to have jobs where initiative and self-direction lead to success. This may be the reason why they are less apt than working-class parents to stress conformity and obedience in their child-rearing, and more apt to teach independent decision making. Kohn tested this theory by dividing up all occupations according to whether they involved *close supervision or autonomy,* work of *high or low complexity,* and work of *high or low routine.* In fact, the middle-class occupations tended to have less routine and more autonomous, more complex work calling for personal decisions; the working-class jobs were the opposite. Moreover, working-class people whose positions did involve more autonomy and complexity and less routine had goals for their children rather similar to those of the middle class. (And conversely, middle-class jobs that actually had little freedom and challenge produced a working-class style of values.) Kohn also found that rural occupations tended to be more like working-class occupations; hence rural child-rearing styles also tended toward the authoritarian, conformity-emphasizing style. These differences in child-rearing styles tend to perpetuate class divisions to the extent that they prepare kids to function best in the class to which they were born.

One more difference between middle-class and working-class child rearing is that the parents in a middle-class family are more likely to agree over how to bring up their kids (Kohn 1977; Komarovsky 1962). This may be partly due to the fact that many wives of working-class males themselves hold middle-class jobs (such as clerical employment). The wives therefore have more middle-class values, which clash with the working-class values held by their husbands. Also, there are more sharply segregated gender roles in the working class; husbands and wives agree less about how to rear their children because they live fairly separate lives. Middle-class parents, on the other hand, are more likely to present a "united front" to their children.

Working-class families also tend to see sharper differences between the gender roles their children should fit into and to insist on conformity to these stereotypes (Lambert, Hamers, and Frasure-Smith 1980). This was especially so for the way they expected their sons to behave. Working-class parents tended to be harsher on misbehavior by their sons than by their daughters, while middle-class parents reacted more sharply against insolence on the part of their daughters. Again, these differences may be related to adult realities. Working-class men often have less chance for expressiveness than their wives do. In the middle-class, in contrast, men often are allowed more initiative, while more conformity is required of women. In general, fathers turn out to be more lenient toward their children's misbehavior than mothers are, perhaps because men get less blame for children's faults.

Placing Children in the Social Mobility Game

The development of mental, psychological, and social faculties is one factor that shapes a person's adult life. Another factor is the social level of the family into which one is born. Children usually end up relatively near their parents' occupational level. This is not to say that they usually have identical positions (though there is a tendency, for instance, for doctors to be sons of doctors, and professors the sons of professors—both professions, with an historically strong gender bias). The further away an occupation is on the prestige scale relative to parents' occupation, the less likely a child is to attain it. This is no surprise even though a fairly small number of people today directly inherit an occupation (such as a family business). The family's main effect on social mobility (or, more likely, social immobility) is its influence on

FEATURE 12.6

IQ: How Much Is Inherited? How Much Difference Does It Make?

The nature-versus-nurture debate has been going on for a long time in the field of IQ. By now, it is generally agreed that there is some biological, hereditary component in IQ. But how great is it, and how much is learned in response to social conditions?

The main research methods for answering this question attempt to compare the IQs of people who have varying degrees of kinship; or they measure changes in IQ over time and in different social circumstances. The classic kind of study is to compare identical twins who were reared apart. These have exactly the same genes (since identical twins are produced by the splitting of a single fertilized egg, whereas nonidentical twins are produced by separately fertilizing two different eggs). If they have been reared apart in different social environments, that should control the "nurture" effect so that whatever correlation there is between their IQs should be what is due to genetics.

As we can see from table 12.1, studies of identical twins reared apart have found correlations between their IQs ranging from .60 to .90. These are the highest correlations in the table except for identical twins who are reared *together*. For many years, such figures were cited as evidence of the importance of heredity. However, the sociologist Howard Taylor (1980) pointed out that, just because the twins were reared in different households, their environments were not necessarily different in important social ways. The psychologists who did these studies ignored patterns of social class or other features of the families and communities in which these twins were brought up. Taylor went back and looked at the primary data on these studies and separated them by degrees of real social difference. Many twins were actually brought up by foster parents whose occupational and educational levels were the same. The twins were usually separated because of some kind of family crisis, and often relatives adopted them. In many

TABLE 12.1 IQ and Kinship

Degree of Kinship	Correlation
Unrelated persons	
Reared apart	.0
Reared together	.18–.30
Foster Parent/foster child	.20–.40
Parent/child	.20–.80
Siblings	
Reared together	.35–.75
Reared apart	.35–.45
Twins	
Nonidentical, opposite sex	.30–.75
Nonidentical, same sex	.40–.90
Identical, reared apart	.60–.90
Identical, reared together	.70–.95

Source: Taylor, 1980, 28–30.

cases, the twins remained in the same town and went to the same school. Moreover, some twins were separated from each other at a relatively advanced age, which gave them a number of years in exactly the same environment.

Taylor's reanalysis had strikingly different results. Those twins who were reared in similar environments had IQ correlations ranging from .66 to .89, with an average of .87. Those who were reared in socially different environments had IQ correlations ranging from .36 to .51, with an average of .43: not particularly striking results, compared to many other figures in the table. Taylor's reanalysis thus showed that *the separated twins studies actually demonstrated greater importance of environment as compared to genetics.*

There is plenty of evidence for the same conclusion. For example (see table 12.1), just ordinary siblings who are reared together have higher IQ correlations than those who are reared apart. Nonidentical twins of the same sex have

(Continued next page)

FEATURE 12.6 (*Continued*)

higher IQ correlations than those of different sex: almost certainly because two girls or two boys are treated more similarly than are a boy and girl. And nonidentical twins have higher IQ correlations than nontwin siblings, even though they are no more closely related genetically: again, because the twins are exposed to more common treatment.

How much of IQ, then, is hereditary? The statistical rule of thumb is that a correlation should be multiplied by itself to arrive at the percentage of variance that is explained; thus, taking the average r of .43 for identical twins who lived in genuinely different environments, we get an r^2 of .18. In other words, something like one-fifth of all differences in IQ seem to be due to heredity.

What Difference Does It Make?

IQ scores thus seem to be more strongly influenced by the family environment and the way the children are treated than by heredity. But whatever its source, how much does IQ affect one's life? IQ is quite strongly related to school achievement and thus has an effect on one's placement in a career. Nevertheless:

- IQ operates mainly within the school system and has little effect in predicting subsequent achievement once one has left school (Bajema 1968; Eckland 1979).
- An individual's IQ score is at its peak during the teen years and declines as one grows older. Hence, children's IQs are much closer to their parents' IQ (average .69 correlation) than adults are to their own parents (average correlation .30) (Taylor 1980).
- There is a wide range of IQs within the same

occupation (Thomas 1956). IQ is not a *sine qua non* for doing any particular job. Even retarded persons having tested IQs of 60 (below the "normal" 100) may turn out later in life to do relatively well in jobs, especially if they have more middle-class patterns of dress and behavior (Baller et al. 1967).

The IQ test thus seems to be a ritual created by our bureaucratic school system, which measures and guides students in placement through the school. It is important in our society only because the educational credentials for procuring jobs have been driven up as a mass, and the competitive school system has expanded (see Collins 1979 for an analysis of this process). However, IQ is not intrinsically related to how well one can do in an actual career.

How Much Can IQ Scores Be Raised?

Major effects on IQ seem to be produced by the home. There have been some efforts to raise the IQ scores of poverty-level children by training them in preschool programs. These were able to raise IQs by seven points, although this difference tended to go away by the time the child had been in the regular school program through the sixth grade (Consortium for Longitudinal Studies 1983). The most notable long-term change took place in Japan after World War II. Japanese born during the 1940s have an average IQ of 104, while those born in 1960 have an IQ of 115: far above the U.S. average of 100 (Lynn 1982). The rapid change shows the variability of IQ scores, in this case probably responding to better nutrition and other environmental factors (such as stress on achievement at school).

how well children do in school. This is ironic because the school system, at least officially, claims to be a "meritocracy" that operates to overcome the advantages and disadvantages that children acquire from their families. Instead, family influence on school achievement turns out to be one of the most important factors researchers have uncovered so far. (There is also IQ; but see feature 12.6.) Children from different social backgrounds come into school with different "cultural capital" and get tracked into different programs that support their achievements, or stigmatize them and lower their levels of aspiration (DiMaggio and Mohr 1985; Rosenbaum 1980; Mercer 1974). In addition, schools in wealthier areas spend more on each pupil, further increasing the educational gap between social classes.

But school only partly determines one's occupational achievement. What else is involved? Many factors are out of an individual's control, such as the state of the economy when one enters the workforce (there are more opportunities during economic booms than during depressions, for instance). Another factor is population patterns: there are fewer opportunities for an age group that enters the work force right behind one that has filled most of the available vacancies. There are also social patterns of bias, such as those that have favored males over females in getting higher level managerial and professional jobs, and that have discriminated in favor of some ethnic and racial groups and against others. The family and the school are only some of the factors that will determine how individuals move through their careers. Structural opportunities and lack of opportunities can thus cut across personality patterns acquired in childhood; and they can also inflate or deflate the value of educational credentials that one has acquired at a particular time (Collins 1979).

Self-Concept and Career Aspirations

As children grow up, they subtly shape their images of themselves and their career aspirations to fit the opportunities and models provided by their social milieu. According to Linda Gottfredson (1981), this happens in stages:

- *Ages Three to Five: Awareness of Size and Power*—Small children at nursery-school and kindergarten level are usually still at an "intuitive" or "magical" stage of thinking. If asked about what they want to be when they grow up, many will express fantasy wishes (such as being a princess or a bunny). Around age five they usually outgrow their beliefs in magic and begin to associate power with being an adult. But they have little conception of what adults actually do and associate adulthood mainly with being large in size.

  One important self-conception that has already developed by this age is *gender self-labeling*, which occurs by around age two or three. By three or four, children prefer playmates of the same sex, and at four or five they pick up some of the male-female stereotypes for adults. But because children before age six do not have a firm notion of the constancy of objects, many of them accept the possibility of females turning into males or vice versa (Kohlberg 1966).

- *Ages Six to Eight: Acquiring Occupational Sex Stereotypes*—At this age children begin to rule out occupations that they consider appropriate only for the opposite sex. Until quite recently, children almost all learned to stereotype the more "macho," physical, or power-wielding jobs as masculine, and positions such as secretary, nurse, manicurist, librarian, and elementary teacher as feminine, along with other socially oriented jobs (such as psychologist and social worker). This sex-role stereotyping is especially strong in the working class (which is not surprising judging by the theory of class cultures: see chapter 6). At this age, though, children have very little understanding of the actual pay, power, or work conditions of various jobs. Both boys and girls accept the stereotypes (or used to, before the recent rise in female career aspirations), and both think that their own sex is superior. Children around age six to eight thus narrow down the range of what jobs are acceptable to them because of their gender, and these preferences tend to remain fixed up through the end of high school.

- *Ages Nine to Thirteen: Awareness of Social Ranking*—At this age children begin to be aware of the class system. Not that Americans are very explicit about social class: we tend not to use the term, and we subscribe to an ideology that everyone is more or less equal. But at the same time, children now begin to recognize that some jobs are too "high" or too hard for them to get, while others are unacceptably low for themselves. Here is where their social-class background begins to shape their images of what they can attain. At the beginning of this stage (around age nine), boys are still aspiring to be policemen, truck drivers, auto mechanics, and athletes, while girls focus on nurses, teachers, and entertainers. They make little distinction between low-prestige and high-prestige jobs.

  As adolescence approaches, these aspirations become more "appropriate." Boys, especially in the higher social classes, begin to shift away from blue-collar jobs and to think of professional and executive levels. Working-class boys, though, do not rule out manual labor and begin only to rule out fantasy occupations (such as professional athlete). Girls' aspirations, on the other hand, tend to *decline*: fewer aspire to be teachers and nurses and more become willing to settle for clerical work. It should be borne in mind that most of these studies reviewed by Gottfredson are from the 1960s and early '70s and hence reflect a much more sexist period than the one we are in now; but even now, there are still plenty of social pressures that stereotype female jobs and apparently especially influence women from working-class backgrounds. Among both sexes, the children adjust their aspirations to the level of their families and the people around them and to their school performance. Those who no longer expect to go much higher occupationally may well stop trying to accomplish anything in school.

- *Fourteen and Up: Acquiring a Unique Self-Image*—Teenagers now begin to narrow down their likely career roles, from a general range of jobs of the "appropriate" gender and social-class level, and to pick out a specific career tailored for themselves. This is partly connected to the move into the Piagetian cognitive stage, at which children have developed an ability to manipulate concepts in their mind, without external social influence. Preadolescents are more likely to have absolute notions of right and wrong, defined as whatever their parents and teachers say, which must always be obeyed. Teenagers develop more self-consciousness and reflectiveness; socially these often take the form of identity problems, in which a youth's self-esteem fluctuates a good deal, and many regard themselves as having very lcw status in the eyes of their peers as well as of adults.

The adolescent usually comes out of this period having settled on some unique self-image. This includes the focus on a specific occupation that the person thinks he would be good at, that is within the range of acceptability and opportunity. Actually attaining this occupation, though, is a matter of compromise. By their late twenties, over 80 percent (of men, at any rate) say they are in the field of work they wanted for themselves, although in fact most of them have changed their aspirations a good deal from what they once wanted. Half of college graduates turn out to be working in areas completely unrelated to the vocational interests they had expressed just a few years earlier.

SUMMARY

1. We tend to think of children as pure, innocent, and deserving of special attention, but this view is a relatively recent historical development. In Western cultures before the seventeenth century, childhood was not distinct from other stages of life, and often children were assumed to be evil. Before the rise of capitalism and industrialization, children were valued primarily for the contributions they could make to the household economy. It was not until the nineteenth and twentieth centuries that children became emotionally "priceless."

2. Compared to modern American child care practices, child rearing in other times and in other cultures was much more collective. Many members of the community shared in watching after children, who spent the majority of their waking hours apart from their mothers. By the twentieth century, production had moved out of the home, the nuclear family had come to be seen as a private domain, and "true womanhood" had come to be associated with being a wife and mother. The growth of the suburbs in the 1950s further isolated families and made mothers individually responsible for their children's welfare.

3. One of the most significant changes in the last half of the twentieth century is that most mothers are now employed. The trend toward increased maternal labor force participation is projected to continue for

some time. When mothers are employed outside the home, young children are cared for by fathers, siblings, relatives, sitters, or in some form of group care. Family day-care homes are typically run by working class women with their own small children. The fastest growing form of child care is day-care centers, but the demand for affordable day care far exceeds its availability. Most day care workers earn wages below the poverty level.

4. There is no evidence of disastrous psychological consequences for children who are not continuously attended by their mother during the first three years of life. Regular, predictable care by any caretaker or set of caretakers is all that is necessary for their psychological well-being. Sharing of caretaking of small children among several persons is favorable for both mother and child, and properly organized day-care centers are psychologically supportive and cognitively enriching.

5. Having young children is associated with higher than average levels of stress and depression. Nevertheless, most of this stress results from lack of money and time, so that mothers with sufficient resources and those living with husbands who contribute to housework and child care enjoy better mental health.

6. Most of the poverty level households in the United States are headed by women. Low-income single-mother families tend to rely on kinship and friendship networks for emotional and material support. Out of necessity, parent-child relationships in single-parent families are often more democratic than those in traditional two-parent families.

7. Because women have traditionally performed most of the child care, researchers have only recently begun to study fathers. While most fathers still play a secondary role and interact with children differently than mothers, a growing number of men are sharing responsibility for parenting. When men perform the routine tasks of child care, their actions and feelings resemble those of mothers more than those of traditional fathers.

8. Children and parents spend much time struggling over control of each other. Small children are highly egotistical and selfishly pursue their self-interest in obtaining love and affection.

9. Parents who control their children by material rewards tend to make the children oriented toward being rewarded rather than concerned with the intrinsic value of the action they did to get rewarded. If social rewards (parent's attention) are used, this results in making the child socially attached to the parent and perhaps to people in general. Control by physical punishment tends to produce children who fight back against parents and others, have a "macho" identification with an authoritarian father, and a sense of right and wrong based on fear of punishment rather than internalized values. Control by shaming produces external conformity to the current standards of the group and control of one's emotional expressions. Control by threat of deprivation of love produces the "strong

superego" personality, with a strongly internalized sense of guilt for trespasses, strong self-discipline, emotional and sexual inhibition. When parents give unconditional love, children become attached to them personally and to social rewards generally.

10. Our models of child development assume that parents' actions are extremely important to children's development but often ignore the fact that children socialize parents as well. We also tend to assume that there is a correct formula for raising optimum adults, but this varies with the individual, the family, the culture, and the historical period. In modern American society, children spend more time in front of the television than they do in school or interacting with parents.

11. Middle-class parents are more likely to value bringing up their children with stress on personal autonomy, happiness, and curiosity; working-class parents stress cleanliness, good manners, obedience, and conformity. These values reflect the occupational structure of the parents' lives, especially the extent to which their work is autonomous or closely supervised, complex and challenging or simple and routine.

12. As they grow up, children become aware of and increasingly define themselves according to gender and social class expectations. By the time they reach teenage years, most have limited their career aspirations to fit the occupational stereotypes for their gender and social class.

FAMILY VIOLENCE

S ince the ideal family provides unique forms of love and support, we tend to think of families as places where people are nurtured and protected. In general, family members do care for each other, provide for each other, and help each other survive in a world that often seems hostile and uncaring. But families also have a "dark side" (Finkelhor, Gelles, Hotaling, and Straus 1983). Families are one of the most common contexts of violence in American society. The likelihood that a man will be assaulted by someone who is a family member is more than twenty times greater than the odds he will be assaulted by someone outside his family. For women the situation is even worse. Women are two hundred times more likely to be assaulted by a family member than by an outsider (Straus 1991, 18). Police receive more requests for help with "domestic disturbances" than with any other problem, and more police personnel are killed trying to settle family fights than in any other line-of-duty affairs. What's more, an individual is far more likely to be murdered by a family member than by anyone else (Straus 1991).

Since we like to think of families as protective and caring, it is sometimes difficult to accept the fact that so many people intentionally inflict pain and suffering on other family members. While it is important to acknowledge that support and gentleness are typical of most families, it is also important to examine violence in families and to begin to understand why it exists. Only then will we be able to work toward reducing violence in American families.

In this chapter, we shall look at some recent research on three major types of family violence: spouse abuse, child abuse, and elder abuse. Besides investigating the incidence of each of these, we will explore some of the myths about family violence and review some of the most common explanations as to why the different types of violence occur. Finally we will look at some attempts to reduce family violence through law enforcement, education, and treatment.

VIOLENCE IN SOCIETY

Before investigating the various forms of family violence, it is useful to turn our attention to the social context in which family violence occurs. Is violence a rarity in our society, or is it widespread? While we might wish it were otherwise, we must conclude that violence is woven into the very fabric of American society. The United States was founded by revolutionaries who fought a violent war to end what they considered to be injustices perpetrated by the English monarchy. White settlers also waged a protracted and violent war against Native Americans, driving them from their homelands and imprisoning them on reservations. White American slaveholders used violence to enslave and control black Africans against their will. Violence was also a common form of justice in the American "Wild West," and contemporary movies continue to idealize rugged individualists who "take the law into their own hands." From

John Wayne westerns to Rambo-type action movies, our popular films tend to feature violent men as heroes.

Not only is violent behavior an integral part of our history and a continuing source of romantic cultural imagery, it is hauntingly present in our everyday lives. Although we fear violence, Americans accept and even condone threatening and abusive behavior that is forbidden in many other cultures. Official crime statistics, which tend to drastically *under*estimate the amount of violent crime, show that an aggravated assault occurs in the United States about once every thirty seconds, a forcible rape every ten minutes, and a murder every twenty minutes. Our murder and rape rates are higher than any other nation that collects such crime statistics. For example, Americans are ten times more likely to die by homicide than the residents of most European countries and Japan (Currie 1985). One reason for our unusually high murder rate is that the ownership of firearms is more common in the United States than in any other country. This is especially so for handguns or pistols, which tend to be used in domestic homicides.

While Americans are not alone in condoning violence, we seem especially concerned that men be tough, competitive, and aggressive. One national survey found that in the United States 70 percent of adults thought it was good for boys to get in a few fistfights while growing up (Stark and McEvoy 1970). In a never-ending barrage of American "action" movies and television shows, male heroes never let other men push them around. What's more, the hero usually overwhelms a beautiful, scantily clad woman, who feigns reluctance while yielding to his aggressive and persistent sexual advances. This one-sided image of male-dominated sexuality is also common in romance novels, popular magazines, video games, and daytime television. One study found that sexual aggression was the second most frequently shown type of sexual interaction on American television soap operas (Lowry, Love, and Kirby 1981). These stereotypes promote the violent rape of women.

The rape rate is ten times higher in the United States than in England or Japan, four times higher than in Australia, and nearly twice as high as in South Africa (Malamuth and Donnerstein 1984). While at least half of all rapes are never reported to the police, the number of reported rapes in the United States increased by about half between the 1970s and 1980s. This is probably due to the increased acceptance of reporting rape rather than an actual increase in incidence, and suggests that we do not really know how widespread rape is. In one study of a random sample of women in San Francisco, one in every four women reported that she had been the victim of at least one rape, and almost one out of every three reported that someone had attempted to rape her (Russell 1984). A 1987 study of 6,000 college students found that 15 percent of women had been the victims of rape, 12 percent of women had been the victims of attempted rape, and one man out of four reported engaging in some type of sexual aggression against women (Koss, Gidycz, and Wisniewski 1987). Another recent study concluded that over one-third of college students were involved in some form of courtship or dating violence (Murphy 1988). Women of all ages are raped, but the highest percentage of rape victims are between the ages of fifteen and twenty-five. While we are just beginning to document the extent of rape, we do know that it most often occurs between people who know each other, rather than between strangers. Myths that women provoke rape or somehow "ask for it" are still widespread, in spite of evidence to the contrary (see feature 13.1).

Nonsexual violence is even more likely to be pictured in the media than

FEATURE 13.1
Rape Myth Acceptance Scale

The following scale (Burt 1980) has been used to assess the extent to which people accept cultural myths about rape. How many do you agree with? A higher acceptance of rape myths has been found to correlate with sexual aggressiveness, traditional attitudes about gender roles, and low educational and occupational achievement.

1. A woman who goes to the home or apartment of a man on their first date implies that she is willing to have sex.
2. Any female can get raped.
3. One reason that women falsely report a rape is that they frequently have a need to call attention to themselves.
4. Any healthy woman can successfully resist a rapist if she really wants to.
5. When women go around braless or wearing short skirts and tight tops, they are just asking for trouble.
6. In the majority of rapes, the victim is promiscuous or has a bad reputation.
7. If a woman engages in necking or petting and she lets things get out of hand, it is her own fault if her partner forces sex on her.
8. Women who get raped while hitchhiking get what they deserve.
9. A woman who is stuck up and thinks she is too good to talk to guys on the street deserves to be taught a lesson.
10. Many women have an unconscious wish to

be raped and may then unconsciously set up a situation in which they are likely to be attacked.
11. If a woman gets drunk at a party and has intercourse with a man she's just met there, she should be considered "fair game" to other males at the party who want to have sex with her too, whether she wants to or not.
12. What percentage of women who report a rape would you say are lying because they are angry and want to get back at the men they accuse?
13. What percentage of reported rapes would you guess were merely invented by women who discovered they were pregnant and wanted to protect their own reputations?
14. A person comes to you and claims he or she was raped. How likely would you be to believe his/her statement if the person were:
    a. Your best friend
    b. An Indian woman
    c. A neighborhood woman
    d. A young boy
    e. A black woman
    f. A white woman

Source: M. R. Burt, "Cultural Myths and Support for Rape," *Journal of Personality and Social Psychology* 38 (1980): 217–230.

sexual aggression. Prime-time television averages over thirteen acts of violence per hour, with even higher rates of occurrence in cartoons. In countless portrayals, frustrated people hit, shoot, and bomb others. By the age of sixteen, the average American child has witnessed a half million violent acts and over 50,000 murders on television (Signorelli 1991). Viewing so much violence on the screen tends to desensitize us to the real violence that is all around us and models aggressive behavior. After studying the effects of television violence on young viewers, the National Institute of Mental Health (1982) concluded that watching violent programs increases the incidence of violent behavior in children and teenagers. While violent media images may serve as a catharsis for some viewers, they send a subtle message that it is all right to use physical force to solve one's problems.

Violence is not just perpetuated by the media, however. It is more likely that movies, television, and magazines are just reflecting some idealized image of what is socially acceptable. If we go back in history, we can see how certain forms of violence have been tolerated and condoned by our religions, schools, and governments. For instance, the Bible encourages stern discipline and admonishes parents who don't punish their children: "He who spares the rod hates his son, but he who loves him disciplines him diligently" (Proverbs 13:24). The U.S. Supreme Court upheld the rights of school teachers to use corporal punishment and recent surveys show that three-quarters of American parents agree with the ruling and want schools to use strict discipline practices (Currie 1985, 42). State governments also tend to legitimate violence by supporting capital punishment for severe crimes, though usually only poor people are actually executed.

SPOUSE ABUSE

Violence has also been institutionalized within the family. As we saw in chapter 8, marriage laws have traditionally given husbands power over their wives. From the days of the Roman Empire, men's right to use physical force against wives has been lawful and even expected (Dobash and Dobash 1979). English common law gave husbands this right, which was modified by the nineteenth century "rule of thumb" allowing a husband to beat his wife with a rod no thicker than his thumb (Davidson 1977). Until the 1870s, wife beating was legal in most of the United States, and remained quite common thereafter (Stets 1988). Aided by governmental neglect and protected by the privacy of their homes, men continued to "keep women in their place" with the threat and use of physical force. In the late 1960s, one in four Americans felt it was acceptable for a husband to hit his wife under certain conditions (Stark and McEvoy 1970). Until it was changed in 1977, the training manual for domestic-disturbance calls published by the International Association of Chiefs of Police essentially recommended that hitting a spouse be treated as a "private matter" and that arrests should be avoided (Straus 1991, 27). It wasn't until the late 1970s that marital violence reached public awareness as a serious social problem that demanded some form of social or governmental action.

How widespread is spouse abuse? Because this form of violence so often goes unreported to the police or other authorities, it is difficult to estimate its true incidence. Asking a random sample of husbands how often they beat their wives may yield some general indication of the extent of family violence, but we can safely assume that the answers to such a survey question would understate the extent of actual marital violence. While wife-beating has been condoned in the past, it carries a social stigma today, so that most men would be reluctant to admit the extent of violence they inflict on their wives. It is also likely that many wives who have been assaulted by their husbands would tend to minimize such instances in an effort to focus on the more positive aspects of the marriage. When the husband is the victim, the social stigma attached to being battered by a woman may reduce men's reports of severe beatings by their wives. However, husbands who beat their wives might be eager to admit that their wives had also assaulted them, because this would serve to justify their own violent actions. These kinds of conflicting social pressures undoubtedly influence people's responses to survey questions about family violence; hence researchers estimate that the true rates might be twice as high as the rates

FEATURE 13.2
Measuring Family Violence: The Conflict Tactics Scale

No matter how well a couple gets along, there are times when they disagree on major decisions, get annoyed about something the other person does, or just have spats or fights because they're in a bad mood or tired or for some other reason. They also use many different ways of trying to settle their differences. I'm going to read a list of some things that you and your (wife/husband/partner) might have done when you had a dispute, and would first like you to tell me for each one how often you did it in the past year, and whether you have *ever* had it happen.

- Discussed the issue calmly
- Got information to back up (your/his/her) side of things
- Brought in or tried to bring in someone to help settle things
- Insulted or swore at the other one

- Sulked and/or refused to talk about it
- Stomped out of the room or house (or yard)
- Cried
- Did or said something to spite the other one
- Threatened to hit or throw something at the other one
- Threw or smashed or hit or kicked something
- Threw something at the other one
- Pushed, grabbed, or shoved the other one
- Slapped the other one
- Kicked, bit, or hit with a fist
- Hit or tried to hit with something
- Beat up the other one
- Threatened with a knife or gun
- Used a knife or gun

Source: Murray A. Straus, "Measuring Intrafamily Conflict and Violence," *Journal of Marriage and the Family*, 41, 1 (1979): 75–88.

reported in most studies (Straus and Gelles 1988, 19). Nevertheless, a surprising number of people admit to committing a wide range of violent acts against other family members.

Researchers at the University of New Hampshire have developed a Conflict Tactics Scale that measures violence between spouses, between parents and children, and between children. (See feature 13.2 for an example of the form used to measure husband-wife violence.) The Conflict Tactics Scale measures three different aspects of conflictual family interaction: (1) reasoning; (2) verbal aggression; and (3) violence or physical aggression. Investigators have used this scale to measure conflictual family interaction, with relatively consistent results for similar populations sampled in similar ways (Straus and Gelles 1988). The violent items in the most recent version of the Conflict Tactics Scale include: threw something at the other; pushed, grabbed, or shoved; slapped or spanked; kicked, bit, or hit with a fist; hit or tried to hit with something; beat up the other; burned or scalded (for children) or choked (for spouses); threatened with knife or gun; used a knife or gun. As you can see, the items are arranged in order of severity, so that if someone has used one of the harsher forms of violence, it is also likely that he or she has used one of the milder forms as well.

Using this scale and other social indicators, researchers have attempted to estimate just how often violence occurs between spouses. At the high end, some estimate that over one-third of all wives will be beaten by their husbands during their lives (Walker 1984). At the low end, Suzanne Steinmetz (1977) reported evidence that 7 percent of all wives and 0.5 percent of all husbands were severely beaten by their spouses at some time during their marriages. Less severe kinds of violence, such as slapping and shoving, tend to be even more common. Evidence from the late 1980s shows a decrease in the reported levels of spouse abuse, but 16 percent of all homes had some kind of spousal violence during the current year, and 28 percent at some time during the entire marriage. Applying these percentages to the approximately 56 million couples in the United States, one can estimate that about 9 million couples experience at least one assault during a single year. Moreover, at least 15 million couples are estimated to have experienced some form of violence during the entire marriage, and in almost 4 million of these households, the violence has a relatively high risk of causing injury. These forms of domestic violence are labeled "battering" or "wife beating," and are the focus of the greatest concern.

Battered Wives, Battered Husbands, or Mutual Combat?

Using national survey data, researchers have found that in homes with couple violence, about one-fourth of the respondents report that men were victims but not offenders; another fourth allege that women were victims but not offenders; and one-half of the respondents report that both husbands and wives were violent (Straus, Gelles and Steinmetz 1980). Such surveys cannot detect whether violence by wives was retaliatory or not, and the reporting of simple estimates of incidence has touched off heated controversies. What appears to be consistent across studies using the Conflict Tactics Scale is that wife-to-husband assault is about as common as husband-to-wife assault. What is at issue are the reasons for, and the consequence of, these different types of assault.

One important difference between men and women is physical size and strength. If a husband assaults his wife, he is much more likely to inflict pain and injury than if she assaults him. In addition, men tend to be more aggressive, so that the same act, such as hitting with a fist, tends to be more violent coming from a man than from a woman. This does not imply that some women aren't bigger, stronger, and more aggressive than some men, but *on average* a husband punching a wife is much more apt to inflict injury than a wife punching a husband. When studies focus on the differences between husbands' and wives' violence, they discover that it is women, not men, who more frequently receive bruises, cuts, broken bones, and internal injuries from the assault. Usually only the woman is injured, but in the minority of cases in which both spouses are injured, the wife's injuries tend to be much more severe than the husband's (Berk, Berk, Loeske, and Rauma 1983). In light of the general pattern of greater harm to wives, it is inappropriate to label examples of women hitting men as "husband abuse," or to consider marital violence simply a case of "mutual combat."

To understand the high rate of marital violence by wives, it is useful to compare women's family violence to women's nonfamily violence. Outside of the family women rarely assault others, and they are victims of assault by outsiders far less often than they are victims within the family. The major reason that women are violent within the family is that, for a typical American woman, home is the place where she is most at risk of being seriously assaulted (Straus and Gelles 1988).

Many of the assaults by women against their husbands should be considered as retaliation or self-defense (Straus 1980). In fact, the most frequent motive for violence reported by battered women is "fighting back" (Saunders 1988). Rarely do women report initiating an attack of severe violence, and they usually plead that their assaults are attempted in self-defense. As to those women who do initiate violence, some researchers speculate that they often sense impending violence from their husbands and initiate the violence themselves, in order to stop the overwhelming build-up of tension (Walker 1984). One author, Kelly (1988, 120), describes a typical situation in which a woman experienced a continual threat of violence:

> What he did wasn't exactly battering but it was *the threat*. I remember one night I spent the whole night in a state of terror, nothing less than terror *all*

night. . . . And that was *worse* to me than getting whacked. . . . That waiting without confrontation is just so frightening. (Emphasis in original)

Saunders (1988, 107) notes that battered women are often convinced that they or their children are in imminent danger of death or great bodily harm, even at times when the husband is not currently attacking or threatening. In these cases, the courts have sometimes held that the women were legally justified in using physical force in a kind of "preemptive strike."

Rape in Marriage

Several studies have recently looked at the incidence of marital rape. Russell (1982) found that 14 percent of the married women in a San Francisco sample had been raped at least once by their husbands. Gelles and Cornell (1985) found that at some time 10 percent of spouses or live-in partners used force or the threat of force to have sex. Shields and Hanneke (1983) found that many battering victims are also the victims of marital rape, and that when rape occurs, it is usually not just an isolated incident, but part of an ongoing pattern of violence and sexual abuse.

In a study of 323 Boston area women, Finkelhor and Yllo (1985) reported that one in ten women had been forced to have sex with their husbands or partners. Violence accompanied the rape in about half of the instances, and Finkelhor and Yllo identified three basic types of rape in marriage. The first, "battering rape," is found in marriages with a high over-all level of violence. Wives were routinely beaten, and rape was an additional element of abuse and humiliation. The second type, "force-only rape," was found in marriages that were otherwise nonviolent. The researchers concluded that in this type of rape, the husband's desire for control led him to use sexual coercion, and he used just enough force to get his wife to comply. The third type, "obsessive rape," was the least common, but most openly sadistic, as the man was obsessed with violent sex and inflicted pain in order to become sexually aroused. While the women's experiences varied greatly in terms of physical pain and injury, all of the women reported being greatly traumatized (Finkelhor and Yllo 1985).

As noted above, rape by intimates is about three times more common than rape by strangers. While the impacts of marital rape are often different from the impacts of stranger rape, they are not less serious, and are often more frightening (Russell 1982). The most profound psychological consequences of marital rape stem from the extreme violation of trust that the sexual violence signifies. Yllo (1988) reports that victims of marital rape experienced great anguish, and many struggled to escape their marriages. For some, it took several years before they would let themselves trust men and develop intimate heterosexual relationships, and others felt that trusting men was now impossible.

When Do Battering Incidents Happen?

It has sometimes been held that marital violence is essentially a problem of the lower class. This is not strictly true, as violence is found in all social classes. Among couples filing for divorce, for instance, more working-class couples than middle-class couples complain of physical abuse as a major cause of the break-up, but even among the middle class, about one-fourth complain of physical abuse. The kinds of violence that are most common are pushing, throwing things, or slapping. These actions often occur in the context of an argument that escalates beyond shouting or name-calling, but spouse battering is not usually an isolated incident. There is a continuum from the normal amount of quarreling through mild violence up to serious battering. Frequently couples treat even rather extreme violence as if it were nothing special, or else dismiss it as if it were an unpredictable aberration. Nevertheless, abused women report repeated attacks, estimates ranging from three to eight times a year (Gelles and Cornell 1985; Giles-Sim 1983).

Gelles and Cornell (1985) summarized the factors that are typically associated with wife beating: (1) The husband is unemployed or employed only part time; the family income is low; the husband's occupation is in the manual working class. (2) Both husband and wife are very worried about economic security, and/or the wife is strongly dissatisfied with the family's standard of living. (3) The wife is a full-time housewife; there are two or more children; disagreements over the children are common. All factors are not always present, but in general they make up a picture that is more common in the lower classes. Similarly, a portrait of men who were woman batterers (Fagan et al. 1983) found that 74 percent did not have more than a high school education, and 30 percent were unemployed. Most were in their early thirties or younger, and the length of the relationship with their wives was less than five years. This looks like the pattern of early marriage strain found in many working-class marriages (chapter 6). Of course, when these conditions are found in other social classes, they also increase the chances of marital violence. In middle- and upper-class marriages there are more incentives for wives to hide the fact that they are being physically abused, and there is less chance that others will report them to the police or to some other government agency. Consequently, we know more about spouse abuse in poor families than in others.

What situations tend to precipitate violent outbreaks? In one study of seriously battered women (Giles-Sim 1983, 141), the first incident of battering was likely to be preceded by some fairly major change in life patterns: 39 percent connected to moving to a new house or apartment; 32 percent connected to pregnancy or birth; 19 percent associated with recent separation or divorce from another partner. Except for the separations and divorces, these do not seem like necessarily traumatic events, but it is just these kinds of shifts in the surface of people's lives that seem to unleash violent impulses that are present for underlying reasons.

While it has not been possible to specify exactly when battering will occur, researchers have identified some of the situations under which batterers feel threatened and attempt to assert their authority. The most common theme is the claim for loyalty. Many battering men demand that wives continually show loyalty by talking in a certain way, cooking certain foods, showing affection in certain ways, and so forth. If an insecure man's wife or partner spends time with other people, even other women friends or relatives, this is sometimes taken to be a sign of rejection. Violent men are typically found to be very possessive (Dobash and Dobash 1979). These men tend to become sullen and irritable about their wives' involvement with others, which tends to exasperate the wife and cause her to withdraw. The husband interprets this as rejection, becomes more possessive, makes more demands that she show loyalty, and eventually becomes violent (Ferraro 1988).

Even children are sometimes viewed as threats to a wife's loyalty to her husband. Several researchers have discovered that the wife's becoming pregnant often precipitates beatings by the husband (Dobash and Dobash 1979; Ferraro 1988; Gelles 1974; Pagelow 1981). One of Walker's informants illustrates this pattern:

> He didn't really start to beat me until I started showing how pregnant I was, until my belly started swelling. Then it was like he was jealous of that baby, before he was even born. He was jealous of my babies even when he went in the Navy. He would come home and beat them and beat me. (Walker 1979, 119)

Sexual jealousy is another form of insecurity experienced by many violent husbands (Dobash and Dobash 1979; Pagelow 1981; Walker 1979). For many, the perceived threats to loyalty are extremely far-fetched, but the consequences are still tragic:

> Sara had a three-year-old child from a prior relationship and was eight months pregnant with the child of her lover. He was so concerned about her sexual fidelity that he locked her in their small trailer each morning before leaving for work. The locked door and the small windows made her escape impossible, even if she had been interested in pursuing a sexual relationship. He nevertheless returned each day to accuse her of being unfaithful, and to beat her for her transgressions. (Ferraro 1988, 133)

As this example illustrates, violent men's consuming preoccupation with sexual fidelity is often irrational. In several studies, battering men were reported to fantasize lovers and develop paranoid perceptions of their wives' intentions, leading to excessively controlling behaviors (Ferraro 1988). For some men, the slightest indication of disloyalty on the part of the wife is taken to be a vicious attack on the man's sense of self. In part, this attitude is supported by attitudes in the general population. About two-thirds of Americans continue to agree that acting violently in a situation involving a spouse's extramarital affairs was justified (Whitehurst 1971).

Another common theme in studies of wife beating is that violence often occurs around issues of control. Deciding how income is to be made or spent, how children are to be cared for, when and how sexual relations should occur, how meals are to be prepared, how leisure should be spent, or how deferential the wife should be, all provide opportunities for exerting control. Violent men often make excessive demands for compliance with their wishes in these areas, and when wives question them, or comply in a manner deemed unsuitable, the men feel threatened and fly into a rage. Often the issues that constitute threats to the men's control appear trivial to others but are interpreted by the men as "failure to fulfill the obligations of a good wife" (Bograd 1984). For example, Ptacek (1988, 147) reports how one batterer threatened his wife by telling her, "I should just smack you for the lousy wife you've been."

Sometimes men justify beating their wives because of things that their wives say to them, implying that wives should not challenge husbands. In one study, the most common explanation men offered for beating their wives was that they had been verbally provoked:

> She was trying to tell me, you know, I'm no fucking good and this and that . . . and she just kept at me, you know. And I couldn't believe it. And finally, I just got real pissed and I said wow, you know. I used to think, you're going to treat me like this? You're going to show me that I'm the scum bag? Whack. Take that. (Ptacek 1988, 144)

Some of the men equated women's verbal aggressiveness with their own physical violence:

> Women can verbally abuse you. They can rip your clothes off, without even touching you, the way women know how to talk, converse. But men don't. Well, they weren't brought up to talk as much as women do, converse as well as women do. So it was a resort to violence, if I couldn't get through to her by words. (Ptacek 1988, 145)

While batterers often try to blame their wives for provoking their anger, researchers find that battered women rarely provoke abuse intentionally. One study found that battered women report relatively minor "misdeeds" as triggering abuse: preparing a casserole instead of fresh meat for dinner; wearing her hair in a ponytail; commenting that she didn't like the pattern on the wallpaper (Martin 1976, 49). Sometimes a marriage can be nonviolent for a long time, and then some relatively minor event, such as cooking the wrong food, becomes a major threat to the man's control and triggers a violent response:

> Until we were married ten years or so there was no violence or anything. But then after a while, it just became, it just became too much. . . . I don't know if I demanded respect as a person or a husband or anything like that, but I certainly, you know, I didn't think it was wrong in asking not to be filled up with fatty foods. (Ptacek 1988, 147)

Once battering has happened, it seems to become more likely, unless something is done to head it off. Battering can become a routine. Hence the factors that set off later incidents, including very severe ones, have no particular pattern, but can include all the factors mentioned above as well as the man losing his job, or the woman wanting to leave the relationship. Many such incidents happen for no apparent reason at all. For example:

> He hit me, and he said, "Now you are going to hit me." I said, "No, I'm not going to hit you. I don't want to hit you." I said, "Please leave me alone." I started crying. He did it again, and I still didn't hit back. Then he just chuckled and walked off. He never slapped me across the face, though.
>
> I wanted to kill him, but I didn't. I just said, "Why, why is this happening to me?" It was like he wanted to punish me.
>
> There was another time when he kicked me. He just kept kicking me. I felt angry, but I couldn't show it. I was too afraid. I don't think I felt. . . . I never felt like I could strike him. I don't know if I ever could. I just didn't know how to give it back. Just couldn't do it. I'm afraid of what I would do if I did. (Giles-Sim 1983, 1)

These incidents had become so normal that many couples had a history of violence with each other long before they were married. In Giles-Sim's sample (1983, 127) 13 percent had their first incident of violence while they were dating and 35 percent while they were living together, yet they still went on to marry. Usually they interpreted these early events as isolated incidents, as something to forgive and forget, rather than the beginning of a pattern. Other studies (summarized in Gelles and Cornell 1985, 65–67) found that violence during a courtship is common. Between 22 and 67 percent of all couples experienced pushing, slapping, or shoving during dates; 38 percent of college dating couples experienced kicking, biting, or hitting with a fist during a date, with the perpetrator of violence usually being the male. Why were these incidents not taken as warning signs? The answer is probably that violence, in a sexual context, may often be interpreted in our society as a sign of passion or of jealousy. Thus, Henton et al. (1983) found that in the cases of dating violence, about 25 percent of the victims and 30 percent of the offenders interpreted the violence as a sign of affection. The reinterpretation of violence as either normal or as affectionate is one of the main processes that keeps violent couples together and hence that allows battering to be repeated.

Causes of Marital Violence

The traditional way of accounting for marital violence was either to ascribe it to lower-class culture or else to describe it as psychologically pathological and

A show of anger may precede family violence. "You're asking for it!" is a typical verbal prelude to physical abuse.

deviant. As we have seen, there is some support for the idea that the lower class is especially prone to violence, but this cannot be the whole story, since violence also occurs in the middle and upper classes. To ascribe it merely to pathological individuals, however, evades the question of explaining it.

One particularly pernicious psychological account is the argument that women are masochistic and hence "ask for it." Actually, violent husbands often use this as an excuse for a description of what happened. What it implies, sociologically, is that there is a widespread cultural belief that women should behave in certain ways; if they do not, it is legitimate to use violence against them. In American society until the twentieth century, laws actually gave a husband the right to physically chastise his wife for nagging or other offenses against her "proper place" (Straus 1991).

There are four main sociological explanations for why violence occurs in marriage today: (1) sexual property rights; (2) economic strains; (3) intergenerational transmission of violence; and (4) social control.

Sexual Property Rights

The traditional patriarchal attitude is that married women are essentially the property of their husbands, economically, sexually, and emotionally. Women who violate the male prerogative are felt to have broken a respectable social tradition and hence to have brought upon themselves a legitimate angry response. This is, in fact, what one would expect given the theory of love and sex as rituals that establish emotional property (discussed in chapters 8 and 9). Dobash and Dobash (1979) show that men who beat their wives tend to regard their action as proper and to justify it as a defense of their traditional rights. This is particularly likely to happen when there is sexual infidelity, or even minor flirtations causing jealousy; it also happens when the husband feels that his own sexual demands are being unjustifiably evaded. The "woman as property" attitude also condones and promotes violence in response to all kinds of actions on women's part that show their lack of respect for male prerogatives. As the theory of rituals shows, these become symbolic violations that call into question the whole structure of the relationship.

Another incident shows how male feelings about women being their sexual property result in wife battering:

> He went down in the cellar to fix the stove and I went down to try to talk to him. He wanted me to bend over so he could have sex downstairs. I said, "I don't drop my pants every time you turn around and want sex." He said, "You do what I tell you to do."
>
> When we went back upstairs to eat, we were sitting at the table and he was saying something—"You're gonna. . . ." do something. I said, "No, I don't want to," and he kicked me under the table. I said, "Don't kick me." He said, "I'll kick you if I want to. I'll do anything I want to you." Then he hit me with a stick and said he'd do anything he wanted to me because I was his property. (Giles-Sim 1983, 2)

As we have seen in chapter 6, married life is most intensely "ritualized," in the sense of creating strongly held traditional beliefs, in the typical living conditions of the working class. It is there that the highly confining social density of the local community creates the most strongly taken-for-granted attitudes. When traditional prerogatives are violated, the result is unthinking anger. Of course, these patriarchal prerogatives can be upheld in the higher social classes as well, although there the culture usually is more ideological and abstract, and men tend to find more indirect and subtle ways of getting what they want.

The fact that the traditional structure of the entire community is involved in generating these beliefs, especially in the working class, means that the male's violence is often regarded as legitimate by persons outside his own family. The police, who tend to share the working-class culture, are generally unwilling to intervene in what they regard as "domestic squabbles" (Pagelow 1981). Hence battered women have frequently had no outside protection or redress.

Economic Strain

Another explanation for family violence draws on the fact that it tends to increase during times of economic strain. This is another reason why lower-class families are especially prone to violence. Such families usually have a very traditional conception of gender roles, in which the husband is supposed to be the provider and head of household. When he fails to provide, he loses his standing both in his wife's eyes and in his own. Sometimes she is able to make up for some of their economic loss by her own work, but frequently this challenges the husband even more. A good deal of lower-class violence may thus be an effort on the part of the male to overcome his feeling of social inferiority and reassert control (O'Brien 1971).

Economic strains also play a role in keeping battered women trapped in their marriages. Surprisingly, battered women are frequently reluctant to leave their husbands. They often express fear that if they leave, their husbands will come after them and punish them even more. But part of the reason is economic, especially for women with small children whom they cannot support on their own.

Intergenerational Transmission of Violence

There is also evidence that violence toward wives (and toward husbands, too) is passed along from the previous generation through parents' child-rearing practices. Straus (1983) shows that violence is contagious in a number of ways. A woman who was abused by her own parents is more likely to stay in a violent relationship with her husband or boyfriend. This is because she tends to perceive violence as normal, or because she has low self-esteem and little sense that she could improve the situation. Perhaps even more importantly, husbands who were subjected to a great deal of physical punishment when they were children are especially likely to assault their wives. As figure 13.1 shows, there is a direct correlation between the amount of physical punishment a boy had when he was about thirteen years old, and the likelihood he will later use violence on his wife. Not only does childhood punishment predict which men will frequently and severely attack their own wives (marked "3" in figure 13.1), but it also predicts which men will occasionally beat their wives (marked "1–2" in figure 13.1) and even which ones will engage in "ordinary" levels of mild physical abuse.

FIGURE 13.1: Violent Husbands Experienced Much Physical Punishment When They Were Children

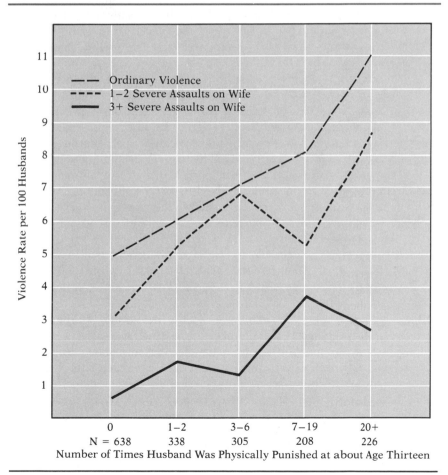

Source: Straus, 1983, 227.

Straus also shows that the more violent a husband is to his wife, the more likely she is to use violent punishment on her children. Violent husbands are also more violent to their children. This closes the circle and sets off the likelihood of children growing up to become spouse abusers in the next generation.

Social Control

One more pattern has to do with how many social pressures and controls there are from the surrounding society. That is to say, many of the factors that lead toward violence (sexual property feelings, economic strains, family backgrounds) can be neutralized or allowed to come forth, depending on how much control is present. In the past, our customs and laws have considered the family to be "private" and not subject to the same controls that operate in more public realms. The patriarchal ideal of "a man's home as his castle" has led to a hands-off attitude on the part of lawmakers and police. This idea of the family as a private "haven" has isolated women and children within families and allowed men to violently abuse them.

Family quarrels do not necessarily escalate into severe violence. This is most likely to happen when the couple has lived in their neighborhood less than two years and when they do not participate in an organized religion or other social groups (Gelles and Cornell 1985). In other words, the family is more socially isolated, with less pressure to restrain what goes on within it. Probably for the same reasons, spouse abuse is most common among couples who are young and who have not been together many years. Women who are full-time housewives are especially likely to be victims of battering, since they are less likely to have social contacts and outside sources of social support. Isolated women are typically unable to find alternatives that would get them out of the victimized situation.

Why does the victim stay? While we should remember that many victims *do* leave when the situation becomes intolerable, a crucial question is why a woman stays in a relationship despite repeated violence. There are several factors involved, some of which have already been mentioned. Her commitment to the relationship is one important factor that makes her stay. This may mean how much she loves her husband, but also how much of her life is contained within her role as traditional wife and mother. Many abused wives blame themselves for their predicament or believe that they deserve the treatment. Some try to minimize the negative impact of abuse through denial, claiming that "it really doesn't happen too often" (Fine 1981, 50). Many are convinced that the violence is a very small part of the relationship, and many cling to the hope that the beatings will stop and the husband will somehow reform. The abused wife may also believe the myth that it is always best for children if parents stay together. The extent to which there are alternative opportunities for her also makes a difference; this includes her education and occupational prospects, which may leave her trapped or give her a chance to support herself, as well as her willingness and ability to go back into the sexual marketplace (chapter 8).

A major choice point usually happens with some particular incident. Frequently a battered woman decides to leave when she fears her children will be hurt. Some women leave when they become resentful toward their husbands for letting the children see their mother beaten, or for exposing their violence to people outside the family (Giles-Sim 1983). Such dramatic incidents can sometimes rearrange an entire worldview.

Intervention and Treatment

There is evidence that battered women can often successfully control violence if they get outside support. Although many women do not call the police to report battering, when they do so the effects are usually positive. A U.S. Justice Department study (*Los Angeles Times*, August 19, 1986) found that when battered wives did not call the police, 41 percent were beaten again by their husbands within six months; but of those who did call the police, the proportion who were beaten again dropped to 15 percent. Typically, women fear reprisal by their husbands if they call the police, and sometimes the police do not take action. But even the threat of police intervention tends to have a deterrent effect, at least on some men.

Sherman and Berk (1984) conducted a field experiment to assess the impact of police intervention in cases of wife abuse. Three types of police response—immediate arrest, removing the husband from the home, and trying to cool down the situation through talk—were used according to a random assignment procedure. They discovered that if the man was arrested, there were fewer subsequent calls for help from the wife and fewer instances of repeated violent behavior. Later studies have shown that arrest is an effective deterrent, and that the risk of retaliation by angry husbands can be reduced in many cases (Ford 1991). As a result of research like this, and pressure by feminist activists, police are becoming more likely to arrest husbands for beating their wives. In addition, it appears that abused wives may now be more likely to call police for help. One reason women do not call the police, however, is that they are attached to their husbands; they love them and do not want to take an action that they feel will make the breach irreparable. These women are caught between opposing forces.

Official responses to social problems are not inevitable consequences of the problems themselves. In fact, most social problems exist long before they are recognized as conditions that demand some form of collective response. Various potential issues are identified, legitimated, and responded to as social problems in response to claims by advocacy groups to redress specific grievances (Spector and Kitsuse 1973). In the case of marital violence, the woman's movement provided the impetus for women to define husbands' use of force against wives as "abuse" and to identify victims of repeated violence as "battered women."

Battered women's shelters first emerged in the 1970s as local efforts by women to provide a supportive and safe environment for victims of abuse. Through volunteer efforts, including consciousness-raising groups, public speaking, lobbying, grant writing, and use of the media, feminists were able to conceptualize husbands' violence against wives as a community responsibility rather than an interpersonal problem limited to a few pathological families (Schechter 1982). Various local groups such as YWCAs, women's clubs, junior leagues, and church organizations solicited funds and lobbied for the establishment of shelters. Eventually, community service agencies, health organizations, child and family service agencies, and local governments began to fund programs for battered women. The shift from grass-roots support to government funding professionalized battered women's shelters and initiated a shift from advocating women's rights to "treating clients." Some argue that the traditional social service agency approach to the problem focuses too much on the personality defects of battered wives, and too little on the problem of men's responsibility for violence (Schechter 1988).

Some of the most recent attempts to address the problem of wife abuse *do* focus on the men. Pro-arrest policies and court-mandated rehabilitation programs are having some success, although treatment approaches vary widely (Lyon and Mace 1991). Probation and mandatory group counseling for batterers generally focus on inducing violent men to accept responsibility for their violence, develop communication skills, and limit their alcohol and drug abuse. Here too, there are debates between activists and professionals over how the problem should be defined and which clinical model to adopt. The following excerpt from an intake session at a Boston-area profeminist antibattering program illustrates how different counseling strategies can define the problem differently:

> *Counselor:* What do you think causes you to hit your wife?
> *Client:* Insecurity. I guess it goes way back. . . . My father was a drinker too. He wasn't just a drinker; he was a mean drunk. Once during Thanksgiving he got mad and threw the whole damned bird on the floor. . . . He'd whack my mother for no reason really. . . . I left home when I was seventeen, got married but it only lasted two months. She was pregnant of course. I've always been insecure with women.
> *Counselor:* This is helping me to understand why you're insecure but not why you hit your wife.
> *Client:* Sometimes I take things the wrong way. . . . I overreact I guess you could say, because of my insecurity. My shrink said I was like a time bomb waiting to go off. She [client's wife] might say something and I don't react at the time, but then the next day or maybe a few hours later I get to really thinking about that and I get really bull shit.

A battered woman who wants to call the police for help may not because she fears that such an action would make the breach between her and her husband irreparable.

Counselor: A lot of people feel insecure but they are not violent. What I'm interested in finding out is how do you make the decision to hit your wife—and to break the law—even if you are feeling insecure?

Client: I never really thought of it that way, as a decision.

Counselor: But you were talking just now as if your violence is the direct result of your insecurity, or of something she says or does.

Client: Yeah, you're right, I do. But I'm still thinking about what you said about the decision. I honestly have to say that I never thought of it that way before. I mean, I'm really dumbfounded! I'm going to have to really think about that.

Counselor: What are you waiting for?

Client: What do you mean?

Counselor: I mean are you waiting to stop feeling insecure before you stop being violent?

Client: Yeah, I guess that's what I've been waiting for. (Adams 1988, 176–77)

This client had been through three years of individual and couple counseling that focused on his violent and abusive behaviors as an outgrowth of his drinking, insecurity, and poor communication. While he had made some progress, the client continued to have "temper tantrums" during which he would grab, shake, slap, or kick his wife. The intent of the new counselor's questioning was to focus on violence as the treatment problem and to induce the client to stop beating his wife. Newer treatment techniques, such as persuading men to take "time out" when they sense they might get violent, avoid the problem of making nonviolence dependent on solving the man's other problems. First and foremost, these techniques teach men how to identify their violent tendencies and make themselves accountable for their violence (Adams 1988). As other researchers have noted, men hit their wives because they can get away with it (Gelles 1987). If we want wife abuse to stop, we must make such violence unacceptable.

CHILD ABUSE

When we got there this baby was crying and we could see his leg was twisted kind of funny to one side. He had bruises on his face. He looked pretty bad.

I ran because I didn't think they cared about me. The night before, my mom told me that they never liked me. She says, "Go live with your friend." And then she goes, "I don't give a damn about you. Just get the hell out of here. I never want to see your face in this house again." (Garbarino and Gilliam 1980, 10, 13)

The amount of violence against children is surprisingly large. It has been estimated from reported cases that 1.5 to 2 million children are abused by their parents, and almost 2,000 children are killed annually (Straus, Gelles, and Steinmetz 1980). A small child has more chance of being killed or severely

TABLE 13.1   Kinds of Family Violence against Children

Type of Violence	Families who reported violence the past year	
	1975	1985
Slapping or spanking child	58%	55%
Pushing, grabbing, or shoving child	32	31
Hitting child with something	13	10
Throwing something at child	5	3
Kicking, biting, or hitting child with fist	3.2	1.3
Beating up child	1.3	0.6
Burned or scalded child	NA	0.5
Threatening to use or using knife or gun	0.1	0.2
Overall violence	63	62
Severe violence	14	11

Source: Gelles and Straus, 1987, 84.

Violence against children includes fairly mild forms such as spanking, slapping, pushing, grabbing, or shoving. More severe forms include hitting a child with something, kicking, biting, or hitting with a fist.

injured by its parents than by anyone else. For children, the home is often the most dangerous place to be.

Most cases of family violence are never made public, since they do not come to the attention of police or welfare authorities. However, national surveys that asked persons how much violence they actually experienced or condoned in their own families found a considerable amount (Straus, Gelles, and Steinmetz 1980; Gelles and Straus 1987). Most parents admitted to using some kind of violence upon their children during the past year. The most common form of violence was fairly mild: over 50 percent said they had spanked or slapped a child, and over 30 percent had pushed, grabbed, or shoved one. But 10 to 13 percent said they had hit a child with something, 3 to 5 percent had thrown something at a child, and 3 percent admitted kicking, biting, or hitting with a fist (see table 13.1). These sound like adults acting more like children. Moreover, when the time range was expanded beyond the past year to ask parents if they had *ever* done certain violent actions, 4 percent admitted beating up a child, and 3 percent had threatened to use or had actually used a knife or a gun on one (Straus, Gelles, and Steinmetz 1980).

Comparing the rates for 1985 to the rates for 1975 shows that there were few changes in the amount of overall violence against young children (see table 13.1). There does appear to be a significant decline of severe violence against children over the decade. This finding may have more to do with public awareness than actual trends in behavior. National media campaigns, new child abuse and neglect laws, reporting hotlines, and almost daily media attention have transformed a problem that was ignored for centuries into a major social issue (Gelles and Straus 1987). Opinion polls conducted in the mid-1970s showed that only about one American in ten considered child abuse a serious problem, but by 1982, nine out of ten did (Magnuson 1983; Gelles and Straus 1987). As a consequence, people responding to questions about violence against children in 1985 may have been less likely to admit punching, biting, kicking, or beating children than they would have been ten years earlier. Alternately, we might assume that all this attention to child abuse could have altered violent parents' disciplinary practices and angry outbursts, resulting in lower actual rates of child abuse.

Whether the likelihood of reporting child abuse is increasing or decreasing, we can assume that reported rates for severe violence against children are well below actual rates. This is because child abuse is typically hidden in the privacy of the home, and most people would be reluctant to admit the full extent of their abuse or neglect. While nine out of ten parents of three-year-olds admitted at least minor forms of physical punishment such as spanking or slapping, about one-third of parents of fifteen- to seventeen-year-olds also admitted using such tactics. When children of any age were more severely beaten, the episodes were reported as being repeated an average of once every two months (Gelles and Straus 1987).

About half of the reported cases of parental physical violence against children involve women, and about half involve men. When we consider the amount of time that children spend with women relative to men in two-parent families, and the fact that most single parents are women, we can see that the rate of violence is much higher for men than for women. Most researchers find that women who are abused by their husbands are most likely to use severe violence against their children, although these women account for only a portion of such cases. While some early researchers speculated that mothers employed outside the home might be more violent toward children than housewives (Fontana 1973), just the opposite seems to be the case. Gelles and Hargreaves (1987) found that women who worked full time had the lowest rate of overall violence toward their children. Fathers with employed wives also had the lowest rates of violence toward children. The rate of child abuse among fathers with wives who did not work outside the home was 2.5 times greater than the rate for fathers married or living with women who worked full-time. Gelles and Hargreaves (1987, 101) speculate that men whose wives do not work maintain traditional beliefs about the man being the head of the household and believe it should be the father who takes responsibility for physically punishing the children ("Wait until your father gets home").

Violence against children is not the only form of violence that takes place in families. We have already seen that spouses often attack one another. Moreover, children often attack their parents when they get old enough to fight back. Almost 2.5 million (or about one out of ten) teenagers every year commit violent acts against their parents, with almost 900,000 adolescents (3 percent) engaging in more severe attacks on their parents—such as punching, kicking, biting, or using a weapon. Mothers are much more likely than fathers to be the victims of both overall and severe acts of violence (Gelles 1987). About 2,000 parents are killed by their children each year, the most publicized example in recent years being the Menendez brothers. It has also been estimated that people over age sixty-five who live with younger family members are also abused, at a rate of about 500,000 persons per year. Violence in the family may start with spouses fighting and parents abusing their children, but it spills over into reciprocal violence later on.

The most common type of family violence, in fact, is scarcely noticed because it is taken for granted. This is violence among siblings. Gelles (1977) found that 80 percent of the children between age three and eighteen who had siblings admitted trying to hurt their brother or sister during the preceding year. Boys were more violent than girls, but the difference was only a matter of degree. Nor was the violence minor. Almost half the children had kicked, punched, or bitten a sibling; 40 percent had hit one with an object; about a sixth had beaten one up. More than half had committed a violent act that would be grounds for legal prosecution if it had been directed against someone outside the family.

Why does this violence occur? One reason is that American culture supports the family use of violence. In national surveys, 90 percent of people agree that it is often desirable to spank or slap children. Many claim that spanking is for the child's own good and say they are glad their own parents used physical punishment on them (Straus 1991). Most parents admit that they hit children who are three or four years old, and over 20 percent also say they spank or slap infants (Straus 1991). This is an almost astounding finding, considering that infants have few cognitive or motor skills that enable them to respond to their environment or control their actions.

Moreover, these tend to be official attitudes as well as private ones. Two-thirds of police, clergy, and educators answering a poll were in favor of spanking; and the Texas legislature, as just one among many, enacted a law in 1974 that stated: "The use of force, but not deadly force, against a child younger than 18 years is justified (1) if the actor is the child's parent or stepparent . . . (2) when and to the degree the actor believes the force is necessary to discipline the child" (Garbarino and Gilliam 1980). Ironically, much of the legislation passed in the last two decades to protect children from assault simultaneously reaffirmed the parents' right to hit their children "when necessary" (Straus 1991, 27). If we add to this the strong belief in the privacy of the household and the right of families to settle their own affairs, we find a strong set of viewpoints that condone and even favor the use of force inside the home.

"Normal" Violence

Most violence against children is ordinary physical punishment carried out by loving and concerned parents. To illustrate, Murray Straus (1991) uses the example of a ten-month old child who picks up a stick and puts it in his mouth. The parent takes the stick away and says, "No, no, don't do that. You'll get sick. Don't put dirty things in your mouth." Unfortunately children crawling on the ground are virtually assured of finding another stick, or rock, or something equally dirty, and promptly put it back in their mouths. Eventually the parent notices and comes over to the child and gently slaps the child's hand, saying, "No, no don't do that." This scene seems quite normal. As

---

FEATURE 13.3
Corporal Punishment as a Ritual of Family Solidarity

Corporal punishment is the background upon which child abuse arises. Many parents regard physical punishment as not only legitimate but desirable, especially for boys (Straus, Gelles, and Steinmetz 1980). For these parents, punishment is not just a necessary evil, but morally proper. We can explain this by using once again the sociological theory of rituals introduced in chapters 8 and 9 to explain the social ties of love and sex. Although it may seem strange that punishment operates in a fashion similar to love, in both cases there can be mixtures of solidarity and conflict; when sexual arousal is mixed in, the result may be clinically a sadomasochistic attachment.

A ritual is an interaction that creates or reinforces feelings of social membership, drawing boundaries between insiders and outsiders, and giving off symbols by which members show their loyalty to the group. The best-known rituals are formal ones, such as flag-salutes, weddings, graduations, and holiday gift-exchanges. But there is also a type of ritual in everyday life that represents the degree of intimacy in personal interactions: greeting and departure ceremonies, kisses, and shared drinks and meals. Since rituals convey membership, the flaunting of an expected ritual carries the implication

that the relationship is being violated. This is the reason why a typical minor family quarrel breaks out when family members do not come to dinner on time; the sitting together at dinner is a ritual, and failing to do so is not just a practical issue but conveys a little break in family solidarity.

Rituals are not necessarily consciously recognized. They can occur as "natural rituals" (Collins 1988, 197–99), whenever the basic ingredients are present: the group assembled face-to-face; awareness of a common focus of attention, excluding outsiders; a shared emotional mood. If these ingredients are present, the emotion grows more intense among all participants, resulting in an additional emotion: the feeling of attunement, which manifests itself in a feeling of membership. It is through this process that actions—such as spankings—can become symbols of relationships. The physical punishment situation brings the punisher and punishee together face-to-face (or at least face to rump), creates a strong focus of attention, and causes an emotional build-up. The conscious contents of the emotion need not be the same on both sides; but there is always some contagion of nervous excitement, anger, and fear. "This is going to hurt me as much as it will hurt

Straus points out (1991, 29), however, the problem is that these actions also teach the child the principle "those who love you are those who hit you." This lesson begins in infancy and continues for most American children until they leave home. Remarkably, one of three parents report that they continue to hit their children even when they are teenagers (Straus 1991, 28).

The fact that the hitting is done by someone who loves and cares for the child probably makes it worse, insofar as the child is more likely to obtain a deep moral sense that it is right to both give and receive physical punishment. The idea that people who love you hit you is also easily generalized to "those you love are those you can hit" (Straus 1991, 29). Because these lessons are learned early in life and ritualized in repeated encounters (see feature 13.3), they tend to carry deep meaning for most people. As with spouse abuse, those who repeatedly or severely beat their children often feel a moral righteousness about it, and justify it to themselves, and to others, on the grounds that the

you" is not a strictly accurate statement, but it does convey an emotional truth that something is shared.

The result is that physical punishment tends to create a ritual bond between parent and child. The bond can be full of anger. But it is typical of rituals that there are two levels of emotion simultaneously. (In a funeral, there is grief but also solidarity; in a political rally against an enemy, anger but also solidarity; in a spanking, there is anger and fear but also a close focus of attention between parent and child.) When rituals have done their work, the surface emotion dissipates as time passes, while the solidarity feeling continues as long as the ritual is periodically repeated. This is how punishment rituals come to carry a moral connotation in families, especially where punishment is the most intense form of interaction. Men who were brought up by physical punishment will often speak of how their fathers would beat them ("Boy would I get a whipping!") with an affectionate and nostalgic tone. Children are often punished for ritualistic pranks; these are deliberate violations of conventional order, as if daring and provoking punishment. Many families and communities in the more traditional sector of society go through a reciprocating cycle of pranks and physical punishment, not unlike the conflictual cycles of tribal vendettas.

The fact that physical punishments so often carry the unconscious message of family solidarity is one reason why the more blatant domestic assaults are often ignored or excused. After all, if a certain degree of violence on children is felt to be a good thing, it is easy to overlook the points at which the violence gets really severe and a child is emotionally or physically hurt. Furthermore, the very success of a family tradition of punishment rituals helps perpetuate it. When children who have been routinely subject to physical punishment go on to use the same methods when they become parents, they are not necessarily displacing their pent-up aggression onto a weaker victim. The former punishee is not merely taking it out on someone else. This is the intergenerational transmission of a tradition, a feeling of carrying on family solidarity from generation to generation. It is because of these kinds of traditions that so many people are resistant to doing anything about controlling corporal punishment, or seeing how it slides over into child abuse.

victim deserved it. As we will see, this tends to have devastating effects on children, who feel responsible for their mistreatment, and who are at risk for treating their own children in similar fashion.

The Causes of Child Battering

In over 90 percent of all cases where children are abused, the aggressor is a member of the immediate family. However, recurrent child abuse cannot go on without the compliance or at least passive acquiescence of other persons besides the abusing parent and the child. Others who know about it must allow it to happen or at least refrain from reporting it. Since there are strong pressures in our society supporting the right of parents to physically punish their child, this is frequently what happens. But often the root causes of the abuse itself come from outside the immediate perpetrator and the victim.

Some parents who abuse their children do so because they themselves are psychotic. They account for a relatively small percentage of all cases of child abuse, although they do tend to be involved in a greater proportion of the cases with severe injuries (Garbarino and Gilliam 1980). More commonly, abuse is perpetrated by "normal" parents, using the "normal" pattern of physical punishment and family authoritarianism, which then gets out of hand. Often the situation starts out with some mild control problem, which is made worse by the parent's use of some extreme form of punishment. A vicious circle is set up, which provokes yet further punishment, until injury occurs. The small child who cries too much (which at an early age may well be due to physiological causes) can provoke a sleepy or harassed mother, father, or mother's boyfriend into spanking, which at that age has little effect except to make the child cry more. As this goes on night after night, the adult may become extremely abusive and even produce physical injuries.

What are abusive families like? The key seems to be a situational match between a child who is a problem and parents who have the wrong skills and motivations to be caretakers (Garbarino and Gilliam 1980).

- Children who are especially demanding are particularly likely to be abused. Thus, children who are ill or handicapped, hyperactive, or suffer emotional problems are especially likely to be victims.
- Parents who have unrealistic expectations about their role are especially likely to become abusive. Becoming a parent requires reordering one's priorities away from the self-centeredness of the courtship stage to caring for the needs of a small child. Many couples who marry especially young, who have children early in their marriage, or who have unwanted children due to carelessness with birth control are likely to be in this category. For them, the arrival of the baby is a shock. Although much attention has been focused on young fathers or boyfriends who abuse children, it is even

more often the mother in this situation who finds her life suddenly made frustrating and confining by her new child and who reacts with child abuse. (Borderline to this situation are parents who spank children younger than twelve months or even six months: ages at which the spanking clearly represents the parents' emotions rather than any rational policy for controlling the child.)

- Parents who have had little opportunity to learn the role of caretaker often run into difficulty. Some in addition were badly treated or abused during their own childhood or grew up without regular caretakers with whom they had affectionate bonds. Child abuse thus tends to be perpetuated across the generations.

- Stress in the parents' lives also acts to precipitate child abuse, just as it affects spousal violence. Particularly fathers who are under economic stress, because of unemployment or career setbacks, tend to increase their violence at home. This seems to occur in tandem with changes in the male-female power situation at home. The husband's relative power vis-à-vis his wife goes down as his contribution to the family income is reduced; hence he may fall back on his remaining resource, domination by sheer force. This results not only in wife abuse but in child abuse as well.

It has been estimated that close to one-quarter of all American families are in danger of abusing their children because of a combination of these factors. This seems to be borne out by surveys of how mothers interact with their infants, in which 25 percent of the mothers are "at risk because they have child-rearing attitudes or experiences characteristic of abusers" (Garbarino and Gilliam 1980, 31).

Abuse of Young Children and Teenagers

> My dad started grounding me for finding dirty spots on the dishes. The last time he grounded me like that, it went on for six months and it got so bad I had to start asking for everything: if I could get up, if I could go to the bathroom, if I could sit down and eat with him, if I could get ready to go to bed, if I could take a bath. You know, everything you take for granted. And his answer to me always was, "do you deserve it, do you think you deserve it?" Well, of course I deserved to eat. I have to eat to live, you know. And it just got really bad. (Garbarino and Gilliam 1980, 12).

There seem to be two rather different patterns of child abuse. There are parents who start abusing their children rather early and who continue it until the children leave home. Garbarino and Gilliam (1980, 127–33) call these "long-term abusers." On the other hand, in some families violence and overstrict discipline against children do not begin until the teenage years. Mothers are more likely to be the abusers of young children, whereas in the teen years, authoritarian fathers are the most likely perpetrators.

Most of the factors that have been reviewed above are especially relevant to the families of long-term abusers. The general pattern is that people who abuse their small children are particularly likely to be suffering from the problems of poverty and to reflect the cultural patterns of the working class: early marriage and child rearing, the tendency to an early marriage crisis, and unrealistic expectation about marriage. Researchers have declared that child abuse is not class based and that it occurs in all social classes (Steinmetz and Straus 1975, 7). But they have often drawn their findings from small samples of college students and have downplayed the overall pattern, which clearly shows poor people to be greatly overrepresented among abusers of small children (Gelles and Cornell 1985). Leroy Pelton (1978) has referred to this as "the myth of classlessness." Hitting small children—as compared to hitting adolescents—is much less likely to be deliberately provoked by the children and more likely to result from environmental stresses. This is not to say that the danger of child abuse cannot exist to some degree in all classes, especially given our prevailing cultural belief that corporal punishment is desirable. It is true that injured children from wealthier families are much less likely than poor children to be labeled as "abused" by attending physicians. But it is misleading to ignore poverty as a major cause of stress. In general, children of the poor are more likely to be abuse victims.

Abusers of *teenage* children have higher incomes and more stable family structures. Less is known about them and about the kind of interaction with their children that leads to the abuse. Teenage rebellion (see chapter 15, especially regarding teenage delinquency rates) may interact with parental authoritarianism in an escalating spiral. For example, we know that children who are emotionally closer to their parents are less likely to become delinquents (Hirschi 1969). On the other hand, children who are severely abused are especially likely to become socially violent. For instance, a survey of violent inmates at San Quentin prison found that all of them had been badly abused during childhood (Maurer 1976).

In general, there is a pattern of the intergenerational transmission of violence. Parents who abuse one another are especially likely to have undergone severe physical punishment (which might now be classified as abuse) when they were children, and these parents are especially likely to abuse their own children. Straus (1983) found the amount of violence husbands used against their wives was correlated with the amount of abuse these fathers perpetrated against their children (see figure 13.2). Wife beating by husbands also correlated with the amount these mothers beat their children. In other words, the battered women tended to pass along the abuse to their kids. Straus also found that children who were beaten by their parents were highly likely to severely attack their siblings (figure 13.3). All the way around, violence is contagious.

FIGURE 13.2: Abusive Husbands and Abused Wives Both Tend to Abuse Their Children

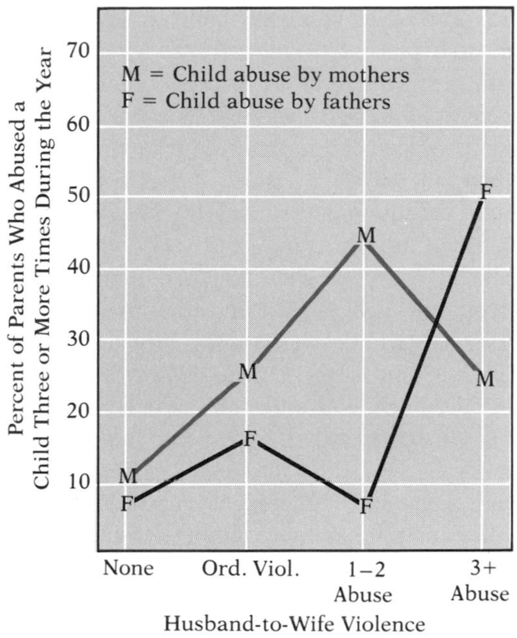

FIGURE 13.3: Children Beaten by Their Parents Tend to Attack Their Siblings

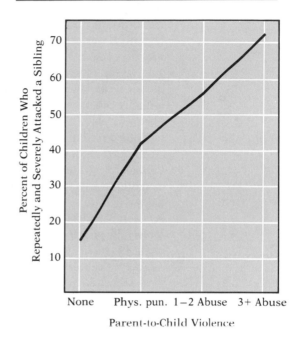

Source: Straus, 1990, 409.

Source: Straus, 1990, 408.

Incest

If I was alone with my dad, he would touch and kiss me. I would try to please him in little ways because it was confusing to me. Then he came into my room one night. I was real scared, but he told me to relax and he wouldn't hurt me. It did hurt, but there were moments I remember enjoying. I mean, I was the ugly one and this was the closest thing I had ever experienced to love. The only affection I can remember were those times with my father. He told me he would kill me if I told, so I never told my mom. I still don't know if she knew. (Garbarino and Gilliam 1980, 151)

Most sexual abuse of children, like violent abuse, is perpetrated by their own families. Because of the extremely scandalous nature of incest, there are great pressures to keep it hidden, and it is difficult to estimate its actual rate of occurrence from the few cases that do come to light. Russell (1986) found that one out of forty-three girls (i.e., 2.3 percent) who had lived with their biological fathers (at least through age fourteen) reported they had been sexually abused; for girls who lived with stepfathers, however, the rate was one in six (16.7 percent). What researchers find out, though, depends on just what questions they ask. Questions that ask women generally about sexual abuse in childhood get a prevalence rate of 6 percent to 14 percent, whereas more specific questions about sexual contacts with fathers or stepfathers will produce a prevalence rate of from 11 percent up to 62 percent (Finkelhor 1986, 43). For instance, the latter, very high, figure (62 percent) is produced by asking women if their fathers ever touched them in ways that made them feel sexually uncomfortable. On a more restricted definition of sexual abuse, however, it still remains a sizable problem.

There are of course many possible types of *incest* (grandparent-grandchild, siblings, etc.), but of all these types one is by far the most common: father-daughter incest. A collection of various studies of incest cases came up with the following pattern (Herman 1981). Of 506 total incest cases, 424 involved incest between parents and their children. Of these, 399 were cases of father-daughter incest, 11 of father-son incest, 12 of mother-son, and 2 of mother-daughter incest.

As we can see, there is a small percent of homosexual incest cases; again, these largely involve the father rather than the mother. Mother-son incest cases were once believed to be so rare as to be nonexistent; in eight out of the twelve cases reported above, the son forced his mother to have intercourse; so the incident could also be reported as rape. In most of these cases, the boy was psychotic or mentally retarded.

Most of the fathers who commit incest, on the other hand, are not psychotic (although psychosis has been described as responsible for 10 to 20 percent of cases [Garbarino and Gilliam 1980, 155]). Nor are these men socially "abnormal" according to the conventional culture. Incestuous fathers are often outstanding citizens to the public, with above-average levels of education and income. In fact, the fathers might be described as ultraconventional: not only are they likely to be churchgoing and "respectable," but they adhere strongly to traditional gender roles. They run authoritarian, male-dominated households, and their habitual violence is at a higher level than the average pattern (Herman 1981). Child sexual abusers themselves tend to have previously been the victims of childhood sexual abuse (32 percent, versus 3 percent of a control group of nonabusers; Finkelhor 1986, 102–3).

There is also sometimes a situational factor. In many cases the mother is ill, withdrawn, or actually separated from her children. Typically, the family roles have been rearranged, and a daughter has taken over the mother's role, either in helping to run the household or at least in being emotionally close to the father. Typically, there are poor relations between the parents, and often between the mother and her daughter. Several studies found that victimized daughters were highly likely to be hostile or to dislike their mothers (79 percent in one study, 60 percent in another), much more than they were to feel this way toward their fathers (52 percent and 40 percent, respectively), even after the incest took place (Finkelhor 1986, 157). Incestuous affairs tend to go on for quite some time, typically beginning around age nine (though sometimes later) and continuing for three or four years. Most incestuous affairs are never reported; they end simply because the daughter grows old enough to move away, often by running away or early marriage. Although daughters may indeed feel considerable emotional attachment to their fathers (at least initially), they almost always have negative feelings about the incest. In about half the cases, the girls physically resisted or gave in under threat of force (Garbarino and Gilliam 1980, 157–58). Moreover, the long-term effects on the victims are often quite severe (Finkelhor 1986). Their self-images tend to be negative. Large numbers of them suffer from depression in their adult lives, almost half of them to the point of contemplating suicide. Drug abuse, alcoholism, and psychosomatic symptoms tend to be much more common than usual among these women, and they also turn out to be much more likely to become victims of spousal beatings (Herman 1981, 93, 99). The chain of victimization thus seems to repeat itself.

Is there any preventative for incest? Several courses of action are feasible. One is for families to become alerted to the conditions that tend to promote it (mother ill, withdrawn, or absent; a conventionally rigid, domineering father who is prone to violence). Another action is to explicitly warn children against keeping secret sexual pacts with adults. It is particularly important for not only the mother but the father to take part in these discussions. "If the father tells the child in a calm, guiding way, 'Never let an adult put his hands down your pants, and please don't keep secrets about that sort of thing from us,' the father has eliminated any chance he ever had of successfully victimizing his children. . . . If the mother alone warns the children, the warning might not clearly extend to adults who are part of the family" (Sanford 1980, 233–34).

ELDER ABUSE

In the 1960s and 1970s, child abuse became a pressing social issue as activists, researchers, service providers, and legislators turned their attention to violence against children. The 1970s and 1980s witnessed a similar concern over spouse abuse. In the 1980s and 1990s violence toward older persons burst onto the scene as a pressing social issue. Until recently, elder abuse was relatively unnoticed. While it is certainly not new, violence against the elderly is now beginning to achieve the status of a full-fledged social problem.

Research and policy on other forms of family violence laid the groundwork for focusing on the problem of elder abuse. In addition, because people are living longer and because a greater percentage of the American population is over sixty-five than ever before, more adult children are caring for their parents. As with other forms of violence, older people are more likely to be abused by family members than by strangers. We do not yet have accurate estimates of the extent of elder abuse. Like other forms of family violence, it tends to be underreported, and few surveys have focused extensively on the older population. Current estimates of the number of abused elders and frequency of abuse are based on reports to social service agencies or taken from relatively small, unrepresentative samples. What social service agencies find is that self-neglect is the most common form of elder maltreatment. Nevertheless, researchers tend to focus on how dependent elderly persons are abused by their caretaker adult children (Sprey and Matthews 1989).

Elder abuse is a general term that can include physical, psychological, and material mistreatment or neglect. While many older people are independent and in good health, the chances of poor health increase with age. Persons over the age of seventy-five are at increased risk of physical, mental, and financial dependency, which, in turn, make them susceptible to the following forms of abuse:

1. Physical violence, such as hitting, slapping, shoving, and other tactics common to spouse or child abuse;
2. Psychological abuse, such as verbal assaults, threats, intimidation, and isolation;
3. Physical maltreatment, such as physical restraint, excessive medication, or the withholding of personal care, food, medication;
4. Material exploitation, such as theft or misuse of money or other personal property; and
5. Violation of personal rights, such as forcing the elderly out of their dwellings or into other settings, such as nursing homes. (Block and Sinnott 1979; Gelles and Cornell 1990; Lau and Kosberg 1979)

The most frequently cited estimates of elder abuse range between 500,000 and 2.5 million cases per year (Gelles and Cornell 1990; Pillemer and Finkelhor 1988). These figures indicate considerably more cases than are actually reported to authorities (Tatara 1989). Pillemer and Finkelhor (1988) suggest that only about one of fourteen cases of elder maltreatment come to the attention of the appropriate authorities. Lau and Kosberg (1979) estimated that one in ten elderly persons living with a family member are abused each year. Using a larger sample, Pillemer and Finkelhor (1988) found that 3.2 percent of the elderly reported being the victims of physical violence, verbal aggression, or neglect. For all these forms of abuse, the victim's spouse was most likely to be reported as the abuser.

Factors Associated with Elder Abuse

Women of very advanced age are the most likely victims of elder abuse. Individuals with physical or mental impairments run the greatest risk of being abused, though major handicaps are not necessary precusors to abuse. For example:

> An elderly husband and wife, aged 81 and 79, respectively, had by their own reports always had a "difficult relationship." The wife reported that her husband had struck her previously, but never repeatedly or severely. The situation became much worse when she suffered a severe stroke. She could not accept her new physical limitations and complained about and insulted her husband. During arguments, she would throw food and objects at her husband, which infuriated him. He responded by pushing and striking her more severely than ever before. (University of Massachusetts Center on Aging, 1984; cited in Pillemer and Finkelhor 1988)

Other than spouses, middle-aged daughters caring for aged parents are the most numerous abusers. The fact that women are overwhelmingly responsible for elder care, however, masks the fact that men tend to be more violent, regardless of the age or gender of the victim. When researchers disaggregate elder abuse according to type, they find that men are more likely than women to use physical force and violence (Filinson 1989).

Elder abuse typically occurs under conditions of high stress. First and foremost on the list of stressors is the enormous financial burden of caring for the elderly: medical, food, housekeeping, and transportation costs can be very large. Second, the physical requirements of tending to an elderly dependent are substantial. The personal care, feeding, cleaning, and transporting tasks associated with elder care can be exhausting. Third, the emotional strain of caring for an aged dependent can be overwhelming. Watching and caring for a spouse who is deteriorating physically and mentally can be devastating. Similarly, caring for a dependent parent can entail special emotional burdens, particularly if there are unresolved issues from the past. Parents may continue to treat their adult caretakers as incompetent children, or conversely, may reverse the traditional roles and treat them as omnipotent parents. In either case, the emotional strains associated with power and dependency issues are often quite intense. Finally, multiple competing obligations often add significant stress to family elder care situations. Simultaneously meeting the needs of spouse, children, ailing parent, and job can prove to be difficult, if not impossible (Block and Sinnott 1979; Gelles and Cornell 1990).

Another factor that makes the elderly vulnerable to abuse is their generally devalued status in society. Older people are subject to what has been labeled "ageism." They are often forced to retire from productive occupational roles and can be isolated socially and geographically. Older people also tend to be discriminated against because of their age and are often judged on the basis of unfavorable stereotypes. Insofar as older Americans accept such stereotypes, they can be more willing to accept abuse. Insofar as adult children and other people define senior citizens as "useless," they may be more likely to perpetuate or condone elder abuse.

More and more individuals are surviving to old age. In fact, the most rapidly increasing portion of the U.S. population consists of individuals over the age of eight-five (Sprey and Matthews 1989). With the aging of the American nation, we will undoubtedly see increased attention paid to the needs of older Americans and the emerging problem of elder abuse.

COMMON FEATURES OF FAMILY ABUSE

David Finkelhor (1983; Pillemer and Finkelhor 1988) has summarized some of the commonalities among the different forms of family abuse. Briefly, he suggests that spouse abuse, child abuse, and elder abuse are similar in the following ways:

1. *Abuse of power*. In general, it is the most powerful family members who abuse the least powerful. Thus, men tend to sexually abuse young girls; parents (especially men) tend to physically abuse young children; husbands tend to physically abuse wives who have few contacts or resources; and healthy caretakers tend to abuse frail elderly dependents.
2. *Response to perceived powerlessness*. Although family abuse is perpetrated by the strong against the weak, the violent acts tend to be acts carried out by abusers to compensate for the perceived lack or loss of power. Thus, men tend to beat their wives when they feel insecure or powerless and to abuse their children when they are unemployed or failing financially. Parents also tend to use physical punishment against children when they feel they have lost control of their children and their own lives.
3. *Shared effects on victims*. Not only are victims of family abuse exploited and physically injured, but they tend to be "brainwashed" into believing it was their own fault. Thus, abused children believe they are bad and unlovable; abused wives are persuaded that they are incompetent, hysterical, and frigid; sexually abused children are made to believe that their father's actions are normal and express love; and abused elders come to see themselves as a worthless burden. Victims of all types of family abuse tend to be intimidated by their abusers, but to remain loyal to them and find it difficult to leave. Long-term effects of all types of family abuse include self-blame, lowering of self-esteem, sense of stigma, despair, depression,

suicidal feelings, self-contempt, and an inability to trust and develop intimate relationships.

4. *Shared characteristics of abusing families.* Many forms of family abuse tend to be more common in the lower socioeconomic strata; under conditions of unemployment or chronic poverty; in families with a male-dominated patriarchal structure; in families with drug or alcohol problems; and in families that are socially isolated. In addition, the various forms of family violence tend to co-occur, and children raised in abusive families are at risk of repeating those patterns in their adult lives.

CONCLUSION: WHEN CONFLICT TURNS INTO FAMILY VIOLENCE

All families have some conditions that cause conflict. When do these conflicts turn into violence? And when do they remain relatively peaceful? The extreme forms of violence and abuse that we have reviewed in this chapter are related to the family realities that we have been analyzing elsewhere in this book. As we have seen, a marriage is created by a process of bargaining and negotiation, even if it takes place at an emotional and symbolic level. Within a marriage, the partners have greater or lesser power, depending on resources such as their occupations, networks, and sexual bargaining position. Power always has the potential for causing conflict. So does stress, which arises in any situation of change or difficulty. On top of this, traditional authoritarian roles that are found in some families exacerbate conflicts and cause them to turn more readily into family violence.

Shifts in Family Power

Power has the potential for creating conflicts, but these conflicts are likely to stay beneath the surface as long as the power situation is stable. No one really likes being controlled by others, and this situation generates at least a latent sense of resentment. But powerless people tend to react passively if they can do nothing about the situation. It is when the power situation is changing that conflict is likely to come to the surface.

Until recently, men have held most of the resources that give individuals power in families. As we have seen in chapter 11, they have generally had more income, more powerful occupations, more education, and wider social networks. These conditions have been changing in the last few decades. It is certainly true that women have not yet caught up with men in income and occupational level, but women have been shifting into the full-time labor force. This has moved their power position, at least some. Women have increased their educational level much more rapidly, and it now exceeds that of men. These changes in education and in employment have no doubt expanded women's social networks, which is an important source of power in the home.

All these changes have been shifting the balance of power in marriages. It should not be surprising that there is an upsurge of family conflict at this time. This may turn out to be a temporary situation. In the long run, if women's occupational position continues to improve, families may stabilize at a point where power resources of men and women are equal. When power resources stop shifting, we might expect that this source of conflict will settle down.

Stress Surrounding Family Life

Another source of conflict and potential violence within the family is the amount of stress that people are undergoing. Research on stress has turned up some points which are somewhat surprising. As we would expect, negative events are stressful: becoming unemployed, having one's car repossessed, having problems with money. It is not surprising that families in these situations, especially working-class and poverty-level families, tend to have outbreaks of family violence. But there are other sources of stress as well, including any major change in life circumstances. Taking a new job is stressful; so is the arrival of a baby. Many events that we might regard as happy events are also stressful and can be associated with an increase in conflict and violence.

It is for this reason that conflict tends to break out around any major family transition. The very process of getting married is a life change that can be quite stressful, particularly when the partners don't know what they are getting into. Similarly, the transition from childlessness to the pressures and responsibilities of being a parent is another stress. Some of these factors interact; for example, a woman's family power tends to go down when she has a baby, since pregnancy and early motherhood usually reduces her economic power and her social networks. Putting these various factors together, we can see that family conflicts can build up when a new baby arrives, and that the shifting-power situation may lead to an outburst of violence that may encompass both wife-battering and child abuse.

Of course, not every family that has a baby responds to the new stress by an outburst of violence. In fact, most families do not. How often these stresses and problems lead to open conflicts depends upon how many resources the family members have for dealing with the situation. If they are economically strapped, or lack social support networks, or are young and inexperienced at coping with household problems, then the rate of violence is likely to be much higher. It is where various stressing conditions intersect that family violence tends to occur.

Traditional Authoritarian Gender Roles

Finally, we come to a social pattern that not only fosters violence but also makes it hard to eradicate. When families are organized around localized, encapsulated networks, there tends to be a sharp segregation of male and female roles. This traditional structure produces a high degree of pressure for conformity. People take their positions as rigidly fixed and immutable. They tend to see the world moralistically, with traditional behavior clearly marked as "right" and any other kind of behavior as "wrong."

These families have a world view in which "men should be men, and women should be women," and the roles and powers of the two sexes should not be altered. They also draw a rigid line between the positions of parents and children and believe the power of the parents should be strongly enforced. *Violence in this type of family often has a symbolic significance.* It is a way of ritually acting out the traditional authority relationships. This is why violence can be set off by incidents that seem trivial: a husband getting furious about the way dinner was prepared, or hitting the children for making noise at the wrong time. To the authority figure, the incident is taken as a provocation, a message that authority is being challenged and that the traditional relationship is threatened. Violence is a ritual for putting tradition back in order.

This is in keeping with ritual interaction theory, since structures of enclosed, high-density interaction fill everyday life with rituals that people are supposed to conform to. (See Collins 1981 for an application of this theory to the patterns of violence throughout world history.) The ritual of spanking children is a way of reinforcing the authority of the parents, but it also reaffirms that the children are a part of the enclosed structure. Both parents and children become firmly attached to these rituals; children even grow up with a sentimental admiration for how tough their father was. In the same way, these traditional families tend to believe that men have the right to control their wives with force, that beating one's wife "is good for her." And since this is a ritual which acts out the marriage relationship, many wives who are deeply embedded in the structure end up interpreting the beating as a sign of affection.

This kind of traditional structure is one that takes sexual property rights seriously. Sex is highly ritualized, and any sign of disloyalty is regarded as a threat to the whole system of family authority. Husbands regard themselves as entirely within their rights to violently beat their wives for adultery or sometimes even for any action provoking jealousy. This is one area in which violence is likely to be more reciprocal. For sexual possession is supposed to go both ways, and wives as well become intensely jealous of a husband's sexual behavior. A woman who is otherwise quite traditional and passive in her marriage may attempt to kill her husband when she feels she has been wronged in sexual matters.

This highly ritualized family structure reinforces violence because it creates a belief that the violence is morally justified. The violent husband—and sometimes the violent wife—is righteously angry and morally outraged at what he or she believes a spouse has done to violate "the way things should be." It is for this reason that the police are reluctant to intervene in family disputes, since they are often confronted by a spouse who believes that (usually) he is acting within his rights, and that no one outside the family has a right to interfere.

The Combination of Violence-Producing Conditions

When the several kinds of conditions we have reviewed here overlap, family violence is most likely to be severe. But much of the time they do not overlap. Not all traditional family structures are abusive: as long as the power relationships remain stable and no one challenges them, as long as there are no important family stresses, as long as there are plenty of economic and social resources for dealing with stress, things may be peaceful. Traditionalists like to point to this as the ideal family situation: the old-fashioned family in which everyone knew his or her place, husbands and parents were tough but loving, and everyone felt secure as a member of the family. The ideal combination of circumstances probably did not happen often, even in earlier historical periods. When conflict did break out in the traditional family structure, life could be extremely cruel.

Most sectors of modern society are moving away from that structure. But elements of it remain. It is most often found in working-class families, but also lingers in middle-class families from traditional backgrounds and communities. The conditions of living in the poverty class may also recreate some of the same kinds of localized, encapsulated, and gender-segregated structures. On the whole, though, the trend is toward a different kind of structure. The newer structure is one in which the family is more of an open network, more connected to the outside society, with more equal power relations between husbands and wives and much less authoritarian control over children. The very fact of transition, of shifting power relationships in the family, does tend to increase the level of conflict. But in the long run, we can expect that family relationships will settle down at a new balance of power. As this comes about, we can look forward to a reduction in the level of family violence.

In the meantime, there are things that educators and social service providers can do to reduce family violence. Although a combination of stressors tends to increase the chances of family abuse occurring, it is also the case that a majority of people who are exposed to several risk conditions do not resort to violence. This makes intervention and education especially important. Social programs can break down the sense of fear and isolation that often accompanies family abuse, and with legal sanctions, treatment programs can help both perpetrators and victims redefine taken-for-granted violence as inappropriate and unacceptable. Interventions aimed at helping those undergoing stress associated with job loss, new babies, or the infirmity of a family member can also help to break the cycle of violence by encouraging people to develop realistic expectations about all aspects of parenting and family life. Just talking about the strains and hassles of everyday life, and knowing that others face similar difficulties, can bring a great sense of relief to family members who might otherwise resort to violence. Learning some simple self-help techniques to use in crisis situations has helped many people avoid recreating ritualized family abuse.

We cannot expect family violence to disappear if the underlying conditions promoting it remain. As long as our culture idolizes violence, our schools and churches teach that physical punishment is moral, our marriages reflect traditional authoritarian gender roles, and our economy keeps a substantial portion of our workers in poverty, then family violence will continue at today's high levels. With change in at least some of these underlying conditions, private and government efforts at intervention and education are likely to meet with some success.

SUMMARY

1. Violence is widespread in American society, as evidenced by homicide and rape rates that are much higher than in most other industrialized countries. We idolize tough guys in the movies, portray aggressive male dominated sexuality on television, and tend to believe that children should be physically punished for wrongdoing.
2. The family is the most common context for violence in American society. While family violence often goes unreported, police still receive more requests for help with domestic disturbances than any other problem. Battering of wives and children is an extension of "normal" patterns of discipline that derive from patriarchal models of family authority. Child abuse, wife abuse, and elder abuse have recently been recognized as important social problems, but efforts to reduce the amount of family violence have been hampered by the assumption that the family is a private domain.
3. While reporting problems make estimation difficult, at least one quarter of American families will experience some violence between family members, and a substantial fraction will experience fairly serious violence. Although women push, shove, and hit men about as often as the reverse, women's violence is often in retaliation or self-defense, and men's violence against women is much more likely to result in serious injury. Wife-beating happens most frequently when the husband is poorly paid, working class, or unemployed; when there is great worry and dissatisfaction over economic

resources; when the woman is a full-time housewife; and when there are many disagreements over children.

4. Men, and often the surrounding community and the police, justify violence against wives as just punishment for violation of male prerogatives of power and especially of sexual property. Economic strains also set off domestic violence, especially when the man is unable to support his family and feels downgraded by his wife's economic contributions. Violence is passed along intergenerationally; men who beat their wives tend to have experienced much physical punishment as children, and battered wives tend to use more violence on their children.

5. Social control does reduce domestic violence. Socially isolated families have more violence, and socially isolated women are less likely to leave an abusive relationship. Calling the police has some deterrent effect, especially when the man is arrested. In general, men hit their wives because they can get away with it.

6. Violence against children includes a very extensive use of mild physical punishment (spanking, slapping), but also injury-causing or even lethal beatings and use of weapons by parents upon children, children upon parents, and children upon each other. Most child battering is not done by psychotic parents but by "normal" ones in particular circumstances. Children who are especially demanding, including handicapped children, are especially likely to be abused. Abusive parents tend to be those who have little commitment to the role of parent, or who have difficulty making the transition from the youth culture themselves; parents who themselves were abused in childhood or otherwise did not have an opportunity to learn the role of a good parent; parents who are under special stress such as economic problems. Nearly 25 percent of all American children are judged to be at risk because of some combination of these factors.

7. Mothers are more likely to be the abusers of small children, fathers of teenage children. Abuse of small children is most likely in the lower social classes; teenage abuse is less well understood. Women who are abused by their husbands are most likely to use severe violence against their children.

8. Most incest is of the father-daughter kind. It is most likely to occur when the mother is withdrawn from the family and especially from her children, due to illness, absence, or family strain. Incestuous fathers tend to be ultraconventional persons in public, highly traditional about gender roles, who run authoritarian households and use above average levels of violence. In about half the cases the incest is forced upon the daughter. Most cases are never reported, but terminate when the daughter runs away or marries to leave home. Long-term effects on the daughters are quite negative, including high tendencies to alcoholism, suicidal depression, and becoming a victim of spousal beatings.

9. Elder abuse, including physical, psychological, and material mistreatment

or neglect, is less studied than wife abuse or child abuse. Women of very advanced age run the greatest risk of abuse, which tends to be perpetrated by spouses or middle-aged daughters. Elder abuse, like other forms of family violence, tends to occur under conditions of high stress. Since the elderly population is growing rapidly, we will undoubtedly see more attention focused on this issue in the future.

10. Violence often occurs within families that believe in traditional authoritarian gender and parent roles. A relatively insignificant event frequently triggers the violence in these kinds of families, and the physical outburst symbolically reaffirms relations of dominance and hierarchy.

11. The common features of family violence include: (1) it is an abuse of the powerless by the more powerful; (2) it is often a response to a perceived lack or loss of power; (3) its victims tend to suffer low self-esteem and blame themselves; and (4) it occurs under conditions of poverty, stress, male dominance, and social isolation.

FAMILY CHANGES

14

DIVORCE AND REMARRIAGE

It is not surprising that there is such a thing as divorce, given what we have seen in previous chapters. Marriage is an arrangement of power, privilege, and conflict, as well as of love and solidarity. Often it is all of these at the same time. When the balance of emotional, economic, and sexual resources changes, the marriage too has to change its arrangement between the partners, or else break up. Some marriages dissolve because they were not too well negotiated in the first place, while others come undone after a longer period of struggle. Not all marriages come apart, of course, although some couples perhaps would be better off if they did separate instead of carrying a de facto emotional divorce inside a hollow frame. For all this, there is also remarriage, which happens to a large majority of divorced people. It appears that we can't quite stand marriage, but we can't stand being without it either. In this chapter, we will attempt to see how and why this is so.

HOW MUCH DIVORCE?

In an average year there are about 1.2 million divorces in the United States. Is this a high number or a low one for a country as large as this one, with our population of some 250 million? Like most things, it depends on your perspective. About 2.4 million marriages take place every year, so one can say that there are more marriages than divorces and that divorces are a little less than half the number of marriages (*Statistical Abstract* 1992). Most of the people who divorce in any one year, however, did not get married the same year but throughout a period that can stretch forty years or more into the past. Hence statisticians seek alternative ways of calculating the divorce rate. One way to do it is to divide the number of divorces by the total number of existing marriages. Since a marriage requires two people, one can count either the married men or the married women; statisticians for some reason usually count by the number of women, although (as already mentioned in chapter 11) there is a small tendency for more women to say they are married than for men to say so. At any rate, these figures are relatively easy to come by from the U.S. Bureau of the Census, and the result is a chart like figure 5.2.

This shows us that the divorce rate is about 21 per 1,000 marriages. It also indicates that the divorce rate has been rising fairly steadily for over 120 years. There is one major exception: between about 1940 and 1947, during World War II and immediately after, the divorce rate skyrocketed and then fell again almost as abruptly. War-related problems seemed to play havoc with the American family. It was the largest war ever fought by the United States, mobilizing over 16 million troops. It is conjectured that many marriages were contracted in haste, as the groom faced possible death in combat. In addition, the temporary separation of families at this time broke many bonds of affection. When couples were reunited, many were not compatible. Hence, the peak of

After spending fifteen months on an aircraft carrier in the South Pacific during
World War II, a GI returns to his wife and six-month-old child he has never
seen. Many marriages contracted in haste because the groom faced possible death in
battle fell apart when the couples were reunited after the war.

the divorce rate was in the two years immediately after the war. But aside from
this episode, the curve has generally been upward; what seemed at the time to
be an unprecedentedly high divorce rate in the 1940s became the normal rate
twenty years later and was surpassed by still higher rates in the 1970s. Only in
the late 1980s, did the divorce rate stabilize again or even dip down a little,
although it is standing at a level that is essentially at its all-time high.

Just how high it is can be seen by looking back a century. In 1860, the
divorce rate was about 1 per 1,000 marriages. It grew steadily to about 8 per
1,000 in 1920 (the "Jazz Age"), more slowly to about 10 per 1,000 in 1960
(ignoring the sudden up-and-down around World War II), and then more than
doubled in the next generation. There is no doubt that we have gone through
the single largest and sharpest rise in divorce in American history and that the

family system has been transformed by divorce as never before. Back in the 1800s, marriages were normally ended by the death of one of the partners: 97 percent were dissolved by death and only 3 percent by divorce (Thwing and Thwing 1887). Today, marriages end more or less equally by death or by divorce.

Another way to calculate divorce rates is to estimate the proportion of all marriages in any given year that will end in divorce (see figure 14.1). In 1890 the chances that a marriage would eventually end in divorce were only about one in ten. By 1930, the chances were about one in four, and by the 1970s, the chances were one in two. This is the rate where we have more or less stabilized today; hence one would have to say that for marriages made today—the marriages of students who are reading this book—the statistical likelihood of divorce is about 50–50.

Divorce rates are higher in the United States than in Canada, Japan, and most of the western European nations. In fact, our rates have been about double the rates of these other countries since 1960 (Sorrentino 1990). But divorce rates have gone up as fast or faster in other industrialized countries than they have here. Whatever has been going on, a unique breakdown in American cultural values is not the culprit.

When Does Divorce Happen?

The average divorce happens about six years after the marriage (U.S. Bureau of the Census 1992e). About 5 percent of all divorces happen within

FIGURE 14.1: The Rising Proportion of Marriages That Will End in Divorce

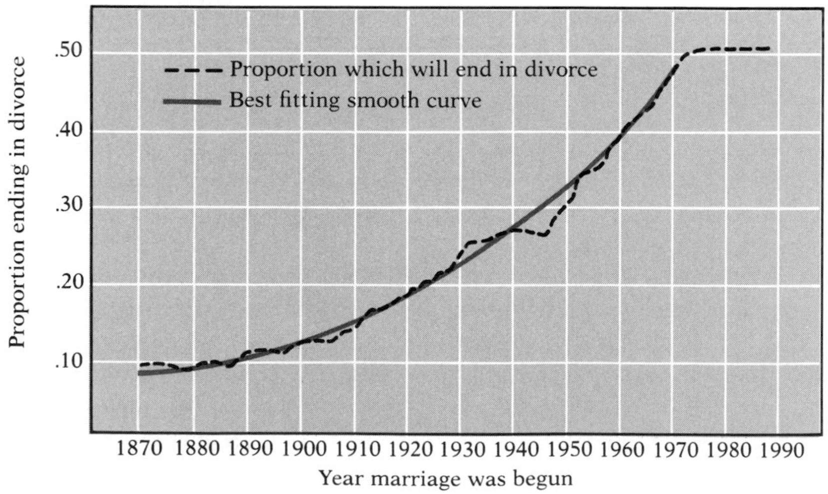

Source: Cherlin, 1992.

FIGURE 14.2: Most Divorces Happen Soon after the Marriage

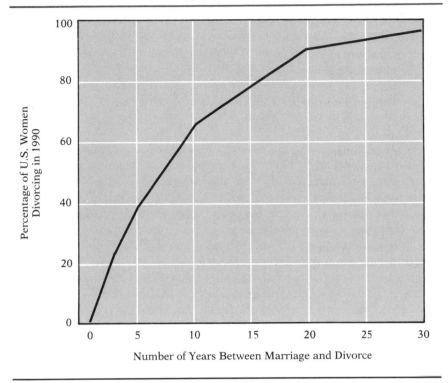

Source: U.S. Bureau of the Census 1992e, 8.

the first year, and the rate peaks at about 9 percent in both the second and third years. Then it starts falling again, until by the tenth year of marriage, two-thirds of those who are going to divorce have done so, and by the twentieth year of marriage, 90 percent of the divorces that are going to happen have already occurred (London and Wilson 1988; U.S. Bureau of the Census 1992e). The first few years of marriage are thus the main danger point (see figure 14.2).

Social Class and Divorce

If one went by the tabloid press, one would get the impression that rich people are the specialists in divorce. Much publicity is given to movie stars like Elizabeth Taylor or Roseanne Arnold. In general, however, the picture is just the opposite. The lower the social class, the more likely the marriage is to break down, whereas the higher the social class, generally speaking, the more likely the marriage is to survive. This is particularly true if we count desertions as well as formal divorces. Especially in the lower social classes, many people break up a marriage by simply leaving, and they never file for a formal divorce decree. This happens most frequently in hard economic times, when the expense of going through the legal arrangements is not felt to be worth it.

There are few twists on this social-class theme. If we use education as a measure of social class (this information is more easily standardized than information on people's occupations, which it closely parallels), we find that generally, as people's educational level goes up, the level of divorce goes down. This is true for both men and women, but educational level has tended to favor marital stability in men much more than it does in women, as shown in table 14.1.

What is going on here? For men, higher education generally brings a higher occupational level. These more affluent men are not only more likely to marry than poorer men, but they are also more likely to keep their marriages together. Apparently being married and staying married are part of the traditional respectability for men in the higher ranks of the class system. Moreover, until very recent years, men were getting most of the graduate education, and hence they usually married women with less education than themselves. (The only place in which this is not likely to be true is in the working and lower classes, where males drop out of high school more than females.) Even now, men tend to monopolize most of the higher paying and prestigious jobs. Thus, high-ranking men have a lot of pull on the marriage market, and once they are

TABLE 14.1

Percentage of Intact Marriages by Educational Level

Level of education	Percentage of marriages intact
Men	
Less than high school education	65%
High school diploma	72
College diploma	79
Post-graduate education	82
Women	
Less than high school education	57
High school diploma	66
College diploma	77
Post-graduate education	63

One might notice that the percentages of intact marriages is lower for women than for men at each educational level. How is this possible, since men and women are married to each other? The answer is that there are different numbers of men and women at each educational level, and they are not necessarily married to someone at the same educational level. Hence different combinations end up producing this pattern.

Source: Glick and Norton, 1977, 9.

married they tend to have the resources to dominate their wives. These men's wives often provide a lot of support for their careers and their egos, and maintain a high level of domestic comfort. As far as affluent men are concerned, marriage is a good deal, and they don't usually want to lose it. From their wives' point of view, things may not be as rosy as they seem. But in terms of hard economic realities, these women tend not to have the job opportunities to keep up their standard of living if they divorce (see figure 14.6). Thus, the economic situation tends to keep the couples together.

Such marriages probably contain a good deal of underlying strain. If the wives are highly educated, they are much more likely to divorce. We see this in the figures below: whereas men with postgraduate degrees are *least* likely to be divorced, women in this group are *very* likely to be divorced. Their level of marital breakup is almost as high as that of the very lowest-educated group. In other words, women at the top and the bottom of the class structure are most alike in terms of marital longevity.

These highly educated women are the minority who are lawyers, physicians, professors, and other highly skilled professionals and administrators. Even given economic gender discrimination (i.e., women professors don't make as much as men professors, and so forth), these are the women who are best off in terms of their own incomes and careers. Such women are among the least likely to marry at all, and if they do, they are among the most likely to divorce. It also turns out that they are least likely to remarry after a divorce. This is probably because they are least willing to put up with traditional male domination of the marriage. They have the most resources to back up their own point of view, and if they cannot reach an egalitarian settlement, they are most willing to go off on their own (see also Cherlin 1979). Also, highly educated women tend to be older when first married and older when divorced, so the pool of eligible remarriage mates is smaller for them than for other women.

Age and Divorce

Another pattern is that divorces happen more frequently the younger the couple were when they married. Among women who marry at age sixteen or seventeen, over a fifth are divorced within the next ten years. In fact the rate is high for all teenage marriages. For example, in 1989 32 percent of the women who had married when they were teenagers had divorced; but women who had waited until age thirty or more before they first married included only 12 percent divorced (see figure 14.3).

The effect of age, however, is mixed up with that of social class. Since persons from the lower social classes tend to marry earlier than those in the higher social classes, it may be that the age effect is secondary. The teenage marriages that show up on the chart may be mainly working-class and lower-class marriages, and that is the reason why the rate is so high. This seems

FIGURE 14.3: People Who Marry Young are More Likely to Divorce

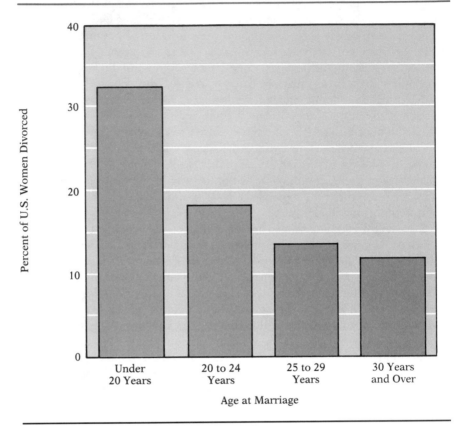

Source: U.S. Bureau of the Census, 1989, table E.

to be true, but it probably accounts for only part of the divorce rate at young ages. Persons from any social class seem somewhat more likely to divorce if they marry young, and the effects of early marriage on divorce seem to be independent of early childbearing (White 1991). We also know that women who divorce tend to have had shorter engagements and were more likely to have married without the approval of their parents (Goode 1956). Thus, one source of the problem may be hasty, spur-of-the-moment youthful marriages— quickly put together, quickly torn apart.

Race and Religion

For a variety of reasons, there tends to be a higher divorce rate among African-Americans than among whites, Asian-Americans, or Hispanics in the

United States. We have examined some of the causes in chapter 7: greater economic strains among the portion of the black population that is poor, but also a particularly independent style among many black women that makes them less willing to put up with male-dominated marriages. Whatever the causes, about 73 percent of the marriages of whites to whites were still intact in 1985, in contrast to only 69 percent of the marriages of blacks to blacks (*Family Planning Perspectives*, 1986).

About 2 percent of U.S. marriages are interracial and about 7 percent of marriages including an African-American are interracial (O'Hare 1992). The more common form of black/white marriages is for a white woman to marry a black man; the less common involves a white man and a black woman. Both of these forms of intermarriage have fairly high divorce rates, but even these divorces seem to fit the "preferred" pattern: when the marriage is between a white female and a black male, the percentage remaining intact (calculated on the same basis as in the previous paragraph) was 63 percent; when it involved a black female and a white male, the percentage still intact was 47 percent. This last figure shows one of the highest rates of marital breakup over this kind of time period for *any* social characteristics yet measured. One of the unsolved mysteries of the sociology of marriage is why these patterns of black/white marriages happen this way.

Religion also affects the divorce rate, but its influence seems to be decreasing. It used to be that Protestants had higher divorce rates than Catholics. This is not surprising, given the Catholic Church's doctrine prohibiting divorce, but the rate of Catholic divorces has nevertheless gone up parallel with Protestant divorces. By the late 1980s, several researchers reported that Catholics were no less likely to divorce than were non-Catholics (Bumpass 1990). Interreligious marriages, like interracial marriages, have also been subject to slightly higher divorce rates, but in general, the influence of religion on marriage dissolution seems to be declining.

CAUSES OF DIVORCE

Why do divorces happen? Obviously, because people are dissatisfied with marriage and feel they would be better off apart. But there is more to it than that. Divorces are somewhat like revolutions: the regime may be oppressive, but it doesn't come tumbling down just because people don't like it; something more specific has to happen that breaks it apart.

Sex and Divorce

In our folklore of soap operas and gossip, the event that breaks a marriage apart is usually an illicit sexual affair. There may be some truth in this. Kinsey

(1953, 455) asked the divorced women and men in his sample who had had an extramarital affair what significance it had for their divorce. About 60 percent said that their extramarital affair was not a factor in their divorce at all. (Presumably this means they did not get divorced because they wanted to leave their husband or wife for another man or woman.) But from the perspective of the *other* spouse (the aggrieved party), the extramarital affair was not so unimportant. Thus, 83 percent of these males said that their wife's sexual infidelity was a fairly important factor in their divorce, and 76 percent of the wives said the same about a husband's infidelity. We see here, incidentally, a bit of the double standard, in which husbands are more likely to treat their wives as exclusive sexual property than vice versa.

Nevertheless, this doesn't mean that 76 to 83 percent of all extramarital affairs result in divorce. As we have already seen (in chapter 9), most extramarital affairs are not discovered, and when they are, about half the time the aggrieved spouse forgives or condones it. Extramarital affairs are a good deal more common than divorces (at least they were in the past). Sexual infidelity, although dramatic, is not really the most important cause of divorce.

Sexual relations of a different sort, though, may be quite important. Chapter 9 indicated that sexual dissatisfaction in marriage is fairly widespread, and that it is correlated with a generally low level of marital happiness. The general unhappiness seems to prevent good sexual relations, which means that if individuals want to improve their sex lives, divorce may be the only route they know how to take. In fact, it does turn out that second marriages are sexually more satisfying than first marriages (Westoff 1977, 126). Not only do people report that their sex lives are improved in quality, but even the frequency of sexual intercourse goes up. This happens even though people are older in their second marriages, and the rate of intercourse tends to decline with age. So it looks as if some people leave their first marriages in search of a better sexual relationship. Fortunately, they tend to find it.

Economics and Divorce

Even though sex and other interpersonal relationships may be a motivation for getting divorced, they aren't the whole story. We can see this from the general economic trend. In times of prosperity, the divorce rate tends to go up; in times of economic depression, it drops (see figure 5.2). During the Great Depression of the 1930s, the divorce rate dipped. This happened for all social classes. To some extent, the reason is that during hard times people are less able to afford a divorce, and instead the couples simply separate, or one spouse (usually the husband) deserts the other. There is also some indication that families feel the need to stick together for economic support when they are in financial straits (Preston and McDonald 1979). Or to put it another way: when job opportunities are bad and the level of unemployment is high, spouses feel

FEATURE 14.1
Dealing with the Legal Entanglements of Divorce

Divorce, like marriage, is a legal condition. It involves one in the complexities of courts and lawyers. Although some divorces are simple, others have considerable legal complications. And recent legal developments, such as the famous *Marvin* case, have begun to extend some of the liabilities of conventional marriage and divorce to the breaking up of cohabitation arrangements.

Grounds for Divorce

Until the early 1970s, divorce was legally very hard to get—at least under the strict letter of the law. At one time couples would head for Las Vegas or Mexico to take advantage of places where one could acquire a quick divorce. In most states, the only grounds for divorce that were legally admissible were adultery, desertion, or extreme cruelty. In the case of well-known people filing for divorce, this led to quite a few lurid headlines. But as the number of divorces began to increase most people simply made up fictional complaints. Most common was the pat formula "extreme mental cruelty" as the ground for divorce.

In 1970, California enacted a *no-fault divorce* law, and more than half the states followed soon after. It was no longer necessary under these laws for one person to be the aggrieved party. Now the marriage could be ended by mutual consent, and it was no longer necessary to lie about the reasons for divorce. Opponents of these liberalized laws argued that they would make divorce all too easy; nevertheless, the rate of divorce in no-fault states has turned out to be no higher than the trend in states that kept the traditional laws (Cherlin 1992).

Who Needs a Lawyer?

With no-fault laws, it has become easier for couples to obtain a divorce without hiring lawyers to represent each side. In many states, it is possible for one spouse simply to obtain papers from the courthouse and file a petition for divorce. If the petition is not contested by the other spouse, the divorce is granted within a certain period of time (usually six months or one year from preliminary decree to final decree). In many large cities, there are "divorce clinics" where a professional will help persons who desire a divorce to file the necessary papers. However, these procedures are feasible only if there is no significant property involved, or if there is no disagreement about custody or support of children. If divorce is complicated by either factor, both parties are well advised to retain their own lawyers to protect their interests. If no lawyers are involved, the judge will decide the property and child-related issues under conventional precedents. If one spouse is represented by a lawyer and the other is not, the latter's chance of getting an equitable settlement, from his or her own point of view, is rather slim. Lawyers usually charge by the hour for their legal services, and a minimum of about $750 for a divorce is typical. If the case is very complicated (i.e., if there is a great deal of property or a difficult custody fight), the cost may be considerably higher.

Property Settlements and Alimony

Different states have somewhat different laws regarding the settlement of marital property. In California and several other states, all property acquired during a marriage is what is legally known as *community property*. This includes all the salaries and other income to either spouse during the time they have been married, with the specific exception of inheritances, which are considered their separate property. Other separate property may consist of a house or a business that was owned before marriage, although these may become converted into community property by

(Continued next page)

FEATURE 14.1 (*Continued*)

gift to the community, by being "mingled" with community property (e.g., mingling old funds with new contributions), or by making continued mortgage payments out of current income on a previously owned house. There are many complicated issues here, which is one reason why it is essential to have a lawyer if there is much property involved in a marriage. All community property is divided 50–50 between the divorcing partners, unless the couple have specifically made an agreement otherwise in the form of a marital contract.

In some other states without community property laws, there continues to be the traditional English common law, which vested all property in the husband, unless explicitly put jointly or singly into the name of the wife. Here divorced women might come away with little or no property at all, although in fact judges have tended to use their discretion to make an equitable distribution. In New York and elsewhere, "equitable distribution" is now the law, and the trend is for this form of law to become more widespread.

Alimony (sometimes called *spousal support*) consists of a fixed income to be paid by one partner to the other after the divorce. Its basis is the inability of one ex-spouse (traditionally the wife) to support herself, due to having given up her career for the domestic responsibilities of marriage. Although one hears about very large alimony payments to the ex-wives of some wealthy individuals, alimony is no longer typical in most divorces. In most states (such as Ohio), alimony is not awarded at all. In others, it is awarded only if the spouse did not work at an outside job at all during the marriage. It has now become common for a court to order spousal support payments for a limited period of time, based on an estimate of how many years of education and retraining it will take for a former housewife to support herself with a paying job.

Marital Contracts

Sophisticated lawyers now point out that one can avoid some of the legal difficulties of a divorce by planning ahead. No one goes into marriage expecting to get divorced, but since the probability is in fact rather high (nearly 50 percent for first marriages, even higher for second marriages), it is rational for a couple to specify what will happen in that event. A *marital contract* (Weitzman 1981) can specify what property is to be communal and what is to be held separately; what spousal support, if any, is to be paid in the event of divorce; and many other matters relating to the rights and duties during the marriage as well as in its dissolution. The only thing a marital contract cannot specify is the question of child custody and support. Since these affect the welfare of the children, the court legally retains jurisdiction to decide these matters after taking input from the contending parties themselves.

---

less confident about going off on their own. It doesn't necessarily mean that they feel more warmly about clinging together and supporting one another in time of need. More likely, wives who would otherwise leave their husbands realize that they just can't afford to do so; and men who might desert in times of prosperity find they are out of work and hang around so that they can live off their wife's salary or other income.

Economics is a very basic motivator in a marriage, and an economic downturn pulls realities closer to that bottom line. Marriage is a kind of trade-off of various resources: income, love and affection, domestic labor, and sex. When the trade-offs are unequal, one person or the other dominates, and this can give rise to considerable dissatisfaction. But as long as the resources are too unequal, the situation may not change. If divorce is like a revolution, an unequal marriage is often like a stable dictatorship. It is only when one partner feels that he or she (very often she) has better opportunities outside the marriage that it falls apart.

Of all these resources, the economic ones are probably most crucial. A marriage is a way of providing a living, a communal economic unit, and that aspect can go on even if the love, common interests, and sexual ties all have dwindled away. The fact that marriages are more apt to stay together during an economic depression is not necessarily a good thing; one might say it is just one more toll that a depression takes on people's lives.

From the mid-1950s through the early 1970s, following the post-World War II deviation in the pattern, the rate of divorce went up in the United States. This was also a period of sustained economic growth. Some economic historians are now saying that it was an economic miracle such as no society ever saw before, and that we may never see again. We became a society in which the great majority of the population—the working class included—could reasonably aspire to owning their own home, as well as a car (or several), television sets, household appliances, stereo system, and many other consumer goods. Near the end of this period, married women began to join the labor force in unprecedented numbers. It was also during this time that the divorce rate rose to a historic peak.

Since the late 1970s, the divorce rate has stabilized and is even inching downward. This is probably not because of some newly awakened surge of traditional morality; all other indicators of sexual behavior tend to indicate otherwise. What we are seeing instead is probably the mirror image of the earlier trend: divorce increases during prosperity and declines during depression. Ever since the oil crisis and the runaway inflation of the mid-1970s, the American economy (like those of other industrial nations) has been stagnant, with at best irregular growth punctuated by downturns. We still remain a very wealthy society; the roller coaster has stalled nearer the top than the bottom. But one result is apparently the end of the long trend of continuously rising rates of divorce, although the level at which the rise has topped off is historically an unprecedentedly high one.

Toward the end of the 1980s and into the 1990s, much attention was paid to the decline of community and the rise of individualism (e.g., Bellah et al. 1985). The basic idea was that Americans were too self-absorbed and that they were no longer able to make the sacrifices that were necessary for their communities and their society to thrive. Religious leaders, politicians, and some family scholars picked up on this theme, as they have periodically for over a hundred years, and lamented the decline of The American Family (e.g., Popenoe 1993). The primary measure of this increasing hedonism, narcissism, and self-absorption was taken to be the rising divorce rate (White 1991). Also buttressing the view was the purported tendency of Americans to talk about

FEATURE 14.2
Marvin versus Marvin: A Divorce Settlement Without a Marriage

Marvin vs. *Marvin* is probably the most famous divorce case of recent times, although it did not actually involve a legal divorce. It was brought by Michelle Triola Marvin against the actor Lee Marvin, with whom she had lived for almost seven years before separating in 1970. During that time Lee Marvin acquired property in his own name (including motion picture rights) worth more than $1 million. When they separated, Lee agreed to pay Michelle $850 a month for the next five years. After eighteen months he refused to make any further payments, and Michelle sued him for breach of contract. The case was eventually taken to the California Supreme Court, which ruled in 1976 that their cohabitation did involve an implicit contract similar to marriage and that Michelle Marvin was entitled to half of the property. Eventually, after the case was returned to a lower court for further hearing of the facts, a judge awarded her $104,000 in rehabilitative alimony.

The newspapers called the award *palimony*, the first time alimony was awarded to a cohabitation partner after the relationship broke up. Did this mean that all couples living together were now liable for the same divorce procedures and property claims as conventional spouses? In effect, the court was adopting a sociological rather than a legal definition of marriage: the exclusive sexual relationship and joint household economy were taken as the key, not the legal marriage document or ceremony.

But in fact the alarms were overstated. The *Marvin* ruling was based on particular circumstances that limit its application to other cases. Michelle Marvin claimed that Lee had made an explicit contract to support her, in return for giving up her career as a singer in order to live with him. She treated this relationship as a marriage, and even changed her name from Michelle Triola to Michelle Marvin. Legally binding contracts may be verbal as well as written, although verbal contracts are harder to prove. In this case, the judge believed Michelle that in fact there was such an oral contract. In the absence of such contracts or promises, a cohabiting couple does not incur any liabilities of this sort.

Practical advice emerging from the *Marvin* case is: (1) if you cohabit with someone, be careful what you say, especially of what can be construed as a promise; (2) if there are any doubts, make an explicit cohabitation agreement. Lenore Weitzman's book *The Marriage Contract* (1981) gives examples of such contracts.

families and express opinions in ways that characterized marriage as a personal, tentative, and fleeting commitment.

It is true that marriage is less obligatory than it once was, and that people are now able to end bad marriages fairly easily. Nevertheless, there is a problem with the argument that individualism, selfishness, or secular humanism caused the divorce rate to go up. For one thing, the divorce rate and attitudes toward divorce seemed to peak about 1980. Since then, the likelihood of divorce has been inching down (see figure 14.1), and people's attitudes toward divorce have become slightly more negative or remained stable (see figure 14.4). What's more, when longitudinal studies are conducted, disapproval of divorce does not seem to affect the divorce rate, and attitudes toward divorce seem to change only after someone goes through it (Thornton 1989). Apparently attitudes toward divorce, or changing views toward marriage, are not the primary forces behind changes in divorce rates. It is much more likely that both attitudes and divorce rates are responding to the same sorts of underlying economic and social pressures.

SINGLE LIFE AFTER DIVORCE

Most people who divorce only do so for strong and pressing reasons: they want out of their marriage. But even with all the incentives to leave, the experience of divorce is often a traumatic one. There tends to be a great deal of anger and resentment, and also anxiety about the future. A psychological support has broken away, and a ritual tie that had been taken for granted has snapped. The results of this structural transformation are flaring emotions not unlike the dangerous spark given off by a piece of electric machinery when it is knocked apart.

The Postmarital Economic Strain

Eventually these feelings, centered on the person who is now one's ex-spouse and on the breakup itself, begin to go away. The formerly married

FIGURE 14.4: Shifting Public Opinion on Divorce

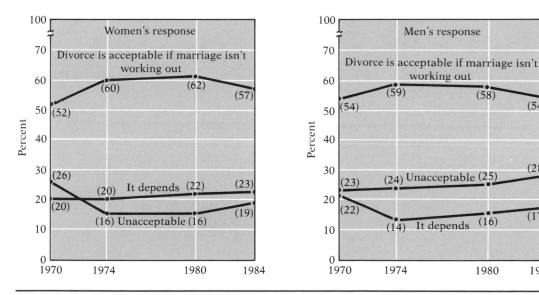

Source: *Public Opinion*, Dec./Jan. 1986, vol. 8, p. 27.

FIGURE 14.5: Few Divorced Women Receive Child Support

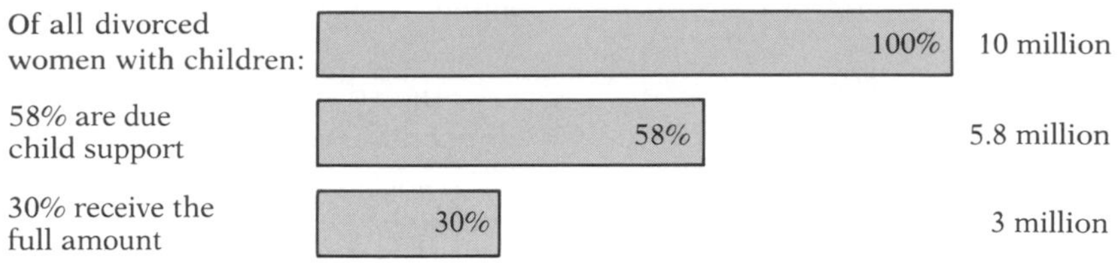

Of all divorced women with children:	100%	10 million
58% are due child support	58%	5.8 million
30% receive the full amount	30%	3 million

Source: *Current Population Reports*, 1992, P-23, No. 181.

person settles into a new life. For all the emotional flare-ups and the social readjustments that must be made, the most important problem is economic. A practical living arrangement has been broken up, and two new ones must be created. Generally, this has been a more severe problem for the woman. If she was a housewife before the divorce, she has the problem of getting a job or securing some other source of income. There may be alimony decreed by the court to be paid by her ex-husband, and if there are children, child support payments. But alimony (where it exists) is rarely adequate to live on, and child support payments are often lower than actual child-related expenditures. What's more, these forms of postdivorce support are often not paid in full. Only about half of divorced women with children received any child support payments from their ex-husbands in 1990, and only about a third received the full amount due (see figure 14.5). The average amount of child support received was under $3,000 per year (Ahlburg and DeVita 1992, 2a).

The lack of payment of child support is one reason why female-headed households are six times more likely to be below the official poverty level than male-headed households (U.S. Bureau of the Census 1992d). Another major reason concerns labor market inequities. Increasingly, divorced women are expected to take jobs, if they do not have them already, and to provide their own support. But the income that a woman makes is on the average only about two-thirds of what a man makes, and hence her household standard of living is very likely to drop sharply.

This phenomenon has been confirmed by several studies. Weitzman (1985) interviewed a sample of persons who had been divorced in California. In the year since their divorce, the standard of living of divorced women declined 73 percent. This was calculated by comparing the women's income with the standard income to support a family of a given number of members. Since in most cases custody of the children was awarded to the women, the budget was calculated in relation to how much was necessary to support an adult and a certain number of children. A terrific decline in the standard of living was seen. In order to survive, some of the women had to change their lives dramatically (14 percent went on welfare; many moved in with their parents or other relatives). Weitzman's sample (from Los Angeles) may be somewhat extreme, however. A larger sample of the entire U.S. population (Duncan and Hoffman 1985) found that the economic status of women fell an average of about 30 percent in the first year after divorce. An earlier study (Hoffman and Holmes 1976, 31) found that over a seven-year period the standard of living of divorced women had declined 29 percent. This is a less substantial decline than the 73 percent Weitzman found, although it is still an important and consequential economic reversal for divorced women (Hoffman and Duncan 1988). More recent studies show that women's incomes tend to remain low for many years after divorce (see figure 14.6).

The position of husbands is more ambiguous. In actual dollar amounts, the income of both husbands and wives declines. In one national survey (Hoffman and Holmes 1976, 27, 31) the income of divorced women went down by 29 percent, whereas the income of divorced men went down by 19 percent. How can both incomes go down? One reason is that the same income now has to be shared by two households. Usually the family home (if there is one) is sold in order to divide up the proceeds so that both ex-spouses must find new accommodations. But since the women usually have custody of the children, the standard of living needed by the ex-husbands and ex-wives is different. In relation to the number of persons to support, then, most studies find that seven years after the divorce the economic position of the divorced women declines while the standard of living of the men improves (Hoffman and Holmes 1976, 31). It is doubtful, though, whether these divorced men actually feel their standard of living is going up. Many must live in an apartment rather than a house, and like the women, they lose whatever economies of scale there had been for two adults living together (such as the cost of appliances and utilities).

One of the most striking findings from the divorce studies is that the difference between men and women is sharper at the higher social class levels. Thus, in families at the poorest income level, divorce can actually improve women's per capita income (the women generally work or receive welfare payments in addition to whatever alimony and child support is awarded). Women in the middle and upper categories, in contrast, depart most from the men in terms of income. In these groups, the husband's per capita income rises while the wife's falls dramatically (see figure 14.6).

Given these economic problems, one wonders how people can afford to get divorced in the first place. But they do, and in large numbers. There are several reasons why it is possible. For one thing, the divorce is usually an emotional blowup, and people do not think much (or, for that matter, know very much) about the economic consequences of what they are doing.

Second, economic factors do enter in covertly and unconsciously, so that people are more likely to divorce if it is economically feasible for them. We can

FIGURE 14.6: Women's Family Income Drops after Marital Disruption

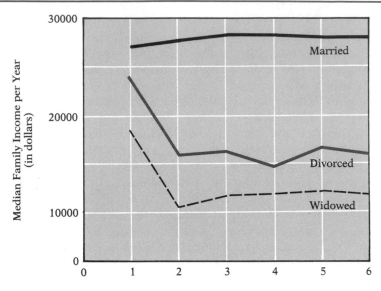

Number of Years after First Interview

When women over 30 experience marital disruption, their incomes decrease significantly and remain low for many years.

Source: Morgan, 1991, table 4.1, p. 73, and figure 4.1, p. 75.

see this in the greater tendency for divorce in dual-income families (which makes the divorce settlement easier on the husband, surely, and also gives the wife some financial means—though generally not enough, realistically—to try living independently). It also shows up in the tendency for divorces to increase during good economic times and decrease in hard times; divorce is one of the luxuries that an economic boom makes possible.

Finally, there is the mitigating fact that divorces tend to occur relatively early in the marriage, when the couple has a lower standard of living. For a young couple, especially if they have not yet bought a home, the drop in standard of living may not be great if they separate. It may well be that a major reason why the divorce rate trails off to a minimum after about ten years of marriage is that couples at this age become too affluent. They have accumulated too many possessions to be able to afford a divorce, no matter what they feel about each other.

A New Social Life

If postdivorce economic realities are usually a down experience, socially and sexually the new divorced person, if childless, often has a more exciting time. At least some of the old friendship networks are broken up, and there is an incentive to make new acquaintances. Often one moves into different kinds of social circles. Married couples tend to socialize with their relatives and with other married couples. Especially in the middle class, women tend to be friends with the wives of their husbands' business or professional friends and vice versa; in other words, the couple system creates the friendship links for each of the persons within it. This arrangement tends to break up with the marriage. Although other people will try to remain friends, they typically find there is too much strain in keeping equal contact with both of the separated individuals. Other married couples find they have to choose which person they will continue to be friends with, and the other ex-partner gradually drifts away. And a single tends to have an awkward time of it in a social environment based on couples. Many divorced women report that married friends became jealous or suspicious of them around their husbands.

The same is true of friendships based on where one lives. The divorce means one person or the other (and sometimes both) moves out of the house or apartment, which breaks up yet another network of friendships. (One indirect way that we know about this, incidentally, is through the fact that the more close friends one has when married, the less likely one is to divorce [Ackerman 1963]. The inference is that people may stay together, not so much because they like each other, but because they don't want to break up their network of friends.)

For the divorced person, then, old friendship networks tend to break up, at least to a certain extent. He or she is less likely to be socializing with other married people. The way our housing is arranged is one reason. Married people (especially those with children) tend to live in certain neighborhoods, the ones with single-family dwellings. These include the typical "bedroom community" suburbs, with their yards, swings, tricycles and bicycles in the driveways, and other signs of conventional family life. Unmarried people usually live in apartments, more likely nearer the central city. These are the same places where one tends to find young married couples without children, and this is often where divorced people move, under the pressure of post-divorce economics, to find a smaller and cheaper dwelling.

Divorced people, then, end up back in the world of singles, only this time with greater sexual experience. It is not surprising that formerly married persons tend to have fairly active sex lives. Older Kinsey data (for the 1940s and early 1950s) found that divorced women below the age of forty were having more sexual intercourse (about three to six times a month) than were most unmarried women at that time, which was less than once a month (Gebhard 1970; Kinsey 1953, 289). For men, the difference was less drastic, but nevertheless postmarital sex was 50 percent higher that premarital (Kinsey 1948, 294). Hunt (1966) estimated that virtually all divorced men and four-fifths of divorced women had postmarital intercourse. These are higher incidence rates than for single people having premarital intercourse (at the time of these data, the early 1960s). In

FEATURE 14.3
Divorce and Downward Mobility

Terry Arendell (1986) interviewed sixty middle-class women with children in Northern California who were divorced in the late 1970s and early 1980s. She found that even if they were employed before and after the divorce, the women experienced financial troubles and "downward mobility" after the divorce. Most could meet the most essential monthly expenses with caution and careful spending, but few had any money for emergencies or unanticipated demands. Some fell farther and farther behind, unable to establish any material security. One single mother, divorced nearly eight years, described her precarious existence:

> I've been living hand to mouth all these years, ever since the divorce. I have no savings account. The notion of having one is as foreign to me as insurance—there's no way I can afford insurance. I have an old pickup that I don't drive very often. In the summertime I don't wear nylons to work because I can cut costs there. Together the kids and I have had to struggle and struggle. Supposedly struggle builds character. Well, some things simply aren't character building. There have been times when we've scoured the shag rug to see if we could find a coin to come up with enough to buy milk so we could have cold cereal for dinner. That's not character building. (P. 37)

Most middle-class wives who divorce expect to recover financially and return to a middle-class standard of living. As Arendell discovered, most cannot realize that ideal. Forced to take low-paying jobs and often unable to keep up with the mortgage payments, many divorced women are forced to sell their houses and move to working-class neighborhoods. Giving up their previous standard of living also means not being able to provide one's children with the middle-class amenities they have come to expect:

> My husband liked really good food and always bought lots and the best. So when he left, it was really hard to cut the kids back. They were used to all that good eating. Now there's no food in the house, and everybody gets really grouchy when there's no food around. . . . I think I've cut back mostly on activities. I don't go to movies anymore with friends. We've lost $150 a month now, because my husband reduced the support. It gets cut from activities—we've stopped doing everything that costs, and there's nowhere else to cut. My phone is shut off. I pay all the bills first and then see what there is for food. . . . I grew up playing the violin, and I'd wanted my kids to have music lessons—piano would be wonderful for them. And my older two kids are very artistic. But lessons are out of the question. (P. 43)

Although the women in Arendell's study suffered through economic hardship, they also reported feeling stronger because of it. One woman whose husband had left her to marry another woman after twenty years of marriage reported realizing that she was better off without him:

> The best thing that's happened out of this divorce experience is acquiring a sense of myself. I've learned that I have a lot of abilities and skills and that I can make it. It's still scary jobwise, but the kids and I will survive. I've done an awful lot of growing, and I feel really good about that. And I feel pretty good about who I am and where I am. . . . The last time I saw my ex-husband, just a few months ago, I said to myself, "Thank God I'm not married to that man." I'd been thinking that for quite a while, but that was the first time I could say it and really mean it. (P. 146)

recent years, the trend is probably to even higher levels of postmarital sex.

Divorced people thus have more active sex lives than unmarried people generally, even though they are biologically older. But this is not to say that the postmarried state is an unending orgy of sexual bliss. There are many readjustments to be made, a lot of strain in ending old relationships with friends and finding new ones. Loneliness and anxiety often accompany this transition, and financial strains in the background intensify the problems. For these reasons the lives of the formerly married, especially for the first year or so, tend to fluctuate from one emotional extreme to the other.

Even though most women suffer economic hardship, the majority report substantial improvements in the quality of their social lives and sexual relationships after divorce. This may be because women are more likely to initiate the divorce in the first place (Kelly 1986, 309). A recent national survey asked individuals who had divorced in the past five years if they had wanted their divorce and whether they were better or worse off after the divorce (Sweet, Bumpass, and Call 1988). Women were two and one-half times more likely than men to report having wanted a divorce. The overwhelming majority of women in the national survey also reported that they were much happier after the divorce (McLanahan 1989). Another long-term study of divorce (Wallerstein and Blakeslee 1989) found that five years after the divorce, two-thirds of the women and half of the men were more content with the quality of their lives.

Few divorces are impulsive. Usually the decision to divorce is preceded by months or years of accumulated grievances and unhappiness. Some studies find a "last straw" phenomenon in which some specific event, such as sexual infidelity, triggers the final decision to divorce. When it happens, few people report that it is a mutual decision, and those who initiate the divorce fare much better in an emotional sense. While spouses who initiate divorce often do so with sadness, guilt, and apprehension, they usually maintain a greater sense of control and do not experience as much humiliation and rejection as those who oppose the divorce (Kelly 1986).

CHILDREN AND DIVORCE

Changing patterns of divorce and remarriage have had a big impact on American families. Perhaps the biggest change is the decreasing stability of children's family situations and the absence of fathers. The likelihood that a child's family structure will change before they leave home is much greater than it was just a few decades ago. Estimates are that between 40 and 60 percent of children born in the 1990s will spend some time in a mother-only family (Furstenberg and Cherlin 1991, 11). As we will see, remarriage is common, but only about half of children in mother-only families are likely to live with a stepfather before they leave home (Bumpass and Sweet 1989).

We should remember, however, that many families continue to be relatively stable. In 1990, almost three-fourths of children under eighteen in the United States were living with two parents (including step and adoptive parents) (U.S. Bureau of the Census 1992e). More than a third of first marriages are likely to remain intact for life, and more than half of the children who begin life in a two-parent family will have intact families throughout their childhood (Bumpass 1990). Thus, while the likelihood of family change continues to increase for children, most still live with two parents. The difference is that they are more likely to have lived through a divorce and to be living with a stepfather than ever before.

About 60 percent of all divorces involve children (Spanier and Glick 1981). The reason the figure is no higher is that divorces tend to happen within the first few years of marriage, before children are born. But quite a few of them occur while a child is on the way, and in fact 10 percent of all births in the United States take place after a divorce and before a remarriage (Rindfuss and Bumpass 1977). For those couples who are already parents when they divorce, the typical number of children is two. In one recent study, Morgan, Lye, and Condran (1988) found that couples with boys were less likely to divorce than couples with girls. Morgan et al. argue that fathers are more involved with sons and this greater parental role integrates fathers more tightly into the family.

In the event of divorce, however, we have single parents with two (usually small) children on their hands. How do they cope with the situation? Usually it is a mother who does the coping, since in 90 percent of the cases today it is still the woman who receives custody. The 10 percent of the fathers who get custody represent a fairly new trend, and they have been experiencing some untraditional problems in their role as single parent (Rosenthal and Keshet 1981; Hanson 1988). For the women in the more familiar role of single parent, these problems are not so foreign (see feature 14.3).

Reacting to the Change

Prior to the 1970s, the prevailing view was that divorce was indicative of individual pathology and that children of divorce would be likely to have major psychological problems. Divorce was viewed as a traumatic event that disrupted "normal" family functioning, and was therefore likely to have negative emotional and psychological impacts on children (Herzog and Sudia 1973). Later research focused on some of the strengths of single-parent families and found that children who experienced divorce were not much different from children in "intact" families. Comparing children whose parents were divorced with children whose parents were still together showed some differences in terms of personality traits, average school grades, test scores, and behavior problems, but when children were followed over time, most of the differences were found to exist before the divorce (Cherlin et al. 1991). Thus, the physical absence of the father appears to be no more harmful than his emotional absence or any number of other problems typically found in "intact" families (Krantz 1988).

One reason why children of divorce may be no worse off than children in two-parent households is that conventionally intact marriages often have major strains within them that are kept beneath the surface, so that breaking up one of these marriages may, as often as not, be an improvement. A parent's personal fulfillment can facilitate a more healthy emotional environment, and sometimes children are better off living with one happy parent than with two unhappy parents. Examining children in various family situations shows that it is the experience of conflict in the home that contributes to the children's positive or negative self-concepts, not whether there has been a divorce (Raschke and Raschke 1979; Demo and Acock 1991). A major national survey found that children living with two parents who persistently quarrel over important areas of family life show higher levels of distress and behavioral problems than do the general class of children from disrupted marriages (Furstenberg and Cherlin 1991). Still, it is misleading to assume that divorce will automatically reduce conflict or produce happiness. Often, parents who divorce continue to have conflict for years after the initial separation.

While at least one parent frequently feels relieved after a divorce, this is not usually the case for the children. Only about one child in ten experiences relief when the parents separate, and that one case most likely involves a threat of violence (Wallerstein and Blakeslee 1989). Most children expect a reconciliation, and tend to reject the parents' initial decision as final. While all children are profoundly affected by the divorce, there is no way to predict the long-term effects for a particular child based on the child's behavior at the time of the divorce. We do know that young children, particularly boys, suffer temporary deleterious effects when their parents divorce, while adolescents tend to be less affected, especially if there is little conflict involved (Demo and Acock 1991, 182). Three factors tend to aid children's adjustment after the divorce: (1) the effectiveness of the custodial parent (usually the mother) in parenting the child; (2) a low level of conflict between the mother and father; and (3) maintenance of a continuing relationship with the noncustodial parent (usually the father) (Furstenberg and Cherlin 1991). Many experts have claimed that a continuing relationship with the father is important to the child's later adjustment, but the result of studies on this are still inconclusive.

While it is difficult to predict the particular responses of any one child to divorce, having one's parents split up carries with it special stresses and problems. Children are often upset by the violent emotions expressed by their parents and miss the parent who moves away. They often feel rejected (consciously or unconsciously) because their father or mother left. The new social and sexual life of their parents brings in unfamiliar people, who may be unpleasant or threatening. The divorce often precipitates a residential move for the custodial parent as well as the noncustodial parent, so that the child's social life may be disrupted. Divorce increases the likelihood that a child will live in a disadvantaged community where jobs are scarce and schools are of low quality. Because the benefits of finishing high school and deferring parenthood are lower in such communities, adolescents are less likely to stay in school and more likely to become teen parents (McLanahan 1989).

Long-Term Impacts

What are the major long-term risks to children of divorce? In general, children who live apart from one or both parents are more likely to drop out of high school and less likely to attend college, more likely to marry and have children in their teens, and more likely to be single parents themselves. These tendencies increase the risk of long-term poverty and economic dependence (McLanahan 1989). While findings are not always consistent, the major differences between children who live with one parent and children who live with two parents can thus be traced to the differing economic situations of the two groups.

Single-parent families, especially female-headed households, have significantly less money than two-parent families. Not surprisingly, poverty causes high levels of physical, emotional, and psychological stress, and lack of money limits children's potential for achieving future upward mobility. One of the most important changes that occurs after a divorce is a decline in the parents' economic investment in children. This is partly the result of having to maintain two households instead of one, and partly the result of the noncustodial parent making smaller monetary contributions to the household. In addition, divorce alters the quantity and quality of the time parents spend with their children. The noncustodial parent (usually the father) tends to spend less time with the children for a variety of reasons, not the least of which are transportation costs and ongoing conflict with the ex-spouse. The custodial parent (usually the mother) tends to have less time to spend with the children as a result of working more hours to make up for lost income. In addition, the quality of the interaction between parents and children often declines because the parents are under stress and in conflict, and parental authority tends to be undermined (McLanahan 1989). Single parents and stepparents are less likely to monitor their children's school work and social activities, and they tend to have lower educational expectations than parents in two-parent families (Astone and McLanahan 1989).

There is another way that divorce affects children's behavior after they have grown up. It seems to give them a more negative attitude toward marriage. Children whose parents were divorced are themselves more likely to go through a divorce (Glenn and Shelton 1983). Not only that, but they are less likely to marry in the first place, or at least to marry as early as everyone else (Kobrin and Waite 1983). This trend shouldn't be exaggerated, though, since these children of divorce are only about 3 to 6 percent less likely to marry in their twenties than other people. But there is a definite tendency in this direction, especially for some groups of people. White women in particular seem to acquire a noticeable disinclination to marriage as a result of their parents' divorce.

Interestingly enough, this antipathy to marriage seems to operate whether the parents eventually remarry or not. In general, remarriages do not seem to affect children's psychological and personality traits very much. Perhaps this is because opposite effects balance out in the statistics: some children thrive in the new situation, while others become involved in new pressures and conflicts. Remarriages do improve the family finances, which has an important background effect upon the level of happiness. On the other hand, we also know that remarriage is a factor in child abuse and especially sexual abuse (see chapter 13).

Child Custody

There are three basic forms of custody: *maternal custody, paternal custody,* and *joint custody.* Traditionally in English common law, legal divorces were extremely rare, and the father retained custody of the children. In the twentieth century in the United States, as divorce became more frequent, the situation switched to the opposite extreme, and courts routinely gave custody to mothers as the "natural" child-rearer. The sexism in this assumption has become more apparent in recent years, but in 90 percent of all cases mothers still get custody. This is partly because of a continued traditionalism on the part of courts, and partly because women are more likely than men to ask for custody.

The courts of many states have recently begun to make two custody determinations: legal custody and physical custody. Physical custody relates to which parent the child will live with on a day-to-day basis, and legal custody relates to the parental power to decide a child's religion, education, and medical treatment. In cases of divorce, the courts respond to requests by the parties involved and grant maternal, joint, or paternal *legal* custody and maternal, joint, or paternal *physical* custody in any combination. Maternal physical and legal custody is the most common arrangement, but maternal physical custody with joint legal custody is becoming increasingly prevalent.

In California, there is evidence that the majority of both mothers and fathers now request joint legal custody when they divorce, and that the courts normally award it. While joint custody only became legally possible in California in 1979, by the late 1980s over three-fourths of divorce decrees provided for either joint legal or joint physical custody (Mnookin, Maccoby, Depner, and Albiston 1990). In most cases the courts continued to award physical custody to mothers. Contrary to media portrayals, most divorcing families with children have very little legal conflict concerning custody or

In 90 percent of all divorce cases involving families with children, the mother is given custody, although joint legal and/or physical custody is becoming more common.

visitation prior to judgment. Over two-thirds of the California cases had either mild or negligible conflict. In 38 percent of the families, the divorce was uncontested and the parental interview information suggested that there was no basic disagreement concerning custodial arrangements (Mnookin et al. 1990). This is primarily because both parties agreed that the mother should have physical custody.

Most divorcing mothers request sole physical custody for themselves. In the California study, more than 80 percent of the mothers said they personally wanted their children to live with them after the divorce. About 15 percent of the mothers indicated a desire for joint physical custody, but less than 2 percent said they wanted the father to have sole physical custody (Mnookin et al. 1990). As might be expected, the fathers' preferences were very different. Roughly equal proportions of the fathers desired joint custody, father custody, and mother custody. Even if all fathers who could not be contacted in this study had indicated a preference for maternal custody, about half the total sample of fathers would still be in favor of joint or paternal custody.

When it came to actual legal requests for custody, mothers were more likely than fathers to act on their preferences. Nearly 80 percent of mothers requested sole custody. Only about half of those fathers who said they wanted joint or paternal custody actually filed the papers requesting such an arrangement. There was a much greater likelihood that the divorcing parties would report intense conflict if the father requested joint or sole physical custody. In about one out of ten cases a parent asked for more physical custody than they

FEATURE 14.4
Joint Custody: What Actually Happens?

Joint physical custody involves sharing the children between two households. There are many different ways in which this is done, ranging between splitting the week (weekdays with one parent, weekends with the other), splitting the day (some hours in one place, other hours in the other), splitting the year (usually summer and other vacations with one parent), or alternating years between the parents.

When joint physical custody is awarded, the actual living arrangements may not reflect this pattern. For instance, in the California study (Mnookin et al. 1990), 20 percent of the total sample were awarded joint physical custody, but about half of these reported a one-parent living arrangement during the interview. Three times as many joint physical custody cases gravitated toward actual residence with the mother (39 percent) as with the father (13 percent). The divorced parents with joint physical custody decrees reported higher levels of conflict than mother-custody families, but those families with a joint decree, and actual residence with the mother, reported the highest conflict of all. In all, a total of 16 percent of the sample were found to have actual joint living arrangements for the children, in part because other physical custody decrees sometimes gravitated toward joint living arrangements (Mnookin et al. 1990). We can see from the California figures

that people don't always file legal papers to get what they want, nor do they always do exactly what the court decided in the divorce decree.

Social welfare professionals are sharply divided on whether joint physical custody is a good idea or not. One camp vociferously claims that joint custody is merely a way of dragging out a divorce, which allows the parents to continue to use their children as pawns in an ongoing fight. Another camp just as vigorously argues that joint custody ought to be normally awarded, unless there are special circumstances dictating otherwise. It turns out that there are advantages and disadvantages. It is *not* true, however, that joint custody is more of a strain on children than other forms of custody. Deborah Luepnitz (1982, 150) concluded that "joint custody at its best is superior to single-parent custody at *its* best."

The main advantages of joint custody are: (1) there are fewer court battles and fewer cases of parental child-snatching; (2) mothers are more likely to receive regular support payments; and (3) both parents have a built-in break from continuous parenting and can rely on each other for "baby-sitting" when they need it. The main disadvantages are: (1) each parent is tied to the ex-spouse and cannot easily leave town for another job or other opportunity; and (2) there tend to be hassles in shuffling children between two houses.

indicated that they wanted in the interview (Mnookin et al. 1990). This might be because they were engaged in strategic bargaining concerning the child custody and child support provisions of the divorce agreement.

Paternal physical custody, when it is awarded, is most likely to be the result of a battle. Deborah Luepnitz (1982) found in her study of divorced families that about half the fathers who got custody went through a bitter struggle, settled either in court or out of court. Some, however, got their children because the mother voluntarily relinquished them out of the feeling that she was less able to care or provide for them. But about half the time fathers got custody only after it had first been awarded to the mother. One father said his custody fight cost $10,000 in legal fees:

> I had to take out a loan, and my parents had to take out a loan. But it was worth it. I would have done anything to get my children back. It was hell for me those eight months while they were with her. I was worried to death about them. They all told the judge they wanted their Dad, and the judge decided because of her drinking and running around that I was the better parent. (Luepnitz 1982, 25)

For the most part, children were rarely asked which parent they wanted to have custody. This was especially the case in the traditional maternal custody arrangements. When paternal physical custody was awarded, as a more unusual situation, about one-third of the children were actually asked by their parents, if not by the judge. Nearly all the children said that it was a hard decision to be asked to make.

Visitation

In single-parent custody, courts usually order some kind of visitation rights for the noncustodial parent with his or her children. This often works out in fact somewhat differently than the court has specified. The custodial parent may use her or his control over when is an appropriate time to visit (since the exact hours are rarely stated in court orders) to punish the ex-spouse. Conversely, some parents rarely use their visitation rights. This is more common in the case of fathers: about a quarter never visit their children if they lose custody of them, and another quarter visit rarely (Luepnitz 1982, 34). Mothers are less likely to do this if they lose custody, but still a quarter of them rarely or never visit. About half of both mothers and fathers visit their children quite frequently.

Visits mean that the noncustodial parent either takes the children out to the movies and ice cream parlor or the like, or has them for a visit at his (her) home. Some children enjoy these visits, others do not. Here's what one seven-year-old girl said.

In single-parent custody situations, courts usually order some kind of visitation rights for the noncustodial parent with the children.

Q: Would you like to see your dad more or less?
A: Never!
Q: Why?
A: Because he makes mean faces and says things about Mom, and you get sick when you go with him.
Q: You get sick?
A: Yeah. I used to like to go before he started asking questions.
Q: What questions?
A: About Mom. Like if she goes out with men, and if she works, and if she leaves us with sitters. (Luepnitz 1982, 35)

Support Payments

Traditionally husbands had the obligation under the law to support their children. The sexist language here has come under attack, but the spouse with the larger income is still obligated to pay for the children's support, and in most cases this is the father. In the past, judges have exercised wide discretion as to the amount of child support awarded. Court records indicate that amounts have ranged from 0 to over 100 percent of the noncustodial father's income (Garfinkel 1988). Child support awards have tended to be regressive, insofar as wealthier fathers are ordered to pay a lesser share of their incomes than poorer fathers. But the level of child support is supposed to reflect the standard of living the family had before the divorce, not merely provide the bare necessities of life. Some states use a sliding percentage of the father's income as a basis for awarding support: e.g., 17 percent for one child, 25 percent for two children, 29 percent for three children, 31 percent for four children, and 34 percent for five or more children (Garfinkel 1988, 333). In dollar amounts, that means a father of two children who earns $20,000 per year is supposed to pay $417 per month in child support, whereas a father of two who earns $45,000 is supposed to pay $938 per month. From the women's viewpoint, these dollar amounts are criticized as being too low. On the other hand, some men feel unjustly deprived when a court awards more than a third of their income, and historically, the courts have been reluctant to award high levels of child support because it may "remove a man's incentive to earn" (Weitzman 1981, 123).

According to recent research, most divorced men could pay more child support without assuming an unfair burden. It is estimated that noncustodial fathers could pay about two and one-half times their current legal obligations and three times what they are actually paying (Garfinkel and Oellerich 1989). But the biggest problem has been divorced men's failure to pay anything at all. As figure 14.5 shows, only about a third of divorced women receive the full amount of child support they are due.

In 1984 and again in 1988 Congress passed legislation concerning awarding and collecting child-support payments. Most of the changes were designed to ensure that the custodial parent (usually the mother) gets an award that is adequate to support the child and that the absent parent (usually the father) can be found and made to pay. Because of high levels of failure to pay in the past (see figure 14.5), the federal government now monitors states' efforts to collect delinquent child-support payments, helps in establishing paternity (pays for blood tests), and locates missing parents who are supposed to pay. Because of the difficulty in bringing nonpaying absent parents into court and getting them to pay, starting in 1994, for all new or modified support orders, child-support payments are to be withheld from absent parents' wages automatically without regard to whether they are in arrears. Thus, child-support payments will be handled like income taxes, and regularly deducted from the absent parent's paycheck.

Effects of Custody Arrangements

All types of custody arrangements, like marriages, place strains on children (and on parents). We do know that children tend to have more postdivorce contact with fathers in joint custody than in mother-custody arrangements (Bowman and Ahrons 1985). Nevertheless, there is no evidence of any difference among children in maternal, paternal, or joint custody in terms of their psychological adjustment or behavior problems (Luepnitz 1982, 149; Furstenberg, Morgan, and Allison 1987; Derdeyn and Scott 1984).

Children in general tend to report being satisfied with whatever custody arrangement happens to exist. There is some evidence, though, that children are more socially competent if they live with the same-sex parent than if they live with the opposite-sex parent. This seems to be particularly true of boys, who do more poorly in their mother's custody than girls do (Luepnitz 1982, 11). There are also strains on the parents in *not* having their children. Psychologists have long used the term *father absence* or *mother absence* to refer to feelings of deprivation experienced by children in the absence of one parent. Now the term *child absence* has been coined to refer to the feelings of deprivation experienced by some parents whose ex-spouse has sole custody of their children. However, we do not have enough evidence as yet to conclude that a particular form of child custody worsens or improves the adjustment of mothers and fathers after divorce (Coysh, Johnston, Tschann, Wallerstein, and Kline 1989).

REMARRIAGE

"Love may be quite compatible with marriage, *if* we are willing to accept the fact that in most people's lives there will be more than one love and more than one marriage," says Constantina Safilios-Rothschild (1977, 67).

Remarriages are now quite common. In fact, over a third of all marriages that take place now are remarriages (Ahlberg and DeVita 1992). Up to 70 percent of all divorced people eventually remarry, although the figure has been dropping in recent years.

Who Remarries Whom?

Men are more likely to remarry than women, and they tend to do it faster as well. More than three out of four divorced men remarry, in contrast to only two of three divorced women. Again, as we saw in chapter 11, men actually find marriage more to their advantage than women do. The average remarriage is within three years after the divorce: a length of time in which people can adjust to a "second-time-around" singles lifestyle and then grow tired of it. It is also enough time for people to find a new partner, if that is what they want to do. The age at which people are remarrying has been dropping, but only if they got divorced at an early age. Back in 1960, the average man remarried at age forty, which by 1985 had dropped to age thirty-seven; for women, the age dropped from thirty-six to thirty-three (London and Wilson 1988). Just as couples have tended to postpone entry into first marriage, they also have stretched the time between marriages. In 1970, the median interval between divorce and remarriage was only about one year. By the late 1980s, the interval was 2.5 years for women and 2.3 years for men (Ahlburg and DeVita 1992).

Younger women remarry more quickly than older women. Women divorced after the age of 40 have a low probability of remarriage, even though remarriage rates among older widows have been rising (Ahlburg and DeVita 1992, 17). Age has little impact on the remarriage of men, again illustrating their favored position in the remarriage market. What's more, men with higher incomes or more education are more likely to remarry than those who are less well-off economically. Just the opposite is true for women: those with more resources are less likely to remarry. In addition, women with children are less likely to remarry than those without.

In remarriages as in first marriages, people tend to match themselves up on the marriage market with someone similar in social class, education, and ethnicity. An additional source of in-marriages among similar people is that divorced persons are most likely to marry another divorced person, just as single people tend (even more overwhelmingly) to marry other singles, and widowed people tend to marry widows. In 1985, almost two-thirds of divorced men and women remarried people who were also divorced, and about a third married single people (London and Wilson 1988). The second choice for single people is to marry a divorced person, which is also the second choice for widows. Thus the divorced seem to be in between two other marriage markets,

In remarriage, the age gap between husband and wife tends to widen. Men who remarry at age forty or older are an average of eight years older than their new brides.

overlapping somewhat with both singles and widows, although each group marries primarily within itself.

Patterns of marriage and remarriage partially derive from people choosing to marry others of similar age. For first marriages, the trend is for men to be about two years older than their brides. However, in remarriages the age gap tends to widen, but differs depending on one's gender and the age at which one remarries. Men who remarry before age thirty tend to be about a year older than their brides, whereas men who remarry at ages thirty to thirty-nine are, on average, four years older. Men who remarry at age forty or older are an average of eight years older than their new brides, and are probably able to marry younger women because they command more economic resources at that stage of their lives. In contrast, women who remarry before age thirty are three and one-half years younger than their grooms, while women remarrying after their thirtieth birthday are about two years younger. More than half of the men and women remarrying before age thirty marry single people, whereas about three-fourths of those remarrying after age forty marry divorced people. These trends are strongly influenced by the number of eligible mates in particular age groups. Because there are more divorced or single women than men in each age group going up from the 30s, men's position in the remarriage market is strengthened even further.

Life in Second Marriages

About a third of all children growing up today will be part of a stepfamily (or "blended" family) before they reach adulthood (Ahrons and Rodgers 1987; Furstenberg 1987). Because of prevailing custody practices, the children usually live with the mother, so if she remarries, her new husband becomes the resident stepfather. Contact between the children and the nonresident biological father often diminishes after remarriage, and in about half of the cases, fathers have little or no contact with their children from a prior marriage. In other cases, nonresident families maintain contact with their children but tend to be more distant and detached than they were before. While relations between noncustodial parents and stepparents are sometimes strained, they are usually polite, and there is no evidence that they are conflict ridden (Pasley and Ihinger-Tallman 1987).

Are remarriages different from first marriages? Negative cultural stereotypes of "wicked stepmothers" and "neglected stepchildren" tend to prejudice us against the potential for remarried families to be similar to "biological" families (Coleman and Ganong 1987). Nevertheless, most research shows that daily life in stepfamilies is more similar than dissimilar to that of first-marriage families (Furstenberg 1987). One difference, however, is that the role of stepparent is different from that of biological parent. Since there is often confusion over what authority the new parent should have, the biological parents, stepparents, and children must negotiate new patterns of relating. In general, stepparents tend to be somewhat more disengaged and are more apt to use an authoritative style of parenting than the resident biological parent. Stepparent-stepchild relations are typically found to be important determinants of marital satisfaction between the new marriage partners as well.

Are remarriages more satisfying than first marriage? We have already seen some hints that they are not always idyllic for children of former marriages, at least when compared to being raised by single parents. But remarriages do seem to be more or less on a par with any previous marriage as far as the children's psychological health is concerned. For the adults, too, there seem to be few differences in spousal reports of marital quality when first and second marriages are compared (Furstenberg 1987). There do seem to be some improvements, as sexual relationships are reported to be better (Westoff 1977). The divorce rate for second marriages is somewhat higher than for first marriages. Third, fourth, and fifth marriages have progressively higher divorce rates (Sweet and Bumpass 1987). But there are not many of such multiple remarriages. For most people, twice is enough.

Conclusion

Divorces are tied to many problems and a great deal of suffering. There is the strain of the divorce itself, the negative effects on many children, and the economic loss suffered—especially by women. Would society be better off, then, if divorces were more difficult to obtain? Social and legal pressure could be applied to keep families together. It is doubtful that this would be a real solution, however. It would amount to turning the clock back to previous times when divorces were virtually impossible to obtain. But family life was also much more authoritarian, based more on economic considerations and social respectability than on personal affection and concern for individual happiness (see the historical material in chapters 3 and 4). The humanitarian movements of the modern period moved in the direction of loosening up those authoritarian family controls and allowing individuals to escape from unsatisfactory marriages. If we went back to restrictions on divorce, people might be better off economically, but the old sources of family conflict would be back again, too. One can predict that, in that event, the pendulum would soon swing the other way, with movements springing up again to allow divorce.

The situation appears to be a dilemma, a trade-off between opposing concerns. Is there any way that the problems can be mitigated, without turning back the clock? Here are several suggestions.

There are two rather different types of divorces: those in which the couple has no children and those in which there are children. Divorces in childless marriages seem to have relatively few negative consequences other than the temporary unpleasantness of breaking up the relationship itself. These divorces tend to happen relatively early in a marriage. They are cases in which the couple discovers they made a mistake in the marriage market. They just weren't matched as well as they believed. Under these circumstances, the best thing for everyone's happiness is the divorce. In fact, such couples ought to be encouraged to get divorced as soon as they recognize their marriage will not work instead of waiting until they have children.

Divorces in which there are children, on the other hand, are full of problems. It is not true that children of divorce necessarily become neurotic or delinquent, but many children of divorce do have serious problems. Even without these, arrangements for custody of the children and visitation by the ex-spouses are full of dilemmas. Nevertheless, married couples who stay together "for the sake of the children" even though they are warring between themselves do not create a good environment for their children either.

The problem would be eased if adults were much more careful about planning when to have children. In particular, they should have every opportunity to gauge whether their own relationship is likely to be permanent before they have children. Cohabitation between unmarried couples as a kind of "trial marriage" is thus a useful thing. The prevalence of contraception, as well as abortion, is directly related to the number of children who are born who are genuinely wanted by their parents. The strain of having children, especially if the parents are not really committed to each other, is one of the factors that makes marriages unhappy and hence more likely to end in divorce. We see, for example, that divorce is higher among women who conceived premaritally (*Family Planning Perspectives*, 1986). Hence one can predict that moves to restrict abortion and otherwise make it more difficult for adults to control childbirth would result in more children of divorce. Conversely, improvements in population control would reduce this problem.

Another potential solution to some of the hardships of divorce would be to provide more practical and economic help to families. Although such solutions are far from easy to implement, many child and family advocates suggest that divorce would be less prevalent, and less devastating when it did occur, if we could provide full employment and decent wages for every American family. A range of child-related services, such as day care and health care, have also been proposed by various advocacy groups, councils, and commissions (Keniston 1977).

Ultimately, a major factor in unhappy marriages is economic and social inequality between men and women. It is men's greater economic resources that give them power within the home, and that in the past has motivated women to take a somewhat calculating attitude toward the marriage market (see chapter 8). More genuine love relationships are fostered by greater equality, since there is less feeling of power on one side and subtle coercion on the other. Many of the issues of spouse abuse, too, are based on the traditions of male domination, which make many men feel justified in using force to control their wives (chapter 13). In the long run, this suggests a potential solution to the problem of divorce. If and when women achieve a breakthrough against economic discrimination in the occupational sphere, they will enter marriages on a level of economic equality with men. A major source of family conflicts, the struggle over power, will have disappeared. When and if that happens, we may expect that the high divorce rates that characterized the period of most intense gender conflict (the 1970s, 1980s, and 1990s) may be a thing of the past.

SUMMARY

1. The divorce rate in the United States climbed steadily for over 120 years. The major exception was during and just after World War II, when the divorce rate suddenly skyrocketed and then fell for a few years, before beginning its climb again to still higher levels. Since the late 1970s the divorce rate has dropped a little, but remains at a very high level.

2. According to current projections, about half of all marriages that occur in the next few years will end in divorce. Divorce is most likely to occur two or three years after the marriage, whereafter the rate begins to fall. Half of all divorces occur within about six years of marriage.

3. Members of the lower social classes tend to divorce more frequently than those in the higher social classes. There is one major exception: highly educated career women (but not men) have a higher rate of divorce than those of lower educational and occupational levels. The reason is probably their greater reluctance to stay in a male-dominated marriage and their greater ability to support themselves independently.

4. Teenage marriages are particularly likely to end in divorce. There is also a tendency for religious and racial intermarriages to have higher divorce rates.

5. Sexual infidelity is sometimes a factor in divorce, but it generally is not the most important factor. Most spouses do not leave marriage for another lover, although men especially are likely to demand a divorce after finding out about their spouse's affair.

6. The most important factor affecting divorce is economic. Divorces generally bring a decline in the standard of living for both partners, but especially for women. Divorce rates tend to decline during economic depressions and to go up during prosperity. The economic downturn of recent years in the United States is one important reason for the leveling out of the divorce rate.

7. After a divorce, both partners tend to experience economic strain, though

the situation is usually worse for women. The smallest drop in the standard of living occurs if the divorce takes place early in the marriage; conversely, couples who have been together longer are less likely to divorce because the economic loss for them would be so great. The greatest economic hardship in a divorce is experienced by women with children. In part, this is because alimony is rarely rewarded, and child support is actually paid in only a minority of cases. Because of widespread failure to pay child support, court-ordered child support will be deducted directly from the absent parent's wages.

8. Formerly married persons tend to have a more active sex life than unmarried persons. They also tend to find their postmarital sexual experiences more satisfying than their marital sex had been.

9. Over one-half of all divorces involve children. While the experience of divorce is difficult for children, the majority suffer no long-term psychological difficulties, and children of divorced parents are virtually indistinguishable psychologically from children with two parents. Nevertheless, two factors associated with divorce can cause problems for children: conflict and poverty. If the divorced parents continue to engage in frequent conflict, children are more likely to experience distress and have more difficulty maintaining a positive self-image. The lack of financial resources often accompanying divorce can cause high levels of physical, emotional, and psychological stress. Children living in poverty are more likely to drop out of school, marry, have children while in their teens, and become single parents themselves. A contrasting long-term effect of divorce on children is that they tend to be more reluctant to marry and are more likely to have divorces of their own after they grow up.

10. Most divorces are uncontested, and most custody arrangements are agreed to by the divorcing spouses. In about 90 percent of all divorces, the mother is granted custody of the children. Many states are now awarding joint legal custody, wherein both mother and father retain rights to make decisions about children's medical treatment or educational endeavors. In some states, like California, about one in five divorcing couples are awarded joint physical custody, wherein the children actually reside for part of the time with each parent. Both types of joint custody arrangements tend to increase the amount of contact fathers have with children after the divorce, but we do not know the long-term impacts of such arrangements.

11. Remarriages are now a typical part of the life cycle. More divorced men eventually remarry than divorced women, probably because being married is more pleasant and advantageous to men. Women with children are less likely to remarry, and so are older and more highly educated women. Although separation is an unpleasant time emotionally, women are happier in the condition of divorce than men. Divorced people are especially likely to remarry other divorced persons.

12. Life in remarried families is not much different from life in first-married families, although spouses, stepparents, stepchildren, and stepsiblings must adjust to each other and negotiate new patterns of relating. There are few differences in spousal reports of marital quality between first and second marriages, although sexual relations are typically reported to improve in second marriages. The chances of divorce in second marriages are higher than in first marriages, and they go up with each subsequent marriage.

15

LIFE TRANSITIONS

The study of adult life transitions is a relatively new field of research for sociologists and psychologists. The theories that have gotten the most attention and upon which there is the greatest agreement are those applying to childhood and especially its first few years (see chapter 12). However, led by Erik Erikson (1959; 1982), students of human development are now realizing that adulthood is not one uniform plateau but tends to be marked by peaks, valleys, and turning points. For instance, Rhona Rapoport (1963) was one of the first to note that individuals encounter a patterned set of family "crises" which she termed "normal critical transitions." Such transitions include leaving home, getting married, having children, launching children, retiring, surviving a spouse, and increasingly, getting a divorce.

LIFE COURSE VARIATIONS

Researchers studying the "life cycle" first attempted to identify regular stages of adult development through which everyone passed. More recent research reveals that life transitions do not occur in the same way or at same time for everyone (Mattessich and Hill 1987). Not only do individuals in the same culture tend to differ somewhat in the timing of the onset of various stages of adult development, but the length, sequence, and composition of life transitions tend to vary according to the period of history and the particular cultural milieu in which one lives. For instance, pastoral (herding) societies tend to socialize children to assume full adult responsibilities sooner than hunting and gathering societies (see chapter 3). This is primarily because the type of subsistence economy—herding large domestic animals—demands many hours of relatively low-intensity labor that children can master. Similarly, societies existing in ecological scarcity tend to utilize child labor in any way that will contribute to the survival of the community. Societies living in scarcity therefore demand that children assume adultlike roles sooner than children in societies that enjoy relatively secure subsistence.

More modern social and economic developments, such as industrialization, urbanization, and the spread of mass education, have similarly impacted the meaning and timing of transitions into and out of childhood and early adult roles. The works of Ariès (1962) and Zelizer (1985) (see chapter 12) illustrate how historical trends in Europe and America influenced our expectations about how children should act and when they should assume adultlike duties. Specific historical events, such as war or economic downturn, can also influence people's family experiences and alter the timing of important life transitions. For example, a lack of money during the depression discouraged young people from marrying or forming their own households and discouraged the already married from having children or getting divorced (Elder 1974). Similarly, the advent of war can influence the nature and timing of the transition to adulthood through recruitment of young men into the armed services. Being in the military, and especially having to fight in a war, makes young men "grow up" faster and also affects decisions about staying in school, pursuing a career, getting married, having children, or getting a divorce (Elder 1986).

When we look at broad historical trends in the United States, we can see that we are returning to timing patterns for getting married and having children that are more like events at the turn of the century than like those of the mid-twentieth century. Between 1890 and the late 1960s, we witnessed a gradual quickening of the pace of life course transitions. Over this time period, people tended to get married, become parents, have children, leave home, and become grandparents sooner than did their own parents (Neugarten and Datan 1973). Since the 1960s, the trend has reversed, with both men and women marrying later and having children later than did their own parents. Returning to patterns of later marriage and later birth has been encouraged by women's increasing chances for going to college and getting jobs (see chapters 5, 10, and 12).

Two other demographic trends are helping to shape the course of adult life. Life expectancy has increased a whopping twenty-seven years since 1900. The average U.S. woman can now expect to live to be seventy-nine years old. Birthrate, on the other hand, has decreased gradually in this century, with only minor reversals about 1908 and 1955 (see chapter 5 and figure 5.4). The average number of children born per female dropped from 3.7 children in 1900 to 1.8 today (Bengtson, Rosenthal, and Burton 1990). These demographic trends are having some major impacts on the processes of growing up and growing old. In this chapter we will look at some of these processes, beginning with the transition from teenage years to adulthood in our own and other societies. We will see how theories try to explain what it means to think like an adult and see how some of this research ignored women. Some other life stages hardly existed in past times. Today many married pairs have a long period as "empty nesters," after their "chicks have flown away." When people commonly bore children into their late thirties and forties, and when one or both parents were more apt to die before reaching a ripe old age, such a life stage was not very common.

The corporate career ladder contributes to another life stage much talked about these days: the male midlife crisis. Whereas men in preindustrial days may have expected to do more or less the same work all their lives, many men today do not. A young man may start out with high ambitions and initially rise rapidly—only to hit a limit to career growth at midlife. His marriage, meanwhile, may have lost its excitement, his kids may be troublesome teenagers, and he may be haunted by the question, "Is this all there is?" At the same time, women may be returning to the work force and feeling that their own lives are expanding. Less attention has been given to two other crises/transitions: the *late life transition*, sometimes called the "retirement crisis," usually in one's sixties; and the final transition, death.

THE TEENAGE YEARS: TRANSITION TO ADULTHOOD

The teenage years, for us, are the period from thirteen to nineteen, though more realistically we usually stop calling someone a "teenager" when he or she has turned eighteen and graduated from high school. Adult status starts at eighteen in certain legal respects: being able to vote, the end of parents' liability for child support (again, in some states). But there are other legal transitions that occur at different ages: sixteen for drivers' licenses, or twenty-one, the old legal age and still drinking age in most places. The fact that the dates float around like this illustrates that the teenage or adolescent years are not merely biologically fixed, but are also social. Of course, there is a biological factor—arriving at adult maturity—but precisely when this happens varies from individual to individual. This is overlaid by our society, which defines the transition in various ways and affects how people experience the teenage transition. Adolescents' self-images and coping abilities are affected by both the coming of puberty and by the shock of moving from a more intimate elementary school to the impersonal organization of junior and senior high school (Simmons and Blyth 1987).

There is, of course, a physical side to the teenage transition. Children go through a growth spurt during their early teens, and most attain their full height by age seventeen. Some mature earlier or later, which creates a temporary lineup of physical advantages and disadvantages. Girls often feel unattractive when they reach their adult height at an early age, sometimes by age twelve or thirteen, and tower over the boys. Boys who attain their full height and weight earlier may have some tendency to push the less-quickly maturing around, or at least dominate in athletics. Hence sheer physical growth has an effect on the teenager's social prestige, though in different ways; since relative positions change during these years, egos can have some special ups and downs. Most of the males' later growth is in weight and musculature, which tends not to come in until the mid-twenties. By that time the adolescent transition, in the fullest sense, is usually over, and these kinds of physical rivalries settle down.

The adolescent transition is a time of sexual maturation as well. Again, different kids go through this at different paces, and the leads and lags can give rise to a status system. Though temporary, it may have a tremendous psychological impact upon young egos. As social life becomes organized around boyfriends and girlfriends, dating, parties, and sex, a person's sexual attractiveness and popularity can be very important for placement in the teenage social networks. Teenagers are not only entering into the adult sexual marketplace; that by itself, as we have seen, creates both emotional excitement and anxieties. At the same time, physical changes are going on in their own bodies, which create additional temporary ups and downs. Putting these factors together, it is no wonder that teenage years are a time of considerable strain and have a reputation for wildness.

Mental illness rates are highest for this age, particularly rates of schizophrenia, a psychosis that involves extreme delusions and withdrawal from social reality (and often commitment to a mental hospital). If a person is going to acquire a homosexual rather than a heterosexual identity (or some mixed, bisexual identity), the teenage years are when this occurs (as we've seen in chapter 9). The delinquency rate is virtually zero at about age twelve, then shoots up rapidly, so that the highest level of juvenile delinquency is at about age sixteen. Thereafter it begins to dip again, and by the time young people are in their twenties, the rate at which they commit (and get caught for) burglaries, assaults, car thefts, and other crimes has dropped to a modest level, where it remains for the rest of their lives. Teenage drivers are the most likely of all to have auto accidents and are most likely to die from them. Perhaps because of these dangers (as well as the social and psychological strains of the sexual and other transitions), teenagers also form the group most likely to undergo a period of intense religious devotion.

ATTAINING ADULTHOOD

We noted above how different societies prescribe somewhat different patterns of life course transitions. The same is true of different social classes. Working-class

The transition from adolescence to adulthood is often delayed by postsecondary education. Here, a fashion design major learns her trade, which eventually will enable her to support herself, and perhaps others, economically.

people typically go through the transition from adolescence to adulthood sooner than middle-class people. Working-class teenagers are more likely to drop out of school, but even when they do get a high school diploma, they are more likely to take a full-time job by the time they are eighteen years old. Entering the adult world of routine employment often signifies the assumption of adult status. In addition, working-class couples are more likely to get married and have children while still in their teens or during their early twenties, again marking the adoption of full adult status. Middle-class children, in contrast, typically delay the full assumption of adult responsibilities by going to college, traveling, taking temporary instead of "permanent" jobs, working part-time instead of full-time, and continuing to receive financial assistance from their parents. Because they typically attend college until about the mid-twenties, and sometimes return home to live with their parents for a while, middle-class children have a later transition to full adult status. This is not to say that they are not considered adults by their peers or by the legal system, but they delay assumption of those occupational and family roles that have traditionally been used to mark the transition from adolescence to adulthood.

THE MIDLIFE TRANSITION

The so-called *midlife crisis* has become an important concept in recent years. This is a period around age thirty-five to forty-five or fifty when an individual may go through several years of psychological and social turmoil. People may become depressed or feel the need to shake up their lives by getting a new job or a divorce. The danger of alcoholism seems to be especially great at this age (Tamir 1982; Farrell and Rosenberg 1981).

Structurally, what is happening at this age is that one's career has reached a plateau or turning point. One has settled into a particular kind of work; the long series of stages of acquiring an occupational self-identity are over by sometime in the twenties, and now the individual has been in a chosen (or compromisingly accepted) field for ten or fifteen years. By now one knows realistically what kind of work it is, and one knows what one's chances are for moving up or into other kinds of work. This is part of the midlife crisis: the realization that you know what you are going to be doing for the rest of your life.

Often there is a feeling of panic connected with this. As Staude (1982) points out in his general theory of life transitions, the midlife period carries with it an aspect of death. What dies during the midlife period is one's youthful view of the world and one's role in it. The youthful period, as we have seen, is unsettled and vague. Although one's exact role in the world is unclear, the range of possibilities seems limitless. The midlife period closes this off. One now has a realistic view of what life is going to be like, and earlier ideals,

The midlife crisis is not a biological phenomenon but a social one, involving the realization that earlier expectations of what one will accomplish in life have not been met.

hopes, and unrealistic dreams have to be put aside. More exactly: you do not usually *decide* at some point to "grow up" and put aside your dreams; instead, you gradually realize that your life has already fallen into a routine, and that life itself has made its decision for you. Psychologically, the midlife crisis involves the realization of the death of one's fantasy self. Connected with this is a real sense of one's mortality. Not that one is close to death (because most people still have another thirty or forty years to live); but for the first time one has the experience of something like death. An idealized part of oneself is gone; what is left, realistically, is a mere mortal being who will eventually face the real biological death. The midlife crisis marks the end of the feeling that there are no limits.

It is not surprising that the midlife crisis has become a popular topic now rather than at some other time in history, in the United States and other wealthy industrial countries. The midlife crisis is not a biological phenomenon but a social one. It is mostly a phenomenon involving white-collar workers, and it occurs among people who live a long time. Even in our own society, working-class people are much less likely to have a midlife crisis. For them, their occupational plateau comes much earlier. Already by his early twenties, a working-class man is likely to know what kind of work he will do for the rest of his life. (The case for women is somewhat different, as we will see below.) His income level, too, usually reaches its peak quite early, whereas middle-class persons usually have more steps on their salary scale and do not reach their peak earnings until they are in their forties or fifties. Working-class men thus seem to have their crisis of "awareness of limits" and "death of youthful dreams" very early. Our culture has not dignified this crisis with a grand label like the "midlife transition," but the working-class equivalent seems to show up in the early marriage crisis that is typical of many working-class families.

The midlife crisis, then, is largely built around the midcareer stage for middle- and upper-middle-class occupations and has the psychological aspect of a growing awareness of where and how one is stuck. Hence the often rather frenzied or anxious efforts to change one's job, to move somewhere else, or to get divorced and start over. Family pressures tend to add to this crisis atmosphere. Parents often go through their midlife crisis at the same time that their children are going through their adolescent crisis. It is very likely that each crisis tends to build up emotional pressures that exacerbate the other.

There is also a third transition or crisis that may be superimposed on the midlife crisis, for a person around forty is likely to have parents who are retiring or even dying. There are the strains of watching one's own parents go through the transitions of later life. For some people in their forties, there is the problem of caring for aging parents. All this adds further pressure to the midlife period.

The Empty Nest Syndrome

It should be obvious that the concept of the midlife crisis has been built up around the experience of men. It is their occupational career plateau that is most often at issue, and the "death of hopes" and the sense of limits are modeled on middle-class masculine experiences. In fact, the studies that first studied midlife crises and developed concepts to explain this time of life were based entirely on interviews with middle-aged men (Levinson 1978; Vaillant 1977).

A comparable turning point for noncareer women happens when their children are grown up and leave home—the *empty nest syndrome*. Lillian Rubin (1979) has produced a sensitive study of what traditionally used to be called "women of a certain age." She captures the upheaval that occurs when one long-standing routine is over and a woman faces the question of what to do for the rest of her life.

> It's unbelievable when I think of it now. I never really saw past about age forty-two, where I am now. I mean, I never thought about what happens to the rest of life. Pretty much the whole of adult life was supposed to be around helping your husband and raising the children. Dammit, what a betrayal! Nobody ever tells you that there's many years of life left after that. He doesn't need your help any more, and the children are raised. Now what? (P. 123)

These women often feel as if they are back in adolescence, going through an earlier transition once more. There are the same feelings of anxiety, the unsettled self-image, the vague trying out of various new alternatives. Often this leads to the decision to make a new career, perhaps by returning for more formal education. Psychologically, the separation from their children may bring back some of the strains and scars of early childhood, when they themselves had to move away from the protective warmth of their own mothers (and perhaps fathers).

> After twenty-five years of raising children, it's like I'm back to being twenty again—maybe only fifteen—and I have to start all over again. Only I'm more scared now than I was then. When you're a kid, you still think you own the world. But I'm not fifteen, I'm forty-five, and I know better now. (P. 123)

On the other hand, some researchers (Stevens-Long 1984) have questioned whether the "empty nest syndrome" always or even usually happens. For many, the departure of grown-up children from home is an event with which people successfully cope and sometimes offers an opportunity to take up new activities or careers. Cross-sectional studies (surveys conducted at one point in time) generally show that families with children in the home report less happiness than couples whose children have moved out (Glenn 1975; Glenn and McLanahan 1982). Those studies that followed a group of families from when children were in the home until after they moved out (longitudinal panel studies) found that women (and sometimes men) were happier after the children were "launched" (Menaghan 1983; McLanahan et al. 1985). One recent large-scale national survey found that marital happiness goes up only when all children are launched, not as each child departs (White and Edwards 1990). Because young adults are also increasingly likely to move back in with their parents for a time, the transition to an "empty nest" is less abrupt and total than it was just a few decades ago. It also appears that the time period immediately after all teenagers depart is the happiest for the married couple and that overall satisfaction with life improves to the extent that the parents continue to see or talk to their launched offspring.

The assumption that middle-age mothers would necessarily feel depressed and aimless after the "nest" was empty appears to be unfounded, particularly if parents and young adult children remain on good terms. This makes sense in light of findings that parenthood is stressful, that it is most stressful when children are infants or teenagers, and that women are most stressed because they do most of the parenting (McLanahan and Adams 1987; see chapter 12). It is also the case that mothers who have other pursuits—as most now do—are least likely to depend totally on mothering activities to bolster their self-worth. For most such women, having teenagers grow up and move out is a liberating experience that provides time for rekindling marital romance and pursuing new opportunities for self-realization.

Menopause

There is also the process of biological change, which occurs at about this time or a little later. *Menopause* typically occurs at about age fifty; it typically closes off the possibility of having more children (although in fact customarily most of the children have long since been born) and also thus affects one's sexual identity.

Women have also feared that at menopause they might cease to be sexual creatures, but in fact this turns out not to be the case. One of Rubin's most striking findings is that the quality of sex had improved considerably for most women by the time they were in their forties.

Feature 15.1
Stereotyping Older Women's Biological Aging

Anne Fausto-Sterling (1985) traces the history of medical interpretations of menopause and comes up with some illuminating findings. In the 1960s, some medical doctors began treating the "disease of menopause" by prescribing pharmaceutically produced estrogen. In medical journals and popular women's magazines, postmenopausal women were described as "unstable estrogen-starved women" who walk the streets stiffly in a "vapid cowlike negative state, seeing little and observing less" (Dr. Robert Wilson and Thelma Wilson 1963, cited in Fausto-Sterling 1989, 298). Another physician painted a bleak picture of what women could expect after menopause: "The vagina begins to shrivel, the breasts atrophy, sexual desire disappears. . . . Increased facial hair, deepening voice, obesity . . . , coarsened features, enlargement of the clitoris, and gradual baldness complete the tragic picture. Not really a man but no longer a functioning woman, these individuals live in the world of intersex" (Dr. David Reuben, 1969, cited in Fausto-Sterling 1989, 299). These poor, aimless, sexless women could be helped, of course, through "estrogen replacement therapy," a procedure that made Premarin (a brand name for estrogen) the fifth most popular drug in the United States in 1975. About that time, studies began to link estrogen treatment to uterine cancer and many physicians became wary of prescribing it. Nevertheless, ten years later, in 1985, 2 million women continued to take Premarin-brand estrogen, which contributed to gross sales of over $70 million per year.

Fausto-Sterling notes that only about a quarter of women going through menopause experience symptoms like hot flashes, and the vast majority of postmenopausal women have almost none of the symptoms that the estrogen-therapy proponents insisted accompanied this "disease." She concludes that "there are no data that support the idea that menopause has any relationship to serious depression in women" (1989, 303). Nevertheless, because our culture tends to define women as mothers and to idealize youthful bodies, estrogen-replacement therapy quickly became a treatment of choice during the 1960s and 1970s. Fausto-Sterling suggests that the negative stereotyping of postmenopausal women came from using men's aging as a model and assuming that there was something "abnormal" about the natural process of aging in women. She reminds us that the labeling of illness is a social process that does not always reflect "objective" biological conditions, and she calls for more and better research on the subject of menopause.

Sex? It's gotten better and better. For the first years of our marriage—maybe nine or ten—it was a very big problem. But it's changed and improved in lots of ways.

Right now, I'm enjoying sex more than I ever did in my life before—maybe even more than I thought I could.

All of a sudden I knew what it was like to feel sexual. I mean, I would get sexual feelings before that when I was stimulated, but they didn't just come by themselves. I used to wonder what people meant when they talked about being horny, but I never knew until after Lisa was born. (Pp. 74, 81)

Why the change? As the last quotation shows, the experience of giving birth itself is sometimes an erotically awakening event (Newton 1973). The whole process of being a mother is in some ways sexually stimulating. On the other hand, as we've also seen, children—especially young ones—direct a woman's attention away from her husband, and their sex life may become less frequent or intense for that reason. The presence of children is often a source of sexual conflict between wives and husbands. Hence it may not be until after the children leave home that this maturer capacity for eroticism can fully blossom for a woman. It is true that the frequency of sexual intercourse declines with age, but the quality of sexual experience when they have it seems to be higher. Fear of pregnancy also is an inhibiting factor in sexual enjoyment for many women; hence the approach of menopause has the hidden benefit of reducing this anxiety as well.

THE LATE LIFE TRANSITION

Retirement may be a crisis for several reasons. One of the most important is that it marks the end of one's active involvement in a career. In contrast, in many cultures there is no such thing as formal retirement. People work until they are no longer able to do so. In our own society, retirement is to a large extent a product of our economic structure. Companies find it expedient to move managers out at a specified age, so that their juniors—having ideas of their own and feeling impatient—can step into the top spots at a predictable time, without too much conflict or unpleasantness. At lower levels in the work force as well, there is pressure to "make room" for the coming generation. Whether this retirement structure can continue remains to be seen as the proportion of the population that is aging increases. Society may not be able to comfortably support more and more people who are not working. How to solve the problem is likely to be an issue in coming decades (see feature 15.2). There are some indications that intergenerational conflict around this point is already shaping up.

Retirement is a plus for many married couples; the vast majority of older people rate their marriages as happy or very happy.

In any case, as matters now stand, some persons die soon after their retirement, because they cannot adjust to doing anything else but working. For others, though, the late life period can be a challenge and an opportunity in its own right. Individuals who adjust to it can have a happy and productive twenty years or more. In fact, many couples report that retirement is good for the marriage, and the vast majority of older people rate their marriages as happy or very happy. This may result in part from the fact that the unhappy marriages have had ample opportunity to be terminated, but most researchers find that marital satisfaction in the later years is much higher than it is in the middle years of marriage. Economic stability contributes to older couples' happiness, as do factors such as viewing one's mate as a best friend, liking one's mate as a person, agreeing on life goals, and maintaining humor and playfulness in marriage (Bengtson, Rosenthal, and Burton 1990).

Gender Trends in Longevity

The structure of the aging population is very gender divided. Women used to outlive men by only one year on the average; now they tend to outlive them by seven years (U.S. Bureau of the Census 1985). The result is that women are much more likely to be widowed then men: at ages fifty-five to sixty-nine, 34 percent of women but only 7 percent of men are widowed; at ages seventy-five to seventy-nine, 60 percent of women are widowed compared to 18 percent of men; over age eighty-five, 82 percent of women but only 43 percent of men are widowed (Sweet and Bumpass 1987).

The explanation for this trend is not quite clear. It may be party due to the decline in the maternal death rate as childbirth techniques have improved, but this can only account for a minor proportion of recent changes.

It is probably not biological, since the difference for the most part has only appeared recently in this century. It is sometimes said that men undergo more strain because of their work, and hence they die more quickly. But there are several reasons to doubt this explanation. For one thing, it is precisely during the period when *the proportion of employed women has increased* that women have built up such a long lead in life expectancy. The "strains" of being in the labor force certainly have not reduced their longevity but exactly the opposite. Moreover, it is probably not true that nonworking women are more "protected" from strains; as we've seen in chapter 11, housewives (and married women in general) have higher rates of mental illness and depression than married men. Whatever it is that has caused women to live longer remains a mystery,

Feature 15.2
The Twenty-First Century: An Aging Population at Zero Growth Rate

Current projections indicate that the U.S. population will reach 300 million persons halfway through the next century and then stay at that level. According to the U.S. Census Bureau, zero population growth will prevail during the twenty-first century. The average life expectancy will have risen to eighty; so many people who are already teenagers or older will be alive to see it happen. The projection is based on assuming the current birthrate will hold at approximately two births per woman and that immigration will provide the balance of population replacement at half a million people per year; most of them are now coming from Asia and Latin America.

The balance between the age groups will shift drastically. Senior citizens, sixty-five and older, will become twice as common, accounting for almost a quarter of the population. The burden on the "working" population, those aged eighteen to sixty-four, will more than double as their ratio to the over sixty-fives will decline from the current 5.4 workers supporting 1 old person to a ratio of 2.6 to 1 in the year 2050.

America will be more crowded, but not impossibly so at 300 million; our current (1991) population is 253 million, and the increase would be about one-fifth. America will become more interracial. The black population will increase to about one in six, the Hispanic population may be larger than that, and the Asian American population substantial. The United States will not be the first country at zero population growth. Several Eastern European countries reached that point during the 1970s; China, with a billion people, is attempting to halt population growth by the year 2000 through a very restrictive policy of allowing only one child per family.

FIGURE 15.1: Living Arrangements of Men and Women over Sixty-Five

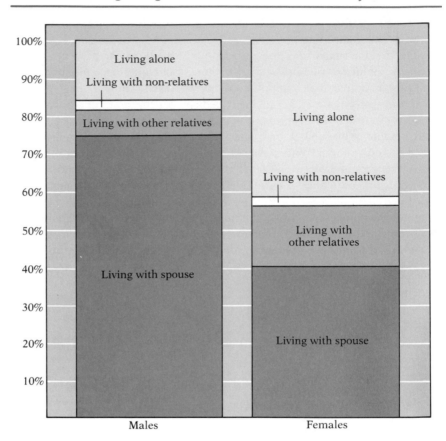

Source: *Statistical Abstract of the United States*, 1992, No. 63.

although it almost certainly has something to do with the gender roles in our society and possibly even involves some subliminal emotional struggle between the sexes that goes on inside families.

One important result of this pattern is that *the aging population is overwhelmingly female.* In the over-sixty-five population, there are about three females for every two males. And the disproportion grows with age. Among people over eighty, the ratio of women to men is almost two to one, and this group is the fastest growing of all segments of our population.

Moreover, these older men and women live in quite different household arrangements (see figure 15.1). Most men over sixty-five are married (about 80 percent of them), whereas less than half of the women are married (40 percent). Elderly women are much more likely to be widowed, and slightly more likely to be either single or divorced. As a result, the vast majority of older men are living with their spouses, while older women are just as likely to be living alone as with their husbands. Why is this? Because of a combination of factors. Women tend to marry men who are on the average three years older in the first place. In addition, women outlive men by almost eight years; the result is that wives outlive their husbands by about eleven years. Not only that, but men over sixty-five are about seven times as likely to remarry (after widowhood or divorce) than are women at that age (Sweet and Bumpass 1987). Thus the development of two predominant populations of the aged: old married men and old widowed women.

Compared to unmarried elderly, older married people are happier and have better health. Married elderly are more likely than their unmarried counterparts to report high levels of morale, life satisfaction, mental and physical health, economic resources, and social support (Verbrugge 1979). Widowhood, in contrast, initially tends to have negative effects on mental and physical health (Pearlin and Johnson 1977), even though these effects usually lessen over the long run (Ferraro 1984).

The major strains of old age, then, are especially experienced by women.

Over one-third of all elderly women live alone, whether by choice or necessity.

There is the problem of living alone—at least this is a problem for persons who may not like to live this way, although of course some people prefer their privacy and independence. Over one-third of all elderly women live alone, whether by choice or necessity. Another 19 percent live with relatives or friends. Only a small proportion (about 5 percent) of either men or women lives in an old-age home or other institution; hence an image of the elderly population as stuck away in formal institutions is not accurate. (And some of these retirement centers have a resortlike atmosphere that hardly resembles the nursing-home stereotype.) For the population as a whole, the biggest problem of old age is the economic strain that often goes along with it, especially for women living alone on a single means of support. Since most elderly women (except in the highest social classes) have relatively little income of their own and are not likely to receive much from their husband's estate (except small Social Security payments), they often can afford only mediocre living quarters. Add to this the cost of health problems, which becomes substantial in later life, and one can see that old age often produces its own, very realistic crises (Lopata 1979).

The Impact of Recent Demographic Trends

Today's elderly have been described as participants in a "quiet revolution" of intergenerational family life. People are growing old in different fashion and in different family contexts than they have in the past. Whereas people used to belong to two- or three-generation families in their later years, today's elderly are likely to be members of four- or even five-generation families (Cherlin and Furstenberg 1986; Bengtson, Rosenthal, and Burton 1990). About half of all people over the age of sixty-five are members of four-generation families, that is, they are great-grandparents (Shanas 1980). One out of five women who die after the age of eighty are great-great-grandmothers—members of five-generation families (Hagestad 1988).

The amount of time that people spend in family roles has increased dramatically since the turn of the century. In the 1800s, parents and children in North America may have shared twenty to thirty years together, and grandparents and grandchildren may have spent only about ten years together (Juster and Vinovskis 1987). Aging parents today may be part of their children's lives for over fifty years. As grandparents, their ties to grandchildren often extend beyond twenty years (Barranti 1985; Preston 1984). Because women live much longer than before, and because they have fewer children, their later-life situations are markedly different from earlier patterns. For example, women in 1980 as compared to those in 1800 spent four times the number of years as a daughter with both parents alive (see figure 15.2). The time they spend as an adult child of one or more parents over the age of sixty-five has increased from about seven years in 1800 to about eighteen years in 1980. Over this period, the amount of time spent in a family with children under eighteen years old has gone down, so that the average number of years spent with parents over sixty-five now exceeds the average number of years spent with children under age eighteen (Watkins, Menken, and Bongaarts 1987). The median age at entry to grandparenthood is now estimated to be about forty-five years old (Sprey and Matthews 1982). Because women can now expect to live to be seventy-nine years old means that most will spend close to half their lives as grandmothers.

Such projections are an oversimplification, of course, because not everyone is "average." In fact there are two contrasting patterns that relate to the early versus late birth patterns (discussed in chapter 12). For those who give birth in

FIGURE 15.2: Women and Their Parents Are Spending More Years Together

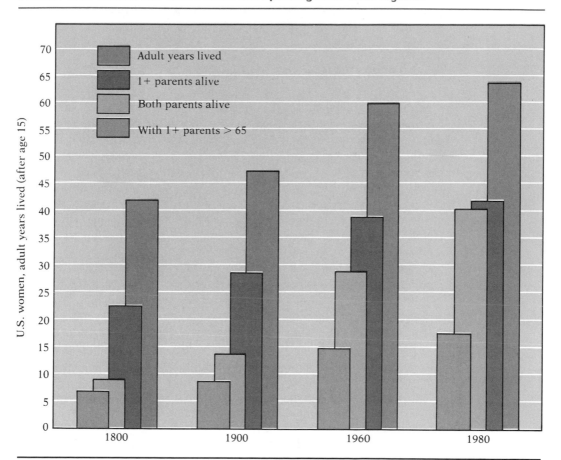

Source: S.C. Watkins, J.A. Menken, and J. Bongaarts, "Demographic Foundations of Family Change," *American Sociological Review 52* (1987): 346-58.

their teens, mothers and grandmothers are quite close in age, and some researchers report that these younger women are not eager to become grandmothers (Elder, Caspi, and Burton 1985). In many instances, substitute child-care responsibilities are then shared with the great-grandmother, often herself about fifty years old. A contrasting pattern results from delayed marriage and birthing. If two consecutive generations delay first birth until age thirty-five, that means that the first parents will be seventy years old before they become grandparents. Clearly, the advanced age of delayed grandparents could limit the quality and quantity of interaction with their grandchildren (Parke 1988). Another important demographic trend that influences family interaction patterns in late life is divorce. When children of elderly parents divorce, the grandparents cannot as easily see their grandchildren, especially if their adult child does not receive custody. Moreover, the elderly parent is often forced to restrict relations with their former daughter-in-law or son-in-law. When divorce is followed by remarriage, multigenerational relationships must be reconstituted and renegotiated.

How Much Contact with Relatives?

It is sometimes said that it is a shame that the old-fashioned family arrangement has disappeared, so that old people are no longer taken care of by their children. As we have seen, this is something of a myth about the past. Many people used to die by the time their children were fully adults. The ideal of the old people sitting by the fireplace surrounded by the younger generations was largely imaginary. Moreover, the myth is not even very accurate *today* as a picture of what people *want*. Surveys show that most parents and their grown-up, married children *do not want to live together* (Shanas et al. 1973). Hence the pattern of elderly people living alone or with roommates is largely their own choice.

That does not mean that parents become completely isolated from their children after they grow up. In fact, a majority of American parents over age sixty-five live *within ten minutes' distance* of one of their children; half of them said they had seen one of their children within the previous twenty-four hours (Shanas 1973, 508–9). This is from the point of view of the parents. If we look at it from the point of view of grown-up children, 68 percent have at least one parent still alive but not living with them. Of this group, 69 percent see their parent at least once a month (*Public Opinion*, Dec./Jan. 1986, p. 34). These are quite astounding figures, completely contrary to the image of old people as isolated and ignored by the younger generation. Other researchers have also found that while few elderly parents live with their adult children, most have frequent contact with them (Treas and Bengtson 1987; Sussman 1985).

There is some evidence that shows, not unexpectedly, that elderly people's morale is higher the more family ties they have (Sauer 1975). But this does not necessarily mean that old people become completely dependent emotionally upon their children. This would be a change from the pattern when they were younger, and we might expect that old people would try to maintain their usual adult pattern. In fact it does turn out that researchers who checked into the importance of nonkinship connections for old people found that *social contacts with friends are even more important for morale* than contacts with relatives (Arling 1976; Wood and Robertson 1978).

There is a social-class pattern to this. Working-class people are more likely to live near their relatives, while middle- and upper-middle-class persons have a greater tendency to move to another town or part of the country. Hence, working-class contacts are more likely to be face-to-face (this is at all ages, not just among old people). In fact, white-collar people are just as concerned to maintain interaction with their relatives as are blue-collar people; but living farther away, they are more likely to make contact by letter or telephone (Adams 1968; Litwak 1960). Somewhat surprisingly, children who are upwardly mobile, moving up and out of their parents' social class, do not thereby break off their family ties. On the contrary, the upwardly mobile are the most likely of all to believe in family ties and to keep up contacts with their aging parents and other relatives. Probably what is happening is that these are the "stars" and "success stories" of their families. Their families are proud of them and like to keep up contact with them; and the successful offspring enjoy the attention and deference they receive from their relatives. The opposite of this pattern is shown in the downwardly mobile: those who are the opposite of a family success story. These persons are most likely to say their relatives are unimportant to them and to know the fewest number of relatives. Even inside the family, then, one's financial and social success has a big effect on one's behavior.

Patterns of contact between elderly parents and adult children also vary by gender. Daughters tend to have more frequent interaction with parents than do sons. Middle-aged daughters most often act as "kinkeepers," regulating the amount and type of contact with parents (Rosenthal 1985). Thus, we can see that women act as emotional managers with aging parents as well as with children. Marital status also influences the amount of contact between elderly parents and children, with unmarried children and widowed parents (usually women) the most likely to have frequent family contact. Ethnic and racial differences also account for some variation in contact between adult children and elderly parents. Hispanics are most likely to have frequent contact, with whites and blacks having lower amounts of contact (Bengtson, Rosenthal, and Burton 1990). Blacks and Hispanics tend to have the highest levels of material support exchanged, which probably has more to do with lack of resources than with race (Mutran 1985; Shanas 1979). Contact with siblings among the elderly also tends to vary by gender and family status, with women and those without children most likely to maintain frequent contact with sisters and brothers (Bengtson, Rosenthal, and Burton 1990).

People over the age of eighty are most at risk for physical and mental frailty

People over age eighty may depend on "children" twenty to thirty years younger for various kinds of support.

and are thus more likely to be dependent on others for help. That help is typically provided by daughters, most of whom would then be in their fifties or sixties. Because families are getting smaller, there are fewer siblings with whom to share the chores of elder care. While much research has focused on the importance of adult children providing support to their elderly parents, recent studies reveal that one's spouse is often a preferred caregiver, and that friendships with same-age peers continue to have beneficial impacts on elderly people's well-being, even if children are available. In addition, in many instances, elders provide more support to adult children than they receive, whether one considers such things as emotional support, financial contributions, provision of living quarters, or routine domestic tasks like cooking and cleaning (Mangen, Bengtson, and Landry 1988).

What happens when parents die? The number of other relatives with whom children keep up contact drops off sharply. Adams (1968) found that the average American in a medium-sized city could name thirty-one relatives that he or she could recognize by sight. But this was only if both the subject's parents were living; if the parents were dead, the number dropped to twenty. This is only the death of two persons, but the number of aunts, uncles, and cousins that one pays attention to seems to hinge strongly upon them. Generally, women knew more relatives than men, and were closer to their own parents. Interestingly enough, women seemed to do this out of a sense of obligation: whereas men were close to their parents to the extent that they felt their parents were affectionate toward them and agreed with their values, women tended to be equally close to their parents whether or not their parents were supportive of them. This implies that men are freer to be close to their parents or not, depending on whether the human relationship is good or bad. But for women this seems not to make much difference. Women seem to feel obligated or even coerced into attending to the parents, whatever their personal preferences.

DEATH AND THE FAMILY

Death is the final transition. At least, so it is from the point of view of the individual. The deeper crisis of the old-age period is facing the inevitable coming of death. In our culture, death is something from which we try to insulate ourselves. Most people die alone, in bureaucratic medical settings, surrounded by nurses and attendants whose professional manner is to emphasize routine and minimize the disruptive or personal quality of what is going on (Sudnow 1967). This creates a superficial calmness, since the overt thought of

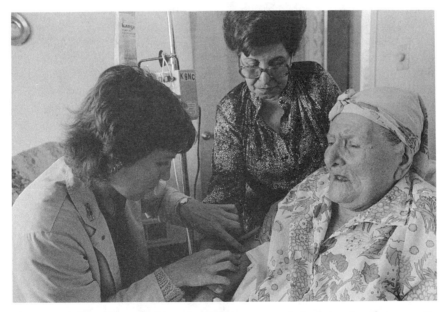

Home nursing provides an alternative to living out one's final days in a bureaucratic medical setting. Here, the patient's daughter is the primary caregiver, enabling this woman to remain at home in familiar surroundings.

death is kept from people's consciousness, or at least from their talk. But repressing the awareness of death, especially when the medical situation indicates that it is quite likely, only creates a split between people's feelings and their public consciousness. This emotional repression can do as much damage as any other aspect of the old-age crisis.

In traditional India or China, the old person was given special honor because he had an institutionalized place in a religious tradition. (In such male-dominated societies, though, old women did not receive the same honor.) Old age was devoted to the cultivation of wisdom, to meditation, to religious study, or to philosophy and art. These roles do not exist in our society, and in fact our public culture tends to be dominated by the extroverted action-oriented ideals of youth (and for that matter, the antiintellectual and antispiritual ideals of working-class and lower-middle-class youth). But as our population becomes increasingly weighted toward the aged, there may be some pressure to redevelop some of these kinds of social roles for developing inward, spiritual, and philosophical interests.

One emerging institution for care of the dying is Hospice. The first hospice appeared in England in the 1960s and was brought to the United States in the 1970s. The term "hospice" refers to compassionate care that is given to terminally ill patients in a variety of settings. Today there are about 2,000 hospices across the country (Knox and Schact 1994). Most provide in-home services for cancer patients, though anyone who is dying is eligible. Since hospice care is given to terminally ill people, the goal is to make the person more comfortable rather than to cure the illness. Hospice stresses companionship and spiritual needs as well as trying to make the patient comfortable and controlling the pain. Hospice staff members work with patients, family, and friends to help them discuss their feelings and expectations. Caring for dying persons at home is generally seen as more humane for the patient, though it can be quite stressful for those who provide care on a daily basis. Though hospice care is becoming more common, rituals associated with its provision have not been fully integrated into our culture.

There is, however, one area of life that is strongly ritualized. This is the funeral. Interestingly enough, the funeral is one of the strongest and most powerful rituals in our relatively deritualized society. Even among people who are not religious, funerals can be important and moving events. Why is this so? The key is to recognize that *a funeral is for those still living, not merely for the dead*. A funeral brings together the people who knew the dead person and assembles them face-to-face. Funerals are effective rituals because they have real work to do for their participants: fitting together the group after one of its members has left it. A death leaves a hole in the group; the purpose of the ritual is to bring everyone together so that they can explicityly recognize that the hole is there and then start to patch it over.

Often there is awkwardness over how to express one's condolences to the spouse or other close relatives and friends of someone recently deceased. You don't quite know what to say, for fear that saying anything at all will make the bereaved persons feel worse. In fact, mourning is a process that people need to go through. They must pass through the grief before they can recover emotionally. This is "working through" the loss. The funeral allows everyone to work this grief through together. At the funeral, condolences are natural; so are tears. Oddly enough, people at a funeral feel sadder than they thought they would; they can become emotionally wrought up over the death of someone they scarcely knew. This is because of the ritual situation: the focus of attention that strengthens the shared emotion in the group. And because of the shared emotion, the group feels itself stronger. The funeral does its work. By attaching a shared ritual onto the end of life, society can assure that death itself ends up strengthening the bonds among those who carry on afterwards.

SUMMARY

1. Individuals encounter a patterned set of family transitions including such events as leaving home, getting married, having children, getting divorced, launching children, retiring, and surviving a spouse. The timing, length, sequence, and composition of life transitions vary according to the historical period and culture in which we live and are influenced by significant events such as war and major shifts in the economy.

2. The teenage or adolescent transition involves biological maturity but also a shifting place in the social structure. Adolescents experience considerable tension as they become ranked in sexual and social markets and in the beginning stages of the adult occupational system. Dangers of pyschosis, delinquency, and accidental death are highest at this time.

3. The midlife crisis in men occurs in upper-middle-class occupations and involves the recognition of limitations on careers. The result is the death of early ideals and of the sense of unlimited possibilities. This crisis often coincides with the teenage crisis of the man's children and possibly with the late life crisis of his own parents.

4. Although early research assumed that the empty nest syndrome represented a traumatic transition for middle-aged wives, recent research shows that women's overall happiness and marital satisfaction improve after the children move out of the house. Earlier depictions of women's reactions to menopause were similarly negative but have given way to a more positive view of women's midlife biological changes.

5. Because men tend to marry women younger than themselves, while women live almost eight years longer than men, wives are especially likely to outlive their husbands. The aging population consists largely of married

men living with their wives, or widowed women living alone or with friends or relatives. The proportion of the aged population that lives in institutions is very small.

6. A very large proportion of old people live near their children and see them quite often. Old people's morale depends partly upon contact with their children, but social contacts with their friends are even more important. Upwardly mobile persons are most likely to maintain contact with a large number of relatives. Women maintain contact with relatives more than men; when men keep up contact, it is more often because the personal relationship is good, whereas women tend to maintain the contact out of a sense of obligation, whether or not they get along personally.

7. Because people live longer and have smaller families, intergenerational family roles have changed dramatically in this century. People spend less time raising small children and more time as adult children of older parents. We are much more likely to be members of families with great-grandparents or great-great-grandparents than ever before. Because U.S. women can now expect to live to be seventy-nine, they are likely to spend close to half of their lives as grandmothers.

8. People over the age of eighty are most at risk for physical and mental frailty and are thus more likely to be dependent on others for help. Wives and daughters are the most likely family members to provide that help. While our stereotypes depict older people as helpless, elders often provide more financial and emotional support to their children than they receive.

9. Death is largely repressed in our society, especially as a result of bureaucratized hospital procedures. This creates many submerged psychological stresses. The aged in our society lack socially esteemed roles for the development of wisdom and spirituality. Funerals, however, do provide powerful rituals for the final life transition, especially for reintegrating the society among those who are left.

THE FUTURE OF
THE FAMILY

Wat can we expect for the future? We know what the current trends have been, and we can extend them through the 1990s or even further. But it is dangerous to assume that trends will just keep on flowing in the same direction. There are famous cases such as the predictions made in the 1930s, when the birth rate was going down and it looked like the U.S. population would level out by the 1960s. Instead, everyone was surprised by the postwar baby boom, which overwhelmed all projections.

We can be a little more confident if we are not just extrapolating a trend but using a theory which explains the causes of social changes. Let us take a look, then, at the main trends we have witnessed throughout this book. In connection with them, we will review the theories which we have used to explain these phenomena. We can estimate whether we should expect that the causes of these trends will continue, or whether we will turn in other directions.

THE MAIN TRENDS: WILL THEY CONTINUE?

- *The marriage rate* has been dropping, although the decline has been leveling out in the last few years. The age at marriage has been going up sharply. Most women do not marry until their late twenties, and the trend may reach the point where a majority will not marry until after thirty. Men have always married later than women, and their age at marriage has been rising even higher (figures 5.2 and 5.7). How far can this trend go? Will people ordinarily be getting married in their thirties or forties? Will we reach the point where most people never marry?
- *Premarital sexual experience* has increased, so that it now occurs for a large majority of both men and women. Women's sexual behavior has tended to converge with that of men, both in premarital and extramarital sex (see figure 5.1 and the data in chapter 9). Does this mean that marriage as a sexual relationship is being displaced?
- *The divorce rate* has risen to a level at which half of all marriages are expected to end in divorce. In the last few years the divorce rate has stopped rising, but it appears to have stabilized near the peak (review figure 14.1). Will it stay at this high level? Or is this just a temporary resting point in a curve whose shape has been uphill for more than a century? On the other hand, are there social conditions which could reverse its direction and bring the divorce rate back down again?
- *The remarriage rate* has been running parallel to the marriage rate. Divorced people tend to get remarried at about the same rate as they married in the first place (figure 5.2). So divorces aren't fatal to the family system. One could even say marriages are popular, because people are having more of them. Nevertheless, as the marriage rate has been dropping, the remarriage rate has been falling off somewhat too. Has the pattern

leveled out, with most people going through several marriages in succession? Or are we in transition to a situation in which most people are going to be either unmarried or permanently divorced, and the marrieds-plus-remarrieds will become a declining majority?

- *Family violence*, including both spouse abuse and child abuse, has come into the spotlight (chapter 13). Here we lack good trend data. The amount of abuse that is *officially reported* has gone up sharply in recent years, but it is likely that there was previously a great deal of family violence that was not reported. In any case, is today's family a violent battleground, and will this situation continue into the future? Can it be that this family conflict is an indication of forces that are in the long run tearing the family apart? Are there any prospects of bringing down the amount of family violence?

- *The birthrate* has been dropping but recently began to rise. For years it has hovered around replacement level, with two children born for every woman (figure 5.4). The *age of childbearing* has also been rising. This parallels the later age of marriage. Women who put off marrying until their late twenties and early thirties are delaying having children for another few years after that. How far can this go? There is a biological upper limit on women's childbearing by the late forties; as women get married later, their number of years of childbearing are narrowed. This may mean that birthrates will drop in the future. What is there to keep the birthrate from falling all the way to zero?

There are a number of side-effects of the falling birthrate:

- *Illegitimacy ratios* are up, as an increasing proportion of the births that do occur happen to unmarried women. In the black population, where the marriage rate is especially low, the proportion of births that are illegitimate has risen to well over half (figure 5.3). Another side effect is that the average *household size* has gotten smaller. There are few families with large numbers of children. On the other hand, there are many singles and childless couples; at the upper end of the age spectrum, we see an increasing number of senior citizens, with a preponderance of widows among the oldest. The mom-and-dad-and-kids family has become a minority.

- The *ethnic composition of families* has been changing in the United States. The trend toward declining birth rates and other features of the modern family structure are strongest in the white population comprised of ethnic groups of European origin. With their birth rates below replacement, recent growth in U.S. population has been due primarily to immigration, especially to groups from Asia and Latin America. Immigrant ethnic groups tend to have more traditional family structures and higher birth rates.

INCREASING ETHNIC DIVERSITY

During most of the twentieth century the minority population was overwhelmingly African-American, and it comprised a relatively stable proportion (13 to 15 percent) of the United States population. Between 1900 and 1960 the minority population doubled, but its rate of growth was only slightly higher than that of the population as a whole. Between 1960 and 1990 the situation changed dramatically. The minority population tripled in size and grew to comprise about a fourth of the total U.S. population. By the year 2000, minorities are projected to account for about a third of the nation's people with Hispanics becoming the most numerous (O'Hare 1992, 9).

With these numbers increasing so rapidly, it will soon become a misnomer

FIGURE 16.1: The Changing Ethnic Composition of American Families

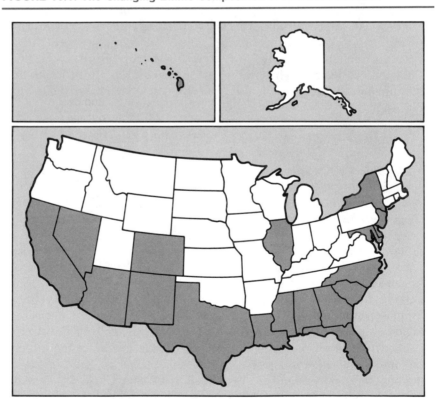

By the year 2010, in the states shaded on this map, more than one-third of the children will be minorities.

Source: Schwartz and Exter, 1989.

to label such groups "minorities." In fact, in some states, "minority" children are already in the majority. In 1990, three-quarters (75.5 percent) of the children in Hawaii and two-thirds (67 percent) of the children in New Mexico were black, Hispanic, or Asian. By the year 2000, over half of those under eighteen in Texas and California will be "minority," and by 2010 over half of the children in Florida, New York, and Louisiana will be "minority." By 2010, nineteen states will be composed of more than one-third minority children (Schwartz and Exter 1989) (see figure 16.1).

Thus we can project that early in the twenty-first century, a substantial proportion of the people in the United States will belong to these ethnic groups. What consequences will this have for the structure of the American family? We will return to this question later in this chapter.

Let us now briefly look at the points of sociological theory that explain the causes driving these trends. Using theory we should be able to guess what directions these trends will take in the future.

LOVE AND MARRIAGE ARE STILL SEXUAL AND EMOTIONAL PROPERTY

The core of the marital relationship is a claim to permanent and exclusive sexual possession of one's partner. Now that there is so much sex outside of marriage, is this core going to fade away, and the family disappear along with it?

The evidence suggests that this is not going to happen. For one thing, although there is more sex outside of marriage than there used to be, *there is still more sex inside of marriage than anywhere else.* Premarital sex is common, in the sense that most people these days have experienced it, but it is not nearly as regular and frequent as marital sex. Many people have premarital sex with the person they eventually marry; in fact, this may be the way in which they negotiate themselves into a relationship that culminates in marriage.

Extramarital sex is also fairly widespread now, in the sense that a majority of persons have at least one adulterous affair at some time or another. But these are mostly rather short episodes. Although they may be dramatic and exciting, they make up a comparatively small part of a person's sex life. In fact, the excitement comes precisely from the fact they are illicit. If we compare arrangements that try to "normalize" extramarital sex, the evidence is that these experiments ultimately show that ordinary sexual possessiveness is more powerful. Communes, we recall (chapter 9), have found that sex is a volatile force and that sexual affairs between individuals tend to tear the group apart. Hence most communes have attempted to institute total celibacy or some other rigid form of control over sex. Open marriages have been advocated and tried (chapter 9), but they seem to be extremely unstable. Open marriages usually terminate within months, either by separation or by a return to exclusive sexual possession. Swinging seems to work better at avoiding personal jealousies because it carefully excludes love from sex; but swinging relationships do not seem to last more than two years at most. None of these is on its way to replacing the conventional marriage pattern tying sex to one primary partner.

Cohabitation

One form of nonmarital union that has become rather common is the cohabiting couple, living together but not married. By 1991, there were 3 million such cohabiting couples in the United States, an increase of 2.5 million couples since 1970, when the idea was still considered new and radical (Ahlburg and DeVita 1992). On the other hand, this is a rather small total compared to the 56 million households composed of married couples. The important point is that *cohabiting couples act much the same as married couples*, except for the fact that they are not legally married. About one-third of all cohabiting opposite sex couples had children in 1991, not remarkably different from the one-half of all married couples living with children. Cohabiting couples have about the same patterns of sexual possessiveness as married couples, and they also have a joint economic relationship as in marriage. Blumstein and Schwartz (1983) found relatively few differences between cohabiting couples and married couples. A cohabitation may be a little easier to break up, since the legal formality of a divorce is not required. But even here there is a tendency for the legal system to treat the partners as having property obligations to each other, if one of them brings the matter to court.

Perhaps the most important reason that sexual possession is not going away is that it is linked to emotional possession. Sex is linked with sexual love; a partner wants an exclusive place in his or her lover's heart, and sexual relationships operate as a symbol of this emotional property. As we have seen (chapter 4), historically the ideal of mutual love and marriage as a sexual possession binding on both partners came into existence with the rise of the individual marriage market. At first impression, the idea of a market seems rather heartless, but in fact love and seeking one's own marriage partner go together. When persons started negotiating their own marriages, rather than serving as pawns in a system of kinship networks or property transactions of patriarchal households, they found themselves in a situation conducive to romance.

How this comes about is explained by the theory of interaction rituals (chapter 8). Having to negotiate one's own fate in an open competition for sexual partners generates the emotional ingredients that create a tie with the person one finds who matches up with oneself. Erving Goffman (1967), who developed the theory of interaction rituals in everyday life, points out that the face-to-face negotiations of modern life produce a cult of individualism. In modern private lives, the self becomes the main sacred object. A love relationship is this cult of the self enhanced to the highest degree, a private ritual in which the lovers idealize each other. At the same time, their cult is a shared one, and sex becomes a symbol of the relationship.

Although there can be sex without love, emotional symbolism tends to grow out of a sexual relationship, and compatible partners make claims on each other for exclusiveness and permanence. Marriage simply ratifies this relationship. This is one important reason why it looks like marriages are not going to disappear. And if we expand the notion of "marriage" to include both legal marriage and cohabitation, we can see that there has been little decline in the institution of marriage.

THE ECONOMICS OF CAREERS ARE CHANGING THE MARRIAGE MARKET

There is no doubt that the marriage market has been changing. Both men and women are waiting longer to get married. One might say they are warier about what kind of deal they are striking. This is spurred by the fact that women have been improving their positions in the marriage market.

Women have become committed to careers as never before. Young women's career aspirations are higher than ever in the past, and women are implementing

A son congratulates his mother at her college graduation with a kiss. As women achieve higher levels of education and experience more success in their careers, they alter their bargaining position in the marriage market.

them by achieving an unprecedented level of education. Throughout most of the twentieth century, males attended college in much greater numbers than females. The popularity of education was increasing throughout this period, as the percentage of young adults who attended college went from 4 percent in 1900 to 53 percent in 1970 (Collins 1979, 4). But after 1970 an important shift took place. College attendance by males leveled off. But the rate of female college attendance has continued to rise and in 1980 there were more women than men enrolled for the first time in history (*Statistical Abstract* 1987, Nos. 265, 267). Now females make up a substantial majority of all college students.

The same thing has been happening in the professional schools, though at a slower rate. Women used to get professional training only in traditional areas such as nursing and primary and secondary education. In 1990, women earned more than half of all master's degrees awarded nationwide. Still, they were less well represented at the highest levels. Only 38 percent of all professional degrees and 37 percent of doctoral degrees went to women in 1990 (*Statistical Abstract* 1993).

Women have not yet made very large inroads into higher management and the more lucrative professions. Only 2 percent of top corporate managers are women, and 20 percent of managers and administrators. But it is clear that women's aspirations have changed.

Women have less incentive to enter into an unequal relationship, and less incentive to stay in a marriage if they are unhappy with it. The result is that the marriage market is now more active than ever. If getting married means going off the market—becoming no longer available to negotiate for sexual and emotional partners—getting divorced puts one back on the market. This has created a large group of divorcees, the most sexually active group of unmarried persons today. But then the tendency of bargaining over sex and love takes over again, and within two to three years, a large proportion of divorced persons have made a new exclusive relationship and remarried. Of those who haven't, many enter the pseudo-marital relationship of cohabitation; in fact, divorced persons make up about a third of the cohabiters.

The marriage market is not disappearing. Men and women are bargaining more equally on it, and are more careful about what they get from it. One might say there is more marital bargaining than ever, more prolonged and with more willingness to reopen negotiations.

THE DUAL-INCOME FAMILY IS THE BASIS OF A NEW FORM OF CLASS STRATIFICATION

Besides sexual and emotional ties, there is one more important incentive for people to get married. The marital relationship has become the most crucial element affecting one's social class position.

As we have seen in chapter 1 (figure 1.2) and in more detail in chapter 6, dual-income households have a huge advantage over most single-income households in their standard of living. To have an upper-middle-class lifestyle today, one usually needs to combine two professional or managerial incomes. Even a moderate middle-class standard of living depends more on the combination of two incomes—even from working-class occupations—than on any single source of income. This becomes not just a matter of keeping up with the Joneses, since these dual-income families have driven up the cost of housing in the United States. As prices go up, there is an incentive to put together two good incomes in a marriage, if only to be able to own a nice place to live.

Family events thus have become a prime determinant of the ups and downs of one's economic position. Whether one marries and who one marries can push income up considerably. Divorce or death of one's partner can send one spinning downwards (chapter 6, chapter 14).

Marriage not only affects the upper part of the income structure. Its absence has an effect—a negative one—on the poverty sector. Both aging families and single-parent families are especially likely to find themselves in this sector. Again the marriage market is involved. In the aging population of seventy and beyond, there are far more women than men, and hence few opportunities to make combined households. In the case of black women who are single parents, there is a severe shortage of marriageable men (chapter 7). When these women have a better source of income than the men who are available, there is no economic advantage in putting together a joint household. This family structure is really the flip side of the dual-income couples who make up the new upper middle class.

We can say, then, that there are strong economic incentives and opportunities to get married, especially in the middle and upper-middle parts of the occupational distribution. The lower-income levels show the opposite pattern. Class stratification has become the bulwark of the modern family system.

SOCIAL CAUSES OF DIVORCE MAY CONTINUE TO RISE

Divorce rates go up when people can afford it. Overall economic prosperity in the last few decades has made divorce a reasonable alternaive for people who are dissatisfied with their marriages. The increasing proportion of women in the labor force has been especially important, since it gives women the resources to be independent. Women tend to suffer economically after divorce, because their incomes are still considerably lower than men's incomes. But compared to the housewives of a generation ago, who had no economic resources whatsoever to fall back on, today's employed women have more room for choice in leaving a marriage.

In the future, even more women will be employed full time, and the proportion who have higher-level jobs with good incomes should increase. The prognosis, then, is that the divorce rate probably will go up still further. As of now, women still have much to lose from a divorce. As their incomes improve, divorce should become even more common.

One of the reasons that couples get divorced is because of one partner's sexual affairs. Extramarital sex violates the bond of sexual possession; when the other partner reacts angrily and demands a divorce, it shows that the ritual of possessiveness is still taken seriously. Sexual affairs thus provide a "push" factor breaking up marriages. On the other hand, such affairs also can provide a "pull" factor, as a husband or wife finds a new sexual partner he or she prefers to the old one, and leaves the marriage for the new lover. How often people have extramarital affairs depends partly upon their opportunities. Men have traditionally had more extramarital affairs than women because they were more likely to work outside the home, where they could come into contact with potential sexual partners. As more women have moved into occupations that bring them into contact with men, their chances for affairs increase. Married women in many jobs are beginning to have as many adulterous affairs as married men.

As women get better jobs in the professions and management, we can expect that there will be more adulterous affairs on both sides, merely because there will be more opportunities. If gender equality increases in the future, so will the number of divorces.

FAMILY CONFLICT: PEAKING AND THEN DECLINING?

Family conflict ranges from quarrels and feelings of dissatisfaction to the extremes of family violence. There are several different sources of conflict: shifts in power relations, stress situations, and traditional authoritarian roles.

Power Shifts

Inequality in power sets the stage for potential conflicts. People who have power feel justified and confident in controlling others; people who are powerless tend to be resentful, but they behave passively if they can do nothing about it. It is when power resources shift that conflict usually breaks out.

Until the last few decades, men have had most of the power resources in families, primarily because of their economic control. One reason for an upsurge of conflict is that women have become mobilized to improve their power position in the home. Women's occupations have given them more economic resources; although these usually do not match the incomes of men, the shift has been enough to at least set off battles over the balance of power.

FIGURE 16.2: Family Conflict as the Balance of Power Shifts

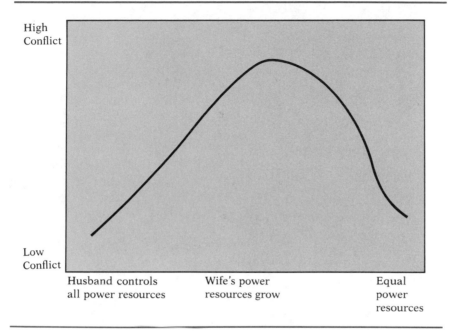

Women's education level has gone up even more rapidly, and this has increased both consciousness around issues of equality and women's aspirations. Changes in occupations and education have probably increased the size of women's networks, which is another important resource for domestic power (chapter 11). We see that the absence of these resources is what makes women most powerless; for instance, battered women who feel unable to leave their husbands are those who lack jobs and networks of their own (chapter 13).

In the long run, we might expect that this source of conflict will rise and fall in a bell-shaped pattern (see figure 16.2). When men had most of the power resources, conflict was relatively low, since women accepted their position passively. Women's power resources increase, and conflict goes up as women challenge men. Eventually, we might expect that the conflict level will fall again and things will stabilize around a more egalitarian situation.

Stress Situations

Family conflict goes up in stressful situations. Families react to these times at different levels of intensity: some merely by feeling unhappy, others by outbursts of violence. All kinds of negative events create stress; but so do events

Working-class marriages are typically more stressful because couples tend to marry and have children earlier, when they have less money and are less equipped emotionally to deal with their situation's demands.

that on the surface may appear only to be a change. Losing one's job is stressful, but so is taking a new job. The initial phase of marriage itself is a shock of adjustment; so is the arrival of children, or moving to a new home. All of these situations are associated with increases in stress, and at the extreme with incidents of violence.

How families react to stress depends on what other resources they have for dealing with stress. For this reason, working-class and poverty-level families are especially likely to have extreme forms of conflict around these incidents. Working-class persons tend to marry earlier, which makes the marital adjustment more of a shock. They also have children earlier, when they have less money and are less equipped emotionally to deal with their demands. These are some of the factors associated with spouse abuse and child abuse.

The long-run picture is that there are fewer early marriages and a declining birth rate, both of which reduce these particular sources of stress. Economic stratification, on the other hand, is getting more severe, and there will be a high level of stress-inducing transitions connected with careers. Here the picture for family conflict is likely to become worse.

Traditional Authoritarian Roles

Probably the most important factor that turns conflicts and stresses into violence is the extent to which men's and women's roles in the family are organized around a traditional pattern of segregated social networks. Here again we can see how ritual density and conformity (chapter 6) help explain class cultures. When families are organized into local, encapsulated networks, with sharp segregation of gender roles, this enforces the tradition of male domination and female passiveness. There are strong pressures for conformity, and violations even of minor matters are taken as symbolic affronts to authority and propriety. In this kind of family, as we saw in chapter 13, men tend to feel that they have the right or even the duty to keep their wives in line; similarly, parents believe that physically punishing their children is "good for them."

These families, then, tend to react to power shifts and to life stresses by a ritualized use of violence. This is also the situation in which ideals of sexual property are most strongly cherished. When there is adultery, or even just a hint of violating sexual possession, the result is likely to be intense jealousy. The jealous husband in particular feels justified in using violence against his wife, since he traditionally has had the power to control her sexually. Wives, too, may react with violence when jealous of their husbands. Often trivial incidents set off violence; we should note that these are symbolic of the whole relationship in high ritual density situations.

On this factor, we can be optimistic about future trends. Traditional encapsulated family structures are giving way to more cosmopolitan networks, as more jobs shift toward the middle-class occupations, and more women have careers. Conflicts will still exist, but the extremely violent reactions to them that are produced by traditional authoritarian roles should be in decline.

CHANGING ECONOMIC AND EMOTIONAL MOTIVATIONS FOR CHILD REARING

The family has always been an economic as well as a sexual/emotional unit. With regard to children, economic payoffs that used to exist in other historical periods have disappeared. Children are simply an economic cost, and the cost has been rising. Housing and medical expenses have gone up drastically. Since children stay in school longer, parents have even higher expenses, continuing for more years. As more mothers are in the labor force, the old source of household labor and childcare is diminishing. It is no surprise that the birthrate declined. As women make even more inroads into professional and managerial careers with their high demands on time and energy, children may become even more of a practical problem. All factors point to the birthrate falling even further in the future.

What's to keep the birthrate from falling all the way to zero? From an economic viewpoint, it is surprising that the birth rate is as high as it is now, near the replacement level of two children per woman.

The countervailing force seems to be that the *emotional side of child rearing* has become more central. There is little reason to have children now except for love. There are fewer children per family, and the births are more carefully planned. This makes children more important to their parents, not less.

From the point of view of sociological theory, this is another instance of the focus on private, personal relationships in modern society. We have a cult of the individual, which reaches a particularly intense form in the search for sexual love between adults. These same conditions act to make children into sacred objects, centers of a great deal of symbolic attention. But practical conditions in modern life are simultaneously undercutting the amount of care parents can give to their children. Children are spending more time in the care of teachers, day care centers, and other forms of surrogate parenting. This is one reason why children are at the center of so many controversies: concerns over child abuse, sexual exposure, school achievement, drugs, athletics, and entertainment. Parents are concerned about their children, at the same time that they are less able to control them. Like love/sex relations among adults, the love relations between parents and children are heavily ritualized and symbolic, and they are undercut by practical realities. But the forces for ritualizing and idealizing these relationships are strong and show every sign of continuing.

A major area of change right now is the pressure to expand the activities of *fathering* along with *mothering*. As women have moved into the labor force and the power balance in the household has shifted, women have adjusted primarily by cutting down the amount of time they spend in housework and child care. So far men have taken up only a small amount of the slack by increasing their own household work. But there is a growing tendency of men to be involved in caring for children, from early infancy on. The main forces operating here are not so much an increasing ideological belief—that men ought to share equally in household and child-rearing responsibilities—but rather practical economic pressures. We have seen (chapter 11) that working-class men are less likely to subscribe to this ideology than are middle-class men, but practical pressures actually cause the former to pitch in more at home. In the upper middle class of intensely career-oriented dual-income families, both men and women are

We will probably become more child-centered symbolically as we become less child-centered in fact.

extremely busy. The pressure there is toward greater equality of responsibilities because their work lives don't allow either one of them to put in much time at home.

As we move from one generation to the next, the effects of greater gender equality in parenting should produce a cumulative effect. There is evidence (chapter 12) that when men participate in child rearing, there is less conflict between the sexes, and less gender stratification in the surrounding society. We can expect that as there is more active fathering, boys and girls will grow up into more gender-egalitarian adults. The shift toward increased fathering should prove to be the beginning of a pattern that will gain even more momentum in the next generation.

The overall picture of parents and children is a combination of increasing ideals, but also numerous strains. We are inching towards parenting equality, and the dual-income pattern will strengthen it even more. Parents idealize the relationship with their children, and devise many symbolic gestures to represent their love. At the same time, the practical strains intrinsic to child rearing are increasing. Paradoxically, these conditions may reinforce one another. For practical economic and career reasons, the number of children will probably decline still further. But as children become more scarce they become even more special, and this increases the tendency of their parents to make them centers of a ritual of parental love. We will probably become more child-centered symbolically as we become less child-centered in fact.

ETHNIC FAMILY PATTERNS ARE APPLICATIONS OF SOCIOLOGICAL PROCESSES

Ethnic changes are particularly visible in the United States at the end of the twentieth century. Nevertheless, the most obvious things about ethnicity are the most superficial: that people have different skin color or hair, or speak with a different language or accent. Such differences have no real effect on social behavior, whether in the family or in economics or politics. If ethnic groups

Intermarriage will continue to contribute to changes in the family. Asian-Americans have particularly high tendencies to marry out of their ethnic group.

have different patterns, it is because *sociological causes are operating there in different proportions.*

We saw one instance of this in the discussion about class stratification and the marriage market. Black women have an extremely high illegitimacy rate. This is not a direct consequence of being black (as we can prove by the fact that this high illegitimacy rate appeared after 1950, but not before). Rather it is due to a combination of several sociological conditions: the shortage of marriageable black men, the class position of black women, and discrimination against blacks that keeps these women in this class position.

Similarly, in chapter 7 we saw that the traditional male-dominated structure of many Hispanic families can be explained by their background in rural societies that are similar to the patriarchal households of earlier agrarian societies. The sociological principles that account for historical differences among societies (see chapter 3) continue to explain differences that we find in the twentieth century as families migrate from agrarian societies into industrial ones. In the United States, the trend for Hispanic women to move into the labor force like other American women is already shifting the power structure of the Hispanic family toward a more egalitarian pattern.

In the long run, it probably will not make much difference in the twenty-first century that the ethnic composition of American families will have changed. The family structure of ethnic groups is not static. There are two processes contributing to change. The first is that social conditions affecting all groups in American society will push family structures toward the forms that are found in the white population.

The second process contributing to change is *intermarriage.* As we noted in table 8.1, most ethnic groups have fairly high proportions of intermarriage with other groups. Asian-Americans, especially those of Japanese and Chinese origin, have particularly high tendencies to marry out of their ethnic group. Hispanics also have a history of considerable intermarriage. The only group with a very low intermarriage rate is blacks. A realistic projection is that most ethnic groups will be widely absorbed into the majority population through intermarriage within the next century. This implies that most ethnic differences in family structure are going to blur, or rather that ethnicity will no longer be as noticeable as class differences. Unless trends change, blacks will be the exception. American families in the twenty-first century apparently will continue to be segregated along lines of white and black.

A KEY PROCESS: GENDER STRATIFICATION AND CLASS STRATIFICATION

The most important process driving almost all of these changes is the movement of women into full-time careers in the labor force. This trend began with the demand for large numbers of white-collar workers, especially secretaries and clerical workers. But these are relatively low-paid jobs, so these workers really constitute a white-collar working class. Another phase of women's careers began after 1970, when women caught up with men in higher education and began to cross over into the professional and managerial occupations. There is a long way to go towards equality in these occupations, but already in the early 1990s it had gone far enough so that we have seen a new upper-middle class, ,consisting of couples that combine two high-paying career occupations.

What will happen in the future? We can imagine two different scenarios. An optimistic scenario for equality suggests that current trends will continue. Chafetz (1990; see also the economic theory of gender stratification in chapter 3) proposes that the demand for female labor will continue to be high. The economy will expand primarily in the tertiary sector of white-collar services, where women already have a strong foothold. Men have dominated in the secondary sector of industrial production, but this is an economic sector in decline. The baby-boom generation, born in the 1950s and crowding the labor market in the 1970s, will begin to retire after 2010, relinquishing many jobs to the post-baby-boomers born after 1965. By around the year 2020, according to this projection, women will have caught up with men in economic positions. And since economic resources are the major source of domestic power, the egalitarian family should finally become predominant.

A pessimistic scenario hinges on the possibility that there will be a major economic downturn, perhaps as severe a depression as that of the 1930s. When jobs are short, men have in the past been able to take positions away from women (Chafetz 1990). If women are driven out of the labor force, their resources for domestic power will decline, and the family may revert to the traditional male-dominated pattern.

Probably the future will be more complex than either scenario. As we have seen, the family structure is now reshaping the class structure. An economic crisis hits different sectors of society unequally. Usually it exacerbates poverty in the lower class and generates unemployment in the working class. People in the higher strata of business and government employment tend to do better. Women are becoming increasingly spread throughout this class structure. The dual-income couples who are beginning to dominate the affluent upper-middle class are likely to be the most impervious to a downturn. Very likely, the single-income families lower down, both male-headed and female-headed, will suffer most. Working-class dual-income couples may just be able to hang on economically, relying on one income to cushion a fall in the other income. So the pessimistic scenario may not so much destroy the new egalitarian family as split society into an upper-income group of families with two high-income wage earners, and a lower-income group whose prospects get steadily worse.

On this score, the optimistic scenario may not be so different. Even in the economic prosperity of the mid-1980s, the only American families that became more prosperous were the upper and upper-middle classes. Inequality has grown. Even if prospects are bright for the dual-income families near the top, there are many economic strains on families below the middle.

In the long run, what we may well be seeing is the renewed importance of class stratification. Gender equality for some women is not gender equality for all; nor is a society of gender equality going to be a society of economic equality across social classes. For the families of the future, social class is likely to shape the biggest differences.

REFERENCES

Ackerman, Charles. 1963. "Affiliations: Structural Determinants of Differential Divorce Rates." *American Journal of Sociology* 69: 13–21.

Adams, Bert N. 1968. *Kinship in an Urban Setting.* Chicago, Ill.: Markham.

Adams, David. 1988. "Treatment Models of Men Who Batter." In Kersti Yllo and Michele Bograd (eds.), *Feminist Perspectives on Wife Abuse.* Newbury Park, Calif.: Sage.

Addiego, F., E. G. Belzer, J. Comolli, W. Moser, J. D. Perry, and B. Whipple. 1981. "Female Ejaculation: A Case Study." *Journal of Sex Research* 17: 13–21.

Agassi, Judith Buber. 1991. "Theories of Gender Equality: Lessons from the Israeli Kibbutz." In Judith Lorber and Susan Farrell (eds.), *The Social Construction of Gender.* Newbury Park, Calif.: Sage.

Aguirre, Adalberto, and R. Martinez. 1984. "Hispanics and the U.S. Occupational Structure." *International Journal of Sociology and Social Policy* 4: 50–59.

Ahlburg, Dennis, and Carol DeVita. 1992. "New Realities of the American Family." *Population Bulletin* vol. 47, no. 2. Washington, D.C.: Population Reference Bureau.

Ahrons, Constance R., and Roy H. Rodgers. 1987. *Divorced Families.* New York: Norton.

Aldous, Joan. 1969. "Wives' Employment Status and Lower-Class Men as Husband-Fathers: Support for the Moynihan Thesis." *Journal of Marriage and the Family* 31: 469–76.

Allen, Walter. 1978. "The Search for Applicable Theories of Black Family Life." *Journal of Marriage and the Family* 40: 111–29.

Alstrom, Carl H. 1961. "A Study of Inheritance of Human Intelligence." *Acta Psychiatrica et Neurologica Scandinavica* 36: 175–202.

Anderson, Elijah. 1985. "The Social Context on Youth Employment Programs." In Charles L. Betsey, Robinson G. Hollister, Jr., and Mary R. Papageorgiou (eds.), *Youth Employment and Training Programs: The YEDPA Years.* Washington, D.C.: National Academy Press.

Arendell, Terry. 1986. *Mothers and Divorce.* Berkeley: University of California Press.

Ariès, Phillipe. 1962. *Centuries of Childhood.* New York: Random House.

———. 1979. "The Family and the City in the Old World and the New." In V. Tufte and B. Myerhoff (eds.), *Changing Images of the Family.* New Haven, Conn.: Yale University Press.

Arling, Greg. 1976. "The Elderly Widow and Her Family, Neighbors, and Friends. *Journal of Marriage and the Family* 38: 757–68.

Arms, Suzanne. 1975. *Immaculate Deception.* New York: Bantam.

Astone, Nan M., and Sara S. McLanahan. 1989. "The Effect of Family Structure on School Completion." Paper presented at the Population Association of America annual meeting, Baltimore, Md.

Atwater, Lynn. 1979. "Getting Involved: Women's Transition to First Extramarital Sex." *Alternative Lifestyles* 2: 33–68.

Bacdayan, Albert S. 1977. "Mechanistic Cooperation and Sexual Equality among the Western Bontoc." In Alice Schlegel (ed.), *Sexual Stratification.* New York: Columbia University Press.

Backman, C. W. 1981. "Attraction in Interpersonal Relationships." In R. Turner and M. Rosenberg (eds.), *Sociological Perspectives on Social Psychology.* New York: Basic Books.

Bajema, C. J. 1968. "Interrelations among Intellectual

Ability, Educational Attainment, and Occupational Achievement." *Sociology of Education* 41: 317–19.

Baldwin, Wendy H., and Christine W. Nord. 1984. "Delayed Childbearing in the U.S.: Facts and Fictions." *Population Bulletin* 39: 1–37.

Baller, W. R., D. C. Charles, and E. L. Miller, 1967. "Midlife Attainment of the Mentally Retarded." *Genetic Psychology Monographs* 42: 235–327.

Bane, Mary Jo. 1986. "Household Composition and Poverty." In Sheldon H. Danzinger and Daniel H. Weinberg (eds.), *Fighting Poverty: What Works and What Doesn't*. Cambridge, Mass.: Harvard University Press.

Banks, Olive. 1986. *Faces of Feminism. A Study of Feminism as a Social Movement*. New York: Blackwell.

Barash, David P. 1977. "Sociobiology of Rape in Mallards." *Science* 197: 788–89.

Bar-Haim, Gabriel. 1988. "Problem-Solvers and Problem-Identifiers: The Making of Research Styles." *International Journal of Education* 10: 135–50.

Barnett, Rosalind C., and Grace K. Baruch. 1987. "Determinants of Fathers' Participation in Family Work." *Journal of Marriage and the Family* 49: 29–40.

Barranti, C.C.R. 1985. "The Grandparent/Grandchild Relationship: Family Resources in an Era of Voluntary Bonds." *Family Relations* 34: 343–52.

Barrera, Mario. 1979. *Race and Class in the South*. Notre Dame, Ind.: University of Notre Dame Press.

Barron, Frank. 1957. "Originality in Relation to Personality and Intellect." *Journal of Personality* 25: 730–42.

Barry, H., and L. Paxson. 1971. "Infancy and Early Childhood." *Ethnology* 10: 466–505.

Bart, Pauline B. 1972. "Depression in Middle-Aged Women." In Vivian Gornick and Barbara K. Moran (eds.), *Woman in Sexist Society*. New York: New American Library.

Bartell, Gilbert. 1971. *Group Sex*. New York: Wyden Books.

Baruch, Grace, and Rosalind Barnett. 1981. "Fathers' Participation in the Care of Their Preschool Children." *Sex Roles* 7: 1043–55.

———. 1986. "Consequences of Fathers' Participation in Family Work: Parents' Role Strain and Well-Being." *Journal of Personality and Social Psychology* 51: 983–92.

Bean, Frank D., Russell L. Curtis, Jr., and John P. Marcum. 1977. "Familism and Marital Satisfaction among Mexican Americans: The Effects of Family Size, Wife's Labor Force Participation, and Conjugal Power." *Journal of Marriage and the Family* (Nov.: 759–67.)

Bean, Frank, and Marta Tienda. 1987. *The Hispanic Population of the United States*. New York: Russell Sage Foundation.

Beer, William R. 1983. *Househusbands: Men and Housework in American Families*. New York: Praeger.

Bell, Alan P., and Martha S. Weinberg. 1978. *Homosexualities*. New York: Simon and Schuster.

Bell, Alan P., Martin S. Weinberg, and Sue Kiefer Hammersmith. 1981. *Sexual Preference: Its Development in Men and Women*. Bloomington: Indiana University Press.

Bell, Robert R., Stanley Turner, and Lawrence Rosen. 1975. "A Multivariate Analysis of Female Extramarital Coitus." *Journal of Marriage and the Family* 36: 375–84.

Bellah, Robert, Richard Madsen, William Sullivan, Ann Swidler, and Steven Tipton. 1985. *Habits of the Heart: Individualism and Commitment in American Life*. Berkeley: University of California Press.

Belsky, J., G. B. Spanier, and M. Rovine. 1983. "Stability and Change in Marriage Across the Transition to Parenthood." *Journal of Marriage and the Family* 45: 567–77.

Bengtson, Vern, Carolyn Rosenthal, and Linda Burton. 1990. "Families and Aging: Diversity and Heterogeneity." In R.H. Binstock and L. George (eds.), *Handbook of Aging and the Social Sciences*, 3d ed. New York: Academic Press.

Bennett, Neil G., David E. Bloom, and Patricia H. Craig. 1989. "The Divergence of Black and White Marriage Patterns." *American Journal of Sociology* 3 (Nov.): 692–722.

Benokraitis, Nijole V. 1993. *Marriages and Families: Changes, Choices, and Constraints*. Englewood Cliffs, N.J.: Prentice Hall.

Benston, Margaret. 1973. "The Political Economy of Women's Liberation." *Monthly Review* 21:13–27.

Berger, Bennett. 1960. *Working-Class Suburb*. Berkeley: University of California Press.

———. 1971. *Looking for America*. Englewood Cliffs, N.J.: Prentice-Hall.

Berheide, Catherine. 1984. "Women's Work in the Home: Seems Like Old Times." *Marriage and Family Review* 7: 37–55.

Berk, Richard A., Sarah F. Berk, Donileen R. Loeske, and David Rauma. 1983. "Mutual Combat and Other Family Violence Myths." In David Finkelhor, Richard J. Gelles, Gerald T. Hotaling, and Murray A. Straus (eds.), *The Dark Side of Families*. Beverly Hills, Calif.: Sage.

Berk, Sarah F. (ed.). 1980. *Women and Household Labor.* Beverly Hills, Calif.: Sage.

———. 1985. *The Gender Factory. The Apportionment of Work in American Households.* New York: Plenum Press.

Berkner, Lutz K. 1972. "The Stem Family and the Developmental Cycle of the Peasant Household: An Eighteenth Century Austrian Example. *American Historical Review* 77: 398–418.

Bermann, E., and D. R. Miller. 1967. "The Matching of Mates." In R. Jesser and S. Fenschback (eds.), *Cognition, Personality and Clinical Psychology.* San Francisco, Calif.: Jossey-Bass.

Bernard, Jessie. 1972. *The Future of Marriage.* New York: World.

———. 1981a. *The Female World.* New York: Free Press.

———. 1981b. "The Good Provider Role: Its Rise and Fall." *American Psychologist* 36: 1–12.

Bernstein, Basil. 1971–75. *Class, Codes, and Control.* 3 vols. London: Routledge and Kegan Paul.

Berscheid, Ellen. 1985. "Interpersonal Attraction." In Gardner Lindzey and Elliot Aronson (eds.), *The Handbook of Social Psychology.* New York: Random House.

Berscheid, E., K. K. Dion, E. Walster, and G. W. Walster. 1971. "Physical Attractiveness and Dating Choice: A Test of the Matching Hypothesis." *Journal of Experimental Social Psychology* 7: 173–89.

Bianchi, Suzanne. 1990. "America's Children: Mixed Prospects." *Population Bulletin* vol. 45, no. 1. Washington, D.C.: Population Reference Bureau.

Bianchi, Suzanne M., and Daphne Spain. 1986. *American Women in Transition.* New York: Russell Sage Foundation.

Billingsley, Andrew. 1968. *Black Families in White America.* Englewood Cliffs, N.J.: Prentice-Hall.

Black, Donald. 1980. *The Manners and Customs of the Police.* New York: Academic Press.

Blake, Judith. 1961. *Family Structure in Jamaica: The Social Context of Reproduction.* New York: Free Press.

Blanchard, R. W., and H. B. Biller. 1971. "Father Availability and Academic Performance among Third-Grade Boys." *Developmental Psychology* 4: 301–5

Block, M. R., and J. D. Sinnott. 1979. *The Battered Elder Syndrome: An Exploratory Study.* College Park: University of Maryland, Center on Aging.

Blood, Robert O. Jr., and Donald M. Wolfe. 1960. *Husbands and Wives.* New York: Free Press.

Bloom-Feshbach, Sally, Jonathan Bloom-Feshbach, and Kirby A. Heller. 1982. "Work, Family, and Children's Perceptions of the World." In Sheila B. Kamerman and Cheryl S. Hayes (eds.), *Families That Work: Children in a Changing World.* Washington, D.C.: National Academy Press.

Blumberg, Rae. 1976. "Kibbutz Women: From the Fields of Revolution to the Laundries of Discontent." In Lynne Iglitzin and Ruth Ross (eds.), *Women in the World: A Comparative Study.* Oxford: ABC Clio.

———. 1978. *Stratification: Socioeconomic and Sexual Inequality.* Dubuque, Iowa: William C. Brown.

———. 1984. "A General Theory of Gender Stratification." In Randall Collins (ed.), *Sociological Theory 1984.* San Francisco, Calif.: Jossey-Bass.

Blumberg, Rae L., and Robert F. Winch. 1972. "Societal Complexity and Familial Complexity: Evidence for Curvilinear Hypothesis." *American Journal of Sociology* 77: 898–920.

Blumstein, Philip, and Pepper Schwartz. 1976. "Bisexual Women." In Josephine P. Wiseman (ed.), *The Social Psychology of Sex.* New York: Harper & Row.

———. 1983. *American Couples: Money/Work/Sex.* New York: William Morrow.

Bograd, M. 1984. "Family Systems Approaches to Wife Battering: A Feminist Critique." *American Journal of Orthopsychiatry* 54: 558–68.

Boserup, Ester. 1970. *The Role of Women in Economic Development.* New York: St. Martin's Press.

Bott, Elizabeth. 1957. *Family and Social Network.* London: Tavistock.

Boulding, Elise. 1976. *The Underside of History.* Boulder, Colo.: Westview Press.

Bourdieu, Pierre. 1984. *Distinction: A Social Critique of the Judgement of Taste.* Richard Nice, trans. Cambridge, Mass.: Harvard University Press.

Bourne, Patricia, and Norma J. Wikler. 1978. "Commitment and the Cultural Mandate: Women in Medicine." *Social Problems* 25: 430–40.

Bowman, Madonna E., and Constance R. Ahrons. 1985. "Impact of Legal Custody Status on Fathers' Parenting Postdivorce. *Journal of Marriage and the Family* 47: 481–88.

Broman, Clifford L. 1988. "Household Work and Family Life Satisfaction of Blacks." *Journal of Marriage and the Family* 550: 45–51.

Bronfenbrenner, U. 1958. "Socialization and Social Class through Time and Space." In E. E. Maccoby, T. M. Newcomb, and E. L. Hartley (eds.), *Readings in Social Psychology.* New York: Holt, Rinehart & Winston.

Bullough, Vern. 1974. *The Subordinate Sex. A History of Attitudes Towards Women.* Baltimore, Md.: Penguin.

Bumpass, Larry L. 1984. "Children and Marital Disruption: A Replication and Update." *Demography* 21 (Feb.): 71–81.

———. 1990. "What's Happening to the Family? Intersection between Demographic and Institutional Change." *Demography* 27: 484–98.

Bumpass, Larry L., and Ronald R. Rindfuss. 1979. "Children's Experience of Marital Disruption." *American Journal of Sociology* 85: 49–65.

Bumpass, Larry, R. R. Rindfuss, and R. B. Janosik. 1978. "Age and Marital Status at First Birth and the Pace of Subsequent Fertility." *Demography* 15: 75–86.

Bumpass, Larry L., and James A. Sweet. 1972. "Differentials in Marital Instability." *American Sociological Review* 37: 754–67.

———. 1989. "Children's Experience in Single-Parent Families: Implications of Cohabitation and Marital Transitions." *Family Planning Perspectives* 21: 256–60.

Burchinal, Lee G. 1964. "Characteristics of Adolescents From Unbroken, Broken, and Reconstituted Families." *Journal of Marriage and the Family* 26: 44–51.

Burchinal, Lee G., and Loren E. Chancellor. 1962. "Survival Rates among Religiously Homogamous and Interreligious Marriages." *Iowa Agricultural and Home Economics Experiment Station Research Bulletin* 512: 743–70.

Burgess, Ernest W., and Paul Wallin. 1953. *Engagement and Marriage.* Philadelphia, Pa.: Lippincott.

Burt, Martha R. 1980. "Cultural Myths and Support for Rape." *Journal of Personality and Social Psychology* 38: 217–30.

Cain, Bruce E., and D. Roderick Kiewiet. 1986. *Minorities in California.* Pasadena, Calif.: The California Institute of Technology, Division of Humanities.

Calahan, Don, and Robin Room. 1974. *Problem Drinking among American Men.* New Brunswick, N.J.: Rutgers Center for Alcohol Studies.

Calderone, Mary S. 1958. *Abortion in the United States.* New York: Harper and Row.

Campbell, Angus. 1981. *The Sense of Well-Being in America: Recent Patterns and Trends.* New York: McGraw-Hill.

Campbell, Elaine. 1985. *The Childless Marriage: An Exploratory Study of Couples Who Do Not Want Children.* London: Tavistock.

Cancian, Francesca. 1985. "Marital Conflict over Intimacy." In G. Handel (ed.), *The Psychosocial Interior of the Family.* New York: Aldine.

Carpenter, C. R. 1964. *Naturalistic Behavior of Nonhuman Primates.* Pennsylvania State University Press.

Carter, Hugh, and Paul C. Glick. 1976. *Marriage and Divorce: A Social and Economic Study.* Cambridge, Mass.: Harvard University Press.

Cates, Willard, Jr. 1982. "Legal Abortion: The Public Health Record." *Science* 215: 1586–90.

Chadwick, Bruce, and Tim Heaton. 1992. *Statistical Handbook on the American Family.* Phoenix, Ariz.: Oryx Press.

Chafetz, Janet S. 1984. *Sex and Advantage: A Comparative, Macro-Structural Theory of Sex Stratification.* Totowa, N.J.: Rowman and Allanheld.

———. 1990. *Gender Equity: An Integrated Theory of Stability and Change.* Newbury Park, Calif.: Sage.

Chafetz, Janet, and A. Gary Dworkin. 1986. *Female Revolt: Women's Movements in World and Historical Perspective.* Totowa, N.J.: Rowman and Allanheld.

———. 1987. "In the Face of Threat: Organized Antifeminism in Comparative Perspective." *Gender and Society* 1: 33–60.

———. 1989. "Action and Reaction: An Integrated, Comparative Perspective on Feminist and Antifeminist Movements." In Melvin Kohn (ed.), *Cross-National Research in Sociology.* Newbury Park, Calif.: Sage.

Chan, Lilly M. 1970. "Foot Binding in Chinese Women and Its Psycho-social Implications." *Canadian Psychiatric Association Journal* 15: 229–31.

Charnov, Eric L. 1982. *The Theory of Sex Allocation.* Princeton, N.J.: Princeton University Press.

Chase, Ivan D. 1975. "A Comparison of Men's and Women's Intergenerational Mobility in the U.S." *American Sociological Review* 40: 483–505.

Chen, Renbao, and S. Philip Morgan. 1991. Recent Trends in the Timing of First Births in the United States. *Demography* 28: 513–33.

Cherlin, Andrew J. 1979. "Worklife and Marital Dissolution." In George Levinger and Oliver C. Moles (eds.), *Divorce and Separation.* New York: Basic Books.

———. 1992. *Marriage, Divorce, Remarriage* (2nd ed.) Cambridge, Mass.: Harvard University Press.

Cherlin, Andrew J., and Frank F. Furstenburg. 1986. *The New American Grandparent.* New York: Basic Books.

Cherlin, Andrew, Frank Furstenberg, P. Lindsay Chase-Landsdale, Kathleen Kiernan, Philip Robins, Donna Morrison, and Julien Teitler. 1991. "Longitudinal Studies of Effects of Divorce on Children in Great Britain and the United States." *Science* 252: 1386–89.

Chesney, Kellow. 1970. *The Victorian Underworld.* New York: Schocken Books.

Chess, Stella, and Alexander Thomas. "Infant Bonding: Mystique and Reality." *American Journal of Orthopsychiatry* 52: 213–22.

Chodorow, Nancy J. 1976. "Oedipal Asymmetries and Heterosexual Knots." *Social Problems* 23: 454–67

———. 1978. *The Reproduction of Mothering.* Berkeley: University of California Press.

Christensen, Harold T., and Kenneth E. Barber. 1967. "Interfaith vs. Intrafaith Marriage in Indiana." *Journal of Marriage and the Family* 33: 461–69.

Ciochon, Russell L., and Robert S. Corruccini. 1983. *New Interpretations of Ape and Human Ancestry.* New York: Plenum.

Clark, M. 1959. *Health in the Mexican-American Culture.* Berkeley: University of California Press.

Clignet, Remi. 1970. *Many Wives, Many Powers.* Evanston, Ill.: Northwestern University Press.

Coleman, Marilyn and Lawrence H. Ganong. 1987. "The Cultural Stereotyping of Stepfamilies." In Kay Pasley and Marilyn Ihinger-Tallman (eds.), *Remarriage and Stepparenting.* New York: Guilford.

Collins, Randall. 1971. "A Conflict Theory of Sexual Stratification." *Social Problems* 19 (Summer): 3–21.

———. 1975. *Conflict Sociology: Toward an Explanatory Science.* New York: Academic Press.

———. 1979. *The Credential Society. An Historical Sociology of Education and Stratification.* New York: Academic Press.

———. 1981. "Three Faces of Cruelty: Towards a Comparative Sociology of Violence." In Randall Collins, *Sociology Since Midcentury.* New York: Academic Press.

———. 1986a. "Weber's Theory of the Family." In *Weberian Sociological Theory.* New York: Cambridge University Press.

———. 1986b. "Courtly Politics and the Status of Women." In *Weberian Sociological Theory.* New York: Cambridge University Press.

———. 1988. *Theoretical Sociology.* San Diego, Calif.: Harcourt Brace Jovanovich.

Coltrane, Scott. 1988. "Father-Child Relationships and the Status of Women." *American Journal of Sociology* 93: 1060–95.

———. 1989. "Household Labor and the Routine Production of Gender." *Social Problems* 36: 473–90.

———. 1990. "Birth Timing and the Division of Labor in Dual-Earner Families." *Journal of Family Issues* 11 (June): 157–81.

———. 1991. "Social Networks and Men's Family Roles." *Men's Studies Review* 8: 8–15.

———. 1992. "The Micro-Politics of Gender in Nonindustrial Societies." *Gender & Society* 6: 86–107.

———. 1994. "Stability and Change in Chicano Men's Family Lives." in Michael Kimmel and Michael Messner (eds.), *Men's Lives,* 3d ed. New York: Macmillan.

———. 1995. *Family Man: Fatherhood, Housework, and Gender Equity.* New York: Oxford University Press.

Coltrane, Scott, and Neal Hickman. 1992. "The Rhetoric of Rights and Needs: Moral Discourse in the Reform of Child Custody and Child Support Laws." *Social Problems* 39: 400–20.

Coltrane, Scott, and Masako Ishii-Kuntz. 1992. "Men's Housework: A Life Course Perspective." *Journal of Marriage and the Family* 54: 43–57.

Coltrane, Scott, and Elsa Valdez. 1993. "Reluctant Compliance: Work/Family Role Allocation in Dual-Earner Chicano Families." Pp. 151–75 in Jane C. Hood (ed.), *Men, Work and Family,* Newbury Park, Calif.: Sage.

Combs, Robert H., and William F. Kenkel. 1966. "Sex Differences in Dating Aspirations and Satisfaction with Computer-Selected Partners." *Journal of Marriage and the Family* 28: 62–66.

Connery, S. J., John. 1977. *Abortion: The Development of the Roman Catholic Perspective.* Chicago, Ill.: Loyola University Press.

Conrad, Peter, and Joseph W. Schneider. 1980. *Deviance and Medicalization.* St. Louis, Mo.: Mosby.

Consortium for Longitudinal Studies. 1983. *As the Twig Is Bent . . . Lasting Effects of Preschool Programs.* Hillsdale, N.J.: Erlbaum.

Constantine, Larry L., and Joan M. Constantine. 1973. *Group Marriage.* New York: Macmillan.

Cooper, John M. 1946. "The Patagonian and Pampean Hunters." In Julian Steward (ed.), *Handbook of South American Indians.* Washington, D.C.: Smithsonian Institution.

Coser, Rose L. 1964. "Authority and Structural Ambivalence in the Middle Class Family." In *The Family: Its Structure and Functions.* New York: St. Martin's Press.

Cott, Nancy F. 1978. "Passionlessness: An Interpretation of Victorian Sexual Ideology, 1790–1850." *Signs* 4: 219–36.

Coverman, Shelly, and Joseph Sheley. 1986. "Change in Men's Housework and Child-Care Time, 1965–1975." *Journal of Marriage and the Family* 48: 413–22.

Cowan, Carolyn P. 1988. "Working with Men Becoming Fathers: The Impact of a Couples Group Intervention." In Phyllis Bronstein and Carolyn P. Cowan (eds.), *Fatherhood Today.* New York: Wiley.

Cowan, Carolyn P., P. A. Cowan, G. Heming, E. V. Garrett, W. S. Coysh, H. Curtis-Boles, and A. J. Boles. 1985. "Transitions to Parenthood: His, Hers, and Theirs." *Journal of Family Issues* 6: 451–81.

Cowan, Philip A. 1988. "Becoming a Father: A Time of Change, An Opportunity for Development." Phyllis Bronstein and Carolyn P. Cowan (eds.), *Fatherhood Today.* New York: Wiley.

———. 1993. "The Sky *IS* Falling, but Popenoe's Analysis Won't Help Us Do Anything About It." *Journal of Marriage and the Family* 55: 548–53.

Cowan, Ruth S. 1983. *More Work for Mother. The Ironies of Household Technology from the Open Hearth to the Microwave.* New York: Basic Books.

Coysh, William S., Janet R. Johnston, Jeanne M. Tschann, Judith S. Wallerstein, and Marsha Kline. 1989. "Parental Postdivorce Adjustment in Joint and Sole Physical Custody Families." *Journal of Family Issues* 10(1): 52–71.

Crain, William C. 1980. *Theories of Development.* Englewood Cliffs, N.J.: Prentice-Hall.

Critelli, J., and L. R. Waid. 1980. "Physical Attractiveness, Romantic Love, and Equity Restoration in Dating Relationships." *Journal of Personality Assessment* 44: 624–29.

Cromwell, R. E., and D. H. Olsen. 1975. "Multidisciplinary Perspectives of Power." In R. E. Cromwell and D. H. Olsen (eds.), *Power in Families.* New York: Halsted Press.

Cromwell, V., and R. E. Cromwell. 1978. "Perceived Dominance in Decision Making and Conflict Resolution among Black and Chicano Couples." *Journal of Marriage and the Family* 40(Nov.): 749–59.

Crouter, Ann, Maureen Perry-Jenkins, Ted Huston, and Susan McHale. 1987. "Processes Underlying Father Involvement in Dual-Earner and Single-Earner Families." *Developmental Psychology* 23: 431–40.

Csikszentmihalyi, Mihaly, and Eugene Rochberg-Halton. 1981. *The Meaning of Things: Domestic Symbols and the Self.* Cambridge, Mass.: Cambridge University Press.

Cuber, John F., and Peggy B. Harroff. 1965. *The Significant Americans: A Study of Sexual Behavior among the Affluent.* Baltimore, Md.: Penguin Books.

Currie, Elliott. 1985. *Confronting Crime: An American Challenge.* New York: Pantheon.

Dallacosta, Mariarosa, and Selma James. 1972. *The Power of Women and the Subversion of the Community.* Briston, Eng.: Falling Wall Press.

Daniels, Arlene K. 1987. "The Hidden Work of Constructing Class and Community: Women Volunteer Leaders in Social Philanthropy." In Naomi Gerstel and Harriet Gross (eds.), *Families and Work.* Philadelphia, Penn.: Temple University Press.

———. 1988. *Invisible Careers.* Chicago: University of Chicago Press.

Daniels, Pamela, and Kathy Weingarten. 1982. *Sooner or Later: The Timing of Parenthood in Adult Lives.* New York: Norton.

David, Henry P., Herbert L. Friedman, Jean van der Tak, and Marylis J. Sevilla. 1978. *Abortion in Psychosocial Perspective: Trends in Transnational Research.* New York: Springer.

Davidson, Laurie, and Laura K. Gordon. 1979. *The Sociology of Gender.* Chicago, Ill.: Rand McNally.

Davidson, T. 1977. "Wifebeating: A Recurring Phenomenon Throughout History." In M. Roy (ed.), *Battered Women: A Psychological Study of Domestic Violence.* New York: Van Nostrand Reinhold.

Davis, Cary, Carl Haub, and JoAnne Willette. 1983. "U.S. Hispanics Changing the Face of America." *Population Bulletin* 38 (June): 3.

Davis, Kingsley. 1936. "Jealousy and Sexual Property." *Social Forces* 14: 395–405.

———. 1941. "Intermarriage in Caste Societies." *American Anthropologist* 43: 388–95.

———. 1949. *Human Society.* New York: Macmillan.

Deboer, Leoburt. 1982. *The Orangutan: Its Biology and Conservation.* The Hague: Dr. W. Junk Pub.

Degler, Carl N. 1980. *At Odds: Women and the Family in America from the Revolution to the Present.* New York: Oxford University Press.

DeLamater, John, and Patricia MacCorquodale. 1979. *Premarital Sexuality.* Madison: University of Wisconsin Press.

Demo, David, and Alan Acock. 1991. "The Impact of Divorce on Children." Pp. 162–91 in Alan Booth (ed.), *Contemporary Families: Looking Forward, Looking Back.* Minneapolis, Minn.: National Council on Family Relations.

Demos, John. 1970. *A Little Commonwealth: Family Life in Plymouth Colony.* New York: Oxford University Press.

Dengler, Ian C. 1978. "Turkish Women in the Ottoman Empire: The Classical Age." In Lois Beck and Nikki Keddie (eds.), *Women in the Moslem World.* Cambridge, Mass.: Harvard University Press.

Derdeyn, A., and E. Scott. 1984. "Joint Custody: A Critical Analysis and Appraisal." *American Journal of Orthopsychiatry* 54(Apr.): 199–209.

DeVault, Marjorie L. 1987. "Doing Housework: Feeding and Family Life." In Naomi Gerstel and Harriet E. Gross (eds.), *Families and Work.* Philadelphia, Pa.: Temple University Press.

Devore, Irven S. 1965. "Male Dominance and Mating Behavior in Baboons." In Frank A. Beach (ed.), *Sex and Behavior.* New York: Wiley.

Dickens, A. G. 1977. *The Courts of Europe, Politics,*

Patronage and Royalty, 1400–1800. New York: McGraw-Hill.

DiMaggio, Paul, and John Mohr. 1985. "Cultural Capital, Educational Attainment, and Marital Selection." *American Journal of Sociology* 90: 1231–61.

DiMaggio, Paul, and Michael Useem. 1978. "Cultural Democracy in a Period of Cultural Expansion: The Social Composition of Arts Audiences in the United States." *Social Problems* 26: 180–97.

———. 1982. "The Arts in Cultural Reproduction." In Michael Apple (ed.), *Cultural and Economic Reproduction in Education.* London: Routledge and Kegan Paul.

Divale, William T. 1984. "Migration, External Warfare and Matrilocal Residence." *Behavior Science Research* 9: 75–133.

Divale, William T., and Marvin Harris. 1976. "Population, Warfare, and the Male Supremacist Complex." *American Anthropologist* 78: 521–38.

Dixon, A. S. 1981. *The Natural History of the Gorilla.* London: Weidenfeld and Nicholson.

Dobash, R. Emerson, and Russell Dobash. 1979. *Violence against Wives.* New York: Free Press.

Dobyns, Henry F. 1983. *Their Numbers Become Thinned: Native American Population Dynamics in Eastern North America.* Knoxville, Tenn.: University of Tennessee Press.

Domhoff, G. William. 1970. *The Higher Circles: The Governing Class in America.* New York: Random House.

———. 1990. *The Power Elite and the State: How Policy is Made in America.* New York: Aldine de Gruyter.

Donegan, Jane B. 1978. *Women and Men Midwives: Medicine, Morality, and Misogyny in Early America.* Westport, Conn.: Greenwood Press.

Dover, K. J. 1978. *Greek Homosexuality.* Cambridge, Mass.: Harvard University Press.

Drake, St. Clair, and Horace R. Cayton. 1945. *Black Metropolis: A Study of Negro Life in a Northern City.* New York: Harcourt and Brace.

Draper, Patricia. 1975. "!Kung Women: Contrasts in Sexual Egalitarianism in Foraging and Sedentary Contexts." In R. Reiter (ed.), *Toward an Anthropology of Women.* New York: Monthly Review Press.

Driver, Harold E. 1969. *Indians of North America.* 2nd ed. Chicago: University of Chicago Press.

Duberman, Lucile, 1975. *The Reconstituted Family: A Study of Remarried Couples and Their Children.* Chicago, Ill.: Nelson-Hall.

Duncan, Greg J. et al. 1984. *Years of Plenty: The Changing Economic Fortunes of American Workers and Families.* Ann Arbor: Institute for Social Research, The University of Michigan.

Duncan, Greg J., Martha S. Hill, and Saul D. Hoffman. 1988. "Welfare Dependence Within and Across Generations." *Science* 239 (Jan. 29): 467–71.

Duncan, Greg J., and Saul D. Hoffman. 1985. "A Reconsideration of the Economic Consequences of Divorce." *Demography* 22: 485–97.

Durden-Smith, Jo, and Diane DeSimone. 1984. "Hidden Threads of Illness." *Science Digest* 92 (Jan.).

Durkheim, Emile. 1893/1947. *The Division of Labor in Society.* New York: Free Press.

———. 1912/1954. *The Elementary Forms of the Religious Life.* New York: Free Press.

Dworkin, Andrea. 1987. *Intercourse.* New York: Free Press.

Easterlin, Richard. 1980. *Birth and Fortune: The Impact of Numbers on Personal Welfare.* New York: Basic Books.

Eckland, B. K. 1979. "Genetic Variance in the SES-IQ Correlation." *Sociology of Education* 52: 191–6.

Edmonds, V. H. 1967. "Marriage Conventionalization: Definition and Measurement." *Journal of Marriage and the Family* 29: 681–88.

Ehrenreich, Barbara, and Deidre English. 1973. *Witches, Midwives, and Nurses.* Old Westbury, N.Y.: Feminist Press.

Ehrensaft, Diane. 1983. "Dual Parenting and the Duel of Intimacy." Paper read at the Annual Meeting of the American Sociological Association, Detroit.

———. 1987. *Parenting Together.* New York: Free Press.

Eitzen, D. Stanley. 1983. *Social Problems*, 2d ed. Boston, Mass.: Allyn and Bacon.

Elder, Glen H., Jr. 1969. "Appearance and Education in Marriage Mobility." *American Sociological Review* 34: 519–33.

———. 1974. *Children of the Great Depression.* Chicago, Ill.: University of Chicago Press.

———. 1986. "Military Times and Turning Points in Men's Lives." *Developmental Psychology* 22: 233–45.

Elder, Glen H., and Avshalom Caspi. 1988. "Economic Stress in Lives: Developmental Perspectives." *Journal of Social Issues* 44: 25–45.

Elder, Glen H., Jr., A. Caspi, and L. M. Burton. 1985. "Adolescent Transitions in Developmental Perspective: Sociological and Historical Insights." In M. Gunnar (ed.), *Minnesota Symposium on Child Psychology*, vol. 21. Hillsdale: N.J.: Erlbaum.

Ellwood, David, and Mary Jo Bane. 1985. "The Impact of AFDC on Family Structure and Living Arrangements." In Ronald Ehrenberg (ed.), *Research in Labor Economics* 7.

Ellwood, David, and Lawrence H. Summers. 1986. "Poverty in America: Is Welfare the Answer or the Problem?" In Sheldon H. Danzinger and Daniel H. Weinberg (eds.), *Fighting Poverty: What Works and What Doesn't*. Cambridge, Mass.: Harvard University Press.

Elmer-Dewitt, P. 1991. "Making Babies." *Time* (September 30): 56–63.

Ember, Melvin, and Carol Ember. 1971. "The Conditions Favoring Matrilocal vs. Patrilocal Residence." *American Anthropologist* 73: 571–94.

Engels, Friedrich, 1884/1972. *The Origin of the Family, Private Property and the State*. New York: International Publishers.

England, Paula. 1989. "A Feminist Critique of Rational Choice Theories." *American Sociologist* 19.

England, Paula, and Irene Browne. 1992. "Trends in Women's Economic Status." *Sociological Perspectives* 35: 17–51.

Entwistle, Doris R., and Susan G. Doering. 1981. *The First Birth. A Family Turning Point*. Baltimore, Md.: Johns Hopkins University Press.

Epstein, Cynthia F. 1971. *Women's Place: Options and Limits in Professional Careers*. Berkeley: University of California Press.

———. 1983. *Women in Law*. New York: Basic Books.

———. 1988. *Deceptive Distinctions: Sex, Gender, and the Social Order*. New Haven: Yale University Press.

Ericksen, Julia A., William L. Yancey and Eugene P. Ericksen. 1979. "The Division of Family Roles." *Journal of Marriage and the Family* 41: 301–13.

Erikson, Erik H. 1959. "Identity and the Life Cycle." *Psychological Issues* 1(1).

———. 1982. *The Life Cycle Completed: A Revision*. New York: Norton.

Erlanger, H. S. 1974. "Social Class and Corporal Punishment in Childrearing: A Reassessment." *American Sociological Review* 39.

Espenshade, Thomas J. 1979. "The Economic Consequences of Divorce." *Journal of Marriage and the Family* 41: 615–25.

———. 1985. "Marriage Trends in America." *Population and Development Review* 11: 193–245.

Etzioni, Amitai. 1993. "Children of the Universe." *Utne Reader*. Vol. 57.

Evans-Pritchard, E. E. 1951. *Kinship and Marriage among the Nuer*. London: Oxford University Press.

Fagan, Jeffrey A., Douglas K. Stewart, and Karen V. Hansen. 1983. "Violent Men or Violent Husbands? Background Factors and Situational Correlates." In David Finkelhor (ed.), *The Dark Side of Families*. Beverly Hills, Calif.: Sage.

Family Planning Perspectives. 1985. "Recently Wed Women More Likely to Have Had Premarital Sex." Vol. 17, Sept./Oct., p. 142.

Family Planning Perspectives. 1986. "Women in 30s Experiencing Record-High Divorce; Level Expected to Decline among Younger Women." Vol. 18, May/June, pp. 133–34.

Farber, Bernard. 1972. *The Guardians of Virtue: Salem Families in 1800*. New York: Basic Books.

Farley, R., and A. I Hermalin. 1971. "Family Stability: A Comparison of Trends between Blacks and Whites." *American Sociological Review* 36: 207–22.

Farrell, Michael P., and Stanley D. Rosenberg. 1981. *Men at Mid-life*. Boston, Mass.: Auburn House.

Fausto-Sterling, Anne. 1985. *Myths of Gender: Biological Theories about Women and Men*. New York: Basic Books.

Fausto-Sterling, Anne. 1989. "Hormonal Hurricanes: Menstruation, Menopause, and Female Behavior." In Laurel Richardson and Verta Taylor (eds.), *Feminist Frontiers II: Rethinking Sex, Gender, and Society*. New York: Random House.

Featherman, David L., and Robert M. Hauser. 1978. *Opportunity and Change*. New York: Academic Press.

Fein, Robert A. 1978. "Research on Fathering: Social Policy and Emergent Perspective." *Journal of Social Issues* 34: 122–35.

Feingold, A. 1982. "Physical Attractiveness and Romantic Involvement." *Psychological Reports* 50: 802.

Feldman, S. S. 1987. "Predicting Strain in Mothers and Fathers of Six-Month-Old Infants: A Short-Term Longitudinal Study." In P. Berman and F. A. Pederson (eds.), *Men's Transitions to Parenthood*. Hillsdale, N.J.: Erlbaum.

Ferraro, Kathleen J. 1984. "Widowhood and Social Participation in Later Life." *Research on Aging* 6: 451–68.

———. 1988. "An Existential Approach to Battering." In Gerald T. Hotaling, David Finkelhor, John T. Kirkpatrick, and Murray A. Straus (eds.), *Family Abuse and Its Consequences*. Newbury Park, Calif.: Sage.

Ferree, Myra M. 1987. "Family and Job for Working-Class Women: Gender and Class Systems Seen from Below." In Naomi Gerstel and Harriet Gross (eds.), *Families and Work*. Philadelphia, Pa.: Temple University Press.

———. 1991a. "Feminism and Family Research." Pp. 103–21 in Alan Booth (ed.), *Contemporary Families: Looking Forward, Looking Back*. Minneapolis, Minn: National Council on Family Relations.

———. 1991b. "The Gender Division of Labor in Two-

Earner Marriages: Dimensions of Variability and Change." *Journal of Family Issues* 12: 158–80.

Field, Tiffany. 1978. "Interaction Behaviors of Primary vs. Secondary Caretaker Fathers." *Developmental Psychology* 14.

Filinson, Rachel. 1989. "Introduction." Rachel Filinson and Stanley R. Ingman (eds.), *Elder Abuse: Practice and Policy*. New York: Human Sciences Press.

Fine, Michelle. 1981. "Injustice by Any Other Name." *Victimology* 6: 1–4; 48–58.

Finkelhor, David. 1979. *Sexually Victimized Children*. New York: Free Press.

———. 1983. "Common Features of Family Abuse." In David Finkelhor, Richard J. Gelles, Gerald T. Hotaling, and Murray A. Straus (eds.), *The Dark Side of Families*. Beverly Hills, Calif.: Sage.

———. 1986. *A Sourcebook on Child Sexual Abuse*. Beverly Hills, Calif.: Sage.

Finkelhor, David, Richard J. Gelles, Gerald T. Hotaling, and Murray A. Straus (eds.), 1983. *The Dark Side of Families*. Beverly Hills, Calif.: Sage.

Finkelhor, David, and Kersti Yllo. 1985. *License to Rape: Sexual Abuse of Wives*. New York: Holt, Rinehart, and Winston.

Firestone, Shulamith. 1970. *The Dialectic of Sex*. New York: William Morrow.

Fischer, Claude S. 1982. *To Dwell among Friends: Personal Networks in Town and City*. Chicago, Ill.: University of Chicago Press.

Fisher, Wesley A. 1980. *The Soviet Marriage Market*. Cambridge, Mass.: Harvard University Press.

Fishman, P. 1978. "Interaction: The Work Women Do." *Social Problems* 25: 398–406.

Fitzgerald, F. Scott. 1931/1956. "Echoes of the Jazz Age." In *The Crack-Up*. New York: New Directions.

Flandrin, Jean-Louis. 1979. *Families in Former Times: Kinship, Household and Sexuality*. Cambridge. Cambridge University Press.

Flango, Eugene, and Carol Flango. 1993. "Adoption Statistics by State." *Child Welfare* 72:311–19.

Flexner, Eleanor. 1959. *Century of Struggle: The Woman's Rights Movement in the United States*. Cambridge, Mass.: Harvard University Press.

Floge, I. 1985. "The Dynamics of Child-Care Use and Some Implications for Women's Employment." *Journal of Marriage and the Family* 47: 143–54.

Fontana, V. 1973. *Somewhere a Child Is Crying: Maltreatment—Causes and Prevention*. New York: Macmillan.

Ford, Clellan S., and Frank A. Beach. 1951. *Patterns of Sexual Behavior*. New York: Harper & Row.

Ford, David. 1991. "Preventing and Provoking Wife Battery Through Criminal Sanctioning." Pp. 191–209 in Dean Knudsen and JoAnn Miller (eds.), *Abused and Battered*. New York: Aldine de Gruyter.

Ford, Kathleen. 1978. "Contraceptive Use in the U.S., 1973–76." *Family Planning Perspectives* 10 (Sept./Oct.): 264–69.

Forrest, Jacqueline D., Christopher Tietze, and Ellen Sullivan. 1978. "Abortion in the U.S., 1976–77." *Family Planning Perspectives* 10: 271–79.

Fowlkes, Martha R. 1987. "The Myth of Merit and Male Professional Careers. The Role of Wives." In Naomi Gerstel and Harriet Gross (eds.), *Families and Work*. Philadelphia, Pa.: Temple University Press.

Freud, Sigmund. 1905/1938. *Three Contributions to the Theory of Sex*. In A. A. Brill (ed.), *The Basic Writings of Sigmund Freud*. New York: Random House.

———. 1913/1938. *Totem and Taboo*. In *The Basic Writings of Sigmund Freud*. New York: Random House.

———. 1914/1957. "On Narcissism: An Introduction." In *A General Selection from the Works of Sigmund Freud*. New York: Doubleday.

———. 1921/1960. *Group Psychology and the Analysis of the Ego*. New York: Bantam.

———. 1924. *A General Introduction to Psychoanalysis*. New York: Boni and Liveright.

———. 1929/1961. *Civilization and Its Discontents*. New York: W. W. Norton.

Friedl, Ernestine. 1975. *Women and Men: An Anthropologist's View*. New York: Holt, Rinehart and Winston.

Fuchs, Victor R. 1986. "Sex Differences in Economic Well-Being." *Science* 232: 459–64.

Furstenberg, Jr., Frank F. 1987. "The New Extended Family: The Experience of Parents and Children after Remarriage." In Kay Pasley and Marilyn Ihinger-Tallman (eds.), *Remarriage and Stepparenting*. New York: Guilford.

Furstenberg, Frank, and Andrew Cherlin. 1991. *Divided Families*. Cambridge, Mass.: Harvard University Press.

Furstenberg, Jr., Frank F., S. Philip Morgan, and Paul Allison. 1987. "Paternal Participation and Children's Well-Being after Marital Dissolution." *American Sociological Review* 52: 695–701.

Fustel de Coulanges, Numa Denis. 1973. *The Ancient City*. Baltimore: Johns Hopkins University Press.

Gallop, Jane. 1982. *The Daughter's Seduction: Feminism and Psychoanalysis*. Ithaca, N.Y.: Cornell University Press.

Gans, Herbert, 1962. *The Urban Villagers*. New York: Free Press.

———. 1967. *The Levittowners*. New York: Random House.

———. 1974. *Popular Culture and High Culture*. New York: Basic Books.

Garbarino, James, and Gwen Gilliam. 1980. *Understanding Abusive Families*. Lexington, Mass.: D.C. Heath.

Garcia, Mario T. 1980. "La Familia: The Mexican Immigrant Family, 1900–1930." In Mario Barrera, Alberto Camarillo, and Francisco Hernandez (eds.), *Work, Family, Sex Roles, Language*. Berkeley, Calif.: Tonatiua-Quinto Sol International.

Garcia-Bahne, Betty. 1977. "La Chicana and the Chicano Family." In R. Sanchez (ed.), *Essays on La Mujer*. Los Angeles, Calif.: UCLA Chicano Studies Center Publications.

Gardner, Robert W., Bryant Robey, and Peter C. Smith. 1985. "Asian Americans: Growth, Change, and Diversity." *Population Bulletin* 4(Oct.): 1–43.

Garfinkel, Irwin. 1988. "Child Support Assurance: A New Tool for Achieving Social Security." In Alfred J. Kahn and Sheila B. Kamerman (eds.), *Child Support: From Debt Collection to Social Policy*. Nesbury Park, Calif.: Sage.

Garfinkel, Irwin, and Sara S. McLanahan. 1986. *Single Mothers and Their Children: A New American Dilemma*. Washington, D.C.: Urban Institute.

Garfinkel, Irwin, and Donald Oellerich. 1989. "Noncustodial Fathers' Ability to Pay Child Support." *Demography* 26: 219–34.

Garland, T. N. 1972. "The Better Half? The Male in the Dual Professional Family." In C. Safilios-Rothschild (ed.), *Toward a Sociology of Women*. Lexington, Mass.: Xerox.

Gaskin, Ina M. 1977. *Spiritual Midwifery*. Summertown, Tenn.: Schocken Books.

Gebhard, Paul H. 1966. "Factors in Marital Orgasm." *Journal of Social Issues* 22: 88–95.

———. 1970. "Postmarital Coitus among Widows and Divorcees." In Paul Bohannan (ed.), *Divorce and After*. New York: Doubleday.

Gebhard, Paul H., and Alan B. Johnson. 1979. *The Kinsey Data: Marginal Tabulations of the 1938–1963 Interviews Conducted by the Institute for Sex Research*. Philadelphia, Pa.: W. B. Saunders.

Geiger, H. Kent. 1968. *The Family in Soviet Russia*. Cambridge, Mass.: Harvard University Press.

Gelles, Richard J. 1972. *The Violent Home. A Study of Physical Aggression between Husbands and Wives*. Beverly Hills, Calif.: Sage.

———. 1974. *The Violent Home: A Study of Physical Aggression between Husbands and Wives*. Beverly Hills, Calif.: Sage.

———. 1977. "Violence in the American Family." In J. P. Martin (ed.), *Violence and the Family*. New York: Wiley.

———. 1980. "Violence in the Family: A Review of Research in the Seventies." *Journal of Marriage and the Family* 42: 873–85.

———. 1987. *Family Violence*. Newbury Park, Calif.: Sage.

Gelles, Richard J., and Claire Pedrich Cornell. 1990. *Intimate Violence in Families*, 2d ed. Beverly Hills, Calif.: Sage.

Gelles, Richard J., and Eileen F. Hargreaves. 1987. "Maternal Employment and Violence Toward Children." In Richard J. Gelles (ed.), *Family Violence*. Newbury Park, Calif.: Sage.

Gelles, Richard J., and M. A. Straus. 1987. "Is Violence Toward Children Increasing? A Comparison of 1975–1985 National Survey Rates." *Journal of Interpersonal Violence* 2: 212–22.

Gennep, Arnold van. 1909/1960. *The Rites of Passage*. Chicago: University of Chicago Press.

Gerson, Kathleen. 1985. *Hard Choices*. Berkeley: University of California Press.

Getzels, Jacob W., and P. W. Jackson. 1962. *Creativity and Intelligence*. New York: Wiley.

Ghiglieri, Michael P. 1985. "The Social Ecology of Chimpanzees." *Scientific American* 252 (June): 102–13.

Gilbert, Lucia. 1985. *Men in Dual-Career Marriages*. Hillsdale, N.J.: Erlbaum.

Giles-Sim, Jean. 1983. *Wife Battering: A Systems Theory Approach*. New York: Guilford Press.

Gillespie, Dair L. 1971. "Who Has the Power? The Marital Struggle." *Journal of Marriage and the Family* 33: 445–58.

Gilligan, Carol. 1982. *In a Different Voice. Psychological Theory and Women's Development*. Cambridge, Mass.: Harvard University Press.

Gilmartin, Brian. 1978. *The Gilmartin Report*. Secaucus, N.J.: Citadel Press.

Girouard, Mark. 1980. *Life in the English Country House*. New Haven, Conn.: Yale University Press.

Gitlin, Todd. 1987. *The Sixties, Years of Hope, Days of Rage*. New York: Basic Books.

Glass, Shirley P., and Thomas L. Wright. 1977. "The Relationship of Extramarital Sex, Length of Marriage, and Sex Differences on Marital Satisfaction and Romanticism." *Journal of Marriage and the Family* 39: 691–703.

Gleicher, Norman. 1984. "Cesarean Section Rates in the

United States: The Short-term Failure of the National Consensus Development Conference in 1980." *Journal of the American Medical Association* 252: 3273–78.

Glenn, Evelyn N. 1983. "Split Household, Small Producer and Dual Wage Earner: An Analysis of Chinese-American Family Strategies." *Journal of Marriage and the Family* 45: 35–46.

Glenn, Norval D. 1975. "Psychological Well-Being in the Postparental Stage: Some Evidence from National Surveys." *Journal of Marriage and the Family* 37 (Feb.): 105–10.

————. 1982. "Interreligious Marriage in the United States." *Journal of Marriage and the Family* 44: 555–68.

————. 1991. "Quantitative Research on Marital Quality in the 1980s." Pp. 28–41 in Alan Booth (ed.), *Contemporary Families: Looking Forward, Looking Back*. Minneapolis, Minn.: National Council on Family Relations.

Glenn, Norval D., and J. P. Alston. 1968. "Cultural Distances among Occupational Categories." *American Sociological Review* 33: 365–82.

Glenn, Norval D., and Sara McLanahan. 1982. "Children and Marital Happiness: A Further Specification of the Relationship." *Journal of Marriage and the Family* 44: 63–72.

Glenn, Norval D., Adreain A. Ross, and Judy C. Tully. 1974. "Patterns of Intergenerational Mobility of Females through Marriage." *American Sociological Review* 39: 683–99.

Glenn, Norval D., and Beth Ann Shelton. 1983. "Pre-Adult Background Variables and Divorce." *Journal of Marriage and the Family* 39: 405–10.

Glenn, Norval D., and Charles Weaver. 1978. "A Multivariate Multisurvey Study of Marital Happiness." *Journal of Marriage and the Family* 40: 269–82.

Glick, Paul C. 1975. "A Demographer Looks at American Families." *Journal of Marriage and the Family* 31: 15–26.

Glick, Paul C., and Arthur Norton. 1977. "Marrying, Divorcing, and Living Together in the U.S. Today," *Population Bulletin* 32 (October): 1–39.

Goffman, Erving. 1967. *Interaction Ritual*. New York: Doubleday.

Goldberg, Marilyn P. 1972. "Women in the Soviet Economy." *Review of Radical Political Economics* 4: 1–15.

Goldman, Noreen, Charles F. Westoff, and Charles Hammerslough. 1984. "Demography of the Marriage Market in the United States." *Population Index* 50(1): 5–25.

Goldstein, M. J., H. S. Kant, and J. J. Hartman. 1973.

Pornography and Sexual Deviance. Berkeley: University of California Press.

Goldstone, Jack A. 1986. "The Demographic Revolution in England: A Re-examination." *Population Studies*. 49: 5–33.

Goodall, Jane. 1986. *The Chimpanzees of Gombe. Patterns of Behavior*. Cambridge, Mass.: Harvard University Press.

Goode, William J. 1956. *After Divorce*. New York: Free Press.

————. 1960. "Illegitimacy in the Caribbean." *American Sociological Review* 25: 21–30.

————. 1969. "Violence between Intimates." In D. J. Mulvihill, Melvin M. Tumin, and Lynn A. Curtis, *Crimes of Violence*. Washington, D.C.: U.S. Government Printing Office.

————. 1971. "Force and Violence in the Family." *Journal of Marriage and the Family* 33: 624–36.

Goode, William J., Elizabeth Hopkins, and Helen M. McClure. 1971. *Social Systems and Family Patterns: A Propositional Inventory*. Indianapolis, Ind.: Bobbs-Merrill.

Goody, Jack. 1976. "Inheritance, Property, and Women: Some Comparative Considerations." In Jack Goody, Joan Thirsk, and E. P. Thompson (eds.), *Family and Inheritance. Rural Society in Western Europe 1200–1800*. Cambridge: Cambridge University Press.

————. 1977. *Production and Reproduction*. Cambridge: Cambridge University Press.

————. 1983. *The Development of the Family and Marriage in Europe*. Cambridge: Cambridge University Press.

Gordon, Linda. 1982. "Why Nineteenth-Century Feminists Did Not Support Birth Control and Twentieth-Century Feminists Do." In Barrie Thorne (ed.), *Rethinking the Family: Some Feminist Questions*. New York: Longman.

Gottfredson, Linda S. 1981. "Circumscription and Compromise: A Developmental Theory of Occupational Aspirations." *Journal of Counseling Psychology Monograph* 28: 545–79.

Gove, Walter R. 1972. "The Relation between Sex Roles, Marital Status, and Mental Illness." *Social Forces* 51: 34–44.

Gove, Walter R., and Michael Geerken. 1977. "The Effect of Children and Employment on the Mental Health of Married Men and Women." *Social Forces* 56: 66–76.

Greeley, Andrew M., W. C. McCready, and K. McCourt. 1976. *Catholic Schools in a Declining Church*. Kansas City, Mo.: Sheed and Ward.

Green, Shirley. 1971. *The Curious History of Contraception*. London: Ebury Press.

Greer, Germaine. 1970. *The Female Eunuch.* London: MacGibbon and Kee.

Greif, G. L. 1985. *Single Fathers.* Lexington, Mass.: Lexington Books.

Greiff, B. S., and P. K. Munter. 1980. *Tradeoffs: Executive, Family, and Organizational Life.* New York: Mentor.

Griffith, Samuel B. 1963. "Introduction." In *Sun Tzu: The Art of War.* Oxford: Oxford University Press.

Grigsby, Jill S. 1992. "Women Change Places." *American Demographics* 14: 46–50.

Griswold del Castillo, Richard. 1984. *La Familia.* Notre Dame, Ind.: University of Notre Dame Press.

Grobstein, Clifford, Michael Flower, and John Mendeloff. 1983. "External Human Fertilization: An Evaluation of Policy." *Science* 222 (14 October): 127–33.

Grossman, Frances K., Lois S. Eichler, and Susan A. Winickoff. 1980. *Pregnancy, Birth, and Parenthood.* San Francisco, Calif.: Jossey-Bass.

Guillemin, Jeanne H., I. David Todres, Dick Batten, and Michael A. Grodin. 1989. "Deciding to Treat Newborns: Changes in Pediatricians' Responses to Treatment Choices." In Linda Whiteford and Marilyn Polan (eds.), *New Approaches to Human Reproduction.* Boulder, Colo.: Westview Press.

Gurin, Gerald, Joseph Veroff, and Alexander MacEachern. 1960. *Americans View their Mental Health.* New York: Basic Books.

Gusfield, Joseph R. 1963. *Symbolic Crusade. Status Politics and the American Temperance Movement.* Urbana: University of Illinois Press.

Gutman, Herbert G. 1976. *The Black Family in Slavery and Freedom.* New York: Pantheon.

Guttentag, Marcia, and Paul F. Secord. 1983. *Too Many Women? The Sex Ratio Question.* Beverly Hills, Calif.: Sage.

Haas, Linda. 1992. *Equal Parenthood and Social Policy.* Albany: State University of New York Press.

Hadas, Moses. 1950. *A History of Greek Literature.* New York: Columbia University Press.

Hagestad, G. O. 1988. "Demographic Change and the Life Course: Some Emerging Trends in the Family Realm." *Family Relations* 37: 405–10.

Haim, Gabriel. 1983. "What Do You Want to Be When You Grow Up: A Cognitive Interpretation of Creativity among Science Career Aspirants." Paper delivered at the Boston Colloquium for the Philosophy of Science, Boston University.

Hajnal, J. 1965. "European Marriage Patterns in Perspective." In D.V. Glass and D.E.C. Eversley (eds.), *Population in History,* pp. 101–43.

Hale, Christine. 1990. "Infant Mortality: An American Tragedy." *Population Trends and Public Policy* 18: 1–16. Washington, D.C.: Population Reference Bureau.

Halle, David. 1984. *America's Working Man, Work, Home, and Politics among Blue-Collar Property Owners.* Chicago, Ill.: University of Chicago Press.

Hamburg, David, and Elizabeth McCown. 1979. *The Great Apes.* Menlo Park, Calif.: Cummings.

Hanson, Shirley M. H. 1988. "Divorced Fathers with Custody." In Phyllis Bronstein and Carolyn P. Cowan (eds.), *Fatherhood Today: Men's Changing Role in the Family.* New York: Wiley.

Hapgood, Fred. 1979. *Why Males Exist. An Inquiry into the Evolution of Sex.* New York: Morrow.

Hareven, Tamara K. 1981. *Family Time and Industrial Time.* New York: Cambridge University Press.

Harlow, Harry F. 1958. "The Nature of Love." *American Psychologist* 13: 673–85.

Harlow, Harry F., and Margaret K. Harlow. 1962. "Social Deprivation in Monkeys." *Scientific American* 207 (Nov.): 137–47.

Harris, Marvin. 1974. "The Savage Male." Pp. 70–93 in *Cows, Pigs, Wars and Witches. The Riddles of Culture.* New York: Random House.

Hartmann, Heidi. 1976. "Capitalism, Patriarchy, and Job Segregation by Sex." *Signs* 1: 137–69.

———. 1981. "The Family as the Locus of Gender, Class, and Political Struggle: The Example of Housework." *Signs* 6: 366–94.

Hartzler, Kaye, and Juan N. Franco. 1985. "Ethnicity, Division of Household Tasks and Equity in Marital Roles: A Comparison of Anglo and Mexican American Couples." *Hispanic Journal of Behavioral Sciences* 7(4): 333–44.

Hatfield, Elaine, and Susan Sprecher. 1986. *Mirror, Mirror: the Importance of Looks in Everyday Life.* Albany: State University of New York Press.

Hauser, Arnold. 1951. *The Social History of Art.* New York: Knopf.

Hawkes, Glenn R., and Minna Taylor. 1975. "Power Structure in Mexican and Mexican-American Farm Labor Families." *Journal of Marriage and the Family* 37(Nov.): 807–11.

Hawkins, Alan, and T. Roberts. 1992. "Designing a Primary Intervention to Help Dual-Earner Couples Share Housework and Childcare. *Family Relations* 41: 169–77.

Hayes, Cheryl D. 1987. *Risking the Future: Adolescent Sexuality, Pregnancy, and Childbearing,* vol. 1. Panel on Adolescent Pregnancy and Childbearing, Committee on Child Development Research and Public

Policy, Commission on Behavioral and Social Sciences and Education, National Research Council. Washington, D.C.: National Academy Press.

Hayes, Cheryl D., and Sheila B. Kamerman (eds.). 1983. *Children of Working Parents: Experiences and Outcomes*. Washington D.C.: National Academy Press.

Hayghe, Howard. 1990. "Family Members in the Workforce." *Monthly Labor Review* 113: 14–19.

Heer, David M. 1958. "Dominance and the Working Wife." *Social Forces* 36: 341–47.

Heiss, Jerold. 1975. *The Case of the Black Family.* New York: Columbia University Press.

Heller, Celia. 1966. *Mexican American Youth: Forgotten Youth at the Crossroads*. New York: Random House.

Henshaw, S. K., and J. Silverman. 1988. "The Characteristics and Prior Contraceptive Use of U.S. Abortion Patients." *Family Planning Perspectives* 20: 158–68.

Henton, J. R., A. Cate, J. Koval, S. Lloyd, and S. Christopher. 1983. "Romance and Violence in Dating Relationships." *Journal of Family Issues* 4: 467–82.

Herman, Judith L. 1981. *Father-Daughter Incest*. Cambridge, Mass.: Harvard University Press.

Hertz, Rosanna. 1986. *More Equal than Others: Women and Men in Dual-Career Marriages*. Berkeley: University of California Press.

Herzog, Elizabeth, and Celia E. Sudia. 1973. "Children in Fatherless Families." In Bette M. Caldwell (ed.), *Child Development and Social Policy*. Chicago: University of Chicago Press.

Hill, Charles T., Zick Rubin, and Letitia A. Peplau. 1976. "Breakups before Marriage: The End of 103 Affairs." *Journal of Marriage and the Family* 32: 147–68.

Hill, Martha S. 1983. "Trends in the Economic Situation of U.S. Families and Children: 1970–1980." Pp. 9–53 in Richard R. Nelson and Felicity Skidmore (eds.), *American Families and the Economy: The High Cost of Living*. Washington, D.C.: National Academy Press.

Hiller, Dana V., and William W. Philliber. 1986. "The Division of Labor in Contemporary Marriage: Expectations, Perceptions, and Performance." *Social Problems* 33: 191–201.

Himes, Norman E. 1963. *Medical History of Contraception*. New York: Gamut Press.

Hirschi, Travis. 1969. *The Causes of Delinquency*. Berkeley: University of California Press.

Historical Statistics of the United States. Series A255–257. 1965. Washington, D.C.: U.S. Government Printing Office.

Hochschild, Arlie R. 1975. "Attending To, Codifying and Managing Feelings: Sex Differences in Love." Paper presented at the Annual Meeting of the American Sociological Association, San Francisco.

———. 1983. *The Managed Heart*. Berkeley: University of California Press.

———. 1989. *The Second Shift: Working Parents and the Revolution at Home*. New York: Viking.

Hodgson, Marshall. 1974. *The Venture of Islam*. Chicago: University of Chicago Press.

Hoebel, E. Adamson. 1954. *The Law of Primitive Man*. Cambridge, Mass.: Harvard University Press.

Hofferth, Sandra, and Cheryl Hayes (eds.), 1992. *Risking the Future: Adolescent Sexuality, Pregnancy, and Childbearing*. Washington, D.C.: National Academy Press.

Hofferth, Sandra, and K. Moore. 1979. Early Childbearing and Later Economic Well-Being. *American Sociological Review* 44: 789–815.

Hofferth, Sandra, and D. A. Phillips. 1987. "Child Care in the United States, 1970–1995." *Journal of Marriage and the Family* 49: 559–71.

Hoffman, Lois W. 1979. "Maternal Employment." *American Psychologist* 34:859–65.

Hoffman, Saul D., and Greg J. Duncan. 1988. "What Are the Economic Consequences of Divorce?" *Demography* 25(4): 641–45.

Hoffman, Saul, and John Holmes. 1976. "Husbands, Wives, and Divorce." In *Five Thousand American Families—Patterns of Economic Progress*. Ann Arbor, Mich.: Institute for Social Research.

Holden, Constance. 1981. "The Politics of Paleoanthropology." *Science* 213: 737–40.

———. 1986. "Depression Research Advances, Treatment Lags." *Science* 233: 723–26.

Holmstrom, Lynda L. 1973. *The Two-Career Family*. Cambridge, Mass.: Shenkman.

Hood, Jane. 1983. *Becoming a Two-Job Family*. New York: Praeger.

Houseknecht, Sharon K. 1987. "Voluntary Childlessness." In Marvin B. Sussman and Suzanne K. Steinmetz (eds.), *Handbook of Marriage and the Family*. New York: Plenum Press.

Houston, Jean W. 1985. *Beyond Manzanar: Views of Asian American Womanhood*. Santa Barbara, Calif.: Capra.

Hrdy, Sarah. 1981. *The Woman That Never Existed*. Cambridge, Mass.: Harvard University Press.

Huber, Joan. 1991. *Macro-Micro Linkages in Sociology*. Newbury Park, Calif.: Sage.

Huber, Joan, and Glenna Spitze. 1983. *Sex Stratification, Children, Housework, and Jobs*. New York: Academic Press.

Humphreys, Laud. 1970. *Tearoom Trade; Impersonal Sex in Public Places*. Chicago, Ill.: Aldine.

Hunt, Janet, and Larry Hunt. 1982. "The Dualities of Careers and Families: New Integrations or New Polarizations?" *Social Problems* 29.

Hunt, Morton M. 1966. *The World of the Formerly Married*. New York: McGraw-Hill.

———. 1969. *The Affair*. New York: World.

———. 1974. *Sexual Behavior in the 1970s*. New York: Dell.

Hutchens, Robert M. 1979. "Welfare, Remarriage, and the Marital Search." *American Economic Review* 69 (June): 369–79.

Isaac, Glynn L., Diahan Harley, James W. Wood, Linda D. Wolfe, Rebecca L. Cann, and C. Owen Lovejoy. 1982. "Models of Human Evolution." *Science* 217: 295–306.

Israel, Joachim, and Rosmari Eliasson. 1972. "Consumption Society, Sex Roles, and Sexual Behavior." In Hans Peter Dreitzel (eds.), *Family, Marriage, and the Struggle of the Sexes*. New York: Macmillan.

Jacobs, J. A., and F. F. Furstenberg, Jr. 1986. "Changing Places: Conjugal Careers and Women's Marital Mobility." *Social Forces* 64: 714–32.

Jacobson, P. H. 1956. "Hospital Care and the Vanishing Midwife." *Milbank Memorial Fund Quarterly* 34: 253–61.

Jarvenpa, Robert. 1988. 'The Political Economy and Political Ethnicity of American Indian Adaptations and Identities." In *Ethnicity and Race in the U.S.A.*, Richard Alba (ed.). New York: Routledge.

Jaynes, Gerald D., and Robin M. Williams, Jr. (eds.) 1989. *A Common Destiny: Blacks and American Society*. Washington, D.C.: National Academy Press.

Jencks, Christopher et al. 1972. *Inequality*. New York: Basic Books.

John, Robert. 1988. "The Native American Family." In *Ethnic families in America: Patterns and Variations*, 3rd ed., ed. C. H. Mindel, R. W. Habenstein, and R. Wright. Pp. 325–66. New York: Elsevier.

Johnson, Robert. 1980. *Religious Assortative Mating in the United States*. New York: Academic Press.

John-Steiner, Vera. 1985. *Notebooks of the Mind. Explorations in Thinking*. Albuquerque: University of New Mexico Press.

Jordan, Brigitte, and Susan L. Irwin. 1989. "The Ultimate Failure: Court-Ordered Cesarean Section." In Linda Whiteford and Marilyn Polan (eds.), *New Approaches to Human Reproduction*. Boulder, Colo.: Westview Press.

Juster, S., and M. Vinovskis. 1987. "Changing Perspectives on the American Family in the Past." *Annual Review of Sociology* 13: 193–216.

Kadushin, Charles. 1964. "Social Class and the Experience of Ill Health." *Sociological Inquiry* 34: 67–80.

Kagan, Jerome. 1976. "The Psychological Requirements for Human Development." In Nathan Talbot (ed.), *Raising Children in Modern America*. Little, Brown.

Kahn, Alfred J., and Sheila B. Kamerman (eds.). 1988. *Child Support*. Newbury Park, Calif.: Sage.

Kamerman, Sheila B. 1989. "Child Care, Women, Work, and the Family." In Jeffery S. Lande, Sandra Scarr, and Nina Gunzenhauser (eds.), *Caring for Children*. Hillsdale, N.J.: Erlbaum.

Kamo, Yoshinori. 1988. "Determinants of Household Labor: Resources, Power, and Ideology." *Journal of Family Issues* 9: 177–200.

Kanin, Eugene J., Karen D. Davidson, and Sonia R. Scheck. 1970. "A Research Note on Male-Female Differentials in the Experience of Heterosexual Love." *Journal of Sex Research* 6: 64–72.

Kanowitz, Leo. 1969. *Women and the Law*. Albuquerque: University of New Mexico Press.

Kanter, Rosabeth Moss. 1977. *Men and Women of the Corporation*. New York: Basic Books.

Katz, A. M., and R. Hill. 1958. "Residential Propinquity and Marital Selection: A Review of Theory, Method, and Fact." *Marriage and Family Living* 20: 27–35.

Keefe, Susan E., and Amado M. Padilla. 1987. *Chicano Ethnicity*. Albuquerque: University of New Mexico Press.

Keller, Eveyln. 1982. "Feminism and Science." *Signs* 7: 589–602.

———. 1985. *Reflections on Gender and Science*. New Haven: Yale University Press.

Kelly, Joan B. 1982. "Divorce: The Adult Perspective." In Benjamin B. Wolman (ed.), *Handbook of Developmental Psychology*. Englewood Cliffs, N.J.: Prentice-Hall.

Kelly, Joan Berlin. 1986. "Divorce: The Adult Perspective." In Arlene S. Skolnick and Jerome H. Skolnick (eds.), *Family in Transition: Rethinking Marriage, Sexuality, Child Rearing, and Family Organization*, Fifth Edition. Boston: Little, Brown.

Kelly, Liz. 1988. "How Women Define Their Experiences of Violence." In Kersti Yllo and Michele Bograd (eds.), *Feminist Perspectives on Wife Abuse*. Newbury Park, Calif.: Sage.

Kemper, Theodore. 1978. *A Social Interactional Theory of Emotions*. New York: Wiley.

————. 1988. "Love and Like and Love and *Love*." In David Franks (ed.), *The Sociology of Emotions*. Greenwich, Conn.: Jai Press.

Keniston, Kenneth. 1977. *All Our Children: The American Family Under Pressure*. New York: Harcourt Brace Jovanovich.

Kephart, William M. 1967. "Some Correlates of Romantic Love." *Journal of Marriage and the Family* 29: 470–74.

Kerckhoff, Alan C., and Keith E. Davis. 1962. "Value Consensus and Need Complementarity in Mate Selection." *American Sociological Review* 27: 295–303.

Kessler, Ronald, and James McRae. 1982. "The Effects of Wives' Employment on the Mental Health of Married Men and Women." *American Sociological Review* 47: 216–27.

Kiely, J. L., J. Kleinmen, and M. Kiely. 1992. "Triplets and Higher-Order Multiple Births." *American Journal of Diseases in Children* 146: 862–68.

Kikumura, A., and Harry Kitano. 1973. "Interracial Marriage: A Picture of the Japanese Americans." *Journal of Social Issues* 29 (Spring): 67–81.

Kimball, Gayle. 1983. *The 50/50 Marriage*. Boston: Beacon.

————. 1988. *50-50 Parenting: Sharing Family Rewards and Responsibilities*. Lexington, Mass.: Lexington Books.

Kimmel, Michael, and Martin P. Levine. 1992. "Men and AIDS." In Michael Kimmel and Michael Messner (eds.), *Men's Lives*. New York: Macmillan.

Kinsey, Alfred C., Wardell B. Pomeroy, and Clyde D. Martin. 1948. *Sexual Behavior in the Human Male*. Philadelphia, Pa.: W. B. Saunders.

Kinsey, Alfred C., Wardell B. Pomeroy, Clyde E. Martin, and Paul H. Gebhard. 1953. *Sexual Behavior in the Human Female*. Philadelphia, Pa.: W. B. Saunders.

Kinsman, Gary. 1992. "Men Loving Men: The Challenge of Gay Liberation." In Michael Kimmel and Michael Messner (eds.), *Men's Lives*. New York: Macmillan.

Kinzer, Nora. 1973. "Priests, Machos, and Babies Or, Latin American Women and the Manichaean Heresy." *Journal of Marriage and the Family* 35: 300–12.

Kitano, Harry H. L. 1976. *Japanese Americans: The Evolution of a Subculture*. Englewood Cliffs, N.J.: Prentice-Hall.

Kitano, Harry, and Roger Daniels. 1988. *Asian Americans: Emerging Minorities*. Englewood Cliffs, N.J.: Prentice-Hall.

Kleinberg, Seymour. 1992. "The New Masculinity of Gay Men." In Michael Kimmel and Michael Messner (eds.), *Men's Lives*. New York: Macmillan.

Klinman, Debra G., and Rhiana Kohl. 1984. *Fatherhood U.S.A.* New York: Garland.

Knox, David, and Caroline Schact. 1994. *Choices in Relationships*. Minneapolis/St. Paul: West Publishing.

Kobrin, Frances E., and Linda J. Waite. 1983. "Effects of Family Stability and Nestleaving Patterns on the Transition to Marriage." Paper presented at the Annual Meeting of the American Sociological Association.

Kohlberg, Lawrence. 1966. "A Cognitive-Developmental Analysis of Children's Sex role Concepts and Attitudes." In Eleanor Maccoby (ed.), *The Development of Sex Differences*. Palo Alto, Calif.: Stanford University Press.

Kohn, Melvin L. 1977. *Class and Conformity*. Chicago, Ill.: University of Chicago Press.

————. 1979. "The Effects of Social Class on Parental Values and Practices." In David Reiss and Howard A. Hoffman (eds.), *The American Family: Dying or Developing?* New York: Plenum Press.

Kohn, Melvin L., and Carmi Schooler. 1978. "The Reciprocal Effects of Substantive Complexity of Work and Intellectual Flexibility: A Longitudinal Assessment." *American Journal of Sociology* 84: 24–52.

Komarovsky, Mirra. 1962. *Blue-Collar Marriage*. New York: Random House.

————. 1976. *Dilemmas of Masculinity. A Study of College Youth*. New York: W. W. Norton.

Kontos, W., and A. J. Stremmel. 1987. "Caregivers' Perceptions of Working Conditions in a Child Care Environment." *Early Childhood Research Quarterly* 3: 100–21.

Kornblum, William. 1974. *Blue-Collar Community*. Chicago, Ill.: University of Chicago Press.

Kortlandt, Adriaan. 1962. "Chimpanzees in the Wild." *Scientific American* 206: 128–38.

Koss, M. P., C. A. Gidycz, and N. Wisniewski. 1987. "The Scope of Rape: Incidence and Prevalence of Sexual Aggression and Victimization in a National Sample of Higher Education Students." *Journal of Consulting and Clinical Psychology* 52: 162–70.

Kotelchuck, M. 1976. "The Infant's Relationship to the Father: Experimental Evidence." In Michael E. Lamb (ed.), *The Role of the Father in Child Development*. New York: Wiley.

Krantowitz, Barbara, and D. Witherspoon. 1987. "The December Dilemma: How to Reconcile Two Faiths in one Household." *Newsweek*, December 28, p. 56.

Krantz, Susan E. 1988. "The Impact of Divorce on Children." In S. M. Dornbush and M. H. Strober (eds.), *Feminism, Children, and the New Families*. New York: Guilford.

Kummer, Hans. 1968. *Social Organization of Hamadryas Baboons*. Chicago, Ill.: University of Chicago Press.

Ladner, Joyce A. 1971. *Tomorrow's Tomorrow: The Black Woman*. New York: Doubleday.

Ladurie, Emmanuel. 1979. *Montaillou*. New York: Vintage Books.

Lamb, Michael. 1981. *The Role of the Father in Child Development*. New York: Wiley.

———. 1986. "The Changing Roles of Fathers." In M. Lamb (ed.), *The Father's Role: Applied Perspectives*. New York: Wiley.

Lamb, Michael E., Joseph H. Pleck, and James A. Levine. 1986. "Effects of Paternal Involvement on Fathers and Mothers." In Robert Lewis and Marvin Sussman (eds.), *Men's Changing Roles in the Family*. New York: Haworth Press.

Lamb, Michael, and Abraham Sagi. 1983. *Fatherhood and Family Policy*. Hillsdale, N.J.: Lawrence Erlbaum.

Lambert, Wallace E., Josiane F. Hamers, and Nancy Frasure-Smith. 1980. *Child-Rearing Values: A Cross National Study*. New York: Praeger.

Langer, W. L. 1972. "Checks on Population Growth: 1750–1850." *Scientific American* 226: 93–100.

LaRossa, Ralph. 1983. "The Transition to Parenthood and the Social Reality of Time." *Journal of Marriage and the Family* 45 (Aug.): 579–89.

LaRossa, Ralph, and Maureen LaRossa. 1981. *Transition to Parenthood*. Beverly Hills, Calif.: Sage.

———. 1989. "Baby Care: Fathers vs. Mothers." In Barbara J. Risman and Pepper Schwartz (eds.), *Gender in Intimate Relationships*. Belmont, Calif.: Wadsworth.

Laslett, Peter. 1971. *The World We Have Lost. England Before the Industrial Age*. New York: Scribners.

———. 1977. *Family Life and Illicit Love in Earlier Generations*. Cambridge: Cambridge University Press.

Lau, E. E., and J. Kosberg. 1979. "Abuse of the Elderly by Informal Care Providers." *Aging* 299: 10–15.

Laumann, Edward O. and James S. House. 1970. "Living Room Styles and Social Attributes: The Patterning of Material Artifacts in a Modern Urban Community." In Edward O. Laumann, Paul M. Siegel and Robert W. Hodge (eds.), *The Logic of Social Hierarchies*. Chicago, Ill.: Markham.

Leavitt, Judith W. 1983. "Science Enters the Birthing Room: Obstetrics in America since the Eighteenth Century." *Journal of American History*. 70: 281–304.

Leibowitz, Arleen, Linda J. Waite, and Christina J. Witsburger. 1988. "Child Care for Preschoolers: Difference by Child's Age." *Demography* 25: 205–20.

Lein, Laura. 1979. "Male Participation in Home Life." *Family Coordinator* 28: 489–95.

Lein, Laura, M. Durham, M. Pratt, M. Schudson, R. Thomas, and H. Weiss. 1974. *Final Report: Work and Family Life*. Cambridge, Mass.: Center for the Study of Public Policy.

LeMasters, E. E. 1957. "Parenthood as Crisis." *Marriage and Family Living* 19: 352–5.

———. 1975. *Working Class Aristocrats: Life Styles at a Working-class Tavern*. Madison: University of Wisconsin Press.

Lenski, Gerhard, E. 1966. *Power and Privilege. A Theory of Social Stratification*. New York: McGraw-Hill.

Lenski, Gerhard E., and Jean Lenski. 1974. *Human Societies*. New York: McGraw-Hill.

Levine, J., J. Pleck and M. Lamb. 1983. "The Fatherhood Project." In A. Sagi and M. Lamb (eds.), *Fatherhood and Family Policy*. Hillsdale, N.J.: Erlbaum.

Levinger, George. 1966. "Sources of Marital Satisfaction among Applicants for Divorce." *American Journal of Orthopsychiatry* 36: 804–6.

Levinger, George et al. 1970. "Progress toward Permanence in Courtship: A Test of the Kerckhoff-Davis Hypothesis." *Sociometry* 33: 427–33.

Levinson, Daniel J. 1978. *The Seasons of a Man's Life*. New York: Knopf.

Lévi-Strauss, Claude. 1969. *The Elementary Structures of Kinship*. Boston, Mass.: Beacon Press.

Levitan, Sarah A., and Richard A. Belous. 1981. *What's Happening to the American Family?* Baltimore, Md.: Johns Hopkins University Press.

Levran, D., J. Dor, E. Rudak, L. Nebel, I Ben-Shlomo, Z. Ben-Rafael, and S. Mashiach. 1990. "Pregnancy Potential of Human Oocytes." *New England Journal of Medicine* 323: 1153–56.

Lewin, Roger. 1980. "Evolutionary Theory under Fire." *Science* 210: 883–87.

———. 1986. "New Fossil Upsets Human Family." *Science* 233: 720–21.

Lewis, Oscar. 1951. *Life in a Mexican Village: Tepoztlan Restudied*. Urbana: University of Illinois Press.

Lewis, Robert A. 1973. "A Longitudinal Test of a Developmental Framework for Premarital Dyadic Formation." *Journal of Marriage and the Family* 35: 16–27.

Lewis, W. H. 1957. *The Splendid Century. Life in the France of Louis XIV.* New York: Doubleday.

Liebman, Robert C., and Robert Wuthnow. 1983. *The New Christian Right*. Chicago, Ill.: Aldine.

Liebow, Elliot. 1967. *Tally's Corner: A Study of Negro Streetcorner Men*. Boston, Mass.: Little, Brown.

Linton, Ralph. 1936. *The Study of Man*. New York: Appleton.

Litwak, Eugene. 1960. "Occupational Mobility and Extended Family Cohesion." *American Sociological Review* 25: 9–21.

Litwak, Eugene, and P. Messer. 1989. Organizational Theory, Social Supports, and Mortality Rates. *American Sociological Review* 54: 49–66.

Lloyd, Peter C. 1965. "The Yoruba of Nigeria." In James L. Gibbs (ed.), *Peoples of Africa*. New York: Holt, Rinehart and Winston.

Locksley, Ann. 1982. "Social Class and Marital Attitudes and Behavior." *Journal of Marriage and the Family* 44: 427–40.

Lockwood, David. 1986. "Class, Gender, and Status." In Rosemary Crompton and Michael Mann (eds.), *Gender and Stratification*. Cambridge: Polity Press.

London, Kathryn A., and Barbara Foley Wilson. 1988. "Divorce." *American Demographics* 10(10): 23–26.

Lopata, Helena Z. 1971. *Occupation: Housewife*. New York: Oxford University Press.

———. 1979. *Women as Widows: Support Systems*. New York: Elsevier.

Lopata, Helena Z., Cheryl Allyn Miller, and Debra Barnewolt. 1984–85. *City Women: Work, Jobs, Occupations, Careers*. 2 vols. New York: Praeger.

Lothrop, Samuel Kirkland. 1928. *The Indians of Tierra del Fuego*. New York: Heye Foundation.

Lovejoy, C. Owen. 1981. "The Origin of Man." *Science* 211: 341–50.

Lowenthal, Marjorie F., Majde Thurnher, and David Chiriboga. 1976. *Four Stages of Life*. San Francisco, Calif.: Jossey-Bass.

Lowry, D. T., G. Love, and M. Kirby. 1981. "Sex on the Soap Operas: Patterns of Intimacy." *Journal of Communication* 31: 90–96.

Luepnitz, Deborah Anna. 1982. *Child Custody. A Study of Families After Divorce*. Lexington, Mass.: D. C. Heath.

Luker, Kristin. 1975. *Taking Chances: Abortion and the Decision to Contracept*. Berkeley: University of California Press.

———. 1984. *Abortion and the Politics of Motherhood*. Berkeley: University of California Press.

Lumsden, C. J., and E. O. Wilson. 1981. *Genes, Mind, and Culture*. Cambridge, Mass.: Harvard University Press.

Lynn, Richard. 1982. "IQ in Japan and the U.S. Shows a Growing Disparity." *Nature* 297 (20 May): 222–23.

Lyon, Eleanor, and Patricia Goth Mace. 1991. "Family Violence and the Courts." Pp. 167–79 in Dean Knudsen and JoAnn Miller (eds.), *Abused and Battered*. New York: Aldine de Gruyter.

Maccoby, Eleanor. 1968. "The Development of Moral Values and Behavior in Childhood." In John A. Clausen (ed.), *Socialization and Society*. Boston, Mass.: Little, Brown.

Macfarlane, Alan. 1979. *The Origins of English Individualism*. New York: Cambridge University Press.

———. 1986. *Marriage and Love in England. Modes of Reproduction 1300–1840*. Oxford: Blackwell.

Macke, Anne S., George W. Bohrstedt, and Ilene N. Bernstein. 1979. "Housewives' Self-Esteem and Their Husband's Success: The Myth of Vicarious Involvement." *Journal of Marriage and the Family* 42: 51–57.

Mackenzie, Gavin. 1973. *The Aristocracy of Labor: the Position of Skilled Craftsmen in the American Class Structure*. Cambridge: Cambridge University Press.

Madsen, William. 1973. *Mexican-Americans of South Texas*, 2d ed. New York: Holt, Rinehart and Winston.

Magnuson, E. 1983. "Child Abuse: The Ultimate Betrayal." *Time*, Sept. 5, pp. 20–22.

Maier, Richard A. 1984. *Human Sexuality in Perspective*. Chicago, Ill.: Nelson-Hall.

Malamuth, M. M., and E. Donnerstein (eds.). 1984. *Pornography and Sexual Aggression*. New York: Academic Press.

Malinoswki, Bronislaw. 1929. *The Sexual Life of Savages in North-Western Melanesia*. New York: Harcourt.

———. 1964. "Parenthood, The Basis of Social Structure." In Rose Laub Coser (ed.), *The Family: Its Structure and Functions*. New York: St. Martin's Press.

Mangen, D. J., V. L. Bengston, and P. H. Landry (eds.). 1988. *Measurement of Intergenerational Relations*. Beverly Hills, Calif.: Sage.

Mann, Jonathan M., James Chin, Peter Piot, and Thomas Quinn. 1988. "The International Epidemiology of AIDS." *Science* 259 (Oct.): 82–89.

Mann, Michael. 1970. "The Social Cohesion of Liberal Democracy." *American Sociological Review* 35: 423–39.

Maples, Terry, and Michael Hoff. 1982. *Gorilla Behavior*. New York: Van Nostrand.

Marcus, Steven. 1964. *The Other Victorians. A Study of Sexuality and Pornography in Mid-Nineteenth Century England*. New York: Basic Books.

Martin, Calvin 1978. *Keepers of the Game*. Berkeley: University of California Press.

Martin, D. 1976. *Battered Wives*. New York: Pocket Books.

Martin, Judith. 1983. "Maternal and Paternal Abuse of Children." In David Finkelhor, Richard J. Gelles, Gerald T. Hotaling, and Murray A. Straus (eds.),

The Dark Side of Families: Current Family Violence Research. Beverly Hills, Calif.: Sage.

Massey, Douglas S. 1986. "The Settlement Process among Mexican Migrants to the United States." *American Sociological Review* 51: 670–84.

Massey, Douglas, and Nancy Denton. 1993. *American Apartheid: Segregation and the Making of the Underclass*. Cambridge, Mass.: Harvard University Press.

Masters, William H., and Virginia E. Johnson. 1966. *Human Sexual Response*. Boston, Mass.: Little, Brown.

———. 1975. *The Pleasure Bond*. New York: Bantam.

Mattessich, Paul, and Reuben Hill. 1987. "Life Cycle and Family Development." In Marvin B. Sussman and Susan K. Steinmetz (eds.), *Handbook of Marriage and the Family*. New York: Plenum.

Maurer, A. 1976. "Physical Punishment of Children." Paper presented at the California State Psychological Convention, Anaheim.

Mauss, Marcel. 1925/1967. *The Gift*. New York: Norton.

May, K. 1982. "Factors Contributing to First-Time Fathers' Readiness for Fatherhood: An Exploratory Study." *Family Relations* 31: 353–61.

May, K., and S. Perrin. 1985. "Prelude: Pregnancy and Birth." In S. Hanson and F. Bozett (eds.), *Dimensions of Fatherhood*. Beverly Hills, Calif.: Sage.

Mayer, Egon. 1985. *Love and Tradition: Marriage between Jews and Christians*. New York: Plenum.

McAdoo, Harriette Pipes. 1986. "Strategies Used by Black Single Mothers Against Stress." *Review of Black Political Economy* 14(2–3): 155–66.

McBride, B. A. 1990. "The Effects of a Parent Education/Play Group Program on Father Involvement in Child Rearing." *Family Relations* 39: 250–56.

McCarthy, J. D., and W. L. Yancey. 1971. "Uncle Tom and Mr. Charlie: Metaphysical Pathos in the Study of Racism and Personal Disorganization." *American Journal of Sociology* 76: 648–72.

McClelland, David C. 1953. *The Achievement Motive*. New York: Appleton.

McLanahan, Sara. 1989. "The Two Faces of Divorce: Women's and Children's Interests." Paper presented at the American Sociological Association, San Francisco, August.

McLanahan, Sara, and Julia Adams. 1987. "Parenthood and Psychological Well-Being." *Annual Review of Immunology* 5: 237–57.

McLanahan, Sara, Julia Adams, and A. B. Sorenson. 1985. "Life Events and Psychological Well-Being over the Life Course." In G. H. Elder (ed.), *Life Course Dynamics*. Ithaca, N.Y.: Cornell University Press.

Mead, George. 1934/1967. *Mind, Self, and Society*. Chicago, Ill.: University of Chicago Press.

Meier, John H. (ed.). 1985. *Assault Against Children: Why It Happens: How to Stop It*. San Diego, Calif.: College-Hill Press.

Menaghan, Elizabeth. 1983. "Marital Stress and Family Transitions: A Panel Analysis." *Journal of Marriage and the Family* 45: 371–86.

Mendoza, Sally P. 1984. "The Psychobiology of Social Relationships." In Patricia R. Barchas and Sally P. Mendoza (eds.), *Social Cohesion: Essays Toward a Sociophysiological Perspective*. Westport, Conn.: Greenwood.

Menken, Jane, James Trussell, and Ulla Larsen. 1986. "Age and Infertility." *Science* 1390: 1389–94.

Mercer, Jane. 1974. *Labelling the Mentally Retarded*. Berkeley: University of California Press.

Merchant, Carolyn. 1980. *The Death of Nature*. San Francisco, Calif.: Harper & Row.

Merriam, E. 1968. "We're Teaching Our Children That Violence Is Fun." In O. N. Larson, *Violence and the Mass Media*. New York: Harper & Row.

Mertes, Kate. 1988. *The English Noble Household. 1250–1600*. Oxford: Blackwell.

Miall, Charlene E. 1987. "The Stigma of Involuntary Childlessness." In Arlene S. Skolnick and Jerome H. Skolnick (eds.), *Family in Transition*. Glenview, Ill.: Scott, Foresman.

Michael, Robert T., Heidi I. Hartmann, and Brigid O'Farrell (eds.). 1989. *Pay Equity: Empirical Inquiries*. Washington, D.C.: National Academy Press.

Michaelson, Evalyn J., and Walter Goldschmidt. 1971. "Female Roles and Male Dominance among Peasants." *Southwestern Journal of Anthropology* 27: 330–52.

Miller, Brent, and Kristin Moore. 1991. "Adolescent Sexual Behavior, Pregnancy, and Parenting." Pp. 307–26 in Alan Booth (ed.), *Contemporary Families: Looking Forward, Looking Back*. Minneapolis, Minn.: National Council on Family Relations.

Miller, Dorothy. 1979. "The Native American Family: The Urban Way." In *Families Today*. edited by Eunice Corfman. Washington, D.C.: U.S. Government Printing Office.

Miller, Joanne, and Howard H. Garrison. 1982. "Sex Roles: The Division of Labor at Home and in the Workplace." *Annual Review of Sociology* 8: 237–62.

Millett, Kate. 1970. *Sexual Politics*. New York: Doubleday.

Millman, Marcia. 1976. *The Unkindest Cut*. New York: Morrow.

Mindel, Charles H. 1980. "Extended Familism among Urban Mexican-Americans, Anglos and Blacks." *Hispanic Journal of Behavioral Sciences* 2: 21–34.

Mirande, Alfredo. 1977. "The Chicano Family: A Reanalysis of Conflicting Views." *Journal of Marriage and the Family* 39: 741–49.

———. 1981. *La Chicana*. Chicago, Ill.: University of Chicago Press.

———. 1988. "Chicano Fathers: Traditional Perceptions and Current Realities." In Bronstein and Cowan (eds.), *Fatherhood Today: Men's Changing Role in the Family*. New York: Wiley.

Mitchell, Juliet. 1974. *Psychoanalysis and Feminism*. New York: Pantheon.

Mitchell, Juliet, and Jacqueline Rose. 1982. *Feminine Sexuality: Jacques Lacan and the Ecole Freudienne*. New York: Norton.

Mnookin, Robert H., Eleanor E. Maccoby, Charlene F. Depner, and Catherine R. Albiston. 1990. "Private Ordering Revisited: What Custodial Arrangements Are Parents Negotiating?" In S. Sugarman and H. Kay (eds.), *Divorce Reform at the Crossroads*. New Haven, Conn.: Yale University Press.

Moghissi, Kamran S. 1989. "The Technology of AID and Surrogacy." In Linda Whiteford and Marilyn Polan (eds.), *New Approaches to Human Reproduction*. Boulder, Colo.: Westview Press.

Mohr, James C. 1978. *Abortion in America: The Origins and Evolution of National Policy, 1800–1900*. New York: Oxford University Press.

Money, John, and Anke Ehrhardt. 1972. *Man and Woman: Boy and Girl*. Baltimore, Md.: Johns Hopkins University Press.

Moore, K. A., and M. R. Burt. 1982. *Private Crisis, Public Cost: Policy Perspectives on Teenage Childbearing*. Washington, D.C.: Urban Institute.

Moore, K. A., and S. B. Caldwell. 1976. "Out-of-Wedlock Pregnancy and Childbearing." Work. Pap. 992-02. Washington, D.C.: Urban Institute.

Moore, Kristin, Daphne Spain, and Suzanne M. Bianchi. 1984. "The Working Wife and Other." *Marriage and Family Review* 7: 77–98.

Morgan, Barrie S. 1981. "A Contribution to the Debate on Homogamy, Propinquity, and Segregation." *Journal of Marriage and the Family* 43: 909–21.

Morgan, Hal, and Kerry Tucker. 1991. *Companies that Care*. New York: Simon & Schuster.

Morgan, Leslie A. 1991. *After Marriage Ends: Economic Consequences for Midlife Women*. Newbury Park, Calif.: Sage.

Morgan, S. Philip. 1991. "Late Nineteenth- and Early Twentieth-Century Childlessness." *American Journal of Sociology* 97: 779–807.

Morgan, S. Philip, Diane Lye, and Gretchen Condran. 1988. "Sons, Daughters, and the Rise of Marital Disruption." *American Journal of Sociology* 94: 110–29.

Morris, Desmond. 1967. *The Naked Ape*. New York: McGraw-Hill.

Morris, Ivan. 1964. *The World of the Shining Prince. Court Life in Ancient Japan*. New York: Oxford University Press.

Mosher, William D. 1990. "Contraceptive Practice in the United States, 1982–1988." *Family Planning Perspectives* 22: 198–205.

Mosher, W. D., and W. Pratt. 1990. "Fecundity and Infertility in the United States, 1965-1988." *Vital Health Statistics*. Hyattsville, MD: National Center for Health Statistics.

Moynihan, Daniel Patrick, Paul Barton, and Ellen Broderick. 1965. *The Negro Family: The Case for National Action*. Washington, D.C.: U.S. Department of Labor.

Mueller, KC. W., and H. Pope. "Divorce and Female Remarriage Mobility." *Social Forces* 58: 726–38.

Muir, Frank, and Simon Brett. 1980. *On Children*. London: Heinemann.

Mulroy, Elizabeth A. (ed.). 1988. *Women as Single Parents*. Dover, Mass.: Auburn House.

Murdock, George P. 1967. *World Ethnographic Atlas*. Pittsburgh, Pa.: University of Pittsburgh Press.

Murphy, John E. 1988. "Date Abuse and Forced Intercourse among College Students." In Gerald T. Hotaling, David Finkelhor, John T. Kirkpatrick, and Murray A. Straus (eds.), *Family Abuse and Its Consequences*. Newbury Park, Calif.: Sage.

Murphy, Robert F. 1957. "Intergroup Hostility and Social Cohesion." *American Anthropologist* 59: 1018–35.

———. 1959. "Social Structure and Sex Antagonism." *Southwestern Journal of Anthropology* 15: 89–98.

Murstein, Bernard. 1967. "Empirical Tests of Role, Complementary Needs, and Homogamy Theories of Marital Choice." *Journal of Marriage and the Family* 29: 689–96.

———. 1972. "Physical Attractiveness and Marital Choice." *Journal of Personality and Social Psychology* 22: 8–12.

Murstein, Bernard I., and P. Christy. 1976. "Physical Attractiveness and Marital Adjustment in Middle Age Couples." *Journal of Personality and Social Psychology* 34: 537–42.

Mutran, E. 1985. "Intergenerational Family Support among Blacks and Whites: A Response to Culture or to

Socioeconomic Difference?" *Journal of Gerontology* 40: 382–89.

Nance, John. 1975. *The Gentle Tasaday.* New York: Harcourt Brace Jovanovich.

Napier, John, and P. H. Napier. 1967. *Handbook of Living Primates.* New York: Academic Press.

Nason, Ellen M., and Margaret M. Poloma. 1976. *Voluntary Childless Couples.* Beverly Hills, Calif.: Sage.

National Academy of Science. 1990. *Developing New Contraceptives.*

National Center for Health Statistics. 1993. "Advance Report of Maternal and Infant Health Data." *Monthly Vital Statistics Report* 42. Hyattsville, Md.: National Center for Health Statistics.

National Commission on Employment Policy. 1982. *Hispanics and Jobs: Barriers to Progress.* Report No. 14. Washington, D.C.: Government Printing Office.

National Institute of Mental Health. 1982. *Television and Behavior: Ten Years of Science Progress and Implications for the Eighties.* Rockville, Md.: National Institute of Mental Health.

National Opinion Research Center. 1986. *Cumulative Codebook for the 1972–86 Surveys.* University of Chicago.

Neugarten, B., and N. Datan. 1973. "Sociological Perspectives on the Life Cycle." In P. Baltes and K. Schaie (eds.), *Life-Span Developmental Psychology: Personality and Socialization.* New York: Academic Press.

Neumann, Franz. 1944. *Behemoth: The Structure and Practice of National Socialism.* New York: Oxford University Press.

Newcomer, Susan, and J. Richard Udry. 1986. "Paternal Marital Status Effects on Adolescent Behavior." Paper presented at Annual Meeting of the American Sociological Association, New York.

Newton, N. R. 1973. "Interrelationships between Sexual Responsiveness, Birth, and Breast Feeding." In J. Zubin and J. Money (eds.), *Contemporary Sexual Behavior.* Baltimore, Md.: Johns Hopkins University Press.

Norton, Arthur J., and Paul C. Glick. 1976. "Marital Instability: Past, Present, and Future." *Journal of Social Issues* 32: 5–20.

Nye, Ivan F. 1978. "Is Choice and Exchange Theory the Key?" *Journal of Marriage and the Family* 40: 219–34.

Oakley, Ann. 1974. *The Sociology of Housework.* New York: Pantheon.

O'Brien, John E. 1971. "Violence in Divorce-Prone Families." *Journal of Marriage and the Family* 33: 692–98.

O'Connell, Martin. 1993. *Where's Pappa: Fathers' Role in Child Care.* No. 20. Washington, D.C.: Population Reference Bureau.

O'Hare, William P. 1992. "America's Minorities: The Demographics of Diversity." *Population Bulletin* Vol. 47, No. 4. Washington, D.C.: Population Reference Bureau.

O'Hare, William P., and Judy Felt. 1991. "Asian Americans: America's Fastest Growing Minority Group." *Population Trends and Public Policy.* No. 11. Washington, D.C.: Population Reference Bureau.

Olsen, M. E. 1970. "Social and Political Participation of Blacks." *American Sociological Review* 35: 682–97.

Olson, J. S., and R. Wilson. 1984. *Native Americans in the Twentieth Century.* Provo. Utah: Brigham Young University Press.

O'Neill, Nena. 1974. *Shifting Gears: Finding Security in a Changing World.* New York: Evans.

O'Neill, Nena, and George O'Neill. 1972. *Open Marriage.* New York: Avon.

O'Neill, William L. 1970. *The Woman Movement: Feminism in the United States and England.* London: Allen and Unwin.

Ortner, Sherry, and Harriet Whitehead (eds.). 1981. *Sexual Meanings: The Cultural Construction of Gender and Sexuality.* Cambridge: Cambridge University Press.

Osherson, S. 1986. *Finding Our Fathers: The Unfinished Business of Manhood.* New York: Free Press.

Ostrander, Susan A. 1984. *Women of the Upper Class.* Philadelphia, Pa.: Temple University Press.

Pagelow, Mildred D. 1981. *Woman-Battering. Victims and Their Experiences.* Beverly Hills, Calif.: Sage.

Paige, Karen E., and Jeffery M. Paige. 1981. *The Politics of Reproductive Ritual.* Berkeley: University of California Press.

Papanek, H. 1973. "Men, Women and Work: Reflections of the Two-person Career." *American Journal of Sociology* 78.

Parenti, Michael. 1983. *Democracy for the Few,* 4th ed. New York: St. Martin's Press.

Parke, Ross D. 1981. *Fathers.* Cambridge, Mass.: Harvard University Press.

———. 1988. "Families in Life-Span Perspective: A Multilevel Developmental Approach." In E. M. Hetherington, R. Lerner, and M. Perlmutter (eds.), *Child Development in Life-Span Perspective.* Hillsdale, N.J.: Erlbaum.

Parsons, Talcott, and Robert Bales. 1955. *Family Socialization and Interaction Process.* Glencoe, Ill.: Free Press.

Pasley, Kay, and Marilyn Ihinger-Tallman (eds.). 1987.

Remarriage and Stepparenting. New York: Guilford Press.

Patterson, Gerald R. 1976. *Living with Children: New Methods for Parents and Teachers.* Champaign, Ill.: Research Press.

Patterson, Orlando. 1982. *Slavery and Social Death: a Comparative Study.* Cambridge, Mass.: Harvard University Press.

Patzer, Gordon L. 1985. *The Physical Attractiveness Phenomena.* New York: Plenum Press.

Pearlin, L. I. 1974. "Sex Roles and Depression." In N. Datan and L. Ginsberg, *Life Span Developmental Psychology Conference: Normative Life Crises.* New York: Academic Press.

Pearlin, L. I., and J. S. Johnson. 1977. "Marital Status, Life-Strains and Depression." *American Sociological Review* 42: 704–15.

Peck, M. Scott. 1978. *The Road Less Traveled.* New York: Simon and Schuster.

Pelton, Leroy. 1978. "The Myth of Classlessness in Child Abuse Cases." *American Journal of Orthopsychiatry* 48: 569–79.

Pepitone-Rockwell, Fran. 1980. *Dual Career Couples.* Beverly Hills, Calif.: Sage.

Petersen, James R., Arthur Kretchmer, Barbara Nellis, Janet Lever, and Rosanna Hertz. 1983. "The Playboy Readers' Sex Survey." *Playboy* 30 (January): 108, 241–50.

Peterson, Gary, and Boyd C. Rollins. 1987. "Parent-Child Socialization." In Marvin B. Sussman and Suzanne K. Steinmetz (eds.), *Handbook of Marriage and the Family.* New York: Plenum Press.

Peterson, James A. 1971. "The Office Wife." *Sexual Behavior* 1(5): 3–10.

Pettigrew, Thomas F. 1989. "The Changing—Not Declining—Significance of Race." In Charles Vert Willie, *Caste and Class Controversy on Race and Poverty.* Dix Hills, N.Y.: General Hall.

Phillips, Deborah. 1989. "Future Directions and Need for Child Care in the United States." In Jeffery S. Lande, Sandra Starr, and Nina Gunzenhauser (eds.), *Caring for Children.* Hillsdale, N.J.: Erlbaum.

Pilbeam, David. 1984. "The Descent of Hominoids and Hominids." *Scientific American* 250 (March): 84–96.

Pillemer, Karl and David Finkelhor. 1988. "The Prevalence of Elder Abuse: A Random Sample Survey." *Gerontologist* 28: 51–57.

Pleck, Joseph H. 1983. "Husband's Paid Work and Family Roles: Current Resarch Issues." In H. Lopata and J. Pleck (eds.), *Research in the Interweave of Social Roles.* Greenwich: Jai Press.

———. 1985. *Working Wives/Working Husbands.* Beverly Hills, Calif.: Sage.

Pleck, Joseph H., and Graham L. Staines. 1985. "Work Schedules and Family Life in Two-Earner Couples." *Journal of Family Issues* 6: 61–82.

Pogrebin, Letty C. 1983. *Family Politics.* New York: McGraw-Hill.

Polatnick, M. 1973. "Why Men Don't Rear Children: A Power Analysis." *Berkeley Journal of Sociology* 18.

Pollock, Linda A. 1984. *Forgotten Children: Parent-Child Relations from 1500 to 1900.* Cambridge: Cambridge University Press.

Poloma, Margaret. 1972. "Role Conflict and the Married Professional Woman." In C. Safilios-Rothschild (ed.), *Toward a Sociology of Women.* Lexington, Mass.: Xerox.

Pomeroy, Sarah B. 1975. *Goddesses, Whores, Wives and Slaves. Women in Classical Antiquity.* New York: Schocken.

Popenoe, David. 1993. "American Family Decline, 1960–1990." *Journal of Marriage and the Family* 55: 527–44.

Power, Eileen. 1975. *Medieval Women.* Cambridge: Cambridge University Press.

Pratt, W. F., W. D. Mosher, C. A. Bachrach, and M. C. Horn. 1984. "Understanding U.S. Fertility: Findings from the National Survey of Family Growth, Cycle III." *Population Bulletin* 39, no. 5

Presser, Harriet B. 1986. "Shift Work among American Women and Child Care." *Journal of Marriage and the Family* 48: 551–62.

———. 1988. "Shift Work and Child Care among Young Dual-Earner American Parents." *Journal of Marriage and the Family* 50: 133–48.

Preston, S. 1984. "Children and the Elderly: Divergent Paths for America's Dependents." *Demography* 21: 435–57.

Preston, Samuel H., and J. McDonald. 1979. "The Incidence of Divorce within Cohorts of American Marriages Contracted since the Civil War." *Demography* 16: 1–25.

Pruett, Kyle. 1983. "Infants of Primary Nurturing Fathers." *The Psychoanalytic Study of the Child.* 38: 257–77.

———. 1987. *The Nurturing Father.* New York: Warner.

Ptacek, James. 1988. "The Clinical Literature on Men Who Batter: A Review and Critique." In Gerald T. Hotaling, David Finkelhor, John T. Kirkpatrick, and Murray A. Straus (eds.), *Family Abuse and Its Consequences.* Newbury Park, Calif.: Sage.

Queen, Stuart A., and Robert W. Habenstein. 1967. *The Family in Various Cultures.* Philadelphia, Pa.: Lippincott.

Radin, Norma. 1982. "Primary Caregiving and Role-Sharing Fathers." In M. Lamb (ed.), *Nontraditional Families: Parenting and Child Development.* Hillsdale, N.J.: Erlbaum.

Radin, Norma, and R. Goldsmith. 1983. "Predictors of Father Involvement in Child Care." Paper presented at the meeting of the Society for Research in Child Development, Detroit, Mich.

Radin, Norma, and Graeme Russell. 1983. "Increased Father Participation and Child Development Outcomes." In Michael Lamb and Abraham Sagi (eds.), *Fatherhood and Family Policy.* Hillsdale, N.J.: Erlbaum.

Rainwater, Lee. 1964. "Marital Sexuality in Four Cultures of Poverty." *Journal of Marriage and the Family* 26: 457–66.

———. 1966. "Some Aspects of Lower-Class Sexual Behavior." *Journal of Social Issues* 22: 96–108.

Rainwater, Lee, R. P. Coleman, and G. Handel. 1962. *Workingman's Wife.* New York: Macfadden.

Ralls, Katherine. 1976. "Mammals in Which Females Are Larger than Males." *Quarterly Review of Biology* 51: 245–76.

Ramirez, Oscar, and Carlos H. Arce. 1981. "The Contemporary Chicano Family: An Empirically Based Review." In Augustine Baron, Jr. (ed.), *Explorations in Chicano Psychology.* New York: Praeger.

Ransford, H. Edward, and John McDonald. 1979. "The Incidence of Divorce within Cohorts of American Marriages Contracted since the Civil War." *Demography* 16: 1–25.

Ransford, H. Edward, and Jon Miller. 1983. "Race, Sex, and Feminist Outlooks." *American Sociological Review* 48: 46–59.

Rapoport, R. 1963. "Normal Crises, Family Structure, and Mental Health." *Family Process* 2: 68–80; 312–27.

Rapoport, Rhona, and Robert Rapoport. 1971. *Dual Career Families.* Baltimore, Md.: Penguin.

Rapoport, Robert, and Rhona Rapoport. 1976. *Dual Career Families Re-examined: New Integrations of Work and Family.* London: Martin Robertson.

Rapp, Rayna. 1982. "Family and Class in Contemporary America: Notes Toward an Understanding of Ideology." In Barrie Thorne and Marilyn Yalom (eds.), *Rethinking the Family.* New York: Longman.

Raschke, Helen, and Vern Raschke. 1979. "Family Conflict and Children's Self-Concept: A Comparison of Intact and Single-Parent Families." *Journal of Marriage and the Family* 41: 367–74.

Red Horse, John. 1980. "Family Structure and Value Orientation in American-Indians." *Social Casework* 61: 462–67.

Reed, James. 1978. *From Private Vice to Public Virtue: The Birth Control Movement and American Society since 1830.* New York: Basic Books.

Reiss, Ira L. 1960. *Premarital Sexual Standards in America.* New York: Free Press.

———. 1967. *The Social Context of Premarital Sexual Permissiveness.* New York: Holt, Rinehart, and Winston.

Reskin, Barbara. 1984. *Sex Segregation in the Workplace.* Washington, D.C.: National Academy Press.

Reynolds, Vernon. 1967. *The Apes.* New York: Dutton.

Richardson, Laurel, and Verta Taylor. 1989. *Feminist Frontiers II.* New York: Random House.

Riesman, David. 1950. *The Lonely Crowd.* New Haven, Conn.: Yale University Press.

Riley, Dave. 1990. "Network Influences on Father Involvement in Childrearing." In Cochran, Larner, Riley, Gunnarsson, Henderson, and Cross (eds.), *Extending Families.* Cambridge: Cambridge University Press.

Rindfuss, Ronald R., and Larry L. Bumpass. 1977. "Fertility during Marital Disruption." *Journal of Marriage and the Family* 39: 517–28.

Rindfuss, Ronald R., S. Philip Morgan, and C. Gray Swicegood. 1988. *First Births in America.* Berkeley: University of California Press.

Risman, Barbara. 1989. "Can Men 'Mother'? Life as a Single Father." In Barbara J. Risman, and Pepper Schwartz, *Gender in Intimate Relationships.* Belmont, Calif.: Wadsworth.

Roberts, Elizabeth. 1986. *A Woman's Place. An Oral History of Working-Class Women, 1890–1940.* New York: Blackwell.

Robinson, John. 1977. *How Americans Use Time.* New York: Praeger.

———. 1988. "Who's Doing the Housework?" *American Demographics* 10: 24–28, 63.

Rohner, Ronald. 1975. *They Love Me, They Love Me Not.* Human Relations Area Files, Inc.

———. 1986. *The Warmth Dimension. Foundations of Parental Acceptance-Rejection Theory.* Beverly Hills, Calif.: Sage.

Rohner, Ronald, and E. Rohner. 1982. "Enculturative Continuity and the Importance of Caretakers." *Behavior Science Research* 17: 91–114.

Rollins, Boyd C., and Kenneth L. Cannon. 1974. "Marital Satisfaction over the Family Life Cycle." *Journal of Marriage and the Family* 36: 271–84.

Rosen, Ellen I. 1987. *Bitter Choices: Blue-Collar Women In and Out of Work.* Chicago, Ill.: University Press of Chicago.

Rosenbaum, James. 1980. "Track Misconceptions and

Frustrated College Plans." *Sociology of Education* 53: 74–88.

Rosenberg, Morris. 1965. *Society and Adolescent Self-Image.* Princeton, N.J.: Princeton University Press.

———. 1981. "The Self-Concept: Social Product and Social Force." In Morris Rosenberg and Ralph H. Turner (eds.), *Social Psychology.* New York: Basic Books.

Rosenberg, M., and E. Gollub. 1992. *American Journal of Public Health* 82: 1473–78.

Rosenberg, Rosalind. 1982. *Beyond Separate Spheres. Intellectual Roots of Modern Feminism.* New Haven, Conn.: Yale University Press.

Rosenthal, C. J. 1985. "Kin-Keeping in the Familial Division of Labor." *Journal of the Marriage and the Family* 45: 509–21.

Rosenthal, K. M., and H. F. Keshet. 1981. *Fathers Without Partners: A Study of Fathers and the Family After Marital Separation.* Totowa, N.J.: Rowman and Littlefield.

Ross, Catherine E. 1987. "The Division of Labor at Home." *Social Forces* 65: 816–33.

Ross, Catherine E., and J. Huber. 1985. "Hardship and Depression." *Journal of Health and Social Behavior* 26: 312–27.

Ross, Catherine E., and John Mirwosky. 1987. "Children, Child Care, and Parents' Psychological Well-Being." Paper presented at the annual meeting of the American Sociological Association, New York.

Ross, Catherine E., John Mirowsky, and Joan Huber. 1983. "Dividing Work, Sharing Work, and In-Between: Marriage Patterns and Depression." *American Sociological Review* 48: 809–23.

Rossi, Alice S. 1968. "Transition to Parenthood." *Journal of Marriage and the Family* 30: 26–39.

———. 1977. "A Biosocial Perspective on Parenting." *Daedalus* 106: 1–31.

———. 1984. "Gender and Parenthood." *American Sociological Review* 49: 1–19.

Rothman, Barbara K. 1986. *The Tentative Pregnancy.* New York: Norton.

Rougemont, Denis de. 1956. *Love in the Western World.* New York: Pantheon.

Rowell, Thelma E. 1967. "Variability in the Social Organization of Primates." In Desmond Morris (ed.), *Primate Ethology.* Chicago, Ill.: Aldine.

Rubel, A. 1966. *Across the Tracks: Mexican-Americans in a Texas City.* Austin: University of Texas Press.

Rubin, Gayle. 1975. "The Traffic in Women: Notes on the 'Political Economy' of Sex." In Rayna Reiter (ed.), *Toward an Anthropology of Women.* New York: Monthly Review Press.

Rubin, Lillian. 1976. *World of Pain. Life in the Working-Class Family.* New York: Basic Books.

———. 1979. *Women of a Certain Age: The Midlife Search for Self.* New York: Harper and Row.

———. 1983. *Intimate Strangers: Men and Women Together.* New York: Harper and Row.

Rubin, Zick. 1970. "Measurement of Romantic Love." *Journal of Personality and Social Psychology* 16(2): 265–73.

———. 1973. *Loving and Liking.* New York: Holt, Rinehart, and Winston.

Russell, Diana. 1982. *Rape in Marriage.* New York: Macmillan.

———. 1984. *Sexual Exploitation.* Newbury Park, Calif.: Sage.

———. 1986. *The Secret Trauma: Incest in the Lives of Girls and Women.* New York: Basic Books.

Russell, Graeme. 1982. "Shared Caregiving Families; An Australian Study." In M. Lamb (ed.), *Nontraditional Families: Parenting and Child Development.* Hillsdale, N.J.: Erlbaum.

———. 1983. *The Changing Role of Fathers?* St. Lucia: University of Queensland Press.

Russell, Graeme, and Norma Radin. 1983. "Increased Paternal Participation: The Father's Perspective." In M. Lamb and A. Sagi (eds.), *Fatherhood and Family Policy.* Hillsdale, N.J.: Erlbaum.

Ryan, Mary P. 1981. *Cradle of the Middle Class: The Family in Oneida County, New York, 1790–1865.* New York: Cambridge University Press.

Ryder, R. G. 1970. "A Topography of Early Marriage." *Family Process* 9: 385–402.

Sacks, Karen. 1979. *Sisters and Wives. The Past and Future of Sexual Equality.* Westport, Conn.: Greenwood.

Safilios-Rothschild, Constantina. 1969. "Family Sociology or Wives' Family Sociology?" *Journal of Marriage and the Family* 31: 290–301.

———. 1970. "The Study of Family Power Structure: A Review 1960–1969." *Journal of Marriage and the Family* 32: 539–52.

———. 1977. *Love, Sex, and Sex Roles.* Englewood Cliffs, N.J.: Prentice-Hall.

Sanday, Peggy Reeves. 1981. *Female Power and Male Dominance: On the Origins of Sexual Inequality.* Cambridge, Eng.: Cambridge University Press.

Sanford, Linda. 1980. *The Silent Children: A Parent's Guide to the Prevention of Child Sexual Abuse.* New York: Doubleday.

Sattel, Jack. 1976. "Men, Inexpressiveness, and Power." *Social Problems* 23: 469–77.

Sauer, William. 1975. "Morale of the Urban Aged." Unpublished Ph.D. dissertation, University of Minnesota.

Saunders, Daniel G. 1988. "Wife Abuse, Husband Abuse, or Mutual Combat?" In Kersti Yllo and Michele Bograd (eds.), *Feminist Perspectives on Wife Abuse.* Newbury Park, Calif.: Sage.

Savage, David G. 1990. "1 in 4 Young Blacks in Jail or in Court Control, Study Says." *Los Angeles Times,* Feb. 27.

Scanzoni, John H. 1970. *Opportunity and the Family.* New York: Free Press.

———. 1975. *Sex Roles, Life Styles, and Childbearing: Changing Patterns in Marriage and Family.* New York: Free Press.

———. 1977. *The Black Family in Modern Society.* Chicago, Ill.: University of Chicago Press.

———. 1979. "Social Processes and Power in Families." In W. R. Burr, R. Hill, F. I. Nye, and I. L. Reiss (eds.), *Contemporary Theories about the Family,* vol. 1. New York: Free Press.

Scanzoni, John H., and Maximiliane Szinovacz. 1980. *Family Decision Making: A Developmental Sex Role Model.* Beverly Hills, Calif.: Sage Publications.

Scanzoni, Letha D., and John Scanzoni. 1988. *Men, Women, and Change.* New York: McGraw-Hill.

Scarr, S., and R. A. Weinberg. 1978. "The Influence of Family Background on Intellectual Attainment." *American Sociological Review* 43: 674–92.

Schacter, Jim. 1989. "The Daddy Track." *Los Angeles Times Magazine,* Oct. 1: 7–16.

Schaller, George B. 1963. *The Mountain Gorilla.* Chicago, Ill.: University of Chicago Press.

Schechter, Susan. 1982. *Women and Male Violence: The Visions and Struggles of the Battered Women's Movement.* Boston, Mass.: South End.

———. 1988. "Building Bridges Between Activists, Professionals, and Researchers." In Kersti Yllo and Michele Bograd (eds.), *Feminist Perspectives on Wife Abuse.* Newbury Park, Calif.: Sage.

Schlegel, Alice. 1972. *Male Dominance and Female Autonomy: Domestic Authority in Matrilineal Societies.* New Haven, Conn.: HRAF Press.

Schooler, Carmi. 1983. *Work and Personality.* Norwood, N.J.: Ablex.

Schooler, Carmi, Joanne Miller, Karen A. Miller, and Carol N. Richtand. 1984. "Work for the Household: Its Nature and Consequences for Husbands and Wives." *American Journal of Sociology* 90: 97–124.

Schwartz, Joe, and Thomas Exter. 1989. "All Our Children." *American Demographics* 5 (May): 34–37.

Scott, Janny. 1990. "U.S. Slips Badly in Infant Mortality Fight, Panel Says." *Los Angeles Times,* Mar. 1.

Scott, Joan W., and Louise A. Tilly. 1975. "Women's Work and the Family in Nineteenth-Century Europe." *Comparative Studies in Society and History* 17: 36–64.

Searle, Eleanor. 1988. *Predatory Kinship and the Creation of the Norman State, 840–1066.* Berkeley: University of California Press.

Sears, Robert R., Eleanor Maccoby, and Harry Levin. 1957. *Patterns of Child Rearing.* Evanston, Ill.: Row, Peterson.

Segura, Denise. 1984. "Labor Market Stratification: The Chicana Experience." *Berkeley Journal of Sociology* 29: 57–91.

Sennett, Richard, and Jonathan Cobb. 1973. *The Hidden Injuries of Class.* New York: Random House.

Shanas, E. 1979. "The Family as a Social Support System in Old Age." *Gerontologist* 19: 169–74.

———. 1980. "Older People and Their Families: The New Pioneers." *Journal of Marriage and the Family* 42: 9–15.

Shanas, Ethel et al. 1973. "Family-Kin Networks and Aging in Cross-Cultural Perspective." *Journal of Marriage and the Family* 35: 505–11.

Shaw, Susan M. 1988. "Gender Differences in the Definition and Perception of Household Labor." *Family Relations* 37: 333–37.

Shehan, Constance L., E. W. Bock, and Gary R. Lee. 1990. "Religious Heterogamy, Religiosity, and Marital Happiness: The Case of Catholics." *Journal of Marriage and the Family* 52: 73–79.

Shelton, Beth A. 1992. *Women, Men, and Time.* New York: Greenwood Press.

Sherman, L. W., and R. A. Berk. 1984. "Deterrant Effects of Arrest for Domestic Violence." *American Sociological Review* 49: 261–72.

Shields, Nancy M., and Christine R. Hanneke. 1983. "Wives' Reactions to Marital Rape." In David Finkelhor, Richard J. Gelles, Gerald T. Hotaling, and Murray A. Straus (eds.), *The Dark Side of Families.* Beverly Hills, Calif.: Sage.

Shorter, Edward. 1975. *The Making of the Modern Family.* New York: Basic Books.

Signorelli, Nancy. 1991. *A Sourcebook on Children and Television.* New York: Greenwood Press.

Silka, Linda, and Sara Kiesler. 1977. "Couples Who Choose to Remain Childless." *Family Planning Perspectives* 9 (Jan./Feb.): 16–25.

Simmons, Roberta G., and Dale A. Blyth. 1987. *Moving Into Adolescence. The Impact of Pubertal Change and School Context.* New York: Aldine de Gruyter.

Sinclair, Andrew. 1965. *The Emancipation of the American Woman*. New York: Harper and Row.

Skinner, B. F. 1969. *Contingencies of Reinforcement*. New York: Appleton.

Skolnick, Arlene S. 1987. *The Intimate Environment: Exploring Marriage and the Family*. Boston, Mass.: Little, Brown.

Skolnick, Arlene S., and Jerome H. Skolnick. 1987. *Family in Transition*. 6th ed. Glenview, Ill.: Scott, Foresman.

Slater, Philip E. 1963. "On Social Regression." *American Journal of Sociology* 28: 339–64.

Smith, Daniel S., and Michael S. Hindus. 1975. "Premarital Pregnancy in America, 1640–1971." *Journal of Interdisciplinary History* 4: 537–70.

Smith, Tom W. 1990. "Adult Sexual Behavior in 1989." Paper presented to American Association for the Advancement of Science.

———. 1991. "Adult Sexual Behavior in 1989: Number of Partners, Frequency of Intercourse and Risk of AIDS." *Family Planning Perspectives* 23: 102–7.

Snipp, C. Matthew. 1989. *American Indians: First of This Land*. New York: Russell Sage Foundation.

Snodgrass, Anthony. 1980. *Archaic Greece*. Berkeley: University of California Press.

Sonenstein, Freya L., Joseph Pleck, and Leighton Ku. 1989. "Sexual Activity, Condom Use, and AIDS Awareness Among Adolescent Males." *Family Planning Perspectives* 23: 102–7.

———. 1991. "Levels of Sexual Activity, Condom Use, and AIDS Awareness among Adolescent Males in the United States." *Family Planning Perspectives* 21: 152–58.

Sorel, Nancy C. 1985. *Ever Since Eve*. London: Michael Joseph.

Sorensen, Robert. 1973. Adolescent Sexuality in Contemporary America. New York: World.

Sorrentino, Constance. 1990. "The Changing Family in International Perspective." *Monthly Labor Review* 113: 41–58.

Spanier, Graham B. 1976. "Measuring Dyadic Adjustment: New Scales for Assessing the Quality of Marriage and Similar Dyads." *Journal of Marriage and the Family* 38(1): 15–28.

Spanier, Graham B., and Paul C. Glick. 1981. "Marital Instability in the United States: Some Correlates and Recent Changes." *Family Relations* 30(3): 329–38.

———. 1986. "Mate Selection Differentials Between Whites and Blacks in the United States." *Social Forces* 58(3): 707–25.

Spanier, Graham B., and Robert A. Lewis. 1980. "Marital Quality: A Review of the Seventies." *Journal of Marriage and the Family* 42(4): 825–38.

Spark, Richard F., Robert White, and Peter Connolly. 1980. "Impotence Is Not Always Psychogenic." *Journal of the American Medical Association* 243: 1,558.

Spector, Malcolm, and John I. Kituse. 1973. "Social Problems: A Re-Formulation." *Social Problems* 21(2): 145–58.

Spiro, Melford E., 1956. *Kibbutz: Venture in Utopia*. Cambridge, Mass.: Harvard University Press.

Spitz, Rene. 1945. "Hospitalism: An Inquiry into the Genesis of Psychiatric Conditions in Early Childood." *Psychoanalytic Studies of the Child* 1: 53–74.

Spitze, Glenna. 1988. "Women's Employment and Family Relations: A Review." *Journal of Marriage and the Family* 50: 595–618.

———. 1991. "Women's Employment and Family Relations." Pp. 381–404 in Alan Booth (ed.), *Contemporary Families: Looking Forward, Looking Back*. Minneapolis, MN: National Council on Family Relations.

Sprey, Jetse, and Sarah H. Matthews. 1982. "Contemporary Grandparenthood: A Systematic Transition." *Annals of the American Academy of Political and Social Sciences* 464: 91–103.

———. 1989. "The Perils of Drawing Policy Implications from Research." In Rachel Filinson and Stanley R. Ingman (eds.), *Elder Abuse: Practice and Policy*. New York: Human Sciences Press.

Stacey, Judith. 1993. "Good Riddance to 'The Family': A Response to David Popenoe." *Journal of Marriage and the Family* 55: 545–47.

Stack, Carol. 1974. *All Our Kin*. New York: Harper and Row.

Staines, Graham L., and Pam L. Libby. 1986. "Men and Women in Role Relationships." In Richard D. Ashmore and Frances K. DelBoca (eds.), *The Social Psychology of Female-Male Relations: A Critical Analysis of Central Concepts*. New York: Academic Press.

Staples, Robert. 1971. "Towards a Sociology of the Black Family: A Theoretical and Methodological Assessment." *Journal of Marriage and the Family* 33: 119–35.

———. 1978. "Race, Liberalism-Conservatism, and Premarital Sexual Permissiveness: A Biracial Comparison." *Journal of Marriage and the Family* 40: 733–42.

———. 1985. "Changes in Black Family Structure: The Conflict between Family Ideology and Structural Conditions." *Journal of Marriage and the Family* 47: 1005–13.

Staples, Robert, and Mirandé, Alfredo. 1980. "Racial and Cultural Variations among American Families: A Decennial Review of the Literature on Minority Families." *Journal of Marriage and the Family* 33: 119–35.

Stark, Rodney, and James McEvoy. 1970. "Middle Class Violence." *Psychology Today* 4: 52–54; 110–12.

Starr, Paul. 1982. *The Social Transformation of American Medicine.* New York: Basic Books.

Statistical Abstract of the.United States 1971, 1981, 1984, 1986, 1987, 1988, 1989, 1990, 1991, 1992, 1993. Washington, D.C.: U.S. Government Printing Office.

Staude, John. 1982. *The Adult Development of C. G. Jung.* London: Routledge and Kegan Paul.

Stebbens, G. L., and F. J. Alaya. 1981. "Is a New Evolutionary Synthesis Necessary?" *Science* 213: 967–71.

Steinmetz, Suzanne K. 1977. *The Cycle of Violence: Assertive, Aggressive, and Abusive Family Interaction.* New York: Praeger.

————. 1978. "Violence between Family Members." *Marriage and Family Review* 1 (May): 1–16.

Steinmetz, Suzanne, Sylvia Clavan, and Karen F. Stein. 1990. *Marriage and Family Realities.* Grand Rapids, Penn.: Harper and Row.

Steinmetz, Suzanne K., and Murray A. Straus. 1975. *Violence in the Family.* New York: Dodd, Mead.

Stendhal. 1967 (1826–42). *On Love.* New York: Grosset and Dunlap.

Stets, Jan E. 1988. *Domestic Violence and Control.* New York: Springer-Verlag.

Stevens-Long, Judith. 1984. *Adult Life: Developmental Processes.* Palo Alto, Calif.: Mayfield.

Stone, Lawrence. 1977. *The Family, Sex and Marriage in England, 1500–1800.* New York: Harper and Row.

Straus, Murray A. 1979. "Measuring Intrafamily Conflict and Violence: The Conflict Tactics (CT) Scales." *Journal of Marriage and the Family* 41(Feb.): 75–88

————. 1980. "Victims and Aggressors in Marital Violence." *American Behavioral Scientist* 23(May/Jun.): 681–704.

————. 1983. "Ordinary Violence, Child Abuse, and Wife-Beating: What Do They Have in Common?" In David Finkelhor, Richard J. Gelles, Gerald T. Hotaling, and Murray A. Straus (eds.), *The Dark Side of Families. Current Family Violence Research.* Beverly Hills, Calif.: Sage.

————. 1990. "Ordinary Violence, Child Abuse, and Wife Beating: What Do They Have in Common?" In Murray A. Straus and Richard J. Gelles (eds.), *Physical Violence in American Families: Risk Factors and Adaptations to Violence in 8,145 Families.* New Brunswick, N.J.: Transaction.

————. 1991. "Physical Violence in American Families: Incidence, Rates, Causes, and Trends." Pp. 17–34 in Dean Knudsen and JoAnn Miller (eds.), *Abused and Battered.* New York: Aldine de Gruyter.

Straus, Murray A., and Richard J. Gelles. 1988. "How Violent Are American Families? Estimates from the National Family Violence Resurvey and Other Studies." In Gerald T. Hotaling, David Finkelhor, John T. Kirkpatrick, and Murray A. Straus (eds.), *Family Abuse and its Consequences.* Newbury Park, Calif.: Sage.

Straus, Murray A., Richard J. Gelles, and Suzanne K. Steinmetz. 1980. *Behind Closed Doors: Violence in the American Family.* New York: Doubleday.

Stroebe, Wolfgang, Chester A. Insko, Vaida D. Thompson, and Bruce Layton. 1971. "Effects of Physical Attractiveness, Attitude Similarity, and Sex on Various Aspects of Interpersonal Attraction." *Journal of Personality and Social Psychology* 18: 79–91.

Stuckey, M. Francine, Paul E. McGhee, and Nancy J. Bell. 1982. "Parent-Child Interaction: The Influence of Maternal Employment." *Developmental Psychology* 18: 635–44.

Sudnow, David. 1967. *Passing On: The Social Organization of Dying.* Englewood Cliffs, N.J.: Prentice-Hall.

Sue, S., and Harry Kitano. 1973. "Asian American Stereotypes." *Journal of Social Issues* 29(Spring): 83–98.

Sullivan, Deborah A., and Rose Weitz. 1988. *Labor Pains.* New Haven, Conn.: Yale University Press.

Surra, Catherine A., 1991. "Research and Theory on Mate Selection and Premarital Relationships in the 1980s." Pp. 54–75 in Alan Booth (ed.), *Contemporary Families: Looking Forward, Looking Back.* Minneapolis, Minn.: National Council on Family Relations.

Sussman, George. 1982. *Selling Mothers' Milk: The Wet Nursing Business in France, 1715–1914.* Urbana: University of Illinois Press.

Sussman, M. B. 1985. "The Family Life of Old People." In R. H. Binstock and E. Shanas (eds.), *Handbook of Aging and the Social Sciences,* 2d ed., New York: Van Nostrand Reinhold.

Swedish Statistical Abstract. 1975. Stockholm: Government Statistical Office.

Sweet, James A., and Larry L. Bumpass. 1987. *American Families and Households.* New York: Russell Sage Foundation.

Sweet, James A., Larry L. Bumpass, and Vaughn Call. 1988. "The Design and Content of the National Survey of Families and Households." NSFH Working Paper No. 1, Center for Demography and Ecology, University of Wisconsin, Madison.

Sykes, Gresham. 1980. *The Future of Crime*. Washington, D.C.: U.S. Government Printing Office.

Synnott, Anthony. 1983. "Little Angels, Little Devils: A Sociology of Children." *Canadian Review of Sociology and Anthropology* 20: 79–95.

Szinovacz, Maximiliane. 1987. "Family Power." In Marvin B. Sussman and Susan K. Steinmetz (eds.), *Handbook of Marriage and the Family*. New York: Plenum.

Tamir, Lois M. 1982. *Men in Their Forties: The Transition to Middle Age*. New York: Springer.

Tatara, Toshio. 1989. "Toward the Development of Estimates of the National Incidence of Reports of Elder Abuse Based on Currently Available State Data." In Rachel Filinson and Stanley R. Ingman (eds.), *Elder Abuse: Practice and Policy*. New York: Human Sciences Press.

Tavris, Carol, and Susan Sadd. 1977. *The Red Book Report on Female Sexuality*. New York: Delacorte.

Taylor, Howard F. 1980. *The IQ Game*. New Brunswick, N.J.: Rutgers University Press.

Taylor, Patricia A., and Norval D. Glenn. 1976. "The Utility of Education and Attractiveness for Female Status Attainment Through Marriage." *American Sociological Review* 41: 484–97.

Taylor, Robert J., Linda M. Chatters, M. Belinda Tucker, and Edith Lewis. 1991. "Developments in Research on Black Families: A Decade Review." Pp. 275–96 in Alan Booth (ed.), *Contemporary Families: Looking Forward, Looking Back*. Minneapolis, Minn.: National Council on Family Relations.

Taylor, Ronald L. 1994. "Minority Families and Social Change" In *Minority Families in the United States: A Multicultral Perspective*. Englewood Cliffs, N.J.: Prentice-Hall.

Teachman, J., K. Polonko, and J. Scanzoni. 1987. "Demography of the Family." In M. Sussman and S. Steinmetz (eds.), *Handbook of Marriage and the Family*. New York: Plenum.

Terman, Lewis M. 1938. *Psychological Factors in Marital Happiness*. New York: McGraw-Hill.

Thomas, Elizabeth M. 1958. *The Harmless People*. New York: Knopf.

Thomas, John L. 1956. *The American Catholic Family*. Englewood Cliffs, N.J.: Prentice-Hall.

Thomas, Lawrence. 1956. *The Occupational Structure and Education*. Englewood Cliffs, N.J.: Prentice-Hall.

Thompson, Linda, and Alexis J. Walker. 1989. "Gender in Families." *Journal of Marriage and the Family* 51: 845–71.

Thornton, Arland. 1989. "Changing Attitudes toward Family Issues in the United States." *Journal of Marriage and the Family* 51: 873–93.

Thwing, C. F., and C. F. B. Thwing. 1887. *The Family: an Historical and Social Study*. Boston, Mass.: Lee and Shepard.

Tietze, Christopher. 1978. "Teenage Pregnancies: Looking Ahead to 1984." *Family Planning Perspective* 10: 205–7.

Tilly, Louise A. 1981. "Women's Collective Action and Feminism in France, 1870–1914." In Louise A. Tilly and Charles Tilly (eds.), *Class Conflict and Collective Action*. Beverly Hills, Calif.: Sage.

Tobias, Sheila. 1973. "What Really Happened to Rosie the Riveter? Demobilization and the Female Labor Force, 1944–47." New York: MSS Modular Publication No. 9.

Torrance, E. Paul. 1962. "Cultural Discontinuities and the Development of Originality in Thinking." *Exceptional Child* 29: 2–13.

Travers, Jeffery, Barbara D. Goodson, Judith D. Singer, and David B. Connell. 1981. *Research Results of the National Day Care Study*. Cambridge, Mass.: Abt Books.

Treas, J., and V. L. Bengtson. 1987. "Family in Later Years." In M. Sussman and S. Steinmetz (eds.), *Handbook on Marriage and the Family*. New York: Plenum.

Trost, Jan E. 1985. "Abandon Adjustment!" *Journal of Marriage and the Family* 47(4): 1072–73.

Trumbach, R. 1978. *The Rise of the Egalitarian Family: Aristocratic Kinship and Domestic Relations in Eighteenth Century England*. New York: Academic Press.

Trunkey, D. D. 1983. "Truama." *Scientific American* 249: 33.

Turner, Barbara F., and Castellano B. Turner. 1974. "The Political Implications of Social Stereotyping of Women and Men among Black and White College Students." *Sociology and Social Research* 58: 155–62.

Turner, Victor. 1977. *The Ritual Process*. Ithaca, N.Y.: Cornell University Press.

Tyree, Andrea, and Judith Treas. 1974. "The Occupational and Marital Mobility of Women." *American Sociological Review* 39: 293–307.

Udry, J. R. 1977. "The Importance of Being Beautiful: A Reexamination and Racial Comparison." *American Journal of Sociology* 83: 154–60.

Udry, J. R., and B. K. Eckland. 1983. "The Benefits of Being Attractive: Differential Payoffs for Men and Women." Unpublished ms., University of North Carolina at Chapel Hill.

Uhlenberg, Peter. 1980. "Death and the Family." *Journal of Family History* 5(3): 313–20.

U.S. Bureau of the Census. 1976a. "Number, Timing, and Duration of Marriages and Divorces in the

U.S." *Current Population Reports* Series P-20, no. 297 (Oct.). Washington, D.C.: Government Printing Office.

U.S. Bureau of the Census. 1976b. "Demographic Aspects of Aging and the Older Population in the U.S." *Current Population Reports,* Series P-23, no. 59. Washington, D.C.: Government Printing Office.

U.S. Bureau of the Census. 1978a. "Perspectives on American Fertility." *Current Population Reports* (July) Series P-23, no. 70. Washington, D.C.: Government Printing Office.

U.S. Bureau of the Census. 1978b. "Perspectives on U.S. Husbands and Wives." *Current Population Reports,* Series P-23, no. 77. Washington, D.C.: Government Printing Office.

U.S. Bureau of the Census. 1978c. "Fertility of American Women: June 1977." *Current Population Reports Series* P-20, no. 325. Washington, D.C.: Government Printing Office.

U.S. Bureau of the Census. 1985a. "Persons of Spanish Origin in the United States: March 1985." *Current Population Reports,* Series P-20, no. 403. Washington, D.C.: Government Printing Office.

U.S. Bureau of the Census. 1985b. *1980 Census of Population.* Washington, D.C.: Government Printing Office.

U.S. Bureau of the Census. 1989. "Studies in Marriage and the Family." *Current Population Reports* P-23, no. 165. See also Series P-20, no. 436. Washington, D.C.: U.S. Government Printing Office.

U.S. Bureau of the Census. 1990. "Work and Family Patterns of American Women." *Current Population Reports* P-23, no. 165. Washington, D.C.: U.S. Government Printing Office.

U.S. Bureau of the Census. 1992a. "Population Projections of the United States, by Age, Sex, Race, and Hispanic Origin." *Current Population Reports* P-25, no. 1092. Washington, D.C.: U.S. Government Printing Office.

U.S. Bureau of the Census. 1992b. "Studies in the Distribution of Income." *Current Population Reports,* P-60, no. 183. Washington, D.C.: U.S. Government Printing Office.

U.S. Bureau of the Census. 1992c. "Workers with Low Earnings, 1964–1990." *Current Population Reports* P-60, no. 178. Washington, D.C.: U.S. Government Printing Office.

U.S. Bureau of the Census. 1992d. "Income, Poverty, and Wealth in the United States." *Current Population Reports* P-60, no. 179. Washington, D.C.: U.S. Government Printing Office.

U.S. Bureau of the Census. 1992e. "Marriage, Divorce, and Remarriage in the 1990s." *Current Population Reports* P-23, no. 180. Washington, D.C.: U.S. Government Printing Office.

U.S. Bureau of the Census. 1992f. "Marital Status and Living Arrangements." *Current Population Reports* P-20, no. 468. Washington, D.C.: U.S. Government Printing Office.

U.S. Bureau of the Census. 1992g. "Demographic State of the Nation." *Current Population Reports* P-23, no. 177. Washington, D.C.: U.S. Government Printing Office.

U.S. Bureau of the Census. 1992h. "Special Studies." *Current Population Reports.* P-23, no. 181. Washington, D.C.: U.S. Government Printing Office.

U.S. Bureau of Labor Statistics. 1992. "Employment and Earnings." Vol. 39, no. 7. Washington, D.C.: U.S. Government Printing Office.

U.S. Congress, House of Representatives. 1990. *Consumer Protection Issues Involving In-Vitro Fertilization Clinics.* Hearings before the Subcommittee on Regulation, Business Opportunities, and Energy of the Committee on Small Business. 101st Congress (March 9, 1989). Washington D.C.: U.S. Government Printing Office.

Valdivieso, Rafael, and Cary Davis. 1988. "U.S. Hispanics: Changing Issues for the 1990s." *Population Trends and Public Policy,* no. 17. Washington, D.C.: Population Reference Bureau.

Valliant, G. 1977. Adaptation to Life. Boston, Mass.: Little, Brown.

Van den Berghe, Pierre L. 1973. *Age and Sex in Human Societies: A Biosocial Perspective.* Belmont, Calif.: Wadsworth.

Vance, Carole S. 1989. *Pleasure and Danger: Exploring Female Sexuality.* London: Routledge.

Vanek, Joann. 1974. "Time Spent in Housework." *Scientific American* 231 (Nov.): 116–20.

Vaughn, Barbara, James Trussel, Jane Menken, and Elise F. Jones. 1977. "Contraceptive Failure among Married Women in the U.S., 1970–73." *Family Planning Perspectives* 9 (Nov./Dec.): 251–58.

Veevers, Jean E. 1980. *Childless by Choice.* Toronto: Butterworths.

Vega, William A. 1991. "Hispanic Families in the 1980s." Pp. 297–306 in Alan Booth (ed.), *Contemporary Families: Looking Forward, Looking Back.* Minneapolis, Minn.: National Council on Family Relations.

Verbrugge, L. M. 1979. "Marital Status and Health." *Journal of Marriage and the Family* 41: 267–85.

Vieille, Paul. 1978. "Iranian Women in Family Alliance and Sexual Politics." In Lois Beck and Nikki Keddie (eds.), *Women in the Moslem World.* Cambridge, Mass.: Harvard University Press.

Voydanoff, Patricia. 1991. "Economic Distress and Family

Relations: A Review of the Eighties." Pp. 429–45 in Alan Booth (ed.), *Contemporary Families: Looking Forward, Looking Back*. Minneapolis, Minn.: National Council on Family Relations.

Walker, Henry A. 1986. "Racial Differences in Patterns of Marriage and Family Maintenance: 1890–1980." In Sanford M. Dornbusch, and Myra H. Strober (eds.), *Feminism, Children and the New Families*. New York: Guilford Press.

Walker, L. E. 1979. *The Battered Woman*. New York: Harper and Row.

——. 1984. *The Battered Woman Syndrome*. New York: Springer.

Wallerstein, Judith S. 1983. "Children of Divorce: Stress and Development Tasks." In N. Garmezy and M. Rutter (eds.), *Stress, Coping and Development in Children*. New York: McGraw-Hill.

Wallerstein, Judith S., and Sandra Blakeslee. 1989. *Second Chances: Men, Women, and Children. A Decade after Divorce*. New York: Trenor and Fields.

Wallerstein, Judith S., and Joan Kelly. 1980. *Surviving the Breakup: How Children and Parents Cope with Divorce*. New York: Basic Books.

Wallin, Paul, and Alexander L. Clark. 1958. "Marital Satisfaction and Husbands' and Wives' Perception of Similarity in their Preferred Frequency of Coitus." *Journal of Abnormal and Social Psychology* 47: 370–73.

Walsh, Robert H. 1970. *A Survey of Parents and Their Own Children's Sexual Attitudes*. Ph.D. dissertation, University of Iowa.

Walster, Elaine. 1974. "Passionate Love." In Arlene Skolnick and Jerome H. Skolnick (eds.), *Intimacy, Family, and Society*. Boston, Mass.: Little, Brown.

Walster, Elaine, and G. William Walster. 1978. *A New Look at Love*. Reading, Mass.: Addison-Wesley.

Walster, Elaine, G. William Walster, and E. Berschied. 1978. *Equity, Theory and Research*. Rockleigh, N.J.: Allyn and Bacon.

Walters, Jennifer. 1983. "The Divorce Experience of Selected Upper-Income Males." Ph.D. dissertation, University of Southern California.

Warren, Carol A. B. 1987. *Madwives: Schizophrenic Women at Mid-Century*. New Brunswick, N.J.: Rutgers University Press.

Washburn, Sherwood L., and Irven S. Devore. 1961. "The Social Life of Baboons." *Scientific American* 204: 62–71.

Watkins, S. C., J. A. Menken, and J. Bongaarts. 1987. "Demographic Foundations of Family Change." *American Sociological Review* 52: 346–58.

Watt, Ian. 1957. *The Rise of the Novel*. Berkeley: University of California Press.

Weber, Max. 1922/68. *Economy and Society*. New York: Bedminister Press.

——. 1923/61. *General Economic History*. New York: Collier-Macmillan.

Weinberg, Martin S., and Colin J. Williams. 1980. "Sexual Embourgeoisment? Social Class and Sexual Activity: 1938–1970." *American Sociological Review* 45: 33–48.

Weinberg, Martin, Colin Williams, and Douglas Pryor. 1993. *Dual Attraction: Understanding Bisexuality*. New York: Oxford University Press.

Weisman, Avery. 1972. *On Dying and Denying*. New York: Behavioral Pubns.

Weiss, Robert S. 1979. "Growing Up a Little Faster: The Experience of Growing Up in a Single-Parent Household." *Journal of Social Issues* 35 (Fall): 97–111.

Weitzman, Lenore J. 1981. *The Marriage Contract. Spouses, Lovers and the Law*. New York: Free Press.

——. 1985. *The Divorce Revolution. The Unexpected Social and Economic Consequences for Women and Children in America*. New York: Free Press.

Weller, Robert H. 1968. "The Employment of Wives, Dominance, and Fertility." *Journal of Marriage and the Family* 30: 437–42.

Weller, Robert, and Frank B. Hobbs. 1978. "Unwanted and Mistimed Births in the U.S.: 1968–73." *Family Planning Perspectives* 10 (May/June): 168–72.

Welter, Barbara. 1966. "The Cult of True Womanhood: 1820–1860." *American Quarterly* 18: 151–74.

Wertz, Dorothy C. 1983. "What Birth Has Done for Doctors: A Historical View." *Women and Health* (Spring): 7–24.

Wertz, Richard W., and Dorothy C. Wertz. 1977. *Lying-In: A History of Childbirth in America*. New York: Free Press.

West, Candace, and Don H. Zimmerman. 1987. "Doing Gender." *Gender and Society* 1: 125–51.

Westoff, Charles F. 1974. "Coital Frequencies and Contraception." *Family Planning Perspectives* 3: 136–41.

——. 1986. "Fertility in the United States." *Science* 234: 554–59.

Westoff, Leslie. 1977. *The Second Time Around*. New York: Viking.

White, Lynn. 1991. "Determinants of Divorce: A Review of Research in the Eighties." Pp. 141–49 in Alan Booth (ed.), *Contemporary Families: Looking Forward, Looking Back*. Minneapolis, Minn.: National Council on Family Relations.

White, Lynn, and John N. Edwards. 1990. "Emptying the Nest and Parental Well-Being: An Analysis of National Panel Data," *American Sociological Review* 55 (Apr.): 235–42.

Whiteford, Linda M. 1989. "Commercial Surrogacy: Social Issues Behind the Controversy." In Linda Whiteford and Marilyn Polan (eds.), *New Approaches to Human Reproduction*. Boulder, Colo.: Westview.

Whitehurst, R. N. 1971. "Violence Potential in Extramarital Sexual Response." *Journal of Marriage and the Family* 33: 683–91.

Whiting, Beatrice. 1963. *Six Cultures: Studies in Child Rearing*. New York: Wiley.

Whiting, John, and Irving Child. 1953. *Child Training and Personality*. New Haven, Conn.: Yale University Press.

Whyte, William H. 1956. *The Organization Man*. New York: Doubleday.

Wilensky, Harold W. 1961. "The Uneven Distribution of Leisure." *Social Problems* 9: 33–44.

———. 1964. "Mass Society and Mass Culture." *American Sociological Review* 29: 173–97.

Wilkie, Jane R. 1987. "Marriage, Family Life, and Women's Employment." In Ann Helton Stromberg and Shirley Harkess (eds.), *Women Working*, 2nd ed. Mountain View, Calif.: Mayfield.

Wilkinson, Melvin. 1976. "Romantic Love: The Great Equalizer? Sexism in Popular Music." *The Family Coordinator* 25: 161–66.

Williams, Norma. 1988. "Role Making among Married Mexican American Women: Issues of Class and Ethnicity." *Journal of Applied Behavioral Science* 24(2): 203–17.

Willie, Charles V. 1981. *A New Look at Black Families*. Bayside, N.Y.: General Hall.

———. 1985. *Black and White Families*. Bayside, N.Y.: General Hall.

Wilson, E. O. 1971. *The Insect Societies*. Cambridge, Mass.: Harvard University Press.

———. 1975. *Sociobiology: The New Synthesis*. Cambridge, Mass.: Harvard University Press.

Wilson, William J. 1978. *The Declining Significance of Race*. Chicago, Ill.: University of Chicago Press.

———. 1987. *The Truly Disadvantaged: The Inner City, the Underclass, and Public Policy*. Chicago, Ill.: University of Chicago Press.

Wilson, William J., and Kathryn M. Neckerman. 1986. "Poverty and Family Structure: The Widening Gap Between Evidence and Public Policy Issues." In Sheldon H. Danzinger and Daniel H. Weinberg (eds.), *Fighting Poverty: What Works and What Doesn't*. Cambridge, Mass.: Harvard University Press.

Winch, Robert F. 1958. *Mate-Selection: A Study of Complementary Needs*. New York: Harper and Row.

———. 1967. "Another Look at the Theory of Comple-mentary Needs in Mate Selection." *Journal of Marriage and the Family* 29: 756–62.

Winick, Charles, and Paul M. Kinsie. 1971. *The Lively Commerce: Prostitution in the United States*. Chicago, Ill.: Aldine.

Wittgenstein, Ludwig. 1953. *Philosophical Investigations*. New York: Macmillan.

Wood, Viviana, and Joan F. Robertson. 1978. "Friendship and Kinship Interaction: Differential Effect on the Morale of the Elderly." *Journal of Marriage and the Family* 40: 367–75.

World Health Organization. "Appropriate Technology for Birth." *Lancet* 8452: 436–37.

Wright, Erik O. 1979. *Class Structure and Income Determination*. New York: Academic Press.

Wright, Erik O., and Joachim Singleman. 1982. "Proletarianization in the Changing American Class Structure." *American Journal of Sociology, Supplement: Marxist Inquiries*.

Wylie, Lawrence. 1964. *Village in the Vaucluse*. Cambridge, Mass.: Harvard University Press.

Ybarra, Lea. 1982. "When Wives Work: The Impact on the Chicano Family." *Journal of Marriage and the Family* (Feb.): 169–78.

Yellowbird, Michael, and C. Matthew Snipp. 1994. "American Indian Families." In *Minority Families in the United States: A Multicultural Perspective*. Englewood Cliffs, N.J.: Prentice-Hall.

Yllo, Kersti. 1988. "Political and Methodological Debates in Wife Abuse Research." In Kersti Yllo and Michele Bograd (eds.), *Feminist Perspectives on Wife Abuse*. Newbury Park, Calif.: Sage.

Young, Michael, and Peter Willmott. 1957. *Family and Kinship in East London*. London: Routledge and Kegan Paul.

Zablocki, Benjamin. 1980. *Alienation and Charisma. A Study of Contemporary American Communes*. New York: Free Press.

Zarsky, Lyuba, and Samuel Bowels (eds.). 1986. *Economic Report of the People*. Boston, Mass.: South End Press.

Zaslow, Martha. 1981. "Depressed Mood in New Fathers." Paper presented at the Society for Research in Child Development, Boston (April).

Zavella, Patricia. 1987. *Women's Work and Chicano Families: Cannery Workers of the Santa Clara Valley*. Ithaca, N.Y.: Cornell University Press.

Zeitlin, Maurice. 1978. "Who Owns America: The Same Old Gang?" *Progressive* 42 (June).

Zelazo, P., M. Kotelchuck, L. Barber, and J. David. 1977. "Fathers and Sons: An Experimental Facilitation of Attachment Behaviors." Paper presented at the meeting of the Society for Research in Child Development, New Orleans.

Zelizer, Viviana. 1985. *Pricing the Priceless Child*. New York: Basic Books.

Zelnik, Melvin, John F. Kantner, and Kathleen Ford. 1981. *Sex and Pregnancy in Adolescence*. Beverly Hills, Calif.: Sage.

Zinn, Maxine B. 1980. "Employment and Education of Mexican-American Women: The Interplay of Mo- dernity and Ethnicity in Eight Families." *Harvard Educational Review* 50(1): 47–62.

———. 1982. "Qualitative Methods in Family Research: A Look Inside the Chicano Families." *California Sociologist* (Summr): 58–79.

Zinn, Maxine B., and D. Stanley Eitzen. 1987. *Diversity in American Families*. New York: Harper and Row.

Ziskin, Jay, and Mae Ziskin. 1973. *The Extramarital Contract*. Los Angeles, Calif.: Nash.

NAME INDEX

SUBJECT INDEX

CREDITS